Friendly Guides to Technolog

Friendly Guides to Technology is designed to explore important and popular topics, tools and methods within the tech industry to help those with and without technical backgrounds come together on a more equal playing field and bridge underlying knowledge gaps to ensure teams, regardless of background, can fully work together with confidence.

This series is for people with a variety of goals, including those without a technical background who work closely with developers and engineers, those who want to transition into the industry but don't know where to start, as well as people who are looking for a clear and friendly introduction to a particular topic.

Focusing on all areas of the modern tech industry from development, product and design to management and operations, this series aims to provide a better, more well-rounded, understanding of the ecosystem as a whole and remove as many barriers-of-access to the industry, be it cultural or financial, to ensure everyone has the ability to successfully learn and pursue a career in technology, whether they want to be a developer or not.

More information about this series at `https://link.springer.com/ bookseries/17128`.

A Friendly Guide to Data Science

Everything You Should Know About the Hottest Field in Tech

Kelly P. Vincent

Apress®

A Friendly Guide to Data Science: Everything You Should Know About the Hottest Field in Tech

Kelly P. Vincent
Renton, WA, USA

ISBN-13 (pbk): 979-8-8688-1168-5   ISBN-13 (electronic): 979-8-8688-1169-2
https://doi.org/10.1007/979-8-8688-1169-2

Managing Director, Apress Media LLC: Welmoed Spahr
Acquisitions Editor: Shaul Elson
Development Editor: Laura Berendson
Editorial Assistant: Gryffin Winkler

Cover designed by eStudioCalamar

Distributed to the book trade worldwide by Springer Science+Business Media New York, 1 New York Plaza, New York, NY 10004. Phone 1-800-SPRINGER, fax (201) 348-4505, e-mail orders-ny@springer-sbm.com, or visit www.springeronline.com. Apress Media, LLC is a Delaware LLC and the sole member (owner) is Springer Science + Business Media Finance Inc (SSBM Finance Inc). SSBM Finance Inc is a **Delaware** corporation.

For information on translations, please e-mail booktranslations@springernature.com; for reprint, paperback, or audio rights, please e-mail bookpermissions@springernature.com.

Apress titles may be purchased in bulk for academic, corporate, or promotional use. eBook versions and licenses are also available for most titles. For more information, reference our Print and eBook Bulk Sales web page at http://www.apress.com/bulk-sales.

Any source code or other supplementary material referenced by the author in this book is available to readers on GitHub. For more detailed information, please visit https://www.apress. com/gp/services/source-code.

If disposing of this product, please recycle the paper

To my dad for bringing me along to work on those quiet Saturday mornings, turning on that extra computer so I could type lines copied from 3-2-1 Contact magazine into a BASIC interpreter.

Table of Contents

About the Author

Kelly P. Vincent is a data nerd. As soon as they saw their first spreadsheet, they knew they had to fill it with data and figure out how to analyze it. After doing software engineering work in data science and natural language processing spaces, Kelly landed their dream job—data scientist—at a Fortune 500 company in 2017, before moving on in 2022 to another Fortune 500 company. They have specialized in the at-first-barely-used programming language Python for nearly 20 years. Kelly has a BA degree in Mathematical Sciences, an MSc degree in Speech and Language Processing, and an MS degree in Information Systems. Kelly is also pursuing a Doctor of Technology at Purdue University. They keep their skills up to date with continuing education. They have worked in many different industries that have given them a range of domain knowledge, including education, bioinformatics, microfinancing, B2B tech, and retail.

Kelly hasn't let their love of data and programming get in the way of their other love—writing. They're a novelist in multiple genres and have won several awards for their novels. Kelly considered how they could combine writing and data science and finally spotted an untapped market with the growth of undergraduate data science and analytics degrees.

About the Technical Reviewer

Hardev Ranglani is a seasoned data scientist with over nine years of experience in data science and analytics, currently working as a lead data scientist at EXL Service Inc. in Chicago since 2021. Hardev holds a master's degree in Data Science from the Illinois Institute of Technology (2021) and a bachelor's in Engineering from NIT Nagpur (2013). Throughout his career, Hardev has worked with clients across diverse industries, including retail, insurance, and technology, solving complex business challenges with data-driven insights. His expertise is in machine learning, leveraging advanced techniques to build predictive models and drive impactful decision-making, with proficiency in Python and R. He is passionate about using data science to uncover actionable insights and drive business impact.

Acknowledgments

I am indebted to a handful of mentors and other people I've learned from while growing as a data scientist, most notably Scott Tucker and Sas Neogi. I also want to thank the members of the "Dream Team," David Smith and Lauren Jensen, who along with Sas helped me see what a great data science team can do.

Introduction

Data science has been celebrated as the sexiest job of the twenty-first century because it has so much potential to help organizations understand themselves and their functions better through insights from data, enabling them to do whatever it is they do, just better. It's easy to get excited about these possibilities, as many organization leaders have been doing. However, setting aside the fact that it's probably a little early to be making pronouncements about an entire century that's not even a quarter over, the two little words "data science" don't really convey the amount of planning and work that goes into getting those helpful insights.

It's common, especially among nontechnical leaders (the majority in the corporate world), to hear the buzz terms "data science," "machine learning," and "AI" and think they will solve all of their problems. Often, these people will spin up a data science team—or sometimes just hire one data scientist—and expect the insights to start pouring out. More often than not, the hapless data scientists they've hired will discover that there's insufficient data to do data science, especially good data science. Even the most skilled data scientist can't turn water into wine—the mantra "garbage in/garbage out" is 100% true in the data world. This is unfortunate and often means that the work simply can't get done. But one even more dangerous thing that can happen in the face of garbage in/garbage out is that inexperienced data scientists will produce garbage results—things that look insightful but are simply wrong. They may pass these faux insights along, and leaders may use them to make completely wrong-headed business decisions. An awareness of the requirements and limitations of data science is crucial to avoid this nightmare scenario and get meaningful insights.

Understanding data science requirements and limitations also serves as a reminder of the time–quality–cost triad in all technical work. You can't get high-quality data insights fast for cheap. At least one of the three will suffer when the other two are prioritized. There are always a lot of moving parts in a data science project.

This book is intended to help you understand why that is, by showing the incredible breadth of topics data science involves. These include statistics, data analysis, programming, ethics, data security, data privacy, data engineering, machine learning (ML), natural language processing (NLP), data visualization, big data, cloud computing, and project management. Most projects only involve some of these areas, but it can be difficult to know far in advance exactly which ones will be relevant. So having a real appreciation for everything that should be considered when doing data science will help temper your expectations to more realistic levels.

The last part of the book focuses on practical aspects of what it takes to become a data scientist. It follows that being an expert in all the above areas is unnecessary to be a good data scientist. It is still important to know what areas data science touches and the areas it operates in, because sometimes an expert in one of those other areas may be needed. Or perhaps you need to develop those skills yourself. Data science is a constantly changing field, and working in it means you never stop learning.

Who This Book Is For

University Students: If you're a student early in a degree program intending to go into data science or one of the many adjacent fields, this book will give you foundational knowledge of everything that you will learn about during the degree. Sometimes when you work your way through a degree program, you can have very myopic views of each major topic (here's a class about unsupervised learning, here's one about data cleansing, etc.), and this book will keep you from forgetting the larger context for anything you're studying. It's like a roadmap for your degree.

Career Changers: If you're planning a move into data science, you will want to understand where everything sits in relation to everything else in the data world. As mentioned, you don't need to make yourself an expert in all areas of data science, but reading this book may help you choose some areas to focus on in your educational journey to a new field.

Organization Leaders: If you're thinking about adding data science to your organization, or if you already have data scientists there, having a good understanding of the entire discipline of data science will be invaluable and help you understand what's being done, why it takes some time, and why it isn't cheap. This is a more technical book than a lot of data books written for leaders. You can obviously skip or skim some of the most technical parts, but this greater depth will help you understand why things are how they are.

Citizen Data Scientists: A lot of people have cracked their knuckles and started writing data science code through sites like Kaggle. If you're one of them and you only have that kind of experience, this book can give you the bigger picture of the field you're dabbling in. It can help you understand the need to develop rigor and identify areas you want to learn more about, similar to career changers.

What This Book Covers

The book is divided into three parts—Part I: Foundations, Part II: Doing Data Science, and Part III: The Future. Part I covers the most important concepts and practices to understand before digging into actual data science. Like its name says, Part II looks at the various components that are a part of doing data science. Finally, Part III looks at how you can make data science a part of your future.

Part I: Foundations

Part I contains nine chapters that are crucial to understanding data science in general. Chapter 1 explores the most fundamental aspect of data science, data itself. Chapters 2, 3, and 4 introduce aspects of statistics because statistics is critical in a lot of data science. Chapter 2 covers descriptive statistics, the heart of the exploratory data analysis that occurs early in most data science projects. Chapter 3 dives into probability and the basics of inferential statistics, including probability distributions, sampling, and experiment and study design. Chapter 4 completes the discussion of inferential statistics and covers statistical testing.

Chapters 5, 6, and 7 are all about analytics. Chapter 5 introduces data analysis as a distinct field that has been around longer than data science and forms the foundation for a lot of data science work. Chapter 6 dives into data science itself, defining it and discussing how it fits into organizations. Chapter 7 talks about "The New Data Analytics," basically an umbrella term for any analytics being done using data nowadays, whether it fits under data analysis or data science or any other area.

Chapters 8 and 9 address some of the considerations that are important when doing data science. Chapter 8 talks about data security and privacy, and Chapter 9 looks at ethics as it pertains to data science.

Part II: Doing Data Science

Part II contains the bulk of the chapters and talks about the specific areas of data science that get carried out. Chapter 10 addresses domain knowledge, the understanding of a particular area like medicine or retail that people must have if they're going to do good data science in that domain. Chapter 11 talks about the programming aspect of data science, specifically the languages Python and R.

Chapters 12, 13, and 14 all address data specifically. Chapter 12 looks at both data collection and storage. Collection can be manual or automated, and there are many challenges that can arise in the process. Data storage is most commonly in databases nowadays, but there are other possibilities. Chapter 13 talks about preparing data for data science through preprocessing steps. Chapter 14 dives into feature engineering, the additional work that almost always needs to be done on data before data science can be done.

Chapters 15, 16, and 17 talk about machine learning, performance evaluation, and language-related techniques used in data science. Chapter 15 is a long chapter covering the many areas of machine learning, including some specific common techniques used. Chapter 16 talks about the many ways of measuring how well your machine learning did. Chapter 17 addresses working with language, either as text or speech, covering many processing techniques used to prepare the data for further work.

Chapter 18 dives into the massively important area of visualization and presentation in data science. Visuals and presentations are often all your stakeholders see, so it's important to do this part well.

The remaining chapters in this part address several different specific areas. Chapter 19 looks at the many fields that are using and can use machine learning and language processing. There are many possibilities. Chapter 20 talks about scalability and the cloud, critical to doing data science today on the large datasets so many organizations have. Chapter 21 looks at data science project management and tracking, a notoriously difficult area of the field. Data science projects often get smushed into systems designed for software development, and they don't usually fit. Chapter 22 addresses an important area that relates to ethics but also to doing good data science: human cognitive biases and fallacies, and paradoxes of the field. It talks about what they are, how they can manifest, and how to deal with them.

Part III: The Future

Part III has three chapters that discuss various aspects of going into the field of data science. Chapter 23 talks about how important it is to start doing actual data science and how it's easy to do with the many publicly available datasets out there (or your own). Chapter 24 talks about how to learn more about data science, something that you'll always need to be doing even when you are a data scientist. Finally, Chapter 25 addresses data science and related careers and how to break into them.

Practitioner Profiles

At the end of every chapter, I've included a profile of a professional working in data science or related areas based on an interview I did with each of them. A few of them work in disciplines other than data science and are paired with a given chapter based on its topic. During the interviews, I tried to get each person's "story"—how they got into data science or their current field. These were interesting to hear because data science and tech in general have people with all sorts of backgrounds. This is partially because degree programs in data science are fairly new, so many current practitioners came from other disciplines (physics was one of the most common sources of older data scientists). It's a good reminder of how breadth of knowledge is valuable in data science. The profiles also include the practitioners' thoughts on working in the field, again showing how different people like some areas more than others, and personal preference always comes into that. They're all interesting to read, so don't skip them.

PART I

Foundations

CHAPTER 1

Working with Numbers: What Is Data, Really?

Introduction

Anyone who's heard of data science also knows that data is at the heart of it. Whether they know what data science is or not, everyone's comfortable throwing around the word "data"—we've all heard about privacy, security breaches, and groups collecting, stealing, and selling data about us. But what *is* data, really?

Data is fundamentally abstracted information that represents something from the real world in a simplified way. That word "represents" is key to understanding data. The only reason data exists is because it makes things more tangible, which can help us understand reality in new ways.

Imagine a family reunion with a backyard full of people spread around, chatting, some hanging out by the food table. You know that men are generally taller than women, and you've heard that people are getting taller over the generations. You wonder if these rules hold true in your

© Kelly P. Vincent 2025
K. P. Vincent, *A Friendly Guide to Data Science*, Friendly Guides to Technology,
https://doi.org/10.1007/979-8-8688-1169-2_1

family, which seems to have a lot of tall women, and you've always thought of your grandpa as a giant. You might try to observe people's heights, but with everybody moving around, it's difficult to figure anything out.

But it will be different if you can turn the scene into data. You pull out your notebook and pen and write everyone's name and gender in columns running down on the left side of the pages, leaving room for two more columns: age and height. Now you go around gathering everyone's numbers (don't worry, you've always been the family oddball, so it fazes nobody).

Once you're done, you can do many things with your data, like grouping by age range and taking averages or comparing the average heights of adult men and women. You can run statistical tests, make fancy charts, determine the shortest and tallest people, and on and on. Even though your interest was in height, you can also get the average age of the family reunion because you collected age.

This is obviously a rather silly example, but it exemplifies the utility of data. You can see that people are of different heights, but using a number to represent how tall each person is allows you to do something with that information that's impractical or impossible through only observation.

DATA VS. DATUM

The word *data* is technically the plural form for *datum*, a single point of information. Lots of people treat *data* as plural when talking about it ("the data are..."), but we're using *data* in the singular here.

Make no mistake—although data science is a fairly new field, data itself is not new. We've been collecting data for thousands of years, starting off with recording it on clay tablets. Farmers and merchants tracked their inventory in some of the earliest writing. Governments produced a lot of early data—the earliest known data is census information collected by the ancient Egyptians around 3,000 B.C.E. and the Sumerians 1,000 years later. The data from these and other censuses was used by rulers to assess taxes for their citizens. Egyptians during Cleopatra's time (around 50 B.C.E.) had long been measuring and recording the height of the River Nile daily with structures called Nilometers, in order to compare it to the previous year and even determine taxes and plan harvests. See Figure 1-1 for an example of a Nilometer in the Temple of Isis in Philae. These tools could be ornate or bare bones like the one in the pictures, but they usually have stairs going down, where they could measure where the water level is on the wall.

Figure 1-1. *Nilometer in the Temple of Isis on Philae in Egypt*

GOVERNMENT DATA

It isn't a coincidence that all of these examples involve government, which is still a big data collector. Censuses are usually the largest-scale data collection efforts out there. The United States has been performing a full census of the entire country by getting details on every single household every ten years since 1790. In 2020, there were over 140 million "housing units"—basically any dwelling with an address—that were contacted. The more recent censuses generate an incredible amount of data and are much more complex than the earliest efforts were, and processes and tech have been evolving the whole time. The 2020 Census was the first time it was possible for households to submit their information online. If you are curious about the kind of data that is collected, check out `https://www.census.gov/`.

In today's world, data is even more ubiquitous. Before the computer and Internet age, data mostly represented aspects of the physical world in some way. But now, the digital world generates tons of data about intangible things. Social media gives us huge numbers of posts, follows, likes, and tags. Virtually all of our online activity is tracked somewhere by somebody. Companies have access to our emails; web searches; any files uploaded, downloaded, and saved in the cloud; and on and on. Our click sequences are tracked. There's no real limit to what companies can track about us.

HOW MUCH DATA IS THERE?

Just for perspective, think about how much data there is. Before computers, data was mostly pieces of paper in filing cabinets or hand-recorded tables of data in dusty books. Now, almost everything you do generates some kind of data somewhere, especially if you carry a smart phone. Back in 1997, a researcher estimated there was 12,000 petabytes of data in the world, but that was back when the Web was still a baby and smart devices were nowhere to be found (`https://www.lesk.com/mlesk/ksg97/ksg.html`). To put that in terms we can understand better, 12,000 petabytes is 12,000,000 terabytes. By 2010, the digital world was radically different and now being measured in zettabytes, or a billion terabytes. There was around 2 zettabytes of data in the world that year. In 2023, it was about 120 zettabytes, and by 2025 it's anticipated to be 181 zettabytes. (`https://www.statista.com/statistics/871513/worldwide-data-created/`). Just to see it visually, 181 zettabytes is 181,000,000,000,000,000,000,000 bytes.

But to bring things back to the simple, physical world, think about when you go to the doctor. They usually record your height and weight. These are pieces of data that paint a simple abstract representation of you. Obviously, this is not a complete picture of your physical self, but it is a starting point, and it might be sufficient for determining a dosage of a medication you might need. In reality, doctors gather much more information about you in order to understand your health.

It's worth mentioning that while data underpins data science, the two aren't the same thing. Data science uses data, and almost any data can be used in data science, but not all data is used. For instance, doctors collect a lot of info about their patients, but most of them aren't doing data science on it. Insurance companies are another story, however, as they've been using data science or simply statistical techniques on patient data to forecast outcomes and determine rates to charge for decades. But the reality is that companies are collecting so much data that most don't even utilize half of what they have, and smaller companies use even less. Collecting it is relatively easy, but using it is harder.

This chapter is going to address all the major aspects of what data is so we can understand and talk about it meaningfully. So we'll be focusing on the terminology and concepts people use when they work with data.

The first to learn is *dataset*, which is simply a casual grouping of data. It might be a single database table, a bunch of Excel sheets, or a file containing a bunch of tweets. Datasets are generally relative to whatever work we're doing, and it doesn't have a specific meaning with regard to any tables. Our dataset might be three tables in the database, while our colleague Hector is working with those three tables plus two more, and that's his dataset.

Data Types

There are a lot of different ways to look at data. This means that we often talk about different "types" of data, where "type" can mean different things in different contexts based on what characteristics we are talking about. Here, we'll talk about the simplest form of data—think the kinds of things you'd see on a spreadsheet or in your student record. Text, media, and other complex data are different beasts and will be addressed later in this chapter.

The term *data type* usually refers to a specific aspect of the data, but there can be a lot of different "types." There are four data types in the classical sense: nominal, ordinal, interval, and ratio. Two of these are also grouped together as *categorical* or *qualitative* data, and the other two are called *numeric* or *quantitative* data. The two categorical data types are nominal and ordinal, and the two numeric ones are interval and ratio.

Categorical or Qualitative

Nominal Data

One thing your doctor tracks is your gender. This falls under the simplest type of categorical data—*nominal* (sometimes *nominative*), a word that basically means "name," because this data type represents things that have no mathematical properties and can't even be ordered meaningfully. Other common nominal types would be religion, name, and favorite color. These are obviously not numeric, and there is no natural order to them.

It might surprise you that a number can also be nominal, such as a US zip code, a five-digit number. There is a slight pattern to them, as they increase as you move east to west across the United States. They are not numbers in the mathematical sense—you can't multiply two zip codes together to get something meaningful, and taking an average of them makes no sense. Also, the pattern isn't consistent—it's not true that each subsequent zip code is always more westerly than a smaller zip code (they are jumbled especially in metropolitan areas, and the order isn't perfect because of the distance north and south that has to be covered). There's no way to know just by looking at the number. It's important to remember that just because it looks like a number doesn't mean you can treat it as a number.

One other nominal value that is a number is the Social Security number in the United States. It seems different from a zip code because it is unique—each one is a unique value, and it also can be used to uniquely identify an individual. But it still has no real mathematical properties. It's important to always consider the nature of any number you see in data.

One special case of nominal data worth mentioning is the *binary* (or *Boolean*) type, because it is used very frequently in data science. The old approach to gender could have fallen under this type, but now there are usually other options. The binary type is more common with simple yes-or-no scenarios. For instance, perhaps your doctor wants to track whether you play any kind of sport or not. The answer would be, yes, you play a sport or, no, you do not play a sport. Often, the number 1 is used for yes and 0 for no, but some systems can also store the values True and False. The important thing to remember is that it has exactly two possible values. Table 1-1 shows an example of student data with a Boolean column representing whether the student is a minor or not.

Table 1-1. *A simple dataset with a Boolean column*

Student ID	Age	Is a Minor?
123	17	True
334	18	False
412	16	True
567	19	False
639	17	True

Ordinal Data

Ordinal data has a lot in common with nominal data, which is why they are both called qualitative or categorical. In the case of ordinal, you can think of it as nominal with one special trait: it has a natural order, where each value comes before or after each other value. We intuitively understand that T-shirt size has a natural order that is clear from the value. These would be values like S, M, L, XL, and so on. So there is an order, but there is nothing numeric about these values, and there is no such thing as a zero value in the ordering. There also is no mathematical relationship between the sizes other than relative order—that is, XL is not twice as large as L or even S. Shoe size and grade level in school are additional ordinals. Again, they are numbers, but they aren't related by any mathematical rules.

Figure 1-2 shows an example of nominal and ordinal data in the form of a handwritten record of a few pizza orders. It has nominal data in the form of the pizza name and crust type and ordinal data in the form of pizza size. We need to know that P stands for personal, which we know is smaller than a small, so the order is P, S, M, and L.

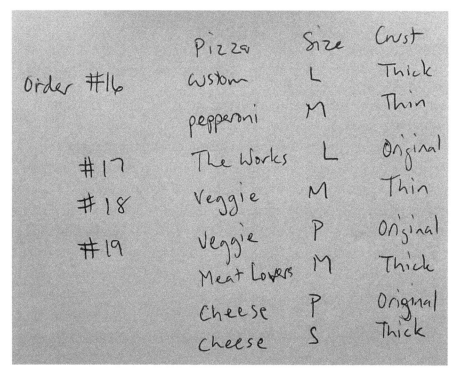

Figure 1-2. *Some pizza orders with categorical data*

Numeric or Quantitative

Ratio Data

The first numeric data type is ratio, where the data is numbers and there is a mathematical relationship between the different values, as well as a natural zero point. Going back to the height and weight your doctor measured, these are both ratio. You can see the mathematical relationship: someone 78 inches tall is twice as tall as someone who is 39 inches tall. Although a real person would never have a height or weight of zero, we can understand what a zero height would mean, so we can say there is a natural zero. Other common ratio data is age, numeric grades on a test, and number of course credits completed.

Interval Data

Interval data is the second numeric data type and has a lot in common with ratio data, with some limitations. Interval data doesn't have a natural zero point, and the mathematical operations you can do on it are significantly limited because of this. For instance, temperature seems like it would be ratio because we have numbers and we can have a temperature of 0. But if you think about what 0 on Fahrenheit and Celsius scales is, those two values mean entirely different things, as a temperature of 0 doesn't mean there is zero heat, so you can't count it as a true zero. Because of this, you can't say that 60 is twice as hot as 30, and summing temperatures would never make sense.

It's interesting to note that unlike Fahrenheit and Celsius values, temperatures on the Kelvin scale would work as ratio data because 0 Kelvin does mean no heat, and the relationships between the different values do follow mathematical rules.

IQ data is another example of interval data, even though at first glance it might seem like ratio data. There are different IQ scales, but they each have a scale that defines the consistent distance between the values. These distances are not arbitrary, unlike the distances between T-shirt sizes, which cannot be quantified without additional information. However, there is no meaningful 0, which means the mathematical operations break down.

One last common example of interval data is a timestamp, which is just a specific date and time. It's very common for these to appear in data, often to indicate when a row first got added to the table. You can measure time between timestamps, but we aren't going to go back to the Big Bang to get a time of 0.

Figure 1-3 shows some sample numeric weather data from the US National Oceanic and Atmospheric Administration, with temperature in Fahrenheit and inches of liquid and solid precipitation. As we've mentioned, temperature is interval in Fahrenheit, but the precipitation measurements are ratio.

U.S. Department of Commerce
National Oceanic & Atmospheric Administration
National Environmental Satellite, Data, and Information Service
Current Location: Elev. 7351 ft. Lat: 44.4568° N Lon: -110.8327° W
Station: **OLD FAITHFUL, WY US USC00486845**

Record of Climatological Observations

These data are quality controlled and may not be identical to the original observations.
Generated on 05/21/2021   Obs

| Year | Month | Day | Temperature (F) 24 Hrs. Ending at Observation Time | | | Precipitation 24 Hour Amounts Ending at Observation Time | | | | At Obs. Time | Evaporation | |
			Max.	Min.	At Obs.	Rain, Melted Snow, Etc. (in)	Flag	Snow, Ice Pellets, Hail (in)	Flag	Snow, Ice Pellets, Hail, Ice on Ground (in)	24 Hour Wind Movement (mi)	Amount of Evap. (in)
2015	01	01	17	-30		0.00		0.0		19.0		
2015	01	02	30	-30		0.00		0.0		19.0		
2015	01	03	19	-1		0.02		1.0		19.0		
2015	01	04	18	8		0.07		2.0		19.0		
2015	01	05	27	15		0.60		12.0		26.0		
2015	01	06	31	27		0.14		7.0		26.0		
2015	01	07	36	16		0.00		0.0		24.0		
2015	01	08	39	13		0.00		0.0		23.0		
2015	01	09	41	-4		0.00		0.0		23.0		
2015	01	10	33	-3		0.00		0.0		23.0		
2015	01	11	36	-3		T		T		23.0		
2015	01	12	33	23		0.02		T		24.0		
2015	01	13	32	-7		0.00		0.0		24.0		
2015	01	14	33	-12		0.00		0.0		24.0		
2015	01	15	34	-12		0.00		0.0		21.0		
2015	01	16	40	-12		0.02		0.5		23.0		
2015	01	17	29	4		0.14		2.0		25.0		
2015	01	18	32	16		0.08		2.0		26.0		
2015	01	19	34	11		T		T		24.0		
2015	01	20	28	0		T		T		24.0		
2015	01	21	28	-3		0.00		0.0		24.0		
2015	01	22	25	-14		0.00		0.0		24.0		
2015	01	23	30	-14		T		T		24.0		
2015	01	24	35	-14		0.00		0.0		23.0		
2015	01	25	29	22		0.00		0.0		23.0		
2015	01	26	36	10		0.00		0.0		22.0		
2015	01	27	54	9		0.00		0.0		22.0		

Figure 1-3. *Example numeric weather data from the US National Oceanic and Atmospheric Administration. Source: Daily Temperature and Precipitation Reports—Data Tables,* `https://www.climate.gov/maps-data/dataset/daily-temperature-and-precipitation-reports-data-tables`

THE FOUR DATA TYPES

All four data types are important. Understanding the differences, which can sometimes be subtle, will help you get better results with data science. Below are all four types in increasing order of complexity.

Categorical (Qualitative):

- **Nominal:**

    - **What It Is:** A name, label, or category

    - **Characteristics:** A name that represents something, with no numeric relationship between possible values

    - **How It Can Be Used in Data Science:**

        - It's the most difficult to use in most data science approaches, often requiring transformation.

        - Often we count the number of occurrences of the different values.

        - A single nominal variable may be converted to multiple variables, with one Boolean variable for each possible value, usually called *dummy variables*.

    - **Examples:** Hair color, religion, hometown, Social Security number

- **Ordinal:**

    - **What It Is:** Similar to a nominal type except that there is a natural order between the values

    - **Characteristics:** Data that has natural and meaningful order

- **How It Can Be Used in Data Science:**

  - This type is a little more useful than nominal, but it's still quite limited.

  - It's common to use counts of occurrences of different values.

  - We can also rank the data.

- **Examples:** Letter grade, medals awarded in a race, a high school or college student's class, T-shirt size

Numeric (Quantitative):

- **Interval:**

  - **What It Is:** Similar to ordinal except we have a little bit more information that is mathematically useful

  - **Characteristics:**

    - Has a natural and meaningful order.

    - Distances between values can be measured.

    - There is no natural zero for the range of possible values.

  - **How It Can Be Used in Data Science:**

    - This is easier to use than categorical data.

    - Many arithmetic operations can be done, so we can quantify interactions between two different pieces of data.

    - Measures like average can be computed.

    - Many statistical measures can be calculated.

  - **Examples:** SAT score, IQ score, shoe size, temperature in Fahrenheit or Celsius

- **Ratio:**

  - **What It Is:** Similar to an interval type, but we have a natural zero as a possible value

  - **Characteristics:**

    - Has a natural and meaningful order.

    - Distances between values can be measured.

    - There is a natural zero.

  - **How It Can Be Used in Data Science:**

    - Considered the most useful type for data science.

    - All of the arithmetic operations can be performed.

    - Measures like average can be computed.

    - All of the statistical measures can be calculated.

  - **Examples:** Height, weight, distance between two points, number of textbooks you own, temperature on the Kelvin scale

Data Transformations

The different data types can be used in different ways, and it's common to transform data from one data type to another to make it easier to work with. This usually involves losing some information as we step down from the most restrictive type to one less restrictive—we can go from ratio to any other type, from interval to nominal or ordinal, and from ordinal to nominal, but not the other direction. But losing information isn't necessarily a problem—in one project I worked on, we found that classifying (or *bucketing*) temperatures into five distinct groups of temperature ranges (freezing, cold, temperate, hot, and very hot) made it ordinal but also improved our predictions over using temperature as a ratio type. We also could have made it interval by dividing it into ten-degree blocks (0–9.99 degrees, 10–19.99, 20–9.99, etc.). What you can do with data is determined both by what data you have and how it can be transformed. Additionally, you may use additional knowledge when deciding how to do this, like you might not use the same bucket ranges for Seattle as in Miami. What people see as cold in Florida is probably considered quite comfortable in Seattle.

BUCKETING DATA

We use the term *bucketing* to describe the process of breaking up a large range of values into a smaller number of values. This can be convenient and also make a better representation of reality. Looking at temperatures, do we really care if its 55° F or 56° F out? Will that one-degree difference make people behave differently, for instance? Generally, it wouldn't, but the difference between those and 30° F would make for different behavior. So a "freezing" bucket might be anything 32° F and below, "cold" might be between 33° F and 45° F, and so on. Obviously, this does mean there's a big difference between 32° F and 33° F, but not so much between 40° F and 41° F. There are always tradeoffs, and that's why data science tests things before making these kinds of decisions on transformations.

One other common transformation that I'll talk about more in
Chapter 14 is a technique for creating what are called dummy variables,
where we split a categorical column into separate "dummy" columns
that are all Boolean. Each column would be one of the possible values of
the column, and each row would have a 1 for the column that matches
the value and a 0 in all of the other new columns. Sticking with the
temperature bucketing example, we start with a bit of temperature data
over a series of Sundays, as seen in Table 1-2. Splitting the temperature
bucket into separate columns would look like Table 1-3.

Table 1-2. *Data with bucketed temperatures*

Date	Temperature Bucket at Noon
2024-03-03	Freezing
2024-03-10	Cold
2024-03-17	Cold
2024-03-24	Temperate
2024-03-31	Cold
2024-04-07	Cold
2024-04-14	Temperate
2024-04-21	Temperate
2024-04-28	Hot
2024-05-05	Very Hot

Table 1-3. *Data with bucketed temperatures transformed*

Date	Temperature Bucket at Noon	Is Freezing	Is Cold	Is Temperate	Is Hot	Is Very Hot
2024-03-03	Freezing	1	0	0	0	0
2024-03-10	Cold	0	1	0	0	0
2024-03-17	Cold	0	1	0	0	0
2024-03-24	Temperate	0	0	1	0	0
2024-03-31	Cold	0	1	0	0	0
2024-04-07	Cold	0	1	0	0	0
2024-04-14	Temperate	0	0	1	0	0
2024-04-21	Temperate	0	0	1	0	0
2024-04-28	Hot	0	0	0	1	0
2024-05-05	Very Hot	0	0	0	0	1

This can be very useful in a lot of data science, especially because many of machine learning models can't handle categorical values at all, and this gives us a way to use it in some. Knowing when this is appropriate comes with practice and experience.

Final Thoughts on the Four Data Types

It's not always easy to figure out what type a given piece of data is. It feels very tempting to treat a zip code as ordinal since it is a number and we know there is a pattern of progression—but it just doesn't work.

The most important things to consider are what information a given piece of data actually holds and what it represents—for instance, someone's height and weight tell you nothing about level of fitness. Someone who weighs 300 pounds could be obese, extremely tall, a muscular athlete, a bodybuilder, or some combinations of these. Remember that data is always an abstraction, and there are many ways to abstract something in the real world. For instance, if you wanted your T-shirt data to be ratio, you can't do that with typical sizes. But if you instead abstracted T-shirts by taking several measurements, like height, width, sleeve length, and so on, those would each be a separate ratio data type. But that level of information is simply not inherent in general T-shirt sizes (each size means something different in different brands). Remember that just because you could theoretically represent something differently doesn't change the way it is currently represented or mean that it can safely be converted to a different type of data.

Figuring Out the Data Type

When you are faced with some data, you can figure out what data type it is by looking at its characteristics and what you plan to do with it. If you think of the data on a hierarchy in terms of how many rules it has to satisfy, that would look like this (starting with the most complicated on top):

> Ratio
>
> Interval
>
> Ordinal
>
> Nominal

Ask yourself if it qualifies as ratio data first, and then if it doesn't, ask yourself if it qualifies as interval, and so on. Often you can skip ratio and interval if it isn't numbers, but when you do have numbers as data, always make sure to go through the hierarchy to ensure that it is truly numeric. This way of looking at data becomes easier with experience.

Structured vs. Unstructured vs. Semi-structured Data

Most people think of tables when they think of data. A lot of data fits easily into tabular form, and this is what most data scientists work with, but it's not the only way to see data. *Structured data* naturally fits into a table structure, where all the records or rows have the same attributes or columns. With this type of data, some of the attributes might be blank, but all the data points have the same basic structure, hence the name.

The opposite, unstructured data, is inherently messier and harder to store and deal with. *Unstructured data* comes in many forms, but all it means is that it can't be easily organized into a clear structure that applies to all data points. Text, video, audio, and images are all common types of unstructured data. While it is possible to enforce a tabular structure on some of these types of data, it often requires quite a bit of manipulation. The main point is that there isn't a straightforward, obvious organizing principle that all of the items have in common—there isn't a natural structure, which is why we call the two types structured and unstructured.

Note that the unstructured aspect of this data applies to the raw data, like the actual text or the image, and not *metadata* about the raw data. For instance, you could track metadata about the videos in a table with columns like video length, file format, genre, language, rating, etc., but this does not represent the video data itself. Several subfields of artificial intelligence (AI), especially natural language processing (NLP), have been trying to put structure on text in a variety of ways. There is a later chapter on NLP. Other disciplines deal with other types of unstructured data.

There's one other type of data called *semi-structured data*, where there are elements of structure, but it's not organized enough to be considered true structured data. Common examples of that, that are seen in the tech world a lot but less in the real world are XML and JSON formatted files. XML is a file with tags that surround text to give information about what it is. If you've seen HTML, it's a specific type of XML, where the tags give formatting information. JSON, pronounced jay-sahn, is another text format with blocks of values within curly braces. I'll include some examples of these types below.

To illustrate the differences between these types of data, we'll look at a few small datasets. First, we'll look at structured data. Table 1-4 shows some basic data on patients in list form.

Table 1-4. *Three patient details*

Patient 1	Patient 2	Patient 3
Name: Julie	Name: Samuel	Name: Veena
Height: 65 in.	Height: 68 in.	Height: 70 in.
Weight: 105 lbs.	Weight: 165 lbs.	Weight: 150 lbs.
Age: 16	Age: 17	Age: 18
Gender: Female	Gender: Male	Gender: Female
T-shirt Size: S	T-shirt Size: L	T-shirt Size: M
Plays Sports: Yes	Plays Sports: No	Plays Sports: Yes

This data isn't structured in its current form, but it absolutely cries out to be put in a table. We even have all the data for each patient. Each patient will be represented in a *row* (all the cells from left to right for one patient), and each characteristic will be a *column* (all the values of a particular characteristic from top to bottom across all patients). Table 1-5 shows what this data in tabular form would look like.

Table 1-5. *Three patient details in a table*

Patient #	Name	Height (in.)	Weight (lbs.)	Age	Gender	T-shirt Size	Plays Sports
1	Julie	65	105	16	Female	S	Yes
2	Samuel	68	165	17	Male	L	No
3	Veena	70	150	18	Female	M	Yes

This structured data is easy to understand—in fact, it's even clearer than the lists in Table 1-4. We can easily see all the values for a given patient by looking across the table rows, and we can also look down the columns and get an idea of what the different values are for the different patients.

In data science, it is common to refer to data stored in columns by a variety of other names: *features*, *variables*, and *attributes*. Similarly, rows are called other things in data science: *records*, *instances*, *entities*, and *cases*. Often the terms used depend on the context or domain, but it is helpful to recognize all of them.

MATRICES

A table is more mathematically considered to be an *n* * *m* matrix (plural "matrices"), with *n* rows and *m* columns, with a header row not considered part of the matrix. This example table would therefore be considered a 3 * 8 matrix.

So what does unstructured data look like? We have two snippets from books. The first is from Charles Dickens's *Nicholas Nickleby*:

> *Mr. Ralph Nickleby receives Sad Tidings about his Brother, but bears up nobly against the Intelligence communicated to him. The Reader is informed how he liked Nicholas, who is herein introduced, and how kindly he proposed to make his Fortune at once.*

The second is from Chapter 2 in a book called *Struck by Lightning: The Curious World of Probabilities* by Jeffrey S. Rosenthal:

> *We are often struck by seemingly astounding coincidences. You meet three friends for dinner and all four of you are wearing dresses of the same color. You dream about your grandson the day before he phones you out of the blue.*

How would you organize this? There is no obvious way to do it that helps us make it clearer. We could put it in a table, as we do in Table 1-6.

Table 1-6. *Two book snippets in a table*

Title	Author	Snippet
Nicholas Nickleby	Charles Dickens	Mr. Ralph Nickleby receives Sad Tidings about his Brother, but bears up nobly against the Intelligence communicated to him. The Reader is informed how he liked Nicholas, who is herein introduced, and how kindly he proposed to make his Fortune at once.
Struck by Lightning: The Curious World of Probabilities	Jeffrey S. Rosenthal	We are often struck by seemingly astounding coincidences. You meet three friends for dinner and all four of you are wearing dresses of the same color. You dream about your grandson the day before he phones you out of the blue.

We've made a table, but we haven't really gained anything. Data *about* books—including the title and author included in Table 1-6—does have obvious structure to it. The book snippet, on the other hand, doesn't have any inherent structure. Techniques in natural language processing, which I'll talk about in a later chapter, can be used to give text some structure, but it requires transformation to be stored in a structured way. For instance, we might break the text into sentences or words and store each of those in a separate row. The types of transformations we might do will depend on what we want to do with the data.

If you are thinking of the wider world, it may have occurred to you that any of the unstructured data I've mentioned has to be converted to a computer-friendly form in order to be stored on a computer—everything ultimately comes down to 1s and 0s, after all. This means that there must be some kind of structure, but it isn't necessarily a structure that makes sense to humans or is useful in analysis. For instance, a computer could encode an image by storing information about each pixel, like numbers for color and location. Data in this form is often used in image recognition tasks, but humans can't look at that data and know what the image looks like.

One of the ways people have tried to deal with unstructured data is to store *metadata*—this is the data about data I mentioned above with the data about books. But an interesting aspect of metadata is that even recording it can be more challenging than other structured data. Imagine a dataset of scientific papers published in journals that stores the word count of each section of each paper. There are conventions in how a scientific paper is organized, but there's a lot of leeway. But most would have Introduction, Methods, and Conclusion sections. Others will have section headings that are not in most of the other papers, and some might be missing even one of those three common sections. If we have a lot of these articles, a table representation can quickly become unwieldy. For instance, it would make sense to have a column for each section, where we store the word count for that section in each paper. Let's say we start going through the papers and only the first four have all of those three sections. Imagine the first four papers we look at all have only those three sections mentioned above, and we record their word counts. That would look like Table 1-7.

Table 1-7. *Simple table with metadata on journal articles*

Paper #	Introduction	Methods	Conclusion
1	100	250	25
2	110	400	150
3	75	350	320
4	55	90	150

But let's say when we get to paper #5, it doesn't have a Methods section and instead has two new sections, Participants and Process. Then, paper #6 has a Summary instead of a Conclusion. We'd have to add those as new columns. Alternatively, we could put the count for the Summary section in the Conclusion column, but then we'd record that somewhere so we wouldn't forget if we needed to know later. The table would start to get ugly in Table 1-8.

Table 1-8. *Table with metadata on journal articles with differing sections*

Paper #	Introduction	Methods	Conclusion	Participants	Process	Summary
1	100	250	250			
2	110	400	150			
3	75	350	320			
4	55	90	150			
5	85		125	230	100	
6	170	350				325

Now we have a bunch of columns with missing values. We'll address the challenges of missing data later, as the choices you make about how to handle it can have consequences down the line. But for Table 1-8, putting a 0 in would probably be a reasonable solution to avoiding missing data—there are no words in a section that isn't present in the paper. But as we look at more and more papers, the number of columns would only increase, eventually being unwieldy and with a significant number of 0s. This is more difficult to work with than most structured data (although real-world structured data also has missing data).

Finally, representing the data this way means that we've lost a lot of information—such as the order that the sections appear in, in a given paper. We'll address unstructured data in later chapters, but the takeaway here is that some data has more inherent structure than other data.

It's usually pretty clear if data is structured or not, but semi-structured data is in an in-between spot. XML, HTML, and JSON have some structure, but not enough that it can be considered true structured data. However, it's easier and more intuitive to convert it into tabular form than unstructured data is. See Figure 1-4 for some examples of these three formats.

XML	HTML	JSON
`<animals>`	`<html>`	`{`

`<animal>`
`<id>1</id>`
`<species>cat</species>`
`<name>Marvin</name>`
`<age>14</age>`
`<age_unit>years</age_unit>`
`<sex>neutered male</sex>`
`<breed>domestic short hair</breed>`
`<colors>`
`<color1>tabby</color1>`
`<color2>brown</color2>`
`</colors>`
`<deceased/>`
`</animal>`
`<animal>`
`<id>2</id>`
`<species>cat</species>`
`<name>Maddox</name>`
`<age>8</age>`
`<age_unit>years</age_unit>`
`<sex>neutered male</sex>`
`<breed>Siamese</breed>`
`<colors>`
`<color1>seal-point</color1>`
`</colors>`
`</animal>`
`<animal>`
`<id>3</id>`
`<name>Pelusa</name>`
`</animal>`
`</animals>`
	`<head>`
	`<title>Two Cats</title>`
	`</head>`
	`<body>`
	`<h1>Get to Know the Cats</h1>`
	`<h2>Marvin</h2>`
	`<paragraph>`
	Marvin was a great cat even though he had some health problems. He was 14 when he died.
	`</paragraph>`
	`<h2>Maddox</h2>`
	`<paragraph>`
	Maddox is a slightly obnoxious 8-year-old seal-point Siamese cat who likes to bite hard things like metal chair legs.
	`</paragraph>`
	`</body>`
	`</html>`
		`'animals': {`
		`'id': 1,`
		`'species': 'cat',`
		`'name': 'Marvin',`
		`'age': 14,`
		`'age_unit': 'years',`
		`'sex': 'neutered male',`
		`'breed': 'domestic short hair',`
		`'colors': {`
		`'color1': 'tabby',`
		`'color2': 'brown'`
		`},`
		`'deceased': True`
		`},`
		`{`
		`'id': 2,`
		`'species': 'cat',`
		`'name': 'Maddox',`
		`'age': 8,`
		`'age_unit': 'years',`
		`'sex': 'neutered male',`
		`'breed': 'Siamese',`
		`'colors': {`
		`'color1': seal-point',`
		`}`
		`},`
		`{`
		`'id': 3,`
		`'name': 'Pelusa'`
		`}`
		`}`

Figure 1-4. *XML, HTML, and JSON*

The example XML and JSON in Figure 1-4 represent the exact same info on three animals in two different formats. HTML is XML-style formatting that serves a specific purpose, telling a web browser how to display things. These examples do look structured, but the reason they are considered only semi-structured is that each of the tags (in XML) or entries (in JSON) is optional in terms of the format itself. In this case, a cat could have multiple colors, but also could have one or even none. The tag for colors could be left out altogether, and it would still be valid XML. Additionally, there's a tag/element that is present only if it's true, which is the deceased element. It exists for the first cat but not the second. But most importantly, we only have an ID and a name for the third animal, and these are still both valid XML and JSON formatting. People using this data may put their own requirements on specific tags in their applications, but the fundamental formats are loose and therefore semi-structured.

Just like it was easy to convert the data in Table 1-4 to tabular form, it's easy from XML and JSON, to a point. There are even coding packages that will do it automatically. But one thing that is common with semi-structured data is that although each animal record here looks very similar to the others, they don't all have the same entries, as we talked about above. Table 1-9 shows the animal data converted from the XML and JSON format (they yield the same table).

Table 1-9. *Three animal details in a table from XML and JSON*

Animal #	Species	Name	Age	Age Unit	Sex	Breed	Color 1	Color 2	Deceased
1	cat	Marvin	14	years	neutered male	domestic short hair	tabby	brown	True
2	cat	Maddox	8	years	neutered male	Siamese	seal-point		
3		Pelusa							

Like with unstructured data, putting semi-structured data into a structured format can yield different results based on how similar the elements are.

Raw and Derived Data

With the exception of the journal articles example, the data we have discussed so far has all been *raw*—direct facts about something in the real world. This can be entered into a computer and stored in some form. But it's still just data that directly reflects the real world. We also often talk about *derived* data, which means it comes from other data in some way.

Deriving new data is incredibly powerful and important in data science. Two common ways data is derived are *aggregations* and *interactions*. In structured data, aggregations are performed in one column across rows: we might take the *average* of all students' heights broken down by gender. Table 1-10 shows an aggregation on height.

Table 1-10. Aggregated data on height by gender

Gender	Average Height (in.)
Female	65
Male	69
Nonbinary	67

Interactions are performed on one row across columns: in the scientific paper dataset, after we've filled the empty fields with 0, we might create a new column called Intro Conclusion Sum, which is the *sum* of the length of the Introduction section and the length of the Conclusion section together. Table 1-11 shows the table with the interaction variable.

Table 1-11. *Table of journal article section lengths with an interaction variable*

Paper #	Intro-duction	Methods	Conclusion	Participants	Process	Summary	Intro Conclusion Sum
1	100	250	250	0	0	0	350
2	110	400	150	0	0	0	260
3	75	350	320	0	0	0	395
4	55	90	150	0	0	0	205
5	85	0	125	230	100	0	210
6	170	350	0	0	0	325	170

As we move along each row in Table 1-11, it's easy to sum the Introduction column and the Conclusion column. With the 0s in place, it's a trivial task.

MORE ON MISSING DATA

If we had left the empty cells alone instead of putting a 0 in, it wouldn't have been as obvious what to do. Empty doesn't necessarily mean 0—it usually means that it's unknown, which means it could actually be a number other than 0, just not recorded properly. It's important to remember that the data alone doesn't tell you what null values mean, despite how "obvious" it can feel as a human. In database systems, empty values (called *nulls*) can really mess up calculations because they don't have the human intuition to know whether we should substitute a 0 or ignore the data point or do something else. In many cases, a sum of 100 and null will come out as null, because it is impossible for the database system to know what null means.

Although interactions are often recorded as new columns in the original dataset, aggregations create a totally new dataset. The columns are usually different from those in the original tabular data, and the rows almost always are. For instance, in the aggregated values of height by gender in Table 1-10, the only columns are Gender and Height, as opposed to Name, Height, Weight, Age, Gender, T-shirt Size, and Plays Sports of the original table. These datasets store two different things—one is a representation of individual patients and the other is a representation of attributes of groupings of patients.

Metadata

I've already mentioned metadata, data that describes other data. It can be quite important in the data world.

METADATA

The word "meta" is casually thrown around a lot, generally to indicate a higher layer of awareness to an experience or facts. It has the same meaning with data—data about data, or data about the information being represented in the data.

In a lot of ways, it doesn't matter if something is metadata or plain data. Both can be used in data science. For instance, imagine you collect all the tweets that mention a particular movie in the week of release because you're interested to see what people are saying. There will be a lot of data associated with the tweets that would largely be considered metadata, like the hashtags, timestamp of the tweet, the username who posted, and the location of the IP address, and if you ran some automatic semantic analysis, a score representing *sentiment* (how positive or negative the post is). Most people will consider the text of the tweet the data and all the other things metadata. But if you're studying hashtags and the sentiment they're associated with, that may be what you focus on, without even looking at the tweets directly.

Data about media is some of the most common metadata. Table 1-12 shows common metadata stored for different types of media (definitely not a comprehensive list).

Table 1-12. *Metadata fields associated with different types of media*

Movie	Video Game	Song	Book	Art
Title	Title	Title	Title	Title
Release date	Release date	Album name	Author	Artist
Director	Studio	Artist name	Illustrator	Series
Producer	Genre	Release date	Editor	Creation date
Editor	Platform	Studio	Series	Medium
Actors	Score composer	Album release date	Series book #	Style
Score composer	Artists	Running time	Publisher	Dimensions
Running time	Writers	Track number	Genre	Monetary value
Studio	Target age	Producer	Target age	
Rating			ISBN	

We can even have more metadata about some of this metadata. For instance, we would likely store information about all the people on these lists—the directors, artists, authors, etc.—such as their names, ages, gender, and agents.

Note that metadata can be stored in the same table—data and metadata can exist together, as long as it makes sense in terms of reality. Some other common metadata you see in databases includes timestamps for when that record was created, a timestamp for when it was modified, and the creator of the record. Note that not all timestamps are metadata. If we record the time of a sales transaction on a cash register, that timestamp would not be metadata because it represents a thing that happened in the real world—the sale itself. But a "last updated" timestamp on the row for that transaction in a table indicating when that row was last modified in some way would be metadata. For instance, imagine that the row contains a link to a generated PDF invoice, and a process is run nightly that adds the file size to the row and modifies the last updated column with the time it does it. This is data about that row in the table and has nothing directly to do with the real-world transaction.

Data Collection and Storage

How data is collected and stored is hugely important in data science and the data world in general. In modern times, data is usually collected automatically from some types of computer systems and stored in different types of database systems. This is a result of evolution from manual processes and storage (like spreadsheets). Data collection and storage will both be covered in a later chapter.

The Use of Data in Organizations

We hear a lot about *big data*, which is an amorphous term basically used to indicate large quantities of data. Data is usually considered big when processing it stretches computer resources, but that varies over time. What would have been considered a lot of data on a computer in 1995 would be considered minuscule today. Many organizations are collecting so much data that they are starting to stretch their resources, which definitely qualifies as big data.

One aspect of big data that it shares with "normal data" is that all of it exists for a purpose—to increase our understanding of the world. As previously discussed, data is an abstraction of reality. Once we've abstracted the world, we can use the data to better understand that world in some way, through data science or data analysis. To put this in context, people have defined different levels of data. A famous diagram called the DIKW Pyramid in Figure 1-5 shows the progression of data from bottom to top as it becomes increasingly useful.

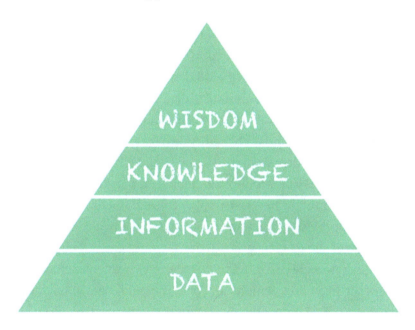

Figure 1-5. *DIKW Pyramid*

If the purpose of data is to help us become wiser about the world, then data analysis and data science are the main methods used to achieve that. Although the word *information* (or just info) is often used synonymously with data, there is an important distinction in this diagram: in this view, information is data that has meaning to it. Beyond that, meaning is clearly useful, but it takes a collection of information and understanding to become *knowledge*. And finally, knowledge doesn't instantly become *wisdom*. It has to build up over time.

Most data analysis and data science in organizations are carried out to help leaders make better decisions. But sometimes data itself is needed simply for processing things, and understanding the various aspects of that plays into the success and efficiency of those sorts of processes. Not all of it is used for analysis. Retail companies store every transaction, organizations that have members store data on their members, and libraries store records of every book and piece of media they have. This is data whether or not it's analyzed in any way.

As an example, we can revisit the US Census. The 1880 Census took eight years to finalize because of the amount of data and the lack of any rapid data processing machines. This was the kind of alarming scale issue that would obviously require some kind of change to keep up in future censuses. As a solution, the Census Bureau commissioned Herman Hollerith to create a "tabulating machine." This proto-computer ran on punch cards and shortened the processing time of the next census by two years. Machines like this were soon in use in businesses, but it wasn't until the 1960s that computing really started taking off, driven largely by a desire to reduce costs by offloading computation from expensive humans to relatively cheaper machines.

One of the bigger revolutions that occurred with data in businesses was the advent of relational database systems, which will be discussed in a later chapter. Although there were other computerized database systems around already, relational database systems made it much easier for analysts to work with data, even though further developments were necessary to keep up, as the amount of data being stored has increased dramatically. Data warehouses were the first step in dealing with this increase in data, gaining traction in the late 1980s, followed by non-relational database systems in the late 1990s. These systems dictate ways of structuring databases, and they made automatic processing and storing large amounts of data much more feasible. Most major database systems started using relational databases principles decades ago and are still following them. This includes airline reservation systems, bank transaction systems, hospitals, libraries, educational institution systems, and more.

While database systems improved the storage and retrieval of data, further developments helped people with processing and displaying data. The 1960s saw the creation of statistical software that allowed people to easily perform statistics and data analysis, another revolution. These tools also enabled easier visualization of data, especially as they were further developed over the years. Most companies make analyzing their data a regular part of their day-to-day operations, and it's all because of how easy it is now. Chapter 5 will dive deeper into this in discussing data analysis.

Why Data Is So Important

We know we live in a time of a data explosion. Almost all companies collect it, sometimes without a clear purpose. But everyone knows that data—digital in general—is the future. "Digital native" companies like Uber, Amazon, and Netflix are all tech companies before being anything else. This is because they've leveraged data from the beginning. To stay relevant and competitive today, established companies need to recognize that they will have to become tech companies themselves. What that really means is learning to get the right data and how to use it to improve decision-making. But it also means that they will have to change their business models, which is why there's such a variety in the types of old companies that are doing well vs. those that aren't. Not everyone is trying, and not all that are trying are succeeding. A lot of brick-and-mortar retailers have struggled with this. But Walmart has modernized itself, mimicking a lot of the innovations that Amazon has done to create such loyalty, like an easy-to-use app, a massive delivery network, and a membership program that allows free shipping (like Amazon Prime).

Data scientists are the ones making much of this possible, by helping companies set their new directions. But data scientists can't do anything without data, and knowing how to work with it is something that comes with experience. Some of the questions and problems data scientists can use data to answer or solve can be seen in Table 1-13.

Table 1-13. *Business problems data scientists can use data to solve*

Industry	Business Problem
Airlines	What is the best way to dynamically change the prices of tickets based on days remaining to travel to maximize revenue?
Banking	What is the optimum credit limit for a new credit card customer? What is the probability that the applicant will default on the credit card payments?
Insurance	Which insurance claims are more likely to be fraudulent and should be investigated further?
Social media	How to personalize the content (posts, ads, etc.) for all the users in real time?
Retail	What is the most optimum inventory to have for all products to minimize overstocking as well as products being out of stock?
Ecommerce	How to measure the effect of marketing spend on the sales that can be directly attributed to marketing?
Streaming media	Which customers are most likely to not renew their subscription services and how to address the same?

Data Governance and Data Quality

One last facet of data that we will mention is the often-neglected fields of data governance and data quality. *Data governance* is basically the idea of taking a big-picture view of the data at an organization and managing it to ensure that data quality and usefulness is maintained at a high level. This means having a comprehensive framework of policies, processes, standards, and metrics that are all used to manage the data. It will include setting guidelines for data access, usage, and sharing. Basically, data governance provides guardrails to company data, and ensuring data quality is one of its important goals.

The meaning of the term *data quality* is pretty intuitive—it tries to address how good the data is. Basically, do its abstractions represent the real world as accurately as possible? But it turns out that this can be way more difficult than it seems like it should be. There are several specific dimensions to data that contribute to overall data quality, such as correctness, completeness, and currency. These may seem similar, but each considers a different aspect of the data and can be useful when diagnosing problems. For instance, we may have a column that has a lot of missing values, but the data that is there is known to be correct. The correctness would be high, while the completeness would be low. If we look at the missing data and notice they are all recently added to the table, maybe there is also a currency problem—only records more than a few days old are good.

The field of data quality is fairly robust, even though most companies don't put enough focus on it. Many don't have any formal data governance or data quality teams. People just kind of muddle through using the data, dealing with it as best they can. Data scientists have to deal with data whether its quality is good or not, so being aware of quality issues can be critical to using the data properly and getting meaningful results.

Part of data governance is identifying data owners, which is important in the context of data quality. This can be a surprisingly difficult thing to do at a lot of organizations. Ownership itself is a nebulous idea, but usually it means that owners should be people who have a deep understanding of the data and ensure that it is accurate to a degree. Sometimes the teams who work with the data most will own that part of the data—for instance, the people who run a college's dorms may own the data on all dorm residents. In other cases, the data may be owned by "Information Technology" (IT), which is itself a rather amorphous concept, but in general it means the technical people who are responsible for maintaining and administering the tools for data storage and other technology. This usually isn't a good idea since they often don't have a good understanding of what aspects of reality the data itself represents, which always requires business knowledge. The main reason this is a challenge in organizations is that people often resist becoming data owners because it is a big responsibility—maybe even an overwhelming one—and they don't want to be held accountable if something goes wrong.

Key Takeaways and Next Up

The word "data" is casually thrown around all the time, but when we want to look at it as a source of information and wisdom, we need to understand what it really is and the variety of ways of looking at it. There is a lot of terminology around it, and it's important to understand the different definitions. Data represents the real world and can come in many forms, but most of the time it can be considered structured and can easily be represented in tables, where columns represent aspects of an object that is stored in each row of the table. There are four primary data types that increase in rigor and restrictiveness, including nominal (for things that have no numeric or ordering characteristics), ordinal (for things that can be ordered, but not really mathematically), interval (numeric data that doesn't quite reach the rigor of the next type), and ratio (numeric that follows certain mathematical rules). You can do the most with ratio data, and nominal is often the hardest to work with in data science. Unstructured data (like text) is another important type that can't be easily put into tabular form. Data can be raw (in a base form similar to how it's originally collected) or derived (transformed, combined, or summed or aggregated in some way). Data is usually organized into datasets (usually one or more tables representing cohesive things), and we often work with metadata (data about data). Data collection and storage is important and will be discussed in detail in Chapter 12. Organizations use data in a variety of ways, but in general the purpose of it is to help them understand their business better and make better, data-informed decisions. Finally, there is a whole discipline around data governance and data quality that is important to keep the data in good shape.

Understanding data's fundamental nature and how it relates to the real world is critical because it underpins everything else in data science. The next chapter is the first of three chapters addressing statistics, another critical part of data science. The chapter will introduce the basics of statistics and cover the metrics and visualizations of descriptive statistics.

PRACTITIONER PROFILE: SAEED JAFARI

Name: Saeed Jafari
Job Title: Data engineer
Industries: Retail, previously energy
Years of Experience: 3 years as a data scientist, 2.5 years as a data engineer
Education:

- PhD Petroleum Engineering

- MS Petroleum

- BE Petroleum

The opinions expressed here are Saeed's and not any of his employers', past or present.

Background

Saeed was an overachiever when younger and had many interests, although programming wasn't one of them. When it was time to go to college, he did well enough on Iran's tough entrance exam to pursue anything he wanted, but his dad encouraged him to go into petroleum engineering, so that's what he did. He encountered programming again while working on his bachelor's and found that he really liked it. He did well on his degree and started a master's to avoid the mandatory military service he was facing. He got to do more interesting programming in his master's, including solving partial differential equations in Matlab. After completing the master's, he started a PhD in Canada, still in petroleum engineering, but his interests had definitely widened by that time. He continued programming during that degree, this time picking up Python, and also started hearing rumblings about AI and machine learning—there was interest in ML in the oil and gas field, but nobody really knew how to use it yet.

Work

Once Saeed started his first job as a petroleum engineer, ML was still top of mind even though he felt like he knew nothing about it and even data in general. But he was still able to do some time series work to predict pressure changes in a gas pipeline. He now knows that it wasn't done well—everything was overfitted, and he had no idea that was a problem. But eventually he learned, and his models got better. He'd learned enough to know that one problem was that the company didn't have very much data, and he really needed more to make great models. He also had truly discovered his love of working with data because of how interesting it is.

Saeed got a new job and again worked on predicting anomalies from sensor data. This work had to do with trailers delivering gas to gas stations. Life threw a wrench in things because on his second day on the job, the data engineer left and Saeed ended up filling in a lot of those responsibilities. So he learned some data engineering at that point and also liked it. But his main job was still the data science work. He was still working exclusively with Pandas, and his code kept crashing, so he started picking up Spark. He was doing his own ETL and then got in trouble one day for using too much compute, so it was learning after learning after learning. Overall, the project was very successful, but Saeed did end up shifting to more data engineering responsibilities and soon started using tools like DBT and Airflow. He began seeing many data engineer openings and landed a data engineering job in a new industry, retail.

Sound Bites

Favorite Parts of the Job: Saeed loves seeing data go from end to end, starting with an API and into a bucket or table for data scientists or business intelligence (BI) to use. He also loves using AI when appropriate. One nice thing about data engineering specifically is that when you're done with the work, you're mostly done (except for production support), which is better than what can happen with data science.

Least Favorite Parts of the Job: One negative with data engineering is that you're always in the background, somewhat siloed, without a lot of visibility or appreciation. Data science gets more attention, which is both good and bad. One other downside to data engineering is production support, which can be very stressful.

Favorite Project: Saeed's favorite project was the one with trailers unloading gas at gas stations mentioned above. The gas stations in question were actually storage stations, not for distribution to customers. Before, they were providing numbers themselves about what had been unloaded. The company hired six engineers to check the amounts being unloaded on each trailer, and after a year they calculated the amounts, but they could only do some, not all, because of human limitations. So they wanted to automate it, and that's when Saeed started using time series analysis on pressure data. When the unloading starts, the pressure drops, and they figured out how to detect when unloading stops, so they could calculate how much gas is actually unloaded. This was a successful project, and they were soon able to do this on thousands of trailers a day.

Skills Used Most: In everything, communication with stakeholders because if you don't understand what they want, they will never use your product. It can be hard because they often don't really know what they want. Also, he uses many technical skills every day.

Primary Tools Used Currently: SQL, Python, Pandas, PySpark, Airflow, Google Cloud Platform, Snowflake, DBT, GitHub, Terraform

Future of Data Science: There will be a lot of GenAI, even in traditional ML models. It will be communicating between ML pipelines, APIs, and orchestration in general. But it will also never be perfect, so we will need data engineers and data scientists (especially because we'll always need human creativity for unusual cases).

What Makes a Good Data Scientist: Understanding the problem and business need. You can always figure out the technical parts of how to solve a problem, but you have to identify it first.

His Tip for Prospective Data Scientists: First, figure out what you really want to do—some jobs are closer to customers than others. Understand data engineering and data science at a high level and remember data engineering is very back-end while data science is more customer-facing (but not always). Be constantly developing your Python and SQL skills (and always comment your code well—your future self will appreciate you).

Saeed is a data engineer with a range of additional experience in data science and petroleum engineering.

CHAPTER 2

Figuring Out What's Going on in the Data: Descriptive Statistics

Introduction

A lot of people have heard the quote popularized by Mark Twain, "There are three kinds of lies: lies, damned lies, and statistics." Statistics isn't really a well-loved field, and the term itself can have different meanings to different people and in different contexts—for most researchers and data professionals, it isn't a bad word. But it certainly doesn't help its reputation that the giants of modern statistics in the centuries leading up to World War II used it to drive the devastating pseudoscience eugenics.

It's true that for many statisticians, data analysts, data scientists, and academics, inferential statistics is the name of the game, where a subset of data can be used to generalize about the larger group that it's pulled from, which we'll dig deeper into in Chapters 3 and 4.

However, for most people, "statistics" means descriptive statistics, which primarily involves tables, charts, and a few calculations like the average, all of which are used to make something better understood by

© Kelly P. Vincent 2025
K. P. Vincent, *A Friendly Guide to Data Science*, Friendly Guides to Technology,
https://doi.org/10.1007/979-8-8688-1169-2_2

describing aspects it. This is a legitimate part of statistics and is what will be discussed in this chapter. We'll first learn about the origins of statistics because it helps us understand how it's used today. The remainder of the chapter will dive into descriptive statistics, describing several basic measurements like mean and median and then introducing six different basic charts commonly used in industry.

The Beginnings of Statistics

Most people wouldn't expect that a field as valuable and rigorous as modern statistics emerged from some curious men trying to figure out how to gamble most effectively. But that's exactly what happened—in 1654, two famous French mathematicians, Blaise Pascal and Pierre de Fermat, exchanged a series of letters about the math around dice and other games of chance, where Fermat provided a simple but clear solution to "the problem of points," which is essentially about how to divide the prize in a fair game when it's interrupted before it's completed. Pascal built on that to lay out the first calculation of something we now call the expected value, which is fundamental to statistics today. We'll talk in detail about the problem they discussed in the next chapter, but first we'll cover the beginnings of the whole field of statistics.

Early Statistics

Statistics as a recognized discipline existed hundreds of years ago, encompassing the data-gathering practice of governments—the term "statistics" even comes from that usage, from the word "state." The industrial revolution inspired faster-paced data collection, on top of the existing interest in demographic and economic data. But probability had very little to do with statistics even into the eighteenth century, which was really what we would now call data analysis. Developments over the centuries eventually led to the rigorous field that is modern statistics.

Statistics developed as a science of measurement and empiricism. As mentioned in the last chapter, data is a representation of something in the real world, but it can be difficult to measure anything in the real world perfectly. For instance, even something as simplistic as measuring a wall in the days before laser measuring devices was difficult. If you have a long measuring tape that stretches long enough, and someone to hold the end, you can get a pretty accurate measure—you might be off by a tiny bit, like a couple millimeters, and it wouldn't be a huge issue on a 15-foot-long wall. But if your tool is a 12-inch ruler with the zero starting a little bit in from the left, you have a much more error-prone process. Measurement errors can add up quickly, or, alternatively, all your little errors could cancel each other out; you can't be sure how much—or how important—the error is. How do you account for that when talking about your results?

Dealing with errors made in observations motivated quite a bit of the progress in statistics in the eighteenth century, leading to the ideas of the median and the normal distribution, among other concepts. There was some progress in quantifying statistical significance (like today's p-value). Late in the century, a statistician invented several charts that are still used today in descriptive statistics, bringing standardized visualization in for the first time.

In the nineteenth century, rumblings of statistics as a truly mathematical field began, laying the foundations for inferential statistics. But it was also a time when more and more disciplines started using statistics in some way to apply rigor and understanding to their data. Astronomers had already been regularly using techniques from both probability and statistics by the beginning of the century and continued throughout. But astronomers were working with well-understood and defined laws and foundational assumptions, where sciences dealing with humans were more difficult to quantify. Still, psychologists were able to start using developments in experiment design to make more accurate conclusions. Social scientists struggled more because of the limitations of the data available to them (and of the collection methods).

Progress was made late in the nineteenth century, but it really wasn't until the twentieth that social scientists really learned how to make their science truly rigorous.

The twentieth century did bring further developments in descriptive statistics (like the median, standard deviation, correlation, and Pearson correlation coefficient and Pearson's chi-squared test, all of which will be explained below or in the next two chapters), as well as computers, which proved invaluable as the field grew and refined itself.

ACTUARIAL SCIENCE

Actuarial science, which is simply the use of statistics and quantitative methods in insurance and similar fields to assess and manage financial risks and expected outcomes, can be thought of as an applied offshoot of traditional statistics that emerged while the modern field of statistics was still developing, but they've evolved as somewhat separate fields. Actuarial science is a discipline that's most associated with insurance and finance, where rigorous techniques from math and other areas are used to assess risk. It first emerged as there was a need for people to handle large expenses down the road, and insurance was the solution. But knowing how much an individual needed to pay wasn't obvious, so they had to develop ways of quantifying risk by predicting future events (like death) and how to handle the money being paid in.

A London man named John Graunt made the first life expectancy tables in the mid-seventeenth century (there's more info on his work below), and the field of actuarial science grew from there. There were slow further developments in this vein, but it wasn't until the mid-eighteenth century that a life insurance company first set their rates based on modern calculations based on the anticipated entire life of the insured. These calculations were based on deterministic models but were still revolutionary in the industry.

The field grew tremendously in the twentieth century, gaining some of the same rigor that statistics was getting. With the advent of computers, actuaries' forecasting ability was expanded significantly.

Examples of Statistics in the Real World

Example 1: Demographics in Seventeenth-Century London

The John Graunt who created the first life expectancy tables had a shop that sold fabric in London in the mid-1600s, but he had a sharp mind loaded with curiosity for things beyond drapery. This led him to study the London Bills of Mortality, a record of burials of people in the Church of England in London. You can see an example of one from 1606 in Figure 2-1. This data source was not perfect, as it recorded burials rather than actual deaths, so if a Londoner was buried outside of London, there would be no record. Additionally, members of faiths other than the Church of England were not recorded, and it also included very limited attributes. In that century, cause of death was recorded, but age was not. Graunt was aware of this, but it was not a hindrance to him. He simply studied the Bills of Mortality and used other information to determine what the age of death was in each burial and recorded that in his own table. One of the interesting things about Graunt's interest in the Bills was that he scrutinized how they were created in order to understand the quality of the data. He believed that cause of death was frequently misreported. This attention to detail is invaluable in data analysis work even today. Despite the obstacles, Graunt was still able to work with the data to discover fascinating insights.

He's widely considered the first human demographer and one of the first epidemiologists. He created the first life table, which gave the probability of surviving to a given age. From his study of the causes of death in the records, he was able to determine basic morbidity and mortality of the common diseases that were reported, despite the fact that he knew the data wasn't perfectly accurate. This work allowed him to conclude that death rates were higher in urban locations than in rural ones. He also observed that more girls were born than boys and that the mortality rate was higher for males than females.

His work is especially interesting because he dealt with one of the things that all people who work with data will face—messy, messy data. But he used ingenuity and creativity to work around that and still come up with significant findings, something that is often required of data analysts (and scientists).

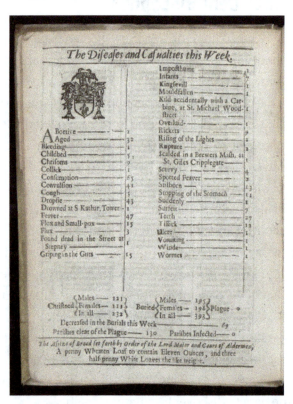

Figure 2-1. *Example Bill of Mortality from 1664 London. Source: "Bills of Mortality Feb 21 - 28 1664,"* `https://commons.wikimedia.` `org/wiki/File:Bills_of_Mortality_Feb_21_-_28_1664_` `Wellcome_L0043358.jpg,` *from Wellcome Collection,* `https://` `wellcomecollection.org/search/works,` *operated by Wellcome Trust, a global charitable foundation based in the United Kingdom*

Example 2: A Desperate and Unheeded Warning

I was in my fifth-grade reading class when my teacher left the room to watch a bit of the US Space Shuttle *Challenger* launch. When she returned, her face was ashen, and it was the first time I'd ever seen an adult that shaken. She told us that the shuttle had exploded and all the astronauts had died, including the very hyped-up first Teacher in Space. There was so much joy and national pride in the space program in the 1980s, and this event destroyed the nation's confidence.

Was it avoidable? Like many technological or industrial disasters, the disaster on that particular day was 100% avoidable. But what makes this one stand out is that while most disasters come about because of negligence—somebody slacking on safety protocols, for instance—there were a handful of engineers who knew it was too cold for a safe launch who begged management to stop the launch. But management ignored the pleas and went ahead with it. The launch had already been delayed once, and there was a feeling that the country was counting on NASA.

A lot of people have written about this disaster, including the often-derisive Nobel Prize–winning physicist Richard Feynman. It all came down to two rubber O-rings that kept the hot, compressed gas of each rocket booster contained where it should be. It was known that a failure in an O-ring could be catastrophic, but there were actually two, so they were considered to have a built-in fail-safe. But although there were two, they were the same. Engineers knew that if one failed because of something environmental, so could the other.

The morning of the launch was supposed to be cold. In the days before, four engineers were deeply concerned about the O-rings failing. They had seen the O-rings lose flexibility in cool temperatures, even in the 60s (Fahrenheit). This can allow hot gas to hit the O-rings and degrade them further. Launch morning, it was actually below freezing, an extreme difference from any testing and real launches they had data on. The engineers tried desperately to convince the decision-makers, but they failed to convey why they were so concerned.

A government report on the disaster highlighted one place things had gone wrong. The engineers shared a particular chart showing temperatures and counts of O-ring failures with management, but it simply didn't convey the most important information. The engineers understood it, but couldn't make the management understand. This is basically a failure of communication between technical and nontechnical people.

Engineers are used to looking at data and understanding that lack of evidence of something doesn't mean it's definitely not present. Figure 2-2 is the chart that was shared pre-launch.

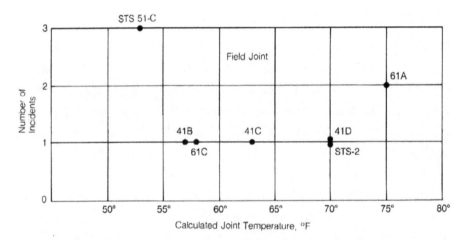

Figure 6
Plot of flights with incidents of O-ring thermal
distress as function of temperature

Figure 2-2. *US Space Shuttle flights with O-ring thermal incidents in the 1980s. Source: "Report of the PRESIDENTIAL COMMISSION on the Space Shuttle Challenger Accident,"* https://www.nasa.gov/ history/rogersrep/v1p146.htm

Each black dot represents a flight with an O-ring that experienced thermal distress on actual flights, showing the temperature the incident(s) happened at (the X-axis) and the number of incidents associated with it (the Y-axis). The astute observer will have noticed that the lowest temperature of any incident on the chart is about 53°. There's nothing anywhere near freezing. There hadn't been any launches below that temperature. This made the engineers nervous, but not the managers. But there is one point that even the managers should have understood—it probably wasn't a coincidence that the most incidents were observed at the lower temperatures.

But they didn't get the severity of the risk of flying into the unknown. Later, in the government report on the disaster, the authors proposed a chart that would have made things much clearer to anyone, technical or not. Figure 2-3 shows the same chart with some additional info—the data points on all the launches that had no incidents.

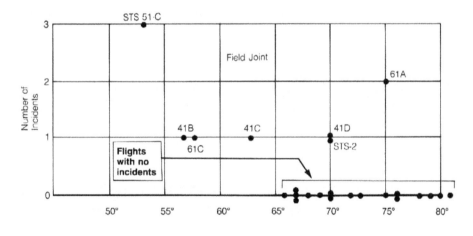

Figure 7
Plot of flights with and without incidents of O-ring
thermal distress

Figure 2-3. *US Space Shuttle flights with and without O-ring events in the 1980s. Source: "Report of the PRESIDENTIAL COMMISSION on the Space Shuttle Challenger Accident,"* `https://www.nasa.gov/history/rogersrep/v1p146.htm`

This chart makes it abundantly clear that all of the launches that had no incidents happened about 66° or higher. You don't have to be a rocket scientist to conclude that lower temperatures are inherently riskier. And although it's common to limit a chart's axes to just outside the range of your data, perhaps if they had started the X-axis at 30° and emphasized how very far 66° was from 31°, their warnings would have been heeded. Instead, seven astronauts died, a nation of people watching the launch live were traumatized, and a low point in the US space program began.

Descriptive Statistics

Descriptive statistics is the science and art of understanding the general characteristics of a dataset and identifying anything special or out of the ordinary about it. Descriptive statistics is a core part of data analysis and data science, and we would almost never undertake a project without doing descriptive statistics—referred to as EDA, or exploratory data analysis, in data analysis and science—first. The simple fact is that you can't do meaningful work with data if you don't understand it. Descriptive statistics helps you get there.

Descriptive statistics primarily involves calculating some basic metrics and creating a few charts and other visualizations, which provides a good overview of your data. The nice thing about it is that it's fairly intuitive once you've learned how to carry it out. We'll look at some different datasets in this section to understand the metrics and visualizations that can be done. We'll use a dataset of video game scores to look at the different metrics and some other datasets for the visualizations.

It's also important to be aware that almost all of the metrics and visualizations in descriptive statistics require numeric data—either interval or ratio. Some descriptive statistics can be done on categorical data, but it is minimal and less informative.

An Initial Look and a Histogram

It's pretty common to look at some plots to get a good sense of your data. This can help you understand aspects of it that might influence decisions you make later. We often talk about the distribution of the data, which generally refers to a few characteristics of a single field. For example, if we have some data about video games and we're curious what the ratings are of all the games, we could do a chart to see what the ratings look like. A first instinct might be to just plot all the values in a bar chart like the one in Figure 2-4, which shows a column for each individual rating with the rating value on the Y-axis.

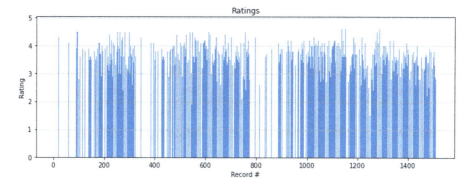

Figure 2-4. *Arbitrarily ordered ratings of video games*

The X-axis is simply the record number, so the order is not related to the actual rating. We'll talk more about bar charts later, but the main thing that's clear from this image is that it's impossible to get a sense for anything about the data, except that the ratings seem all over the place, but on the higher end. Our first instinct at clarifying things might be to add some order to it. Sorting it will help you understand it a little better, as you can see in Figure 2-5.

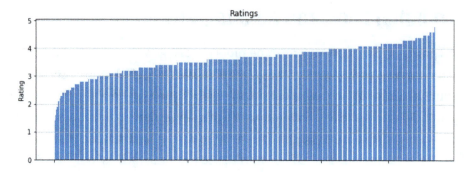

Figure 2-5. *Video game ratings in numerical order*

Figure 2-5 lets you see pretty quickly that there are only a handful of ratings under 3, with most in the 3–4 range. This is useful, but not super useful, as it's still hard to really say what's going on in the overall dataset with all of these individual values.

Fortunately, there's an easy solution: the histogram. We're going to cover how histograms are created in a later section in this chapter, but as a quick intro, the *histogram* summarizes the shape or distribution of the data by showing frequency counts of bucketed data, which in this case basically means we'll look at how many of each average rating we have (ratings can be any value between 0 and 5 including decimals). It reveals a lot that helps us understand the statistical calculations that will be discussed below, but Figure 2-6 shows a simple histogram of the ratings data.

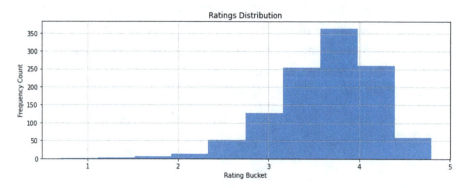

Figure 2-6. *Histogram of video game ratings*

This time the X-axis shows the range of ratings in each bucket (buckets are usually automatically determined by the software we use), and the Y-axis shows the total number in each corresponding bucket. This makes it clear that most ratings are in the 3.5–4.5 range, which was harder to see than in Figure 2-5.

This distribution actually looks a little like the familiar normal distribution (also called the bell curve) that a lot of us will have seen before. But because the median is relatively high and the score is capped at 5, the right side doesn't stretch out as much as the normal distribution does. We'll be talking more about the normal distribution in the next chapter, but for now we have a better understanding of ratings, which are mostly on the higher end of the scale.

Try to keep this histogram in mind as we cover the many statistical measures we calculate below.

Metrics

We can measure a variety of things in a dataset, starting with summary statistics, which can be broken down into measures of location and measures of spread. Summary statistics give us a picture of what's going on in a particular column in a dataset. We'd never include multiple columns in any of these calculations.

Measures of Location

Measures of location are also often called measures of centrality, because they really are trying to give us a sense of where the "middle" of the data is, which helps us understand what is typical of our dataset. If we have some heights of American people and the mean is around four feet, we'd probably guess that the data records heights of children rather than adults.

The primary measures of location include mean (and its variation trimmed mean), median, and mode. With the exception of the mode, all of these require numeric data. The mode can be determined for numeric or categorical data, but in this case numeric is more difficult to work with and usually requires bucketing. We'll see why below.

Most people know how to calculate the *mean* (or *average*), which is simply the sum of all the individual data points divided by the total number of data points. This does give us a sense for where the middle of the data is, but it can be distorted—or skewed—by *outliers*, values that are different from most of the rest of the data. The *trimmed mean* is an adjusted calculation, with a certain percentage of extreme values at either end removed, intended to keep outliers from giving a nonrepresentative mean. A 5% trimmed mean cuts off the top 5% and bottom 5% of all the data, meaning that only the middle 90% is used. We can also trim off specific counts at either end—for instance, in Olympic scoring by judges, they remove the minimum and maximum scores before calculating the mean (this is also sometimes called the *modified mean*).

OUTLIERS

The casual definition of an outlier is basically a data point that is very different from the vast majority of the other data. For example, we all know that a basketball player who is 7'2" is definitely an outlier in human height. But what about someone who's 6'4"? If you come from a short family, that might be extreme, but there are plenty of tall families who have several members that tall. Although there is no universally agreed upon way to define outliers statistically, there are some techniques that can be used to identify them. But sometimes when the dataset is small or if there are values that are just extremely far from the others, just looking at the values is enough to identify them.

The *median* is another key measure of location, as it can show a truer middle of a dataset than mean sometimes does, especially when the data is very skewed (skew will be addressed in the next section). The median is simply the exact middle of the data once it's been placed in numeric order. If there are an odd number of data points, it's the exact middle data point; otherwise, it's the mean of two middle data points. The median is considered to tolerate outliers better than the mean.

The *mode* is simply the value that appears the most in the dataset. This is one that works on both numeric and nominal data. It is another way of giving us the sense of what is typical of the data. It's a simple calculation with categorical and ordinal data, plus any numeric data that is evenly spaced, like whole numbers. If we're working with numeric data that has many possible values that can be close to each other, it usually makes sense to bucket or truncate them before calculating the mode. For example, if student grades can be any calculation, it might be more valuable to group values like 92.1, 92.25, 94.0, 94.25, and 94.33 into just two buckets by cutting of the decimal, leaving you with two 92s and three 94s, with 94 being the mode.

Although it's not always important which measure of location you should use, they can be best in different situations. The mean is good in situations where the data is fairly symmetrical and without many outliers. Things like human height tend to have this characteristic. The median is good when the data is skewed one direction or the other. House prices can be like that, because most houses are on the relatively low end, but there's really no upper limit on how expensive a house can be. If you have the following five house prices $125,000, $150,000, $175,000, $200,000, and $17 million, your dataset is very skewed. The mean is over $3 million, even though most of the prices are way lower. The median here would be $175,000, which is obviously intuitively more in line with the actual values.

CALCULATING MEASURES OF LOCATION

We'll start with a subset of our exam scores with the following ten data points:

67, 11, 93, 53, 79, 60, 74, 93, 99, 78

The first thing we'll want to do is put them in order:

11, 53, 60, 67, 78, 79, 74, 93, 93, 99

We can see that the middle two values are 78 and 79, and there's an intuitively clear low outlier of 11.

We can calculate the **mean** by summing all values and dividing by the count, giving us 70.7

If we want to calculate a **trimmed mean**, we can get a 10% trimmed mean by taking off the top and bottom values (both 10% of the total values) to get 74.6.

The **median** is easy to calculate as we observed above that the middle two values are 78 and 79, making the median the mean of those two, or 78.5.

The **mode** is also easy to calculate here because we're working with a small number of whole numbers. The only number that appears more than once is 93, so that's the mode. If we'd had no repeats here, we wouldn't have a mode.

Data analysts and data scientists rarely have any reason to calculate these things manually because we use software or code, but it's important to understand what these labels truly mean.

Measures of Spread

Understanding the middle of your data is useful, but you also want a sense for how spread out your data is. Additionally, if you have a situation where there are a lot of values at different ends of the scale, the "mean" isn't typical at all. Imagine a family where the men are very tall and the women unusually short—if we take the average of all of their heights, we'll get something in the middle, which doesn't represent a typical family member at all.

Looking beyond central measures can help us understand how different the various data points are from each other. *Measures of spread* help with this and include the range, interquartile range (IQR), variance, standard deviation, skew, and kurtosis.

When working with spread, you usually start by noting the *minimum* (the lowest data point) and *maximum* (the highest data point). The most basic measure of spread is the *range*, which is simply the minimum subtracted from the maximum.

Data can also be divided into quarters by ordering the data and splitting it half at the median and then splitting each of those in half by their median, making four equal parts. Each quarter part is called a *quartile*. Notably, we refer to the quartiles in a certain way, by referring to the max point of each one, so quartile 1 is at about 25% of the data points, quartile 2 at 50% (the median), and quartile 3 at 75%, leaving quartile 4 as the top one at 100%. Note that quartile 2 is simply the median of the entire dataset. A related measure is called the *interquartile range*, which is simply quartile 1 subtracted from quartile 3. If you've ever seen a box plot, you will have some familiarity with quartiles. There will be more info on the box plot below.

Note that we also have a measure similar to the quartile called the percentile. These are whole numbers between 0 and 100 that represent what percentage of the data is below that value. The 90th percentile indicates the number at which 90% of all data values are below. The four quartiles correspond to percentiles 25, 50 (the median), 75, and 100.

Another two measures of spread are the variance and the related standard deviation. *Variance* is a single number that gives a sense of the spread of the data around the mean. Calculating it involves summing the square of each data point's difference from the mean and then dividing the total by the number of data points. The squaring ensures that opposite direction differences from the mean don't cancel each other out. *Standard deviation* is just the square root of variance. It's often preferred because it's in the same units as the data points, so it's more intuitive.

Two other measures that aren't calculated as much in practice are skewness and kurtosis. Data is considered skewed when the spread isn't symmetric, which occurs when the mean and median are quite different. When data is skewed, the top of the histogram is not centered. One simple method of estimating the *skewness* is by calculating the difference between the mean and median, dividing that value by the standard deviation, and then finally multiplying that by 3. If the median is larger than the mean, the data is skewed negative and positive if the median is smaller than the mean. Only when the mean and median are the same can the data be said to be unskewed.

Kurtosis also looks at the tail, but it looks at how far out the tails (the low counts) on the histogram stretch, so it's sometimes said to be a measure of "tailedness." Calculating it requires several steps. First, for each data point, raise the difference between it and the mean to the power of 4. Then take the average of that value, and divide that by the standard deviation raised to the power of 4. Sometimes when we want to look at kurtosis in the context of a normal distribution, 3 (the normal distribution kurtosis) is subtracted from this number to get the *excess kurtosis*, or the amount attributed to the difference of the distribution from the normal.

CALCULATING MEASURES OF SPREAD

We can start with the same exam score data as above with the following ten data points, already sorted:

11, 53, 60, 67, 78, 79, 74, 93, 93, 99

We can clearly see the **minimum** is 11 and the **maximum** is 99.

The **range** is simply the difference between the minimum and maximum, or 88.

Calculating the **quartiles** isn't difficult with this set. First, we split the data in half around the median, 78.5, which we calculated above. Then each half is split at the median of that half, leaving us with the following quartiles: quartile 1 is at 60, quartile 2 at 78.5, and quartile 3 at 93.

The **interquartile range** (IQR) is simply quartile 1 subtracted from quartile 3, or 33.

Variance involves summing the squares of the difference between each value and the mean (70.7 from above) and dividing that sum by the number of values, 10, which gives us 583.1. That's obviously hard to really understand, so we take the square root to get the **standard deviation**, 24.2. That's easier to understand since it's in the same units as the exam scores.

Skewness is one of the less common values, but we can still calculate it by taking the difference of the mean and median, dividing that by the standard deviation, and multiplying that by 3, giving us a **skewness** of –0.97.

Kurtosis is also less common and not very intuitive, but we can still calculate it. We calculate the difference of each point from the mean, then raise each of those to the power of 4, then take the mean of those values, and finally divide by the standard deviation raised to the power of 4, to get a **kurtosis** of 4.1.

These measures all give us a better sense for how our data is spread out.

Visualizations

There are a variety of charts and other visualizations that are considered a part of descriptive statistics. Some of these aid in understanding the measures of location and spread visually, but others can reveal information about relationships between different fields.

Scatterplots

Even though the scatterplot is one of the most basic plots out there, it can be incredibly useful. It involves plotting one variable against another, with one on the X-axis and the other on the Y-axis. This type of plot can only be used with two columns of numeric data. A dot is placed on the plot for every row based on the values in the two columns. It can look like a bunch of dots haphazardly filling the plot area, but at other times patterns can be seen, such as correlation. And looking for a correlation or other patterns is one of the most common reasons for looking at a scatterplot.

One of the easiest ways to understand a scatterplot is to look at one with a correlation. We have a dataset with heights, weights, and gender of a group of kids with asthma, plus some measurements related to their airways (how well they work under different conditions). This dataset isn't going to tell us anything about how their height and weight compare to kids who don't have asthma, but we can look at how the values we have interact with each other.

If we make a scatterplot of their ages and height, we can see a clear pattern in Figure 2-7.

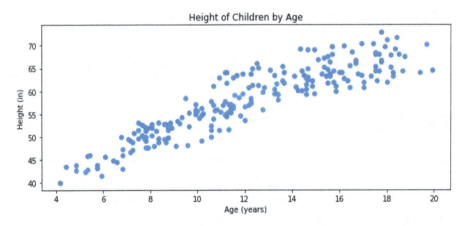

Figure 2-7. *Scatterplot of heights of a group of children*

It's easy to see that as the kids get older, they also get taller. There's a pretty clear trend of the two variables increasing together, which is one relationship variables have together when we say they are correlated. We'll talk about correlation and how to measure it later in the chapter, but this quick visual inspection is one of the main reasons we use scatterplots.

However, while the basic scatterplot can be great for identifying correlation and other patterns, the fact that we only plot two variables at a time can be a significant limitation when we have many variables, which is common. There are a couple ways around this: we can plot different classes on the same chart in different colors, and we can get a third variable in there by changing the size of the point to correspond with a third variable.

In the case of our dataset, we also have gender. Everyone knows there are general differences between girls and boys in height. It's easy to create a plot that shows this, simply by making the dots different colors based on gender, which we can see in Figure 2-8.

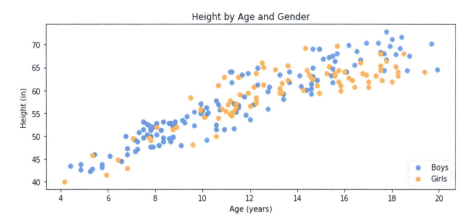

Figure 2-8. *Scatterplot of heights and genders of a group of children*

This simple modification reveals something—girls seem to reach a max height earlier than boys and stop growing, while boys keep growing longer.

The other way we can add a variable is by changing the size of the point by creating what's now called a bubble chart. It's quite common with bubble charts to also make the bubbles slightly transparent since the bigger ones can cover the other bubbles. Our dataset has some asthma-related measurements, so we included one of them, a number that indicates the measured resonant frequency each kid is exposed to when their airway goes from working to not working well, which can be seen in Figure 2-9.

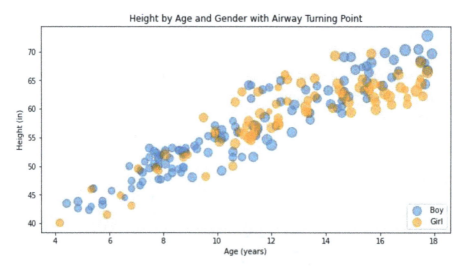

Figure 2-9. *Scatterplot of heights, genders, and asthma metrics of a group of children*

Although there aren't huge differences in the bubbles in this example, it's still clear that some bubbles are bigger than others. The differences in the bubble sizes depend on the range of values in the variable. Bubble charts are valuable mostly because we can get more values on the chart than a basic scatterplot. With color, we've got four variables in a two-dimensional chart.

Bar Charts

The *bar chart* is another simple visualization, and because it is generally intuitive, it gets used a lot in the media. We saw one type above in Figures 2-4, 2-5, and 2-6, but we were focused on understanding the distribution of ratings, and there's a lot more that a bar chart can show. It's used most with nominal data, often displaying counts of values per single variable. As we saw above, a histogram is a type of bar chart, but it is for continuous data and will be discussed separately below. The simplest bar charts show values of a single variable across the X-axis and usually counts or percentages on the Y-axis. The bars in bar charts can be shown vertically or horizontally, although vertically is most common.

If we look again at our video game dataset, we can see what the average rating is each year since 2015, which is shown in Figure 2-10.

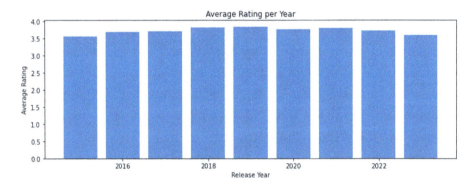

Figure 2-10. *Simple bar chart with average video game ratings over nine years*

We can see that the ratings don't vary much year to year, but there was a small trend up until 2019 and then a small trend back down. It just happens to be very symmetrical.

This sort of chart only gives minimal info, and we have a lot more in the dataset, including the development company of each game. Maybe Nintendo is wondering how they're doing compared to EA, with all other companies clumped together into a third group. We can look at that easily. *Grouped bar charts* show several related columns next to each other. Figure 2-11 is a grouped bar chart showing the same data as in Figure 2-10, but broken down by company.

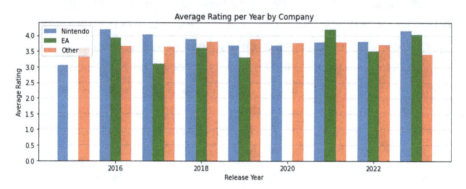

Figure 2-11. *Grouped bar chart of average video game ratings over nine years by company*

Figure 2-11 allows us to see that Nintendo and EA have different trends and that the "Other" group still follows the trend we saw in Figure 2-10 when all the companies were grouped together. But if we look only at the Nintendo bars, we can see that 2015 was a bit of an outlier, at just over 3. The next year the average shot up to over 4; then over the years it declined and started rising again. EA seems all over the place, and there are two years in this range that they didn't have any releases with ratings.

There's another important type of bar chart, called a *stacked bar chart*, which is perfect for looking at how different values proportionally relate to each other. This involves putting each company's bar on top of the other for the same year, so the height of the combined bar would be the sum of all the bars in that year. This doesn't make any sense to do with ratings, but we also have the total number of reviews per game, so we created Figure 2-12 to show a stacked bar chart with Nintendo and EA.

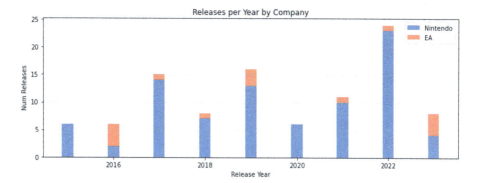

Figure 2-12. *Stacked bar chart with number of video game releases per year from Nintendo and EA*

We can easily see that Nintendo releases many more games than EA does and generally Nintendo has a much higher proportion of the total games released by both together. There's one final type of bar chart that would make the specific proportion even clearer. A *segmented bar chart* shows the same data as a stacked bar chart but as literal proportions, where all the full bars sum to 100%. Figure 2-13 shows the same data as in Figure 2-12, converted to percentages.

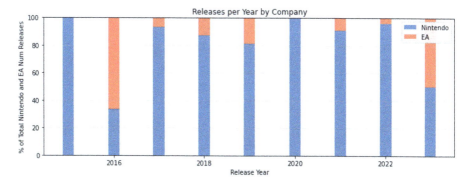

Figure 2-13. *Segmented bar chart with number of video game releases per year from Nintendo and EA*

Nintendo released a much higher proportion of the games than EA did in every year except 2016, like we saw above. This type of chart would be more interesting if we looked at every company in the data rather than just these two, but it would also get unwieldy quick. There would be a lot of thin lines, but the most productive companies would stand out clearly.

Histograms

We talked a bit about the histogram above. A *histogram* is a special type of bar chart that's used with interval or ratio data. The reason it's considered a separate type of chart is that it gives a specific view of data and requires an extra step called *binning*, which splits numbers into batches of similar values. This also means that we won't have too many columns because there's only one per bin. Histograms show counts of occurrences, although we usually talk about this as frequency rather than counts.

We can look at the video game data again, this time focusing on the number of reviews. Games have between 0 and around 4,000 reviews, and if we just look at them all again in Figure 2-14, we can see those large numbers of reviews are rare and most are below 1,000.

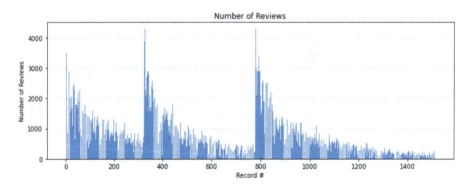

Figure 2-14. *Bar chart with each game and the number of reviews it has, arbitrarily ordered*

A histogram will tell much more about the most common number of reviews, as it's hard to really understand that from this chart. We can create a histogram with bins automatically set, as is shown in Figure 2-15.

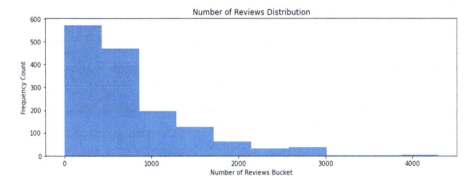

Figure 2-15. *Histogram of video games by number of reviews*

Figure 2-15 makes it clear just how unusual it is to have more than 1,000 reviews. Since so many are under 1,000, we thought it might be worth looking at the distribution of the number of reviews at or below 1,000. We can see this in Figure 2-16.

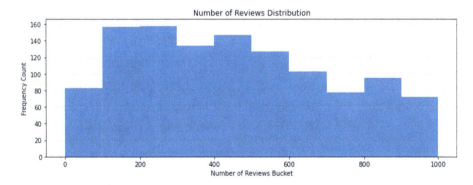

Figure 2-16. *Histogram of video games by number of reviews (with up to 1,000 reviews)*

One nice thing about Figure 2-16 is that the bins are clear, as there are 10 out of the 1,000 total reviews, so each is 100 wide. This distribution is interesting—it doesn't really look like a normal distribution because the tails aren't that different from the middle. There are more occurrences of between 100 and 700 reviews, but otherwise it looks there isn't a huge amount of variability in the number of reviews each game has.

Something to keep in mind is that most of the software we use to create histograms can automatically determine the bin widths, but sometimes different widths can yield very different-looking histograms. It's never a bad idea to try different numbers of bins to see if anything interesting pops up. This is also a good reminder of why it's important to try different things when you are doing EDA. Sometimes you make one chart that reveals nothing, but a little variation on that one can unveil valuable info.

Here we've looked only at histograms that have bins of equal widths, which is the most common type, but they can have different widths. For instance, there's a method called Bayesian blocks that creates optimally sized bins based on certain criteria. It's good to know about these, but we don't often need them.

The Box Plot

You've probably seen a *box plot* before, a chart with rectangles with a line across the middle or so and lines ("whiskers") extending from the ends of the rectangles. Box plots are constructed from several different spread and location measures, which you'll have calculated when doing your summary statistics. The components of a typical box plot include

- First quartile
- Median

- Third quartile

- Minimum

- Maximum

Some box plots also account for outliers and adjust the whiskers accordingly. If outliers are to be shown, the traditional way is to limit the whisker to no more than 1.5 multiplied by the interquartile range (Q3 – Q1) and then outliers are shown as circles farther out from the ends of the whiskers.

For an example box plot, we can return to the children's height and weight data from above, looking at BMI for boys and girls at different ages in Figure 2-17. Healthy BMIs change based on kids' ages, so the age groupings were determined based on ranges provided by the CDC.

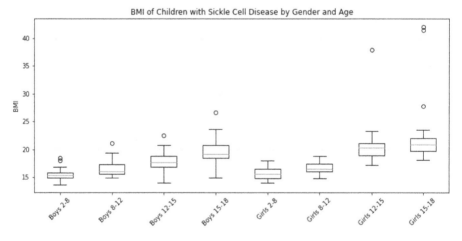

Figure 2-17. *Box plot with BMI of children with sickle cell disease by gender and age*

In this box plot, the box for each group shows the median in orange and the first quartile as the bottom bound of the main box and the third quartile as the top bound of the box. The bottom horizontal line is the minimum and the top line the maximum (the max value within 1.5 * the interquartile range). Finally, the dots above the max are outliers.

One thing that stands out on this chart is how much more extreme the outliers are for the girls than the boys. Also, girls' BMI seems to quickly increase after the age typical of puberty, while boys' BMI tends to increase more steadily. Additionally, both girls and boys are mostly in the healthy range of BMI (according to the CDC guidelines), and in fact they're on the low side. It's possible that having sickle cell disease keeps kids ill enough that they don't grow as much as other kids their age.

Line Charts

One of the most common charts created when the data has a time element is the *line chart*. Although a line chart could be created with different types of data, it's really only suitable for numeric data, and there should be a natural way the data connects. Time is commonly on the X-axis and the numeric value on the Y-axis. There are different variations possible with line charts, and we'll look at some with some pizza sales data. Figure 2-18 shows the total number of pizzas ordered in the dataset by day of the week.

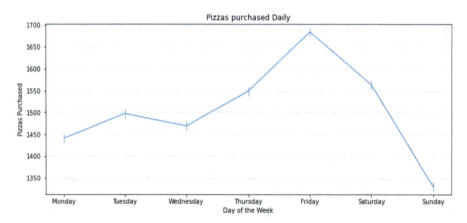

Figure 2-18. *Line chart with number of pizzas purchased by day of week*

We can see that Thursday, Friday, and Saturday are the biggest days. But beyond that, this isn't the most informative plot. One thing that is great about line charts is how easy it is to have multiple lines. Although there certainly is a limit to how many can be added before the chart becomes unwieldy, two to four is generally a good guideline. Figure 2-19 shows the same data broken down by mealtime (determined by time of day of the order).

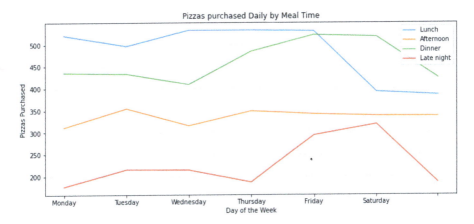

Figure 2-19.   *Line chart with number of pizzas purchased by mealtime and day of week*

It's not surprising that lunch and dinner are the busiest times, and it's also clear that they sort of mirror each other—lunch is busiest Monday through Friday but slow on the weekends, and dinner is less busy at the beginning of the week but picks up later in the week. Afternoon sales are pretty steady, and the late-night crowd is buying more pizzas on the weekend than during the week.

Multiple lines on a chart can be very illuminating, often revealing a relationship between the two plotted variables. But there is another basic but powerful thing that can be done with line charts: adding a second Y-axis. This can be useful when you want to plot two variables but they are on completely different scales. Figure 2-20 shows the total number of pizzas sold on the left Y-axis and the total amount of sales of only large pizzas on the right Y-axis. It's not too surprising that they track fairly close together, but it is interesting that Thursday deviates from the all-pizzas-sold line.

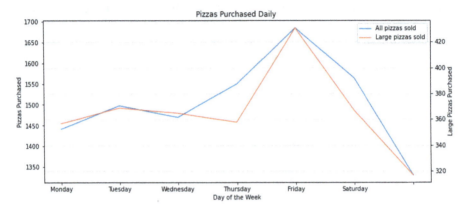

Figure 2-20. *Dual-axis line chart with all pizzas sold by day of week and total sales of large pizzas by day of week*

All of these charts give us good information. It's especially important with charts like these that appropriate labels are added to the chart, so that it's clear what each line represents and, if there is a second Y-axis, which line goes with which axis.

The Pie Chart

Most data analysts and data scientists don't like *pie charts* very much, regarding them as too simplistic. But there are times when they reveal proportions very quickly and easily, especially for our nontechnical stakeholders. We don't always share EDA with stakeholders, but there are times we do, so using one may be in order. Pie charts should also be limited to the number of values they show, and three to six is often a good range. Otherwise, they can get unwieldy fast, just like line charts with too many lines. Figure 2-21 shows the breakdown of the four different types of pizza sold.

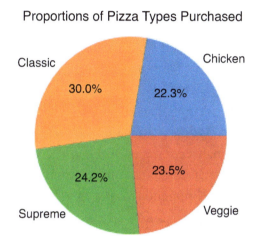

Figure 2-21. *Pie chart of proportions of pizza pie types purchased*

This makes several proportions clear. It's obvious that there are more classic pizzas sold, but otherwise the others are sold in close proportions. But you can't get much else out of this chart. While it can be visually impactful, it isn't that informative, and it even becomes hard to read if there are more than a handful of values. Pie charts are usually more useful for showing nontechnical people than for your own EDA. But every once in a while, a quick pie chart can make something clear.

Key Takeaways and Next Up

In this chapter, we looked at descriptive statistics, which is crucial in data analysis and data science, because it helps us understand our data and know its limits. It primarily encompasses measures of location, measures of spread, and visualizations. Measures of location tell us about where the center of our data is, or what is typical of it. Measures of spread help us understand how much the data varies—are most values close to the center or are they all over the place? Visualizations are another window into data, allowing us to see important relationships, trends, and other revealing information about it. Understanding the data we're working with is critical to doing good data analysis and data science.

We'll continue our discussion of statistics in the next chapter, moving on to the foundations of inferential statistics. First, we'll look at the history of that area and then at probability and probability distributions, sampling, and experiment design. All of these topics are critical for setting things up to do inferential statistics, which we'll talk about in Chapter 4.

PRACTITIONER PROFILE: CINDY MINERA

Name: Cindy Elizabeth Chavarria Minera

Job Title: Business intelligence analyst

Industries: Healthcare, military

Years of Experience: 5 years in data roles

Education:

- MA School Psychology

- BA Psychology

- AA Liberal Arts

The opinions expressed here are Cindy's and not any of her employers', past or present.

Background

Cindy Chavarria was a first-generation college student and hadn't been exposed to a lot of career options when she was in college. The first time she took a stats course for her psychology degree, she felt like she had no idea what was going on and ended up having to retake it. The second time around she was determined not to fail and sought out a tutor, and then something almost magical happened—the tutor was really good, and Cindy understood stats for the first time and discovered she loved it. From then on, she sought opportunities and found mentors who helped her develop her skills. She was able to work with a researcher and even had her first published paper while she was still an undergrad. She continued on to grad school but found that her program wasn't as research-focused as she'd wanted, but she found a short-term job with another researcher and was able to get more experience doing statistical analysis for the researcher. She was able to start learning R in her classes. After finishing the degree, she entered the workforce.

81

Work

After finishing this degree, she knew she wanted to work in a research role and that she wanted to be in a position to help people. The job market was tough, but she found a role at a nonprofit doing a combination of data analysis and data management. She enjoyed that but soon wanted to expand her technical experience and move into a business intelligence role. In that role she wasn't doing as much data analysis as she'd hoped, but she was getting valuable experience with different tools and learning more about the kind of work she really likes to do, which will inform her entire career.

Sound Bites

Favorite Parts of the Job: Cindy loves working with one team at work that is trying to shift to making truly data-informed decisions. They are asking for data and data analysis in their area, and they listen to the findings she comes up with. It's made her realize she's interested in prescriptive analytics.

Least Favorite Parts of the Job: Meetings, so many meetings that could have been an email. It's a real problem in the corporate world.

Favorite Project: She did a fairly simple statistical test that showed that the results of a particular program weren't significant, which led them to dig into why. In the process, they discovered that many of the participants in the program didn't want to be there. After adjusting the program and focusing on the people who did want to participate, they found that the results were significant, which was very gratifying.

How Education Ties to the Real World: Nothing in the real world is as neat and tidy as in school. She especially found that in the real world, nonparametric statistics are needed a lot more because the many nice statistical distributions aren't as common in the real world as in statistics textbooks.

Skills Used Most: Soft skills in general—she has to be easy to work with and gain people's trust for her and her work.

Primary Tools Used Currently: SQL and Excel mostly, occasionally Power BI and R

Future of Data Analysis and Data Science: Rapid change is going to continue, and things are getting more and more specialized.

What Makes a Good Data Analyst: Patience, good social skills, and curiosity. Because you're working with people from all sorts of different backgrounds and skill levels, it helps to know how to talk to people. Curiosity and a just-try-it attitude are important for learning and problem-solving. You can't let fear of the unknown hold you back.

Her Tip for Prospective Data Analysts and Data Scientists: This applies to everyone, but especially girls and women, don't let fear and competition get in the way of learning and achieving what you want. If people tell you that you can't learn something, prove them wrong.

Cindy is a data scientist and business intelligence developer with experience in several fields including health and education.

CHAPTER 3

Setting Us Up for Success: The Inferential Statistics Framework and Experiments

Introduction

We know that statistics was a field that developed a lot over the years, but the twentieth century was a huge year for modern statistics. Some of the biggest developments were in inferential statistics, where you can take a subset (sample) of a large population and do some math on that subset and then generalize to the larger population. This is critical to so many fields, where it's impossible to measure every single thing. Probability, sampling, and experiment design were all critical to allowing for good inference. The work done in the late nineteenth and early twentieth set us up for data in the computer age.

© Kelly P. Vincent 2025
K. P. Vincent, *A Friendly Guide to Data Science*, Friendly Guides to Technology,
https://doi.org/10.1007/979-8-8688-1169-2_3

This chapter will first go a bit into the history of modern statistics and give some interesting examples of statistics in real life. I'll talk about methods of counting, which are critical to understanding probability, and probability theory. Then I'll cover several important probability distributions that are used extensively in statistics. I'll give a quick explanation of the central limit theorem before moving on to sampling, or selecting your data, and finally experiment design.

The History of Modern Statistics: Part 1

The twentieth century saw the fields of probability and statistics finally come together when scientists formalized the math and created modern mathematical statistics. Statistics spread widely into many fields: the social sciences, economics, physics, and chemistry, among others. Unfortunately, early-twentieth-century work in statistics is tainted by the association of many individual statisticians with eugenics, the thoroughly debunked pseudoscience that drove Naziism and a lot of racist, sexist, and ableist policies in the first few decades of the century—many of which are still in place today. Eugenicists used the language of statistics and science to drive and support the creation of these policies along with inhumane actions like forced sterilizations.

Wrapping up racism and other "isms" in the veneer of science has meant that some people still believe in the tenets of eugenics. Other people wonder how anything good like statistics could emerge from eugenics if it wasn't the real deal—but all you have to remember is that a lot of scientific and philosophical development came out in the futile pursuits of alchemy and attempts to prove the existence of God. What were really revolutionized in statistics in the early decades of the twentieth century were methods and techniques and the math they were built on. These methods can be applied to anything, including studies based on atrocious premises as well as properly designed experiments.

Early-twentieth-century statisticians were prolific, developing a lot of the descriptive statistics we previously talked about, along with important concepts like the idea of probability distributions that could be used both to describe data and to make inferences. Statisticians were surprised to discover that many measurable things fit the normal distribution (also known as the bell curve), but definitely not everything. Several additional distributions were discovered as the century wore on.

Modern statisticians developed new ways of looking at experiment design and improved on some of the existing ways of setting up experiments, which is still massively important today. Experiment design influences almost every science, medicine, and social science discipline.

Another change that occurred in the twentieth century was an entirely new paradigm for looking at probability. In the early period, statisticians often looked at probability the way modern Bayesians do, who basically believe that a probability quantifies how much something can be believed, based either on a basic belief or assumption or perhaps the outcome of a previous experiment. This means that there is a prior probability assumed as part of the final probability computation.

But the big guns in statistics at the time rejected that paradigm in favor of one that came to be called frequentist, which holds that including a prior (an additional probability for a beginning state) in the calculation taints the rigor of any probability calculation. There are still Bayesian statisticians, and there are machine learning techniques that are partially based on Bayesian principles. There are also other paradigms that some statisticians favor. Basically, while most statisticians fall in the frequentist camp, paradigm selection can still get their blood pressure up, and some believe that different situations demand different paradigms.

BAYESIAN PROBABILITY AND BAYES' THEOREM

Bayesian probability primarily differs from frequentist because frequentists believe probability comes from counting all the possible outcomes (they care about frequency of occurrences) and Bayesians believe that probability is considered a "reasonable expectation," where someone has quantified a belief (like that this particular video game player has an 80% chance of winning the next game) or otherwise quantified something based on knowledge other than frequency of occurrence. We'll learn more about the frequentist approach later in the chapter, and I'll just talk a bit about the Bayesian paradigm here.

Bayes' Theorem is part of Bayesian probability, serving as a way to determine conditional probability (specifically, what's called the posterior probability). The *posterior probability* is basically the chance of Event A occurring given that Event B has already occurred. It relies on having three different probabilities calculated: the probability of Event A occurring, the probability of Event B occurring, and the *prior probability*, or the probability of Event B occurring given that Event A has occurred. The prior probability is the part frequentists have a problem with, since this is the one that can be based on belief, rather than rigorous calculations—it's attached to something happening before an experiment is run. You can then run the values through the formula and get the posterior. It basically allows you to update a probability based on new information.

Note that you don't have to be a die-hard Bayesian to use Bayes' Theorem in data science—we care about using techniques that work, whether they perfectly represent the real world or not.

A final important area of modern statistics is sampling. The importance of random sampling became clear after a highly publicized debacle where a major US magazine predicted the wrong winner in the 1936 US presidential election based on a deeply biased sample of their own readers, who skewed Republican. Sampling methods improved dramatically over the last century-plus, and now reputable people doing surveys and polls know how to select a representative sample.

Modern statistics is all about rigor and proper planning, but it took a while to get to the good spot we are at today.

Example of Statistics in the Real World

Example 1: The Problem of Points

Our famous mathematicians Fermat and Pascal's problem of points was a specific scenario involving an interruption of a dice game where two players put the same amount of money into the pot and then roll a die until one player wins a predetermined number of times (i.e., they rolled the highest number). They keep a tally of each player's wins, and the first to reach the designated number wins the whole game. You can imagine a game where Fermat has won six rolls and Pascal four, with the target of ten wins. Everything would be great if they continued their game, with Pascal coming from behind to win his tenth time while Fermat was stuck at eight. Pascal walks away with the pot, a smile on his face.

But what happens if the game is interrupted? Perhaps the candle burns out. What is the correct way to split the pot? An obvious answer might be to go based on the proportion of wins each has—in the above scenario, Fermat would get 60% of the pot and Pascal 40%. Mathematicians in the seventeenth century were pretty sure that wasn't right, and in Fermat and Pascal's letter, Fermat lays out the fair solution, explanation, and all. In a game of chance, it's not what has happened that matters—but what *could* happen in the future. So splitting the pot based on previous rolls doesn't account for all the ways the game might have ended—like Pascal coming from behind to win. But how could the many possible futures be modeled?

Another mathematician had already figured out that what needed
to be considered was the number of different ways each play could win.
Pascal first tried to figure this out and came up with a possible solution
based partially on his famous triangle of sums, sending it off to his friend
Fermat for a second pair of mathematical eyes. But Fermat was such
a natural that he came up with a simpler solution almost effortlessly.
Much of the exchange was him trying to get Pascal to understand it, but
after Pascal did understand, he made improvements on the approach by
simplifying some things.

The solution is a little tricky, but a key point is to understand that from
the perspective they took here, a target-10 game split 6-4 is the same as a
target-20 one at 16-14. There is the same number of ways each player could
win in both games: player 1 needs four more wins and player 2 needs six. It
becomes an issue of counting all the possible outcomes, and an unintuitive
aspect is that Fermat's method involved counting cases that wouldn't have
happened because one play would have already reached the target wins.
The counting starts by recognizing that the game will be concluded within a
known number of rounds—the sum of the number each player needs to win
minus 1, or $6 + 4 - 1 = 9$ in the scenarios here. You take 2 to the power of that
calculated number to count the total number of possible outcomes. Then
you count the number of ways each player can win and divide that by the
total, and that's the proportion of the pot each player should walk away with.

If you found this hard to follow, don't worry—so did one of the best
mathematicians in history. The last section of this chapter will address
counting because it's fundamental to the probability and inferential
statistics we'll address in the next chapter.

Example 2: Professional Gambling

Poker's been around a long time, but in recent decades it's become a
spectator sport and a job for a handful of professional gamblers, with
famous players with nicknames like Kid Poker and Poker Brat winning

hundreds of thousands of dollars regularly. All gambling is based on "beating the odds," and most gamblers enjoy taking some risk when there's a chance of a reward. But professionals play smarter than that.

Professional gamblers have incomplete information, so they deal in definite probability, expected value, and volatility. The *expected value* is basically the most likely value given a set of possible outcomes, often similar to the mean. It's calculated by looking at all possible outcomes and their likelihood and then summing them. Players are looking for games where the expected value is greater than 0, so that if they play multiple times, they will come out ahead overall. But the expected value only gives you the result given the basic likelihood of outcomes. In the real world, the results vary wildly. In a coin flip game, the player wouldn't win exactly half the time. The volatility helps quantify this variability, as it is simply the standard deviation, so it gives a sense of the spread of realistic values. Higher volatility brings greater risk of a loss, but a bigger gain if the game is won.

But the catch with gambling at a casino is that the games are configured so that the odds are in the casino's favor by at least a little (if not a lot). So the expected value for the player in a game based purely on chance is going to be less than 0. This is one reason that poker is so popular—luck obviously plays a significant role, but strategy and psychology are hugely important. Poker is partially a mind game. For each poker player at a table, all the other players are in the same boat, where luck dictates the particular cards they get dealt, but good strategy, reading people well, and bluffing convincingly mean players can minimize the damage if they don't have a good hand and maximize the win when they do have a good hand.

Despite the popularity of contests like the World Series of Poker, only a tiny fraction of players make money playing poker, and only a small proportion of those make a lot of money. And because winning at poker requires significant skill, players must study and practice before they can win consistently. It's probably not the wisest career choice for most people, but it makes for good TV.

Inferential Terminology and Notation

One thing that can throw people when studying inferential statistics is the notation and terminology. In inferential statistics, we're usually thinking about two groups. The first is the *population*, which is the "whole thing" that we're interested in and want to understand and we don't have (and can't get) all the values. The second is the *sample* dataset, which is a subset of the population that we can (at least theoretically) get values for. The core idea is that you can take characteristics of the sample you have and generalize to the population, with a recognition that there is going to be some error in your inferred estimates.

We talk incessantly about variables in statistics and data science, although they can go by different names. A *variable* is simply something that can be measured, quantified, or described in some way. *Dependent variables*, also called *outcome variables*, are the things we're looking at in the results. If we want to compare the results of an online advertising campaign, we might measure total clicks between two different ads— clicks are the dependent variable, because it depends on everything that happens before. *Independent variables*, also called *predictors*, are the variables that affect the outcome, the value of the dependent variable. In this case, the ad placement and design attributes would be the independent variables. The dependency between the variables is that the dependent variable depends on the values of the independent variables, which should be independent of each other.

We also need to get used to the notation. There are conventions of using Greek and other specific letters sometimes with diacritics to represent certain things, and in general the trend is to use a Latin letter for sample parameters and a Greek letter for population parameters. Table 3-1 shows an example of some of the notation.

Table 3-1. *Notation in inferential statistics*

Metric	Notation	Verbal
Sample mean	$\bar{x}$	x bar
Population mean	μ	Mu
Sample standard deviation	s	S
Population standard deviation	σ	Sigma

Basic Probability

Most people know enough about probability to understand that it deals with the likelihood of specific things happening and helping us understand risk. But it goes beyond that, and aspects of probability underlie statistical techniques that data scientists use every day.

Probability Theory

There are several concepts in probability theory that rely on specific terms that we'll define in this context.

An event in normal usage is just something that happens, but in probability it's defined in relation to a trial. A *trial* can be thought of as a single attempt, play, or round of something that will have some kind of result. For instance, in a dice rolling game, each play involves a player rolling their dice to generate a result. The trial is the roll of the dice, and the outcome—the combination of dice numbers—is called the *event*, the outcome or thing that happens.

We refer to all possible outcomes as the *sample space*. The sample space depends on what we are actually tracking. In the case of rolling two dice in order to generate a sum, the sum is the outcome we care about (as opposed to the two specific numbers on the dice). So all of the possible sums are the sample space, even though there are multiple ways to get many of the sums.

As an illustration, imagine Pascal rolling two dice in Trial 1 and getting a 3 and a 4, totaling 7. Fermat rolls his in Trial 2 and gets a 1 and a 6, also totaling 7. The two outcomes are the same even though the way they were achieved is different. The sample space of this game is all the numbers between 2 and 12, inclusive, but they aren't equally likely—it's much easier to get a 7 than a 2.

Another important term to understand is independence. We say that trials are *independent* if the outcome of one cannot affect the outcome of the other. This is true for the dice game we're talking about, but it might be clearer to look at a case where trials wouldn't be independent. We could have a game where if the player rolled a six, their next roll would be doubled. In that case, the second roll (trial) would not be independent of the previous trial.

One idea related to independence is independent and identically distributed variables, referred to as iid variables. *iid variables* are those that are all independent from each other and also identical in terms of probability. In a dice game where we're rolling six dice trying to get as many 6s as possible, each die represents a single variable and has a one-sixth chance of getting a 6. So these are iid variables.

One more set of concepts important to probability come from *set theory*, the math behind Venn diagrams. In set theory, a *union* of events means that at least one of the conditions is satisfied. An *intersection* is when all of the conditions are satisfied. The *complement* is the opposite of the desired condition.

Imagine if we're concerned with the exact rolls of two dice in a single trial. We care about the actual configuration, and there are different payouts for certain ones. Let's say the player gets a payout of half of what they put in if they roll two even numbers (call this Event A) and a free second roll if they roll double 5s or 6s (call this Event B). The union of these events is when something happens in A, B, or both, which is all rolls with two even numbers or two 5s (the two 6s are covered in both events).

The intersection of these events is simply double 6s. Finally, the
complement of Event B is all rolls that are not double 5s or 6s, and the
complement of Event A is any roll that has at least one odd number in it.

Counting, Permutations, and Combinations

As we saw in the problem of points, counting was known centuries ago to be
the foundation of calculating probabilities. We must figure out all possible
outcomes and the different ways specific outcomes could happen. This is
foundational in figuring out how likely things are, but counting isn't always
as easy as it sounds, especially when the number of possibilities explodes.

It's often rather instinctive to literally lay out all possible combinations.
For instance, we might roll two dice and sum them. The sample space
is every number between 2 and 12, but some sums are more likely than
others. Table 3-2 shows one way of writing out all the different die rolls.

Table 3-2. *Two-dice sum game outcomes*

Sum	Configurations	Count of Ways to Get Sum
2	1+1	1
3	1+2	1
4	1+3, 2+2	2
5	1+4, 2+3	2
6	1+5, 2+4, 3+3	3
7	1+6, 2+5, 3+4	3
8	2+6, 3+5, 4+4	3
9	3+6, 4+5	2
10	4+6, 5+5	2
11	5+6	1
12	6+6	1

At first glance, this might look good, but we've left something important out—the fact that you could get all sums except 2 and 12 multiple ways because you've got two distinct dice. If you're looking for a sum of 3, you would get there either with die 1 being a 1 and die 2 being a 2 or the reverse (die 1 being a 2 and die 2 being a 1). These are actually two different outcomes. It turns out that counting the number of distinct possibilities in a two-dice sum game is simple: there are six possible values for die 1, and each of those possibilities has six possible values for die 2, so it's 36 (6 for die 1 * 6 for die 2).

But we could modify the game so that if the second die matches the first, it's an automatic reroll until a nonmatching number is rolled (this is similar to a lotto draw). In that case, the first die could still be any number, but there would only be five possible values for die 2, so 6 * 5 = 30. Imagine adding a third roll that also can't match. That would be 6 * 5 * 4 = 120. What we're calculating here is also called a *permutation*, a way of choosing k things that have n possible values where order matters and the values can't repeat.

There's a related calculation called a *combination* that is similar to a permutation except order does not matter. This is usually said "n choose k." There might be a basket containing five balls, each a different color. A combination would tell you how many possible outcomes if you reached in and pulled three balls out. In this case, we don't care which came first because it's one selection activity (as opposed to rolling one die and then another). In this case it would be 5 choose 3, which comes out to 10. This is much smaller than the number of permutations (5 * 4 * 3 = 60).

Now that we've done all this counting, you might wonder what the point of it all is. Counting is critical to calculating probabilities. In the colored ball example, what are the chances that when you pull out three balls at once, one is a blue ball, one is red, and another is yellow? There's exactly one way to do this since there is only one of each color, so the numerator is 1, and that is divided by the total number of possible combinations we counted above, 10, so our probability is 0.1. In this case, every particular combination is unique, so each set is equally probable. But what if you ask what the chances are that you pull out two primary colors (red, blue, yellow) and one secondary (only green and purple in this basket). If we calculate the number of ways we can have two primary colors, it's 3 (red + blue, red + yellow, or blue + yellow). For each of those pairs, there are two possible values for the secondary-color ball, so we have the simple 3 * 2 calculation of possible outcomes matching our requirements, leading to a probability of 6/10, or 0.6.

Whether you've used combinations or permutations to get the total number of possibilities, probability is always calculated by dividing the number of selections that meet your criteria by the total number of ways the items could be selected. Instincts for picking the right way to count come with practice.

Bonus Example: The Monty Hall Problem

One of the most famous examples of misunderstood probability came out of a game show called *Let's Make a Deal*, which was hosted by Monty Hall, whom the problem is named after. In the game show, contestants were shown three doors that were each hiding something—one was something desirable like a car, and the other two were goats nobody would want to win. The contestant picked one of the doors, and Hall would open one of the other two doors to reveal a poor goat (obviously he would pick the one with the goat, not the car if it was there). In the real game, Hall didn't give the contestant a chance to switch doors, but in the reimagined Monty Hall Problem, he did. The contestant was offered a chance to stick with their choice or switch to the other unopened door. Most people's instincts say that it doesn't matter probabilistically whether they switch or not.

But in 1990, a statistician posed this question in a magazine with an answer that shocked everyone, including arrogant statisticians and mathematicians who refused to believe it. Her answer was that the contestant should switch, because they have a two-thirds chance of winning if they switch and only a one-third chance if they don't. The story blew up, with 10,000 people (apparently including 1,000 academics) writing the magazine to say this was wrong.

If you're paying attention, you might have recognized something here: it seems like the probability is changing based on new information, which would be a Bayesian thing, but the reality is that nothing is changing with the probabilities and this is handled with frequentist rules. So how does it work?

When there are three unopened doors, there is a probability of the car being behind each door of exactly 1/3. Once the contestant picks a door, that doesn't change. Their door has a one-third chance of being the fun one. This also means that there's a two-thirds chance of the car being behind one of the doors that the contestant didn't pick. When Hall reveals the goat behind one of the unpicked doors—and this is where it gets unintuitive for most people—there's still a two-thirds chance that the car is behind one of those doors (this doesn't change because of new information per the frequentist paradigm). But now the contestant knows which one has a goat, and that means that the one unpicked and unopened door still has (the unchanged) two-thirds chance of being the one with the car. Obviously, the contestant would be wise to switch, as two-thirds odds are better than one-third.

If you're thinking this can't be right, you're not alone—but you're still wrong. If you're wanting to see it for yourself, bring up a Python interpreter and try simulating it. Simulations prove that the switching strategy is superior. But simulation does remove the human element, and if you picked one door and switched to the next door, you're probably going to be more upset with yourself for switching than if you'd kept your door and ended up with the goat.

One More Bonus Example: The Birthday Problem

There's a famous scenario in probability that tends to blow people's minds, but it drives home the importance of understanding the risk of something happening to anyone vs. the same thing happening to you specifically. "Something rare" happening in the wider sense can occur in so many possible ways that it can be likely, which is counterintuitive. The chance of you dying in a car wreck tomorrow is tiny, but the chance of someone somewhere dying in a car crash tomorrow is pretty much 100%.

The famous scenario is called The Birthday Problem. It says that we only need 23 people in a room together for there to be slightly more than a 50% chance that two of them share the same birthday. This is where perspective matters—the chance of *you* sharing a birthday (month and day only) with one other person is still very low (5.8% if you're curious). But if you don't care which two people have the same birthday, the more people there are, the more opportunities there are for a match. Usually when this problem is worked out, we simplify it by ignoring leap years, the possible presence of twins, and any other trends that might make birthdays cluster. This means that we assume there are 365 possible birthdays.

One simple way to see how this happens is to calculate the probability that everyone has a different birthday and then subtract that from 1, which gives us the probability that at least two people have the same birthday. If we number the people from 1 to 23, it's a matter of multiplying all the probabilities for each person to have a different birthday. The probability that a single person out of a total of 1 has a distinct birthday is simply 365/365, because it could be any of the 365 days out of 365 possible days. But when we add a second person into this calculation, because they cannot have the same birthday as the first person, there are only 364 possible days. So Person 2's probability is 364/365. This continues with each person having one less possible day, with Person 23 being 343/365. Then we multiply all of these values together, which comes out to about

0.4927. Remember that this is the chance that there are no people who have the same birthday, so the chance that at least two people share a birthday is the complement of that, or 1 – 0.4927, or 0.5073.

The likelihoods of two people sharing a birthday just keep going up with more people. With 40 people, it's over 89% and with 70 it's 99.9%. But for perspective, in a room of 23 people, the likelihood of anyone else having the same birthday as *you* is much lower. There need to be at least 253 people in the room for at least a 50% chance that a particular individual shares a birthday with someone else—a lot more than 23.

Casinos and lottery runners rake in the money by relying on people's inability to understand chance. People see other people winning tens of thousands of dollars on the slot machine or on the news posing with a giant check for many millions of dollars, and they think, "Hey, that could be me." If you drop your money in a slot machine and pull the lever or pick up a lottery ticket at the corner store, technically you could win. But you won't. If you're lucky, you'll walk away with what you spent.

Probability Distributions

Probability distributions are a theoretical construct that represents the range and shape of values of a particular thing. They are important in inferential statistics, because if you can identify the distribution of a set of data you have, you can infer other reasonable values outside of the data you have at hand. The most famous distribution is probably the normal distribution, but it's just one of many probability distributions out there that describe different kinds of things.

It's much easier to understand distributions when you look at specific ones, so we'll look at a few different distributions that are important in data science. We'll cover a few more that relate to more statistical tests in a section further below.

Note that there are two fundamental types of distributions: discrete and continuous. *Discrete* distributions encapsulate outcomes that can only hold a finite number of values, like coin flipping—it can only be a head or a tail. *Continuous* distributions have outcomes that can be any value on a continuum, like people's heights.

Binomial Distribution

One of the simplest distributions is the *binomial distribution*, a discrete distribution that is understood to encapsulate the behavior of a series of n independent binary trials where the probability of a success is the same in each trial (like flipping a quarter for heads a bunch of times). Independence here means that one trial is not in any way impacted by any previous trial, which is definitely the case in coin flipping.

The distribution is defined by a known number of trials (n) and a known probability (p). The distribution gives us a formula that allows us to calculate the probability of getting a particular number of successes based on n and p. We also can plot a binomial distribution. Figure 3-1 are a couple binomial plots for particular n's and p's.

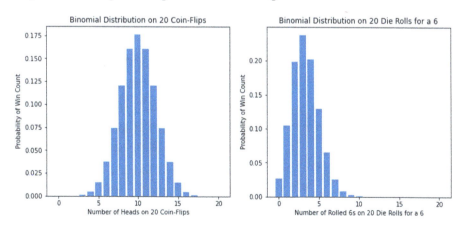

Figure 3-1. *Two binomial distributions representing coin flips and die rolls*

The chart on the left shows the distribution of a sequence of flips of a coin, with the likelihood of winning (getting one particular side) 0–20 times on 20 flips. n is 20 and p is 0.5 because the coin is fair. It's not surprising that ten wins is the most common because it's the halfway point in a distribution with a p of 0.5. If you're familiar with the normal distribution (which we'll look at next), you might notice that this looks a little like that.

But when p is not 0.5, the distribution looks completely different, as we can see in the right chart of Figure 3-1, which shows the chance of getting a specific number on a die in 20 rolls. In this case, n = 20 and p = 1/6 (~0.167), so the plot is skewed to the right. But again, it makes sense because the most common count is 3, which is 0.15, the closest fraction to the p of 1/6.

BINOMIAL DISTRIBUTION IN THE WORLD

Although this is a simple distribution, it can be useful in looking at many things in the world. For instance, if we wanted to gauge how successful a pro player of a particular video game is and we know the probability of success based on their rating, we could measure the probability of winning a certain number of games out of a series of games. If the player has played 100 games with a 0.75 probability of success and won 90 games (and lost 10), intuition would tell us that this player is performing really well. But we can quantify how well. If we want to know how likely it is they'd win exactly 90 games, it would be 0.0094%, very small. A more interesting number might be how likely it is that they'd win at least 90 games. We'd have to add the likelihood of every number of wins between 90 and 100, which turns out to be 0. 013%, a bit higher than exactly 90, but still very low. The player is doing incredibly well.

Since their likelihood of winning is 0.75, you might wonder how likely it is
that they have exactly 75 wins out of 100 plays—it's only 9.18%. But that's
because we're asking if they've had *exactly* 75 wins, when in actuality, 74 and
76 wins are similarly likely and would be considered a typical success rate. In
fact, the likelihood of them winning between 73 and 77 wins is 43.6%

One interesting fact about the binomial distribution is that for a
given p, the higher n is, the closer the distribution gets to meeting the
requirements of the normal distribution, which we'll discuss in the next
section.

Bernoulli Distribution

The Bernoulli distribution is a special case of the binomial, where only one
trial occurs and has two possible outcomes, usually success and failure.

Normal Distribution

Most people have heard of the normal distribution—a continuous
distribution also called the bell curve or Gaussian curve—and would
recognize it plotted with a peak in the middle and curving down to zero
on either side of the center peak similar to what we saw with the binomial
distribution with a probability of 0.5. The *normal distribution* represents
the expected theoretical shape we would see if we plotted all the values of
something that has a natural middle high point with values getting lower
as we move away from the middle. This would include things like exam
scores, where most people do okay, some do slightly worse or better, and
fewer do really bad or really well. It is *unimodal* because it only has one
peak. Many other things fit the normal distribution, including the height of
adult men, ACT scores, birth weight, and measurement error.

You can visually see the normal distribution in a lot of areas of real life, especially in wear patterns of things used a lot over time. Old stone or cement stairs will gradually become sloped, deepest in the middle but curving back up to the original step height. Paint on the part of a door that's pushed to open it will get most worn in one spot and less so the further you get away from that central point. It can even reveal different layers of paint, all the way to the bare wood in the middle, with the different layers showing up the further you get from the center spot. Check out the bar weights on a weight machine at the gym, and you'll see there will be a middle weight that's most worn, with the weights being less worn the further you get out. Figure 3-2 shows an example of the wear pattern on the tiled kitchen floor of my childhood home.

Figure 3-2. *The wear pattern after years of foot traffic on a tiled kitchen floor*

The lighter outside in Figure 3-2 is obviously minimally touched, and there's a clear path of the darker reddish color, with some yellow in the middle zone between the dark and light, with spotty red and tan mixed outside the yellow. No one would doubt what the most common route is from the lower right to the upper left on this floor.

Despite the many things we see in the real world, it's important to realize that most things aren't truly normal. But because things are often close enough to normal, the distribution is incredibly important in statistics. A lot of the tests and techniques that are used require us to assume normality. This works because things are close enough. Several of these tests will be discussed in the next chapter.

The normal distribution is defined by two of the metrics we previously learned about, the mean and the standard deviation. We use the population notation for distribution parameters, so μ is the mean and σ is the standard deviation. With those two values we can plot the curve. The mean is the very top of the peak, and the standard deviation captures how tall and spread out the curve is.

Knowing something is normally distributed helps you understand it. If it's normal, we know that the mean, median, and mode are all the same value and the data is symmetrical—the distribution mirrors itself on either side of the mean. Also, we know that about 68% of all the data will be within 1 standard deviation of the mean, 95% will be within 2 standard deviations, and 99% will be within 3 standard deviations. For instance, if a number is more than 3 standard deviations from the mean, we know that it's quite atypical. Similarly, if it's less than 1 standard deviation from the mean, that's a very common value, not particularly notable.

The normal distribution can come in all sorts of configurations, so it's common to transform our data to fit what's called the *standard normal distribution*, which is simply a normal distribution that has a mean of 0 and a standard deviation of 1.

Figure 3-3 shows two normal distributions, the standard normal distribution and one overlaid on some real data that we know is normally distributed—the height of men (we used the fathers' heights from Galton's famous family height dataset). We left the Y-axis values off because they aren't really important for understanding the concepts of the normal distribution.

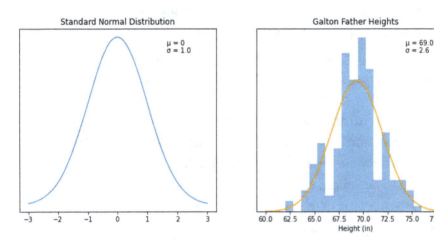

Figure 3-3. *Two normal distributions: the standard normal on the left and heights of men in a classic dataset with a normal curve overlaid on the right*

The plot on the left is the standard normal distribution where μ (the mean) is 0 and σ (standard deviation) is 1. We can see the peak in the middle at 0 and the tails are not too far out, fitting with a standard deviation of 1. It's known that with a big enough sample size, the heights of adult men are normally distributed, so we plotted heights of the fathers from Galton's famous family height dataset. It's not a huge set and there are some deviations, but we can see it's mostly normal from the normal curve that's plotted over it in orange, scaled to this data.

These curves look a lot alike, and one of the tricks to understanding them is to remember how much data lies between different values on the X-axis. We know that about 68% of all the data is within $1 * \sigma$ of the mean, so between $\mu - \sigma$ and $\mu + \sigma$. On the standard normal plot, 68% of the data is between -1 and 1, and on the heights plot, it's between 66.4 and 71.6.

Using the Normal Distribution

This curve and values may be all we need for some things, but often we will use the normal characteristics in some applied way, and that frequently means transforming our data so we can talk about it in a standardized way.

Z-Scores

Because most normal data doesn't line up with the standard normal distribution perfectly, it's common to normalize our data so that we can talk about it in relation to standard normal. This is done by computing a *Z-score*, which is the number of standard deviations that a given value is from the mean. It's simple to compute: take the difference between the value and the distribution mean and divide that by the distribution standard deviation, so it can be positive or negative. On the standard normal distribution, there's no need to divide by the standard deviation since it's one—which is the whole point of it.

The Z-score is a simple but great metric to use for identifying something out of the ordinary, like in a simple anomaly detection task or to define a threshold for reporting an unusual value (like fraud detection or a performance problem on a monitored computer system).

Tests for Normality

Because so many statistical techniques require us to assume normality, it's important to test the data to see if it really is normal. The *QQ plot* allows you to investigate this visually. A QQ plot shows the quantiles of a normal distribution on the X-axis (this is the reference point for normality) and all of the Z-scores in the dataset in numerical order from low to high on the Y-axis (these are the values you're testing). If the data is normal, you will have a near-perfect diagonal line from the bottom left to top right. You can see an example in Figure 3-4, where the dots represent the tested values and the red line is what we expect to see if the data is normal.

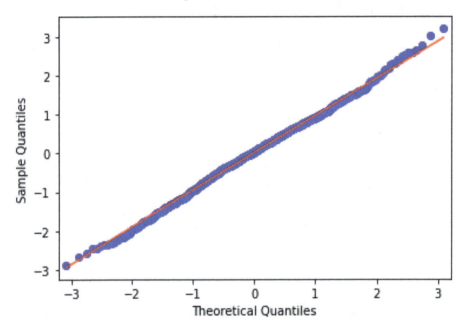

Figure 3-4. *An example QQ plot with random normal data*

Figure 3-4 is how we expect it to look if our data is normally distributed, and a QQ plot will make it clear when the data is not remotely normal. Figure 3-5 shows data from a Poisson distribution, which we'll look at in the next section, and it's clear the data isn't lining up on the diagonal line.

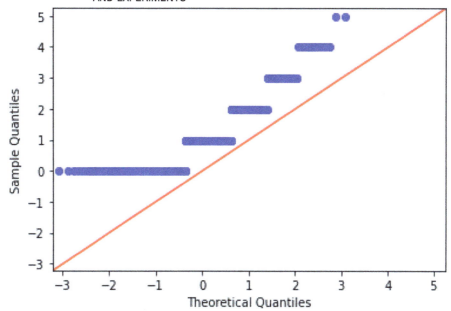

Figure 3-5. *An example QQ plot with random non-normal data
(from the Poisson distribution)*

Another common test is the *Kolmogorov–Smirnov test*, sometimes
simplified to the K–S test. This test compares the actual distribution of
standardized values in a dataset to the standard normal distribution.
It generates a number representing the distance between the normal
distribution and the data's distribution where they are furthest apart.

We already know that most things aren't truly normal, but often they
are close enough that we can still use some of the techniques that require
normality. This is often a judgment call that gets easier with experience,
but it's still important to test rather than to just assume.

Poisson Distribution

The discrete Poisson distribution is interesting because it looks at things
that seem hard to quantify at first. Specifically, the *Poisson distribution*
models irregular events or occurrences that happen repeatedly. Typical
examples are events that are spaced out over time like a customer coming
into a restaurant or someone calling out sick at a job and events spread
out over space like defects on pages in a printed book. Anyone who's ever
worked in a restaurant or a retail store knows that you can always have
a "rush" of customers that's much more than normal or surprising dead
times, with no obvious explanation why. It just happens.

The Poisson distribution uses a rate parameter, λ (lambda), which is
the average number of events per specified amount of time (or of specified
space, like in the book example). So we do have to know something about
the situation we're trying to model. But a restaurant could track customers
coming in over a few weeks and come up with this average and then be
able to use the distribution in the future. Note that λ is also the variance of
the distribution.

One limitation of the Poisson distribution is that it requires λ to be the
same over time. Restaurant workers know that there are more customers
at lunchtime than 3:30 in the afternoon. We could handle this by using one
λ for lunchtime, one for dinner, and another one for between those, with
a fourth for after dinner. But the shorter the periods of time are, the more
difficult it can be in practice.

We can look at the pizza restaurant data we looked at in Chapter 2 to
see if it fits the Poisson distribution. Figure 3-6 shows the distribution of
orders per hour during lunch and dinner over a particular week (weekdays
only), shown as the blue bars. In order to get the Poisson distribution
plotted in the orange line, we calculated λ by taking the average for those
matching time slots over a week in the previous month.

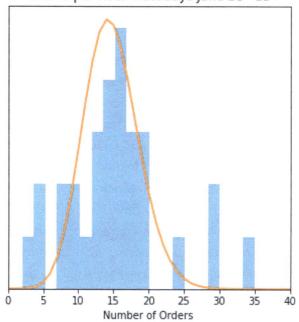

Orders per Hour Weekdays June 21 - 25

Figure 3-6. *Poisson distribution of number of orders per hour over weekdays*

This matches pretty well, which is why the Poisson distribution can help planning in restaurants. But as we'll eventually see, not everything matches so well in the real world.

Exponential and Weibull Distributions

The exponential distribution and Weibull distribution are related to Poisson, but they are both continuous distributions because they track time rather than finite events. The *exponential distribution* considers the same type of scenario used in Poisson and also uses the λ parameter, but it models the time between one event and the next. So when one customer comes in, the exponential distribution could tell us how much longer before the next customer is likely to come in.

111

Sometimes the Poisson distribution seems to be the right one, but when an event becomes more or less likely over time, the *Weibull distribution* is the better option, as it models the amount of time that will pass before a particular event happens, also over time or space like Poisson. A classic example is tracking the failure of a mechanical device. Intuitively, we know that the longer a device is used, the more likely it is to fail because of the continued wear and tear. The Weibull distribution can basically tell us how likely a device is to last a specified amount of time, and it requires two parameters. The first is referred to as the shape parameter and notated with a β (beta), which represents how quickly the probability of failure changes over time. If it's greater than one, then the probability increases over time, and if it's negative, it decreases over time. The second parameter is the expected life of the device, notated with η (eta). One limitation to Weibull is that β itself needs to be constant over time.

Central Limit Theorem

The central limit theorem is another concept that deals with sample statistics, where you're calculating the mean of many samples from the same population. The *central limit theorem* says that if you take the mean (or another statistic) of multiple samples from the same population under appropriate conditions, the distribution of those means will approach a normal distribution as the sample size (the number of means of samples taken) becomes large enough.

This is true regardless of the distribution of the population the samples are taken from and the statistics being measured. Imagine having a dumpster full of nothing but irregular marbles of various sizes and taking a thousand samples of a hundred randomly selected marbles and calculating the mean of all of those. These samples will vary in quality (how well they represent the population), but many will be close

to real representation with fewer deviating from true representation. So
we will see the measured means hovering around a central point—the
ones from the samples that represent the real marbles well—with fewer
measurements spreading right and left from that central point, just like
the normal curve. The peak of the curve is likely an accurate value for the
mean of the marbles in the dumpster.

A concept related to the central limit theorem is the idea of regression
to the mean, which basically says that over time, values become less
extreme. So, if an extreme value pops up—like a player getting a really high
score on their Solitaire game—the score in the next game will probably be
lower (less extreme).

Sampling

Sampling is the act of getting a subset of data from a larger population.
The term can mean a couple of related things depending on context. In
experimental design, it means selecting members of that larger population
to gather data on. In most cases, we won't be able to get data on the entire
population of something. A market research company can't literally talk
to every American or even every American woman between 25 and 45.
We have to figure out a way to talk to a smaller number of them so that we
can generalize to the whole population meaningfully. Then we can run
our study and work with them in some way to get data from them. The fact
that we're taking a sample means that how we select the members of the
population to collect can have huge consequences.

The second case is more common in data science, where we
sometimes have the opposite problem—too much data. We usually
aren't creating studies, but we may need to select a subset of data that
we already have to reduce computational load or to create subsets for
specific purposes. For instance, we might have millions of rows of data
on server performance, thousands of rows for each of many thousands

of servers. We could select a subset of servers, or simply a subset of rows, or other possible subsets. This is also called sampling. Choosing which points go into our sample in this case has most of the same issues and considerations as in the first case. However, here, we have the advantage of possibly being able to have a better picture of what's in the full population (sometimes we could do basic summary statistics on the full dataset, just not the more advanced data science we want to do), which can inform our sampling choices.

The rest of this section will focus mostly on the first type of sampling, which is usually done as a part of an experiment or study.

Bias and Representativeness

One of the main goals in sampling in inferential statistics is for the sample to represent the population in a meaningful way, and the best way to ensure representativeness is to avoid bias in our sampling. In this context, *bias* is simply the systematic favoring of a particular subset of the data, which means the sample misrepresents the population, which comes from the sampling method chosen. We can also have bias in measurement that comes from measuring things in a problematic way.

If we want to understand how popular Snapchat and Facebook are with Americans, we would have a huge bias in our sample if we only surveyed teenagers. We would have picked a sample that was completely unrepresentative of "Americans" as a whole.

There's actually a fairly famous example of systemic bias in psychology. So many psychology experiments are done by academics at their universities on college students who volunteer (or are voluntold if they're taking a psychology class). The prefrontal cortex, the part of the brain associated with impulse control and decision-making, is not fully developed until around age 25, so most college students are still developing. Drawing scientific conclusions about how their brains work when they're still developing will lead to a biased sample. There's nothing inherently wrong with these studies on their own—and they could be really helpful in designing a bigger study of randomly selected adults. The problem is that these results are sometimes reported in a way that generalizes them to the larger population of "people" (especially by journalists).

To better understand bias in sampling, let's go back to the dumpster filled with marbles. Imagine we have a machine that will reach into the dumpster, mix it a bit, and squeeze until it picks a marble and pulls it out. Let's say that the gripper malfunctions and doesn't close all the way, so it rarely pulls out the smaller marbles because they fall out. The sample of selected marbles is going to be different from what's in the dumpster overall. If we take the average of the sampled marbles, it will be higher than the true mean of the marbles in the dumpster.

If we take all our marbles out of our dumpster (this one's dollhouse-sized) and set them on the table on the upper left, we can see the range in size of the marbles. But if our automatic picker can't completely close, it's not going to be able to hold on to the smaller marbles. It's easier to grasp this when you can see it visually. Figure 3-7 shows what these samples could look like. The top image shows the marbles laid out on a table, spread out a bit. If we put our marbles back in the dumpster and tell our broken picker to pull out ten marbles and then repeat that a couple times, the three samples on the bottom left would be typical. If we fix the picker and rerun the process to get three new samples, we'd expect something like the samples on the bottom right.

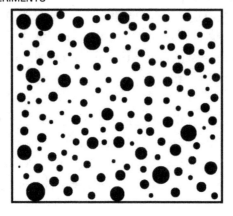

Broken Picker Samples   **Fixed Picker Samples**

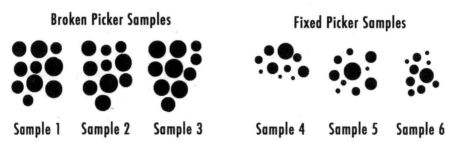

Sample 1   Sample 2   Sample 3   Sample 4   Sample 5   Sample 6

Figure 3-7. *Six example samples from a table full of marbles, three
with a biased picker and three with a fair picker*

The broken picker can't hold the smaller marbles, so all the samples
have only medium-sized and bigger marbles. Those samples don't look
anything like what's on the table. The samples on the right, taken after the
picker was fixed, look far more balanced.

It's also important to remember that humans can fairly easily
eyeball balance in this scenario to a degree, but it's critical to pick
more systematically, because humans don't have the big-picture
precision needed.

Sampling with Replacement

We often talk about sampling with or without replacement. This really is only relevant when you are sampling data to use from a larger, complete dataset. *Sampling with replacement* would allow some data points in the complete dataset to be selected more than once—basically, the data point is put back into the dataset after being "removed" in selection. So *sampling without replacement* is just the opposite, where each time a data point is selected, it is effectively removed from the complete dataset. The most natural way to sample from the marble dumpster would be to pull one marble out and set it down, then reach in for another to set aside, and so on, which would be sampling without replacement. But you could also toss the marble back in after selection and stir it up a bit, which would be sampling with replacement. Having people fill out a survey would normally be sampling without replacement (you wouldn't ask someone to fill it out twice). If you want a truly random sample of numbers, where each number has an equally likely chance of being selected, you'd want to sample with replacement (otherwise, the probability of each number being selected goes up, and it's no longer random).

Sample Size

Like with many things, what we are trying to accomplish should dictate the amount of data needed for a study. Most of us have a general sense that more data is better than less data. But often, the opposite is true. Especially in the era of big data, we often want to use as much of it as possible—maybe even the whole population. It's definitely true that we're more likely to miss exceptional cases if we take a subset of a population. But working with the full population is not feasible if we want to understand something about all adult citizens of the United Kingdom, for example. One surprising thing is that in both cases, it's often the case that a smaller but carefully selected sample is actually better than a huge dataset. In general, it is easier to work with small datasets, especially in terms of computational resources.

There are obviously situations where a large amount of data is truly required, but they are few and far between. A couple obvious ones are Internet search and large language models. We can't skimp on data on those because they are supposed to be complete in terms of breadth and depth. But if we want to know what British people's favorite hobbies are, we don't need to ask every single one.

There are some general rules that can help with identifying a good sample size. When the population doesn't have a lot of variability (most of the members of the population are fairly similar), smaller samples are good as long as the sampling method is rigorous. The opposite is true when the population does have a lot of variability—in that case, a larger sample is more likely to be representative, because it will capture more of the differences. Additionally, large sample sizes are preferred when we're studying rare events (like computer failures or fraudulent transactions on credit cards).

Figure 3-8 shows the three scenarios and how sampling can be impacted. There's a population for each scenario, and then a small sample and a large sample from each. On the left is a population of marbles with minimal variability, so a small sample still looks pretty representative (the large sample doesn't give us much more info). The middle shows a much more varied population, and we can see that the small sample isn't very representative, so we need a larger sample. The rightmost population is one with one very rare event—the big marbles. We will need a larger sample to make sure we get some of the big marbles in it. Alternatively, our small sample could end up with multiple large marbles simply by random chance, so that sample would be incredibly nonrepresentative.

Populations

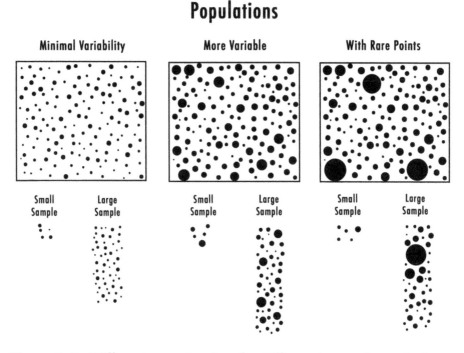

Figure 3-8. *Different samples sizes for different types of populations*

One last consideration when considering sample size is what the consequences of the work could be. If it's potentially damaging, it's usually good to go with the bigger sample to ensure better representation. This might apply if it's part of a study that will be used to write public health policy or involve a company making significant financial decisions. This plays into ethics, which we'll discuss in a later chapter.

Random Sampling

Most of us instinctively know that randomness is usually the key to avoiding bias and unfairness. But in practice, this isn't always straightforward and doesn't always bring the intended results, especially any time we're dealing with people in some way. If we manage to randomly select a group of people to interview or collect data on, we would still have to depend on all of those people actually participating—whether it be being willing to sign up for an interview or to fill out a questionnaire.

Also note that to do any kind of random sampling, we have to have a way to select truly randomly, which means we need some kind of random number generator. This means involving a computer in some way.

Simple Random Sampling

Simple random sampling really is as its name implies—random and the simplest method there is. In *simple random sampling*, every point in the population has an equal chance of being selected, and points are selected one after the other until the desired sample size has been reached. This is often trivial to do in existing datasets, but if you are selecting people or something else to be part of a study, it can be more difficult.

Stratified Random Sampling

Stratified random sampling is similar to simple random sampling, where we split the population into groups (*stratas*) based on characteristics we care about and then apply simple random sampling over each of those groups. For instance, imagine that we were going to study video game players and were interested in understanding the different experiences male and female players have. There are more male players than female over the entire video game player population, so if we just sampled randomly over the whole population, our sample would have many more men than women. But our question is what the experience of each group is, so we would actually want a sample with about the same number of each gender. The random sampling within the groups ensures that each strata is representative of that group, which is what we care about here. Obviously one limitation of this approach is that you have to know something about the larger population to know which strata to put which data point in.

Cluster Sampling

Cluster sampling splits the population into groups called clusters, similar to stratified random sampling, but it differs in that we usually split the population into a large number of clusters with a hierarchy. We then select specific clusters to focus on based on convenience in terms of data collection. It's common to do this kind of sampling when it's logistically difficult to use the other random sampling methods. The most common scenario for this is geographical—for example, a hierarchy could be country, state, county, city, and neighborhood. We'd select a limited number of neighborhoods and randomly select people within that neighborhood, and the advantage here is that it's more practical to send one person out to collect data on many people in a particular neighborhood than a bunch of people to collect data on a few people in a slew of neighborhoods. This method is usually chosen because of practicality and cost.

Nonrandom Sampling

Randomness is generally the best way to avoid bias and ensure representativeness, but it isn't always practical, especially when dealing with tangible things like people. Sometimes we need to simplify things and go with data points that are available to us in some way.

The lack of randomness doesn't mean these methods are worthless. We have to remember that we can't generalize from a sample to a population as confidently with a nonrandom method. But it can still be useful if we just want to try things out, perhaps test some questions on a survey to work out kinks before sending it off to our larger, random sample.

There are many different ways of sampling without randomness, but we'll talk about just a few here.

Volunteer and Snowball Sampling

Volunteer and snowball sampling are methods that are often used when working with people and tend to produce very biased samples. *Volunteer sampling* is a method of asking a huge number of people to participate (like filling out a survey) and then using the data from those who chose to participate—these people self-selected or volunteered. *Snowball sampling* is statistics' answer to multilevel marketing—individuals are selected and then they are asked to provide names for further participants. In college, I got a job selling knives, which involved doing presentations to people I knew, who were supposed to give me names of people they knew who I could contact and so on. I was no social butterfly, so this went nowhere, and it's not a coincidence that I only lasted about a month. The quality of your snowball sampling is entirely dependent on how willing your sampled people are to share their own network. Both volunteer and snowball sampling are likely to be biased because of the self-selection aspect, but it can be a good way to get participants who have a big stake in your questions because they'll be more interested in participating.

Convenience Sampling

As its name implies, *convenience sampling* is a method of selecting from a population based on how easy it is to do. If a student wants to do a study of college students, they might just survey everyone in their class. This is obviously biased based on what the subject is, what class level it is, and what time of day the class is or whether it's remote or in person.

There's nothing wrong with doing a study like this as long as we remember it is not generalizable to any other population, not even to all college students.

Experiment and Study Design

Most of us have participated in a quick survey online, and while they can
be fun, they are almost never actually meaningful because they're full
of bias and poorly framed questions. It turns out that creating a rigorous
experiment or study isn't a trivial thing, and there has been a great deal of
work done in this area over the last 200 years to make it better.

Even if a data scientist doesn't directly design or run experiments,
understanding how they are designed and run can be very beneficial to
understanding the data and knowing how to work with it.

Experiments

Experiments are a staple in the physical sciences, with some social
scientists—especially psychologists—joining the fun. This is because when
done well, they produce reliable and rigorous data that can be used to
draw real, meaningful conclusions.

To be an experiment in the scientific sense, it needs to be testing a
single variable while controlling for all others. There need to be two groups
at a minimum, a *control group* that has no change and an *experimental
group* that encapsulates the change. There can be multiple experimental
groups, but they should be different in the same variable. Study objects
(whether human participants or something else) must be assigned
randomly to one of these groups. There also must be at least one *response
variable*, which is what the experiment is trying to measure—did the
experimental groups respond differently from the control group?

There are other things that must be considered, including sample
size, how many response variables there will be, what type of statistical
techniques will be used to analyze the data, and what level of significance
will be used as a threshold for statistical significance. These things are all
intertwined in order to ensure that differences that appear significant are
truly statistically significant. There are several methods that you can use to

calculate a minimum sample size depending on what kind of statistics you will be using to analyze the data. For instance, there may be a minimum size for a scenario where you want to be able to calculate a confidence interval around a particular metric. Another common calculation is called *power*, which tells us how likely we are to see a significant result for a given sample size.

As an example, Meta might want to know which of three ads works best in the feeds of a certain group, perhaps men between 25 and 35. First, they need to pick response variables, which could be many things, including number of clicks, number of impressions (times shown) before they click the first time, and even a follow-up to clicking that they care about (like purchasing after clicking). The experiment would be split into four groups, one control group that gets no ad and three experimental groups, one per ad. Men who meet those criteria would be assigned randomly to these experimental groups. Then Meta would show the assigned ad to each participant (or no ad for the control group). Depending on what they're looking for, they might show the ad only one time or multiple times (the same number of times per user) and then measure the two response variables—clicks and purchases.

A/B Testing

A/B testing is probably the most common type of experiment that data scientists are involved in running. It's basically an experiment with a control group and exactly one experimental group. It's really popular in digital marketing and in web design because it's fairly easy to implement. A company may want to see if changing the font on their newsletter headline gets better read-through or not, so they set it up to randomly pick which headline to show a user.

Quasi-experiments

Quasi-experiments have a lot in common with true experiments, but they do not have the same rigor because at least one of the criteria for an experiment is not met, most commonly either by not having a control group or not assigning participants randomly. There may be practical or ethical reasons for this, which means that the study cannot be considered a true experiment, even if it meets the other criteria. These studies can still be valuable, but we can't generalize to other situations as safely as with experiments.

Observational Studies

Although experiments and quasi-experiments are both considered the gold standard in research design because they minimize bias and improve accuracy, sometimes it's not possible to have even that much flexibility in design. In some situations, we can't randomly assign participants because of ethical or legal constraints. For instance, if we wanted to study the impact on kids of having a parent in prison, we can't exactly throw one parent in jail—we have to just work with families we can find. *Observational studies* enable us to study something even when we can't control all the parameters by just using data points we have access to.

In this kind of study, one way to mimic experiment-like variable control is to do a *case–control design*, which involves matching the cases (in the above example, the families with a parent in prison) and the controls (families with no parent in prison) by ensuring that as many characteristics are otherwise the same or very similar. We would identify the attributes we think are important to match, which in this case might be things like neighborhood, number of kids, age of kid(s), race of family members, and family income.

Surveys

For most people, when they think of research, they probably think of surveys, which we constantly hear about and sometimes even participate in. *Surveys* are basically just a list of questions to be answered by a participant. They are also called polls and questionnaires, but they're just different names for the same thing (polls are often associated with political topics). We see survey results all the time on the news, in magazines, or on the Internet. It can be really easy to get some snazzy visuals that look good in color from surveys, and compared with most experiments, they're simple and fast.

Casual web surveys aren't scientific at all, but more serious surveys can be well-designed to yield meaningful conclusions. As a starting point, a survey needs to be focused, and we have to identify a clear population—which also needs to be reachable and sampleable. We need to decide what delivery method to use for the survey, which can be online, the mail, the phone, or door-to-door. Sometimes timing is an important consideration, especially around political issues.

Serious surveys also need to ensure operating within the law and ethical guidelines. That generally means explaining the purpose of the study, reminding participants that filling it out is voluntary, and specifying if results will be anonymous or confidential. It's also good practice for researchers to include their contact info on the survey. If the survey is associated with an organization, there might be additional requirements from that organization.

Writing questions for a survey is an art that's been made more scientific from trial and error. The wording can dramatically influence answers people give, which can lead to bias and inaccurate results. The biggest risk is creating leading questions—ones that subtly (or sometimes not so subtly) guide a respondent to a particular answer. There are many ways to phrase a leading question, but a clear one is to start with something like, "Do you agree that ..." For instance, "Do you agree that

there are enough women in the tech industry?" The question-writer's
bias is clear and could influence how the participant responds. A better
way to phrase it might be "Do you think women are fairly represented
in the tech industry?" One highly recommended method for ensuring
that your questions are good is to do a small pilot study where you have
a small number of people complete the survey and ask the respondents
for feedback. This can help you identify confusing, misleading, or simply
biased questions.

Sampling for Surveys and Response Rate

Once we've identified a target population, we have to get a good sample,
which can be especially difficult with surveys because of generally
low response rates. The *response rate* is simply the percentage of
responses received out of all surveys sent. While it is possible to select
target participants randomly, unless a survey is mandatory or highly
incentivized, many people won't complete it. So it's possible that the
final sample wouldn't be random even if the people you sent it to were
randomly selected.

For instance, if we want to learn about what people think about the
behind-schedule roadwork that's been snarling traffic for months, we
could randomly pick people out of an address list and mail them the
survey, but most of them wouldn't fill it out, so we'd end up with a sample
of the kind of people who like to fill out surveys—which isn't all people.
Participation is always a problem, and even incentives could bias the
sample because lower-income people are probably more willing to fill
out a survey for a $25 gift card than a Mercedes-driving corporate lawyer.
We've again found people who aren't necessarily representative of the
population we want to learn about.

Additionally, we have to consider the size of the sample that we need to
ensure potentially significant results. We need the number of participants
to be big enough, not just the people we send survey to, which means we

have to estimate the response rate in order to send it to enough people. Response rates vary widely depending on delivery method, who the target population is, and content. Most companies receive participation rates in employee surveys well over 50%, while web surveys can be in the teens or even lower. One of the unfortunate facts is that the lower the response rate, the more likely it is that the results will be biased. Additionally, in some situations, a high response rate is required, for instance, by some scientific or medical journals.

Key Takeaways and Next Up

This chapter focused on the foundations of inferential statistics— probability, probability distributions, sampling, and experiment design. All of these will play into the inferential statistics discussed in the next chapter. Probability distributions, especially the normal distribution, are fundamental to many of the ideas and statistical tests used in inferential statistics. We talked about sampling, which is hugely important because how we choose a sample influences how good the information we can glean from it is. It's important for a sample to be as unbiased and as representative as possible, and we covered a variety of different techniques for creating a sample. The chapter wrapped up with a discussion of experiment design, a topic most data scientists won't work with much, but understanding different designs and the limitations on them can help us do better work on the results of studies we didn't design.

In Chapter 4, we will conclude the explanation of statistics, starting by finishing the history of inferential statistics. Then we'll dive into the concepts that underlie common statistical tests, including hypothesis testing, p-values, Type I and Type II error, margin of error, confidence intervals, and one- and two-tailed tests. Then we will cover the major statistical tests, including ones that require us to assume the data fits a distribution like the Z-test, t-tests, Analysis of Variance (ANOVA), and chi-squared tests, as well as ones that don't require a distribution.

PRACTITIONER PROFILE: DANNY VACHALEK

Name: Danny Vachalek

Job Title: Data scientist

Industry: Retail

Years of Experience in Data Analysis and Data Science: 8

Education:

- MS Data Analytics and Data Science

- BS Business, Marketing, and Advertising

The opinions expressed here are Danny's and any of his employers', past or present.

Background

Danny Vachalek took a fairly practical route in college and focused on getting a business degree, learning everything he could about marketing and advertising, so it wasn't a surprise that he ended up working in advertising after graduating. But he had no idea that this first job would eventually lead him in a different direction.

Work

Much of Danny's first job involved relatively simple analysis on different advertising campaigns, looking at success rates and other indicators of performance. He was very curious to learn more, and his stakeholders all wanted to know more than he could deliver, but the tools he had at his disposal were basic and he could only look at a few variables at a time for any campaign. He knew there had to be more he could do, so he looked into the options and soon started a master's in Data Analytics.

While working on the degree, Danny saw an opportunity and cofounded an ad agency focused on automating branding communication. But soon he realized that this was a dead-end venture because large language models were becoming increasingly mainstream and available (and they'd soon make his business obsolete). He landed in a job at an advertising startup where he was thrown into the deep end and learned SQL and also met a bunch of old-school machine learning gurus who introduced him to the wider world of data science. That job was rewarding because he learned a lot and did some valuable work, but it was also time-consuming.

After he and his wife learned they had a baby on the way, Danny changed jobs and became a consultant. He often felt over his head in the beginning of several of his assignments, and he realized there was a lot for him to learn before he'd feel comfortable at the start of any new assignment. Specifically, he needed to better understand how to manage the full lifecycle of a project, from gathering requirements to doing a full implementation. He also found that his stakeholders peppered him with fairly random questions every week and again wanted more than his primary tools could provide. Everyone wanted data science, but there was rarely enough time with all the ad hoc queries. Still, his appetite for doing more advanced analytics grew. A lot of his time went to working on his master's and learning the academic side, but he still found himself squeezing in some more business-focused and applied study when possible.

All that education combined with his work on a project that got him some accolades at a consulting gig helped him eventually turn that into a permanent job at his current company, where he's one of several data scientists on a dedicated data science team. Since then, his knowledge of data science and software design has grown, and he loves the work he's doing.

Sound Bites

Favorite Parts of the Job: Danny likes working with and learning from a variety of people at work, many of whom have very different backgrounds and perspectives from him. He especially learns from engineers because he appreciates their systems thinking. He also likes dealing with stakeholders of varying technical expertise and trying to find clear and accessible explanations to simplify complicated concepts, so they can all figure out the best solution. He also likes solving a problem and contributing to important initiatives.

Least Favorite Parts of Job: The public perception of data science perpetuated by the media and in Wall Street is so off base, which means stakeholders often have very unrealistic expectations for what data scientists can accomplish. In some jobs it can be difficult to get the basic tools to do everyday data science because leaders don't understand what's technically necessary. Sometimes kludgy workarounds are the only way to get things done.

Favorite Project: There were several great projects when he worked at the startup, but in one, Danny figured out which product feature flags correlated with the highest revenue and discovered that the users generating the most revenue were ones who'd installed multiple versions of their software. This and further analysis and testing led to doubling revenue based on the same input.

How Education Ties to the Real World: He found that the problems and projects he worked on during his master's were nothing like those in the real world in terms of data quality (it was too high) and the projects weren't realistic because they weren't end-to-end. Learning how to take a project from concept and planning through execution and into production isn't something that programs seem to teach, and it's hugely important in the work world.

Skills Used Most: Analytical and critical thinking are both hugely important, but so is a curious mind that's constantly generating questions. Danny has found that it's especially important to think about questions that can help him understand the real problems stakeholders are trying to solve and then knowing what real solution options there are. It's important to have the ability to recognize when a problem really doesn't have a feasible solution because of data or infrastructure (tool) limitations.

Primary Tools Used Currently: SQL with CTEs, Snowflake, Python, JupyterHub, VS Code, GitHub, DBT

Future of Data Science: Danny thinks that there is a lot of risk in terms of privacy and security (and ethics) because of the way data is collected and handled and the way data science is done nowadays. He worries that something bad will have to happen before people push for more regulation, which he thinks we need. He thinks it would be good if data scientists had to qualify for a license or certification that would require ethical training and more.

What Makes a Good Data Scientist: What separates a good one from a great one is the ability to ask interesting and incisive questions that dive into important business needs. Domain knowledge is incredibly powerful.

His Tip for Prospective Data Scientists: Find your area of focus (the domain you want to specialize in) as early as possible. Basically, pick an industry and learn everything you can about it to make yourself stand out from other entry-level candidates.

Danny is a data scientist working in retail with a strong background in marketing.

CHAPTER 4

Coming to Complex Conclusions: Inferential Statistics and Statistical Testing

Introduction

The importance of experiment design, good sampling, and probability cannot be overstated, but the twentieth century also brought us huge advancements in statistical testing and our ability to evaluate the quality of estimates, both of which power inferential statistics. Correlation and covariance are other important ideas that were developed in the twentieth century. The computer delivered us from small datasets into "big data." They learned many different things they could do under the umbrella of "statistics," and data scientists rely heavily on these new techniques in their day-to-day work, even though they also do a lot of things that fall outside of statistics.

This chapter will cover the remaining history of modern statistics. We'll look at a couple examples of statistics in the real world, with one example from World War II and the other from COVID-19 modeling. We'll then

K. P. Vincent, *A Friendly Guide to Data Science*, Friendly Guides to Technology, https://doi.org/10.1007/979-8-8688-1169-2_4

go over the basics of modern statistical testing to lay the foundation for
learning about specific tests. This includes hypothesis testing, p-values,
Type I and Type II errors, test power, margin of error, confidence intervals,
and one- and two-tailed tests. Then we'll dive into some of the most
common statistical tests—Z-test, t-tests, ANOVA, chi-squared tests, and
some nonparametric tests. We'll look at how testing for significance works
and then look at correlation and covariance.

The History of Modern Statistics: Part 2

Methods for statistical testing exploded in the twentieth century. As
probability distributions were discovered, many methods of forecasting
and hypothesis testing cropped up. The normal distribution and the
Student's t-distribution (which is related to the normal) were both crucial
in the early days, and many tests derive from these, including the Z-test,
the multiple variants of t-tests, ANOVA, and the chi-squared test. But
because of the limitations of these tests (we have to assume the data is
normally distributed and statisticians knew we could not assume that in all
cases), other tests called nonparametric tests were also developed.

Several more important techniques and concepts were defined—Type
II error, confidence intervals, and test power, all of which are related to
testing hypotheses—and we'll discuss them below. Then, the techniques
maximum likelihood estimation (MLE) and linear discriminant analysis
(LDA) were both developed. MLE allows us to estimate the parameters of a
distribution (a population) based on some known data points. LDA allows
us to find a line that separates two different groups of data. Then, the
t-distribution got even more attention and development, and another way
of measuring correlation was defined. We'll also talk about this one below.
All these things were crucial for growth of the field.

Most of the work in the first half of the century was done on relatively
small datasets. When students are learning statistics nowadays, they're

still learning the calculations on small datasets, often small enough that they can do the calculations by hand. The advent of the computer clearly brought substantial change to the field, allowing for more complex and time-consuming computations that could never have been done by hand. Several statistical software programs like SPSS, Minitab, and Statistica were developed and continue to be used today, especially by social scientists, whereas data scientists, many actuaries, and a lot of other computational scientists are using Python, R, or SAS.

Not all the work that data analysts and data scientists do relies on statistics, but a huge chunk of it does. The fact that the field is still developing promises further benefits in the form of new techniques that might make tasks like forecasting even more accurate. Data scientists are wise to keep themselves informed about what's going on in statistics.

Example of Statistics in the Real World

Project 1: German Tank Problem

During World War II, the Allies needed to know how many tanks and other weaponry the Germans had. Spies in Axis areas gathered intelligence to try to estimate the numbers produced each month, but their estimates were consistently too high. So the Allies' mathematicians used a statistical approach to estimate the number and infer other qualities of German tanks and weaponry.

German tanks and other weapons were printed with serial numbers, which the Allies were able to see on tanks that were damaged or captured. It's pretty natural to make some assumptions about the serial numbers, including that they are sequential. In other words, we can assume that each serial number is one number larger than the previous one. The list of all serial numbers represents a sample of all the serial numbers on all tanks (the "population").

They wanted to infer the population size based on a sample, even though they had no way of knowing how the sample related to the population. There are a few overly simple ways to do that (like doubling the max or doubling the mean or median), but none of these is very accurate. Instead, they used a calculation called the Minimum-Variance Unbiased Estimator (MVUE), which involved dividing the max serial number by the total number of tanks and adding the max serial number to that, and then subtracting 1.

Although it wasn't possible to confirm until after the war ended, the statisticians' numbers turned out to be quite accurate, especially when compared with the numbers gathered through traditional intelligence— the differences would often be off by almost an order of magnitude, like 250 tanks being produced in a month vs. 1,500, when the lower number was close to the real number. The Allies were able to use the lower estimates in their planning. An understanding of the real number of tanks the Germans had at their disposal made them more confident in taking the risk of an invasion, specifically with D-Day, which was a major turning point in the war.

Project 2: COVID-19 Modeling and Forecasting

Soon after COVID first broke out in China and started spreading around the world in early 2020, organizations started collecting data on cases and trying to understand spread patterns and forecast new cases. This work was critical in informing public policies like lockdown, social distancing, and mask mandates, especially in the months before vaccines were available.

But there were many challenges in modeling, and a lot of us will remember hearing how bad many of the forecasts were in the early days. This is a real example of the problem of low data quality as talked about in Chapter 1. You'll hear the mantra "garbage in/garbage out" in data science because it is 100% true. So little was known about the illness that its characteristics were totally not understood (we made lots of invalid assumptions), and testing was inconsistent and limited, so it was underreported. We didn't even know how it was spread or how long it could survive before infecting someone in the early days. Since we didn't even understand how COVID worked in the real world, it's no surprise that we couldn't turn the information we did have into usable data. Remember that all data is simply an abstraction of aspects of the real world.

Disease modeling has long been a part of epidemiology, but it relies on a lot of known factors like transmission rate, how long infected people are contagious, and mortality rate. Many of the diseases that crop up in different areas are reasonably well-known from data on previous outbreaks, like the measles and the plague. So, with a relatively new disease like COVID, a lot of guesswork was necessary even once we started getting better data coming in. Scientists obviously looked for parallels with other diseases (especially those in the same coronavirus family, including MERS) to define starting points for models, but COVID behaved differently (for instance, the fact that it didn't impact kids more than non-elderly adults was unusual). Scientists just did their best with limited information.

Figure 4-1 is a good example of some early efforts at predicting COVID in Sweden, with five different predictions made between April and July 2020. They all include the historical data, which is the peak you see on the left side. But you can see how they differ after April. One has it petering out in June, and another has it still going strong a year later. One of the biggest challenges in forecasting is that the further out you go, the less confidence you have in the forecast. And they only had a few months to work with, and they were forecasting out much longer than that, something that is generally not done. We'll talk about this more in Chapter 15.

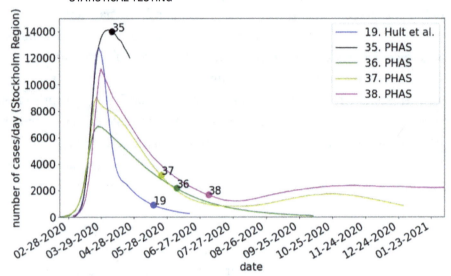

Figure 4-1. *An early effort at predicting COVID-19 in Stockholm,
Sweden. Source: "Computational models predicting the early
development of the COVID-19 pandemic in Sweden: systematic
review, data synthesis, and secondary validation of accuracy" by
Philip Gerlee, Anna Jöud, Armin Spreco, and Toomas Timpka in
Scientific Reports 12: 13256 (2022), available at* https://www.
nature.com/articles/s41598-022-16159-6

As soon as a variety of models from different groups appeared in
the first few months, the US Center for Disease Control and Prevention
(CDC) started creating ensemble forecasts from the many models and
forecasts. *Ensemble machine learning* is when multiple algorithms are used
to make one forecast or result. It can be a powerful way to compensate
for one model's weaknesses while still getting benefits from it (we'll
discuss it more in Chapter 15). There were many types of models being
used in forecasting—some traditional epidemiological ones and others
statistics-based.

There were several different types of epidemiological models. These
models tend to require more assumptions than purely statistical ones do

because they use information about disease behavior and other things like the availability of medical care and human behavior. So with all the unknowns, these were hard to get right. Many of the early ones were dramatically wrong, usually overestimating COVID cases. These models were also very sensitive to data collection problems like underreporting.

Statistical modeling seemed promising, but it requires a lot of data, so it couldn't really get underway until there was enough data. These approaches can be simpler because they look directly at disease case counts and don't really require a lot of outside information. Some early ones included time series approaches and exponential smoothing. Both of these will be covered later in the book, but *time series analysis* is a classical statistics approach where values of a variable are tracked over time (picture a line chart with dates on the X-axis and a line stretching across the center of the chart). *Exponential smoothing* is another forecasting approach that relies on a weighted average of previous data (the average modified by some other factors). Despite the long history of using these techniques, these early COVID models weren't very accurate, but they still tended to outperform the epidemiological ones.

One major drawback to statistical modeling is that outside changes (like lockdowns and the introduction of the vaccine) mean that the previous trend doesn't really apply anymore because circumstances are totally different. Another important disadvantage of traditional statistical approaches that are dealing with occurrences over time is that they can't reliably predict out very far, usually only a handful of weeks (though they can sometimes go a little further with more data).

Scientists eventually realized that a hybrid approach was better, with statistical models best for very short-term forecasts and epidemiological models best for longer-term projections. Scientists did the best they could under the circumstances, even though we now see that most of the models overestimated disease counts in the early days. But this is a reminder that when data science is going to impact people, it's important to be as rigorous as possible.

Figure 4-2 shows one of the more sophisticated models predicting R_t, the average number of new infections caused by one person with the disease on a given date. The screenshot was taken mid-September, which is why the light-blue area fans out from there (it's a 95% confidence interval, something we'll talk about more later in the chapter).

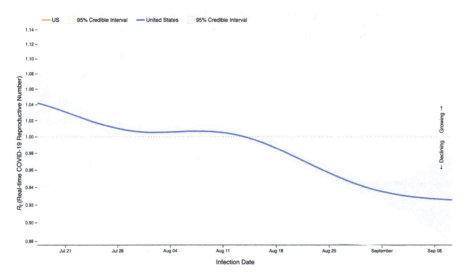

Figure 4-2. *A later, much improved model of COVID-19 transmission rates. Source: From "Current Epidemic Trends (Based on R_t) for States" on CDC CFA: Modeling and Forecasting,* `https://www.cdc.gov/cfa-modeling-and-forecasting/rt-estimates/index.html`

Statistical Testing Concepts

In statistics, the most common goal is to determine if an effect of a treatment is strong enough to claim that it has made a genuine difference. This strength is measured as statistical significance, which technically tells us if something is likely to have occurred simply by random chance or if it's a real result of the thing we're studying. Traditionally, this is measured with p-values and confidence intervals, both of which we'll discuss shortly. It might be testing the performance of an ad on a website against a different ad. Or it could be looking at if a tutoring program improved the grades of high school students struggling in math. We use the term *treatments* in experiments or studies to refer to the different test cases—usually a control group and one or more others. Treatments are technically an independent variable that is manipulated. For instance, in the study of the tutoring program, we would have a control group of kids who aren't enrolled in the tutoring program and a group enrolled in the tutoring program (the treatment). We refer to difference between the results of the control and the treatment(s) as the *effect*.

A mistake a lot of people make when looking for an improvement in something—did making a budget mean we spent less money this month than last?—is to calculate a change in the before and after and presume that it means something. One of the fundamental ideas in statistics is that there is always a dollop of randomness in anything. If people running the tutoring program have 100 students in program at the start and 90 at the end and they discover that 3 students went from a D grade to C and there were no other changes, most of us can instinctively tell that's not enough of a change to really say the program made a difference. They should not call the tutoring program a success, even though technically there was improvement. Additionally, if they have a control group, they would probably find that some of those kids in the control group also went up a letter grade. Statistical tests allow us to say that a difference is meaningful (statistically significant). Without even running the numbers, we know that the tutoring program did not improve things enough to invest time and money in it.

There are many things wrong with the plan to test the budget idea by comparing two months next to each other. First, a sample size of one—one control and one treatment (the use of a budget)—is never going to yield a statistically significant result. Additionally, this is an easily confounded study—what if we pay our car insurance quarterly rather than monthly and it came out last month? That would be a large expense we wouldn't have this month. Other expenses are the same every month and can't easily

be adjusted—the rent, car payment, cable bill. You would need to narrow
your spending down to categories that are adjustable, like groceries or
entertainment.

Hypothesis Testing

Historically, testing for statistical significance has been done by
formulating the problem being studied in a particular way that can seem
a little odd at first, but establishes a common language for statistical
analysis. This is called hypothesis testing because we start with the
null hypothesis, written H_0, and the alternative hypothesis, written H_1.
The *alternative hypothesis* represents the effect we are looking for, like
whether the tutoring program improved grades. The *null hypothesis* is the
opposite—it says there is no significant effect, or the tutoring program did
not make a difference in grades. The way we talk about these is to say that
the goal of hypothesis testing is to see if we can reject the null hypothesis,
which would mean that we believe that the alternative hypothesis is true
and there is a significant effect.

p-Values and Effect Size

One of the most well-known statistical measures is the p-value, even
though it isn't necessarily well-understood. The p-value is a probability,
or a number between 0 and 1, that is used to indicate the likelihood of
obtaining a particular result by chance alone. If it's a low number, that
means it is unlikely to have occurred "by chance"—that is, without a
logical reason (such as the hypothesis likely being true). People sometimes
misunderstand it as a percentage meaning how likely it is to be wrong. We
mentioned that randomness is relevant in statistics, and the p-value helps
us deal with this. A common threshold number for the p-value is 0.05,
which basically means that if you run the study 20 times, we wouldn't see

the effect appear in more than one of the runs due to randomness if the
null hypothesis was actually true.

Some statisticians are starting to question the p-value as such a
universal measure. For instance, assume we want to understand how good
a surgeon is and we set up the following test:

- H_0: The surgeon is not a superstar.

- H_1: The surgeon is a superstar.

- After measuring their performance over dozens of
  surgeries, we calculate a p-value of 0.025, which is
  lower than the common cutoff of 0.05, meaning we
  can reject the null hypothesis and conclude that the
  surgeon is a superstar.

This doesn't mean that the surgeon has a 97.5% success rate. Instead, it
means that if the surgeon really isn't a top performer and we measure their
performance 20 times, they wouldn't show up as a top performer more
than one time.

But this idea that we can incorrectly see an effect as successful 1 time
out of 20 concerns some people. The p-value also doesn't say whether the
specific test even made sense for the problem we're looking at. And one
other major problem is that multiple testing for p-values of slight variants
of the original alternative hypothesis without adjusting the study can lead
to spurious results.

Because of the limitations of the p-value, some people are starting to
report the *effect size*, which is simply a measure of the size or impact of the
effect that you have found to be statistically significant. This is a metric that
will allow you to step away from the p-value to some degree, so it is often
valuable to share.

Type I and Type II Errors

In the context of hypothesis testing, there are two possible errors. The first, *Type I error*, is when we reject the null hypothesis when we shouldn't—in other words, we think the thing we're testing for is true when it actually isn't. In data science, we'll usually see this called a *false positive*. The second, *Type II error*, is when we don't reject the null hypothesis when we should, so the thing we're testing for actually is true but we don't detect that. This would be a *false negative* in data science.

There are commonly accepted levels of risk for both types of errors in statistics (generally 5% for Type I and 20% for Type II), but in the real world, we sometimes have to make adjustments. For instance, in cancer screening, a false positive isn't as concerning as a false negative. A false positive would lead to further tests that may rule out cancer, which the insurance company won't like, but it would be even more costly to treat the cancer at a later stage if it's missed in this first screening. Alternatively, if the judge in a criminal case uses a number from an automated system that gives the likelihood of a prisoner reoffending on release as a major factor in their sentencing decision, false positives would be damaging to people because it could mean they spend more time in jail than they should. See Table 4-1 for a summary of possible results during hypothesis testing.

Table 4-1. *Hypothesis testing and Type I and Type II errors*

		Decision	
		Accept H_0	Reject H_0 (Accept H_1)
Reality	H_0 is true.	Correct	Type I error
	H_0 is false (H_1 is true).	Type II error	Correct

Margin of Error and Confidence Intervals

Any time you take a sample of a population, you're missing some information—you don't know anything about the points not selected, and they are bound to be different in some ways from those in the sample. Obviously, the bigger your sample, the more likely it is to be representative—up to a point of diminishing returns. But the fact is that unless you literally look at every point in the population, there will be aspects to the unsampled points that are unknown. This means that there will *always* be some error in any inferences you make about the population based on the sample. The presence of error doesn't mean there is a mistake anywhere, just that the likelihood of a sample perfectly matching a population is miniscule.

There are a couple of different ways of handling this error. One is the *margin of error*, which is given as a percentage representing the range from either side of a statistic we're providing. The range is known as a *confidence interval*, and we define a particular confidence level with this. For instance, if we've done a survey that says that 57% of Americans prefer the Marvel-Verse over the DC Universe and we calculate a margin of error of 3%, that would mean that we are claiming that we're fairly certain that the true value is within the confidence interval of 54% and 60%.

There are different ways of calculating the margin of error for different types of data, but they all rely on the sample size and the standard deviation on a normal distribution. Remember that sample statistics are often normally distributed, so we can calculate a z-value based on a desired level of confidence (commonly 95%) and multiply that by the sample standard error to get the margin of error. The z-value is simply the number of standard deviations that holds the proportion of data specified in the confidence level.

One-Tailed and Two-Tailed Tests

When we're testing a hypothesis, we need to decide if we are only going to consider a change in one direction (so the treatment led to a better result or not) or either direction (so the treatment led to either a better result, a worse result, or neither). In the one-direction case, we use a one-tailed test so we can know if there was an improvement or not. In the second, it's a two-tailed test because we just want to know if there is any difference.

It's important to understand these different types of tests for a couple of reasons. First, for a one-tailed test, we are only checking for a significant effect in the direction we defined. If we're asking if the treatment led to a significantly higher result with a one-tailed test, we would not be considering the significance of a lower result—it would not tell us whether a lower score is significant, even if that was the case. The two-tailed test is required to look at both sides.

The second reason it's important to understand the difference between one- and two-tailed testing is that the significance level is split in a two-tailed test. If we look for a p-value of 0.05 or lower in a one-tailed test and we want that same level in a two-tailed test, it would be split across the sides, so it will actually be looking for 0.025 at both sides.

Two-tailed tests are sometimes the better test because often the effect is not in the direction we'd expect—our expectations can be wrong, which is why we have to be scientific about it. However, there are also going to be cases where one-tailed is better. For instance, if we're testing a marketing campaign, we're just trying to decide if we should use it or not. We want to know if it improves things so we can run the campaign or not. It doesn't really matter if it has worse results than the status quo.

Statistical Tests

Statistical tests are well-defined tools used to determine statistical significance—whether observed data supports a particular hypothesis or not. Many statistical tests rely on a distribution with known characters and involve producing a *test statistic*—a single number calculated based on a specific formula—that can be placed somewhere on the distribution to draw conclusions. Although we talk about this in the context of the distribution, if we're calculating this manually, what we do once we have the test statistic is look at tables that show the significance of a particular value. Sometimes this value is based on the *degrees of freedom*, a slightly tricky concept, but it has to do with how many intermediate values were calculated before the current calculation or how many independent pieces of information are in the data in order to determine the shape of the reference distribution. It's usually one less than the number of observations for a single sample and two less for two. Looking at the right column of the table with the test statistic and the degrees of freedom if necessary, we just have to see if the test statistic is significant at the p-value level we have selected (often 0.05). Note that all tests based on a distribution can be one- or two-tailed. Also, most of these tests require numeric data.

The basic testing process is to pick the right statistical test, get all the data you need for the test, calculate the test statistic, and find the significance level in the table for that test. An even better way to do all this is to use Python or R, which has functions that give us all the numbers we care about for a huge number of tests, and we don't have to scour statistics tables manually. We still have to understand the tests well enough to know what to provide in the code and what parameters to specify, but it still reduces the work. There are a couple of tables in a later section that summarize the tests we've talked about along with their requirements, uses, and more (Tables 4-4 and 4-5).

Parametric Tests

The normal distribution revolutionized the field of statistics. Some of the most powerful statistical tests are tied to it, or occasionally to other well-known distributions. We'll talk about several normal-based tests in this section. As mentioned above, all of these involve calculating the test statistic, which we will use to determine the significance level and decide if we should reject the null hypothesis or not.

Z-Test

The *Z-test for mean* can be used to test if the sample is different from the population at a statistically significance level. It relies on the *Z-distribution*, which is simply the standard normal distribution, so it has a mean of 0 and a standard deviation of 1. The Z-test is similar to the t-test discussed in the next section, but the Z-test is used with a larger sample size (30 or more) and the population standard deviation is required.

The process of running the Z-test involves calculating the *Z-statistic*, and because we are dealing with the standard normal distribution here, the Z-statistic is the same as the Z-score (remember from Chapter 3 that that's the number of standard deviations that a given value is from the mean).

The t-Distribution and t-Tests

The t-test is a family of classic statistical tests based on the t-distribution. It is useful in a variety of situations, with dataset size usually on the small side, but it's not required. The distribution and test were developed by a statistician at Guinness and published under the name Student, because the company didn't want people knowing they were using statistics at that time (which is funny given how much companies name-drop the terms AI and machine learning nowadays), so the distribution is sometimes called the Student's t-distribution.

The Z-test and t-test often go hand in hand, and there are situations
where you'd choose one over the other. The primary question is whether
you have the population variance. If you don't, you can't use the Z-test.
If you do have it, if you have at least 30 observations, the Z-test is best.
Otherwise, go with the t-test.

The t-Distribution

The t-distribution is similar to the normal distribution in that it is
symmetrical with thin tails, but the tails are heavier than the normal
distribution and the peak is lower when degrees of freedom are lower. As
you approach 30 degrees of freedom, it starts looking more like the normal
distribution. It's a bit meta because it models distributions of sample
statistics and helps us calculate a confidence interval around the relevant
statistics, basically helping us understand sampling error. For instance, if
we have a bunch of height measurements and have calculated the mean,
we can identify a 90% confidence interval around the mean, meaning that
if we take different samples over and over from the same population, 90%
of the time the sample mean will lie within the range. It was heavily used in
the pre-computer era to understand samples, but not as much nowadays
because we can use computers to understand the error through some
other techniques.

The shape of the distribution is determined by the degrees of
freedom, a number determined for the specific test involved based on the
sample size involved. The higher the degrees of freedom, the closer the
t-distribution gets to the normal.

The distribution is useful in situations where we want to know how
likely it is that the mean of a sample we have is the actual population
mean. We can use this when we have a population that we can assume
is normally distributed, but we don't know the mean or the standard
deviation of the population. We can use the sample mean and standard

distribution with the t-distribution to estimate the distribution of a sample
mean, which can help us understand the population mean.

Figure 4-3 shows a couple t-distribution curves with a standard normal
one overlaid so we can see how increasing the degrees of freedom brings it
closer to normal. At low degrees of freedom, the curve (in blue) is different
from normal (in black), but at 40 degrees of freedom (red), it's virtually
indistinguishable from the normal.

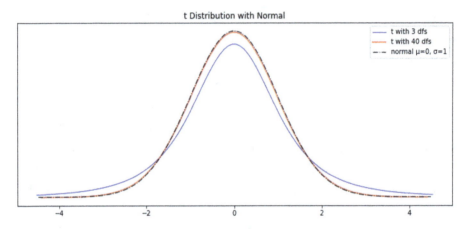

Figure 4-3. *Some example t-distributions compared to a standard
normal overlaid in black*

The t-Tests

t-tests are a family of classic statistical tests and excel at small sample sizes
(generally less than 30). There are several different tests, which will be
described below. All of these have some basic requirements, including that
the data is assumed to be normal. I'm not listing all of them, but if you are
planning to use these tests, you'll need to ensure your data is suited to it.

One-Sample t-Test

The most basic t-test is the *one-sample t-test*, which is usually done to see
if a sample population is different from a population by looking at their
means. We might know that first-generation first-year college students at a
particular school have an average GPA of 3.03. If we want to test whether a
mentoring program that pairs these students with first-generation third-
years helps raise their GPA, we could use the one-sample t-test to see if
the mean of the mentoring group is different from the overall average (the
population of all the first-generation third-years). We might first think to
only test if it improved the mean GPA overall (one-tailed), but it's probably
a good idea to do a two-tailed test to allow for the possibility that it had the
opposite effect (perhaps because of more party invitations).

Independent Samples t-Test

The *independent samples t-test*, also called the *unpaired t-test*, is used to
compare the means of two separate and independent samples to see if
they are different. Independent in this case means that there is nothing
to tie the two samples together like repeated measurement on the same
person or any other relationship between the samples. The samples should
be close in size. An example of this type of study might be to compare the
GPAs of first-generation students vs. legacy students at a college.

Paired t-Test

The *paired t-test*, which also has several other names including the
repeated measures t-test, involves two naturally related samples, such
as a set of before and after measurements with some intervention. For
instance, if we find some professional soccer players whose physical fitness
seems to have plateaued, a trainer might try a new training program and
measure their fitness before the program and again after two months with
the program. Each measure in sample 1 has a pair in sample 2.

Welch's t-Test

Welch's t-test, also called the *unequal variance t-test*, also allows us to compare two populations to see if they have equal means. It's like the independent samples t-test, but it works better than that test if the samples are different sizes or the variance between them is different. Still, we must assume both populations are normally distributed.

ANOVA

The different t-tests only allow the comparison of the means of two samples. Analysis of Variance (shortened to ANOVA) also allows us to compare means, but the advantage is that more than two can be compared at once. The misleading use of the term "variance" in the name of the method refers to how it partitions variance based on different factors, not because we're analyzing the variances. ANOVA relies on the F-distribution, whose curve is defined by two different degrees of freedom.

There are several assumptions that must be met for ANOVA to work reliably. The data must be continuous (interval or ratio) and also independent, which means that the variables cannot affect each other's values. It also should be close to normally distributed within each group, and the variance of each of the groups should also be similar. ANOVA is somewhat robust and some of these assumptions can be a little weak, with normality and similar variance in particular being loose, but the requirements of independent and continuous data really aren't negotiable.

In ANOVA, we often call the variables we are testing *factors* and the possible values each one can have *levels*. We also talk about *groups*, which are the subsets of the data corresponding to each specific combination of variables and levels. For instance, if we have a study that has two factors each with two levels, there would be four possible groups, as seen in Table 4-2.

Table 4-2. *Groups tested on two different factors for ANOVA*

Group	Factor 1 Level	Factor 2 Level
1	1	1
2	1	2
3	2	1
4	2	2

One-Way ANOVA

Like with the others, there are a few different forms of ANOVA. A *one-way ANOVA* involves testing a single independent variable (factor). If that variable has only two levels, then it's equivalent to a t-test. In this test, we are looking at variability both between groups and within groups.

As an example of a study appropriate to this type of ANOVA, we could look at the effect of screen time on ten-year-old kids' ability to concentrate. Concentration is measured by the number of minutes they can focus on a particular task. There can be three levels for daily screen time: none, one hour, and two hours. There are ten kids in each of three groups here, one for each level, and we want to know if there are differences between the mean concentration time of the three groups.

Factorial ANOVA

In the real world, it's somewhat rare to study a single factor. The effort of putting a study together is enough that often it seems more worth it to throw a few things in there at once. There are reasons for caution with this both because factors can interact with each other and compromise the study and because testing too many things can lead to spurious results (see the section "Testing Limits and Significance"), but it's still more common to need to run an ANOVA with a few factors rather than just one. Getting samples large enough to compensate for all the different combinations of factors is important when designing a study (or considering if ANOVA can be run on existing data). A *factorial ANOVA* is simply an ANOVA test involving two or more factors.

A two-way ANOVA is simply a specific case of a factorial ANOVA involving two different factors with multiple levels each. Table 4-2 shows the number of different interactions there are with two factors (four). If we added a nutritional supplement to the study on kids' concentration (where the two levels would be a supplement provided and one not provided), we'd be able to do a two-way ANOVA.

A factorial ANOVA with three factors has even more interactions because each two of the three factors have to be tested for an interaction effect, plus the interaction of all three. A larger sample size will be needed in each group. If some of the kids in the concentration study have ADHD, we could add that as a third factor and be able to do a three-way ANOVA.

Clearly, this can continue on to more factors, but the more factors you study, the more likely you are to see spurious results.

ANCOVA

One further related test is Analysis of Covariance (*ANCOVA*), a statistical method that extends a factorial ANOVA by using *covariates*—interval predictor variables that are known to have an influence on the outcome variable but are not the primary factors under investigation. It's considered a combination of ANOVA and regression, which we'll cover in Chapter 15. For an example of when it could be used, in the kids' concentration study, if we added kids of different ages into the study, we would assume that the age of the child will have an impact on their concentration ability—older kids should have an easier time focusing than younger kids. In this case, the age factor would be called the covariate.

Nonparametric Tests

Statistics has long relied on distributions to understand and test significance. The normal distribution in particular features heavily, as many tests require an assumption of normality. A lot of people doing statistical analysis are cavalier in their assumptions, running tests assuming normality or other requirements without ever testing for those things. This is easy to do, but it's irresponsible and will lead to potentially invalid results.

One of the solutions to the problem of these assumptions—whether we don't know if our data meets the requirements or if we know that it doesn't—is *nonparametric tests*, which are simply tests that don't have all the assumptions required for many of the distribution-based tests. There are many of these, some of which allow us to test for things like normality. Others allow us to do tests on non-continuous outcome variables. We won't go into detail on any of them, but it's important to understand what's available.

This section will look at our first categorical tests, mostly chi-squared tests, and others that work on numeric data. We're again calculating a test statistic and determining significance based on that, which informs our decision to reject the null hypothesis or not.

Categorical Tests

There are several nonparametric tests used to understand categorical variables better, especially to figure out if behavior is different from what was expected. The most well-known categorical tests are the chi-squared tests. There are several different chi-squared tests, but we'll be talking about the Pearson chi-squared test in this section, with two others, McNemar's test for matched pairs and Fisher's exact test. Even with the Pearson chi-squared tests, there are several different ones: the chi-squared test for independence, which tests if two variables in a

study are independent of each other; the chi-squared test for equality of
proportions, which tests whether the distribution of a selected variable
is the same for samples drawn from multiple different populations; and
the chi-squared test of goodness of fit, which tests whether a particular
categorical variable follows a known distribution. This is often a uniform
distribution, meaning we'd see equal proportions across all combinations,
but it can be anything. There are some requirements for these tests to
ensure that the test statistic truly follows the chi-squared distribution,
which is necessary for accuracy of the result.

Chi-Squared Test for Goodness of Fit

We often want to know if a variable follows a known distribution (or a
less rigorous pattern), and this test lets us check for that. With the *chi-squared test for goodness of fit*, we're looking at frequencies to see if the
observed frequencies are the same as what we'd expect from the presumed
distribution. We might look at college students' monthly reading habits
and see if this adheres to an expected breakdown we found in another
study. This is still categorical, so we might have the proportion of 100
students who finish a set number of books per month, with three levels: 55
finish one book, 25 finish two, 15 finish three, and 5 finish three or more.
We work with a table listing these total number of students in each group
to determine if the actual counts fit the expected proportions.

Chi-Squared Test for Independence

With *the chi-squared test for independence*, we want to know if two
categorical variables are independent. The null hypothesis is that they are
independent, so a rejection of the null hypothesis says that they may be
dependent on each other. Imagine a study that looks at whether having
a job impacts high school students' abilities to graduate. We'd have a
count of students for each of the four possible combinations (no job/
graduated, no job/did not graduate, has job/graduated, and has job/did

not graduate). To look at this, we'd first calculate the expected values for
each cell based on the total in each column and row. This time we work
with a contingency listing one variable's values across the top and the
second's values down the side. In our example, it would look like Table 4-3.
We would add counts and column and row totals to calculate the statistic.

Table 4-3. *Setup of a contingency table in a
chi-squared test for independence*

	Graduated	Did Not Graduate
Had a Job		
Did Not Have a Job		

McNemar's Chi-Squared Test

Another chi-squared style of test that relaxes the requirement for
independence in the actual chi-squared test is *McNemar's test for matched
pairs*, which looks at paired-sample data to see whether a particular value
of a variable changes between the two paired measurements. This test is
especially useful when the observations are not independent because it's
looking at the changes within pairs. For instance, if a college requires all
sophomores to take a seminar on social justice issues, they may want to
measure if students' opinions on several issues change after they take the
seminar. They'd measure this at the beginning of the class and then again
perhaps when they are juniors to see if it changed their views long term.
To get the statistic for each issue, they'd create a contingency table similar
to the chi-squared test for independence, but the statistic is calculated
differently.

Fisher's Exact Test

Fisher's exact test is similar to the chi-squared test but doesn't require the same assumptions to be met. It does actually relate to a distribution, the hypergeometric distribution, but it's otherwise less stringent that the chi-square. *Fisher's exact test* measures the probability that the results in the study are as extreme or more extreme than what's observed. For example, in the study looking at if having a job affects high school graduation rates, the test would determine if the risk of not graduating is higher for those with a job by saying that the likelihood that the given number of people who didn't graduate is this value or higher.

Numeric Tests

There are many tests for different situations with numeric data. These also allow ordinal data. Several of them use ranks of the data.

Wilcoxon Rank Tests

The Wilcoxon rank tests for independent data are useful for understanding two usually (but not necessarily) equally sized samples of ordinal data, numerical data that can't be treated as ratio or interval. There are actually two related tests. The first is the *Wilcoxon rank-sum test*, also called the *Mann–Whitney test*, which involves sorting each sample and ranking them in a particular way that accounts for ties. These ranks can then be summed, which allows us to calculate the U statistic and test for significance.

The *Wilcoxon signed-rank test* looks at the magnitude and direction differences between paired observations to tell us whether the two paired samples are different. The test statistic is calculated by first taking the differences between each pair and then summing the multiples of the sign (–1 or +1 based on whether the value is lower or higher than the mean) and the rank. Then the significance can be computed and the null hypothesis rejected or not.

The *sign test* is a variation of the Wilcoxon signed-rank test where the magnitude of the differences between the paired observations isn't needed (especially if the magnitudes aren't very reliable). It can be used on data that has a binary outcome variable and tests whether a sample has a hypothesized median value. It uses the binomial distribution to get the probability of the given outcome if the null hypothesis (that the population median is the one specified) were true.

Kruskall–Wallis Test

The *Kruskall–Wallis test* is analogous to a one-way ANOVA and checks if the median of multiple samples is the same. It's convenient because it doesn't require the samples to be the same size and works with small samples. It's also less sensitive to outliers. Additionally, it's similar to the Wilcoxon tests because it also deals with ranks. The statistic can be calculated using a formula with each sample size, sum of ranks for that sample, and the combined sample size.

Kolmorogov–Smirnov Test

The *Kolmogorov-Smirnov test* is an important test for normality mentioned above in the Tests for Normality section in Chapter 3. It can be used to test if one sample came from a particular distribution or if two samples came from the same one. This one-sample option involves ordering the observations and getting the proportion of observations that are below each observation in the sample (this is referred to as the empirical distribution). This can then be plotted and compared to the distribution we are checking against. With the two-sample version, we compare the two empirical distributions rather than one to a theoretical distribution.

Friedman Test

The *Friedman test* can be used when there are more than two matched
samples to detect differences between repeated measurements. It also
uses ranks, and the test statistic can be calculated with the sample size,
the number of measurements per subject, and the ranked sums for each
measurement round.

Selecting the Right Statistical Test

There are a large number of statistical tests, and we only looked at the most
common. Depending on the work you're doing, you might find you need a
totally separate test. But the ones we've covered will give you a good start.
See Table 4-4 for a summary of the covered parametric tests that assume
normality and Table 4-5 for a summary of the nonparametric tests. So how
do you go about deciding what tests to use?

Table 4-4. *Parametric tests summary*

		Type of Data	Compares to Distribution	Differences Checking For	Num of Samples	Paired/Independent?	Min Sample Size	Assumptions			Required Values					
								Assumes Normality	Assumes Similar Variance	Assumes Similar Sample Size	Population Mean	Population Variance	Sample Mean	Sample Variance	Sample Size	Degrees of Freedom
Z-test	One-sample Z-test for mean	Numeric	Normal	Sample mean and hypothesized (population) mean	1	N/A	30	Yes	N/A	N/A	Yes	Yes	Yes	No	Yes	No
	Two-sample Z-test for mean	Numeric	N/A	Two sample means	2	Independent	30	Yes	No	Yes	Yes	Yes	Yes	No	Yes	No
t-test	One-sample t-test	Numeric	Normal	Sample mean and hypothesized (population) mean	1	N/A	Small+	Yes	N/A	N/A	Yes	No	Yes	Yes	Yes	Yes
	Independent samples t-test - similar variance	Numeric	N/A	Two sample means	2	Independent	Small+	Yes	Yes	Yes	Yes	No	Yes	Yes	Yes	Yes
	Independent samples t-test - different variance	Numeric	N/A	Two sample means	2	Independent	Small+	Yes	No	Yes	Yes	No	Yes	Yes	Yes	Yes
	Paired t-test	Numeric	N/A	Two paired sample means	2	Paired	Small+	Yes	Yes	Yes	Yes	No	Yes	Yes	Yes	Yes
	Welch's t-test	Numeric	N/A	Two sample means	2	Independent	Small+	Yes	No	No	Yes	No	Yes	Yes	Yes	Yes
ANOVA	One-way ANOVA	Numeric	N/A	Multiple sample means with 1 categorical independent variable (factor)	3+	Independent	Factor-based	Yes	Yes	Yes	N/A	N/A	Yes	Yes	Yes	Yes
	Factorial ANOVA	Numeric	N/A	Multiple sample means with 2+ categorical independent variables (factors)	3+	Independent	Factor-based	Yes	Yes	Yes	N/A	N/A	Yes	Yes	Yes	Yes
	ANCOVA	Numeric	N/A	Multiple sample means with 2+ categorical independent variables (factors) + covariates independent from factors	3+	Independent	Factor-based	Yes	Yes	Yes	N/A	N/A	Yes	Yes	Yes	Yes

Table 4-5. *Nonparametric tests summary*

		Type of Data	Compares to Distribution	Differences Checking For	Num of Samples	Paired/Independent?	Min Sample Size
Categorical	Chi-Squared goodness of fit test	Categorical	Selected	Are proportions of groups based on two categorical variables as expected	1	N/A	Large
	Chi-Squared test for independence	Categorical	N/A	Are two categorical variables independent	2	Independent	Large
	McNemar's test for matched pairs	Categorical	N/A	Does an intervention change the outcome for a group	2	Paired	10+
	Fisher's exact test	Categorical	N/A	Are two categorical variables independent similar to Chi-squared test for independence for small samples)	2	Independent	Small+
Numeric	Wilcoxon rank-sum test (Mann-Whitney test)	Ordinal & Numeric	N/A	Are the two samples from the same distribution	2	Independent	Small
	Wilcoxon signed-ranks test	Ordinal & Numeric	N/A	Does an intervention change the outcome between samples	2	Paired	Small
	Sign test	Ordinal & Numeric	N/A	Does an intervention change the outcome between samples	2	Paired	Small
	Kruskall-Wallace test	Ordinal & Numeric	N/A	Do two samples come from the same distribution	2	Independent	5+
	Kolmorogov-Smirnoff test - one sample	Ordinal & Numeric	Selected	Does single sample comes from selected distribution	1	N/A	50+
	Kolmorogov-Smirnoff test - two sample	Ordinal & Numeric	N/A	Do two samples come from the same distribution	2	Independent	50+
	Friedman test	Ordinal & Numeric	N/A	Are there differences between the multiple samples	3+	Independent	5+

There are several steps to follow to select the right test(s):

1. Define the research question and null and alternative hypotheses.

2. Determine the type of data you have (nominal, ordinal, or numeric) and, if numeric, if it's normally distributed.

3. Determine if you have one sample or two or more and, if two or more, if the samples are paired or independent.

4. Consider the possible tests based on #2 and #3 and look at the assumptions of each to see if your sample(s) meet(s) them.

5. Pick your test based on what assumptions your data meets.

Note that there are tests that you may need to do to see if your data does meet the assumptions required.

Deciding the tests to use in statistics is something that is sometimes difficult at first, but comes with practice. You will probably need to spend some time researching a test to learn more about it any time you are considering using one. But one day, your instincts will make it easier.

Testing Limits and Significance

One important final consideration around statistical testing that we need to remember is that every time we run a test looking for a significant find, we are at risk of encountering a Type I error. It's tempting to test for a bunch of things at once, especially when we don't really know what could be important. For instance, remember what a p-value 0.05 really means— basically that 1 time out of 20 (5 out of 100) we could see an apparently significant result that isn't actually significant and just occurred by chance.

If we have ten predictor variables that have no genuine impact and run tests on all of them, it's possible that at least one will come up as significant even though it isn't. We can even compute the probability of one variable testing positive out of ten variables that aren't impactful. Basically, with a significance level of 0.05, that means an insignificant variable has a 0.95 chance to correctly test insignificant, but a 0.05 chance of incorrectly being labeled significant. The probability that all ten tests run for these insignificant variables will accurately test insignificant is 0.95^{10}, or only 0.60. Even with only ten variables tested—a number that doesn't seem too crazy—there's a 40% (1 – 0.60) chance that at least one will incorrectly test significant. If we increase the number of variables to 20, it's actually more likely that at least one will show up as wrongly significant than not (0.95^{20} is 0.36, so there's a 64% chance).

This is why some people advocate stronger p-values like 0.01 or looking at other numbers besides the p-value. But even at a 0.01 level, the likelihood of at least one of ten insignificant variables tested will show up as significant is 10% and 18% if we test twenty. So it doesn't solve the problem, but it does make it less likely to crop up. There's also a technique called the *Bonferroni correction* that reduces the chance of a Type I error (a false positive) by adjusting the p-value threshold based on the number of tests run.

This is another reason why repeating studies is so important for understanding of things. If we test only one thing in a study at a 0.05 level of significance, there is still a 5% chance that test will yield a significant result even if the variable isn't actually impactful. But if someone else tests the same thing in a different study, the chances of them also finding the variable significant are also 5%, but the likelihood of both of those things happening is only 0.25% (0.05 * 0.05), clearly much lower.

Correlation and Covariance

Most people have heard the warning, "Correlation is not causation," even if they don't know exactly what that means. As a concept, *correlation* is simply a measure of the strength and direction of the relationship between two variables. In *positive correlation*, when one variable increases, so does the other one. In *negative correlation*, the variables are inversely related— when one increases, the other decreases. Height and weight of children are positively correlated with each other, and both are correlated with age, whereas the amount of sleep a child needs is negatively correlated with age.

Covariance is a similar concept to correlation, also specifying a relationship between two variables, but the difference between them is basically in the range of values they can possibly have and how they are calculated, which we'll see below.

Data analysts and scientists are in the business of explaining why things happen based on data, so it's probably not surprising that understanding relationships between variables is really important. They calculate correlations and covariance frequently for different reasons, which we will go into in later chapters.

Correlation

Scatterplots are also often created to visually see a correlation, as it's easy to look at a plot and see that the values are positively correlated when the data points hover around a diagonal line going up from the origin to the top right. Similarly, a negative correlation would have a diagonal line going from the top left to the bottom right. But seeing a correlation in a chart is only valuable to a point—if we want to know if the correlation is statistically significant, we need to quantify it. See Figure 4-4 for examples of different types of correlation.

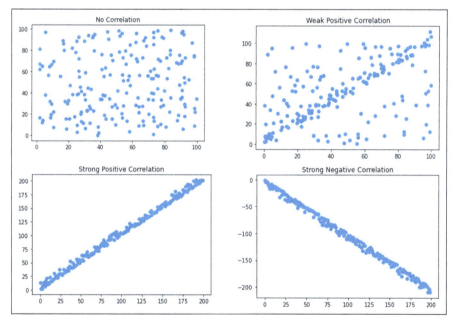

Figure 4-4. *Four views of correlation*

The top left is entirely random data with no correlation. The upper right has some correlation, but it's weak because of a lot of random values. The bottom left is strong positive correlation (as one variable increases, so does the other), and the bottom right is strong negative correlation (as one variable increases, the other decreases at a similar rate).

Quantifying correlation first means calculating the *correlation coefficient*, which is a single value representing the relationship between two specific variables. It's what's called a standardized calculation because it's a product of a measure of the two variables being compared divided by the product of their standard deviations, which makes it unitless and always between –1 and 1. Positive values indicate positive correlation (same direction), and negative indicate negative correlation (opposite direction). Because the value is standardized and unitless, it makes it easy to compare correlation of completely different datasets. It's common to display several correlation coefficients from a single dataset in a *correlation matrix*, which displays them in an easy-to-read grid format.

There are several different options for calculating the correlation coefficient, each of which has pros and cons in particular situations. The *Pearson correlation coefficient* is the most common, and it requires the data to be normally distributed. A couple other correlation measures that we won't discuss here but don't require normality are Spearman rank and Kendall rank. Calculating Pearson requires multiplying each X and Y point's difference from the mean and dividing by the product of the X and Y standard deviations.

However you decide to calculate correlation, it's common to display multiple correlations in a *correlation matrix*. We ran Pearson on the data on sick kids with their height, weight, and age that we looked at in Chapter 2, and Table 4-6 shows the results in a typical view.

Table 4-6. *Correlation matrix of sick child age, height, and weight*

	Age (years)	Height (in)	Weight (Kg)
Age (years)	1.000000	0.910758	0.704657
Height (in)	0.910758	1.000000	0.751830
Weight (Kg)	0.704657	0.751830	1.000000

Each cell shows the correlation of the column and row variables, so they are mirrored across the diagonal. You can also see that the diagonal shows correlations of 1, because a variable is always perfectly correlated with itself. It's also common to only show the correlations on the diagonal and lower-left corner (leaving the upper-right side blank), since the upper right mirrors the lower left. Correlation is one way only, so height is correlated with age exactly as age is correlated with height.

Another common way to look at the correlation matrix is a heatmap, which makes the values a little clearer. Figure 4-5 shows a heatmap on this same data.

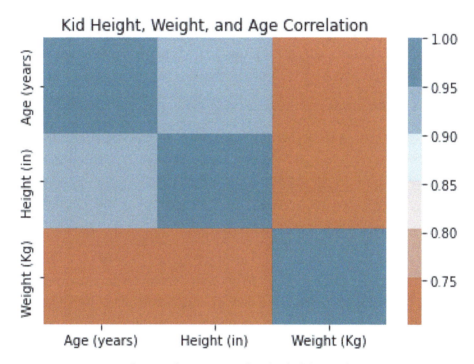

Figure 4-5. *Correlation heatmap of sick child age, height, and weight data*

These variables are all correlated pretty significantly, so the range of colors from low to high all show for positive values from 0.7 to 1.0. It's also common for a heatmap to show negatives in one color and positives in another. These are great for a few variables, up to ten or so, but can get unwieldy like a lot of other visualizations.

Covariance

Covariance is conceptually similar to correlation, but where correlation is always between –1 and 1, covariance depends on the scale of the two variables being compared. Calculating it means multiplying the differences of the means of both variables and summing for each data point and then dividing by degrees of freedom.

It's also common to display the covariance matrix the same way as correlation, but the numbers will be different, sometimes quite different. Table 4-7 shows the results of the covariance calculation on the same data on sick kids.

Table 4-7. *Covariance matrix of sick child age, height, and weight*

	Age (years)	Height (in)	Weight (Kg)
Age (years)	14.692915	25.910734	60.605008
Height (in)	25.910734	55.086591	125.204445
Weight (Kg)	60.605008	125.204445	503.446999

We can see that these numbers are not confined between –1 and 1 like correlation, and it's more difficult to really understand what they mean. Covariance is not often inspected directly and is more often used in other calculations, like ANCOVA as we saw above. Correlation is more intuitive.

Correlation vs. Causation

Let's revisit the adage "Correlation is not causation." The point it's making is that you can't assume that a change in one variable *causes* a change in the other variable. Sometimes things happen to seem correlated, but it's simply coincidental. There's a blog called Spurious Correlations that finds funny data that correlates. For instance, the popularity of the name Thomas between 1985 and 2020 correlates almost perfectly with the number of car thefts in Maine, which can be seen in Figure 4-6.

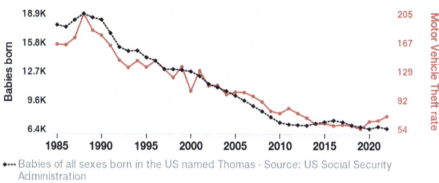

Figure 4-6. *Spurious correlation between two unrelated variables. Copyright: Tyler Vigen. Used with permission. Source:* `https://tylervigen.com/spurious/correlation/1882_popularity-of-the-first-name-thomas_correlates-with_motor-vehicle-thefts-in-maine`

It would be so tempting to think these are related, as they're even both slightly level at the top and bottom of the chart, yet common sense tells us they can't be. But if we had done our own study and were looking at two variables in the study that we really wanted to have a causality relationship, we might be tempted to assume they were genuinely correlated. It's important to stay cautious and skeptical when we're looking at correlation. Doing good statistics means never jumping to conclusions and especially not believing something just because you want to. Data analysts and data scientists must always have the mentality of making sure before claiming that there's a trend or effect.

Key Takeaways and Next Up

Statistics is used to some degree in virtually every field, but the style and perspective varies widely across fields. Psychologists do different things from computational biologists. There isn't one way data science is done, either, because data science techniques can be used in so many fields. Most data scientists work with large amounts of data often generated by computers or transactions of various types, from retail purchases to advertising responses, to banking fraud detection, to computer system monitoring. It's not as common for data scientists to work so much with humans directly like psychologists might. But since data science can actually be done on almost anything, and different aspects of statistics might be used in different fields, this chapter and the previous two took a bit of a step back to look at statistics as a whole, not only the parts that are most commonly used by data scientists.

This chapter focused on the inferential side of statistics—basically trying to generalize from a sample to an entire population. This is done through statistical tests that rely on probability distributions, many of which we covered in the previous chapter. There are several statistical measures of significance and other aspects that need to be understood in the context of statistical tests. We discussed tests for continuous and categorical data, many of which assume that the populations are normally distributed, and some nonparametric tests that don't assume normality. Then correlation and covariance were introduced.

The next chapter will finally dive into the applications of the ideas we've talked about in the book so far. Chapter 5 is going to discuss data analysis. Even if a data scientist isn't going to have the job title data analyst, they will most likely be doing data analysis as part of their regular work. The chapter will describe the origins of data analysis, describe two examples of data analysis in the real world, talk about practical skills needed in data analysis, and finally address the process a data analysis project generally follows. The truth is that most of the technical knowledge and techniques used in data analysis have already been introduced, but the chapter ties it all together.

PRACTITIONER PROFILE: SASWATI NEOGI

Name: Saswati Neogi

Current Title: Associate director of data science

Current Industry: Retail

Years of Experience: 15+ years in statistician or data scientist roles with various titles

Education:

- MS Applied Statistics

- MS Mathematical Statistics

- BS Statistics

The opinions expressed here are Sas's and not any of her employers', past or present.

Background

Saswati Neogi, Sas for short, grew up in India and always excelled in math, especially taking to the more abstract parts of it. So it surprised no one when she majored in Statistics for her undergrad degree, although she does admit she chose it partially because it was the only degree that didn't require physics or chemistry. She loved diving deeper into math and statistics and went straight into a rigorous master's degree in Mathematical Statistics that focused on the theoretical aspects of statistics and deeply understanding the algorithms and methods she studied. While at the University of Delhi, she learned about a great scholarship opportunity for a master's in Applied Statistics at the University of Akron in Ohio and decided to go for it. She knew that it would put her in an even better position to find a good job after graduation, because it would be more practical and tool-focused. She had really enjoyed building models in college and was excited to see them applied in real situations. During this last degree, she learned both SAS and R (two statistical programming languages heavily used in the industry at the time), which made her very employable.

Work

After finishing this degree, she already knew she wanted to work in a statistician role. She loved her work in school and wanted to be able to apply it in the real world, where it could have an impact by helping companies improve their business. Would she be able to run forecasts to make their lives easier or more accurate? Could she help automate things to get rid of time-consuming and error-prone manual processes? There were so many possibilities.

Soon Sas moved on to her first job as a risk analyst for a bank. She worked on interesting problems there and loved diving in to help solve real-world problems. One of the projects she worked on had to do with determining which customers were more sensitive to price changes, which allowed them to do targeted marketing. Another thing she discovered at her first job was that she loved working with people and explaining these complicated models to nontechnical people. Since that first job, she has worked in both the insurance industry and the retail industry. One project in insurance she worked on was helping to determine the pricing for customers. The generalized linear model, a flexible variant of linear regression that is basically the bread and butter of insurance, was used widely in her work. In retail, she worked on forecasting and profiling projects, using a variety of machine learning techniques. She noticed that retail was a very different world from insurance, because in insurance everyone is used to using all sorts of statistics and machine learning methods and even black box techniques (which give results whose inner workings can't be explained) are accepted as normal. But in retail, data science is newer than in insurance, and often techniques whose internal steps can be explained have to be used instead. Additionally, the data in retail is often a bit messier than in insurance.

Favorite Parts of Job: Some of Sas's favorite things about her work are interacting and communicating with other teams and people, working with smart and creative people, coding and applying models, and testing the performance of models with metrics.

Least Favorite Parts of Job: She is not a fan of messy data and not always being able to get all the data you need to build a good model. Also, sometimes your stakeholders are not totally receptive, and they can have unreasonable expectations or be unwilling to apply your models.

Favorite Project: The company had paid an expensive consultancy firm to build a model to predict sales of a particular item in stores. When her team found out, they built a model in a few weeks that beat the consultants' accuracy and then were given ownership of the project going forward.

How Education Ties to the Real World: You'll probably only use a fraction of what you learn in college, but the experience of learning new things in school will help you know how to learn new things in the real world. The ability to improve your skills in data science is very important.

Skills Used Most:

- Communication, machine learning models, coding, and visualization (as a data scientist)

- Project management and communication (as a data science manager)

Primary Tools Used Currently: SQL, Python, R, Snowflake, other cloud tools (AWS, Sagemaker, DataBricks, EMR (AWS PySpark))

Future of Data Science: The hype around data science, AI, ML is going to fade in the future—AI and ML have been these buzzwords expected to solve all problems, but they come with pros and cons, and it's not realistic that they can solve all problems. Experience will always be needed to know what can be applied in a given space, with what data and technology is available.

What Makes a Good Data Scientist: Technical skills, good coding skills, communication, and a collaborative attitude (nobody knows everything, so different people have different expertise, and you will work with them to complement your own skills). Being a team player is huge.

Her Tip for Prospective Data Scientists: Don't get too used to having clean data and the research projects done back in school—they help you work on your skills, but are not reflective of reality (a lot of time is spent cleaning and understanding data in the real world).

Sas is a leader in data science after nearly 20 years of working as a statistician, data scientist, and manager.

CHAPTER 5

Figuring Stuff Out: Data Analysis

Introduction

In previous chapters, we've talked about the most fundamental thing in data analysis and data science—the data itself—along with statistics, which is the foundation for everything else in data science. You obviously can't do anything in data science without data. But you also can't just take some raw data and start pulling out fascinating insights or predictions from it. There are a lot of steps to carry out before the data can yield valuable information. In fact, there are many ways that data can be explored, and the field of data analysis is devoted to understanding and working with data. Data analysis primarily involves slicing and dicing data in well-informed ways to extract meaning from it with analytical tools. These tools include programming languages, spreadsheets, and techniques from the statistics we covered in the previous three chapters, especially charts and other visualizations. Often data analysts stick to descriptive statistics, but some will delve into the more advanced statistics. Data analysts have been around for a long time, looking at what has happened before, explaining what it means, and sometimes using that information to help us understand the world and predict the future.

© Kelly P. Vincent 2025
K. P. Vincent, *A Friendly Guide to Data Science*, Friendly Guides to Technology,
https://doi.org/10.1007/979-8-8688-1169-2_5

This sounds a lot like what data science can do, and that's not a coincidence. It can be hard to say where data analysis ends and data science begins. I think of the two being on a continuum with lots of overlap. Almost always, a data science project involves doing data analysis, especially at the beginning. Most data analysts' work does not dip into the more advanced data science world, but many data scientists do data analysis all the time. In fact, in order to be a good data scientist, you need to be a good data analyst first.

"Data analysis" is a loaded term that means a lot of different things to different people in the business world, but the simplest definition is that it is the process of investigating and analyzing data in order to better understand what that data represents. We will talk more about this in Chapter 7, but the more general label of "analytics" is divided into four types: descriptive, diagnostic, predictive, and prescriptive. As you can guess, descriptive basically just describes what's in the data, while diagnostic seeks to explain things by looking into the data. These two are the basic domain of data analysis, with data science more focused on the last two.

So data analysis work can involve a huge range of techniques, and if you look at job listings with the title "data analyst," you will see a surprising range of skills required. In many cases, they are actually looking for a data scientist, asking for advanced programming skills, machine learning, and more. Perhaps they are trying to avoid paying a data scientist salary by using the other label, or (more likely) they don't know what they want or need. In other cases, they are looking for a high school graduate with basic math skills who knows or can learn to use Excel. Most of the time, the title means something in the middle. This chapter is going to focus on that Goldilocks data analyst—someone who would normally be expected to have a college degree along with several other skills, but not be as technical as a typical data scientist.

You may wonder, if data analysis is its own field, why does it have a chapter in a book on data science? As I mentioned above, being a good data scientist involves being able to do good data analysis. Someone who's a data analyst by job title may be more skilled in visualization or other things than a data scientist might be, but good data analysis done early in almost any data science project is critical to the project's success. This is because of a fundamental fact of data science: you cannot do good data science if you do not understand the data. Data analysis has a set of approaches that bring about that understanding.

While this chapter is about data analysis and will refer to those performing this work as data analysts, this is just the name of the hat they're wearing and equally applies to data scientists performing the data analysis part of their overall process. Additionally, data analysis can involve a lot of different approaches and tools, including those that fall under the label of statistics. This chapter will focus on the concepts, processes, and basic techniques of data analysis. It will start with a history of the field and two examples of data analysis in the real world. Then I'll go over the four critical skill areas needed to be a good data analyst. Finally, I'll cover the process that is generally followed to do data analysis work, a process called CRISP-DM (CRoss Industry Standard Process for Data Mining).

The History of Data Analysis

Data analysis has been around a long time, basically as long as data itself. The whole point of data is generally to understand whatever it represents better, and that's what data analysis helps you do. Of course, the kinds of analyses the ancient Sumerians and Egyptians were doing were incredibly simplistic compared with what we do now, but it was still important. Figure 5-1 shows an account of commodities on an ancient Sumerian tablet from around 2,500 B.C.E. This one shows quantities of barley, flour, bread, and beer. The data from these and other tablets was used by rulers to assess taxes for their citizens.

Figure 5-1. *Ancient Sumerian tablet circa 2,500 B.C.E. tracking quantities of several commodities (barley, flour, bread, and beer). Source: Sumerian economic tablet IAM Š1005.jpg,* https:// commons.wikimedia.org/wiki/File:Sumerian_economic_tablet_ IAM_%C5%A01005.jpg

While basic tasks with data have been done for a long time, data analysis as a modern field really took its first baby steps in the 1600s and didn't really learn to walk until the twentieth century. A lot of the important work in data analysis has been in visualizing the data in ways that help people understand the data and situation better. There are some pretty cool visualizations (especially maps and charts) that came out in the 1800s that we will see later in the book, with one simple original visualization in an example below (Figure 5-2). There is an entire chapter dedicated to visualization later in the book.

I'm going to talk about the computer and how it revolutionized data analysis next, but it's also worth looking at what data analysis can do before we understand how it's done. I'll share a couple of examples of real-world data analysis work. One of these is quite recent, but the other one was done more than 150 years ago and is still impressive today.

The Advent of the Computer

In the early days, analysis was always done on a rather small amount of data. It might be too much for people to keep it all in their heads, which is why visualization can be useful, but it was still a small enough amount that calculations could be done manually. There are definite limits to how much data was too onerous. We saw this in the previous chapters on statistics, where a sample size of 30 was "big enough" to use the more demanding Z-test over the t-test. Thirty data points sounds laughable to a modern data scientist, but things changed with the computer.

The computer has revolutionized a lot of things, and data is no different. The biggest change it enabled in data analysis is the ability to look at large amounts of data. This doesn't necessarily mean "big data," a term we mentioned in Chapter 1 and will revisit later. It simply means that everything doesn't have to be calculated by hand anymore, so we can use a lot more data. Early data analysis often involved collecting data points and running them through simple statistical techniques, perhaps starting by summing the numbers by literally adding the numbers together on paper or later with a simple calculator. Working with even hundreds of data points this way can get unwieldy, and as the data grows, these calculations soon become impossible.

US CENSUS TABULATING MACHINE

The US Census was getting very complicated by the last few decades of the 1800s. Hand counting people was too expensive and also error-prone. A machine was invented that helped speed up the process for the 1870 US Census, but it was still a mostly manual endeavor. The 1880 US Census was so difficult and time-consuming that they did not finish it until 1887. They needed something different. The first solution that came in was called a tabulating machine, and variations of that machine were used through 1940, after which the Census Bureau finally moved on to proper computers.

The main sign of this shift for regular people is the creation of statistics and analysis tools that allowed people to do their own analyses on the larger datasets that were difficult for manual work. One of the earliest was a product called SAS that's still used by some statisticians, especially in the insurance industry. It's a programming language with a custom, proprietary interface. It was first developed in the second half of the 1960s and is still evolving. Another product that came out in the early days is SPSS, which has primarily targeted people working in the social sciences, so it's used a lot in academia. By the 1970s, a lot of statistical work was done in one of the era's workhorse general programming languages, FORTRAN, which was difficult for less technical statisticians. The S programming language emerged in response to this, designed as an alternative to working with FORTRAN directly. People started using it, and then a version of S called S-PLUS came out in the late 1980s. S in general has been superseded by R, the modern open source and free statistical language based on S that was originally developed in the early 1990s. Nowadays, R is used by statisticians and data scientists and some data analysts, although most data scientists are switching to Python, which is a general-purpose programming language with a lot of statistical libraries that is often regarded as better than R because of its ease of use and performance.

OPEN SOURCE VS. PROPRIETARY TOOLS

With the exception of R and Python, the products mentioned above are proprietary, so they are very expensive and not accessible to most individuals. The proprietary tools are falling out of favor as people switch to open source R or Python. But a couple things that keep the proprietary products around are that they include guaranteed security and compliance, critical in many industries, and that companies that have code written in these languages or tools would have to rewrite everything in their chosen open source language. This is expensive, time-consuming, and not without risk (somebody could accidentally introduce a new bug), so it will be some time before these tools are abandoned, if they ever are. An additional reason that organizations sometimes stick to proprietary software is that because they are paying for it, they are entitled to customer support and can also influence future development of the software. Anyone who's gotten stuck with some code that isn't doing what they expect can understand the value of just being able to pick up the phone and call someone for help rather than hitting up Stack Overflow, the Internet's best free spot for technical questions. However, it is worth noting that there is customer support available for open source tools through some private companies (for a fee, of course).

Examples of Data Analysis in the Real World

Examples of data analysis are everywhere, but I've picked a couple interesting and very different examples to look at here. The first has to do with bringing baseball into the data age, and the second is an early win for public health.

Example 1: Moneyball

For decades, American baseball teams were staffed by scouts with "good instincts," who would go out into North American high schools and colleges to find young players with the aid of word of mouth. There was a strong tradition of scouts going by gut instinct and considering the potential of these young men with some consideration of traditional stats, like their hitting average, home runs, or RBIs (runs batted in). They would work with other team officials to prioritize a list for the draft to bring these players on with the intent of developing them into amazing players. The scouts were critical to this process.

In the mid-1990s, the owner of the Oakland A's died. He had been bankrolling expensive players with a philanthropic mindset, but the new owners were more practical and didn't want to spend so much on the team. As a consequence, the most expensive—that is, the best—players left for greener pastures. The team was no longer winning games, and nobody liked that. After a while, somebody in the organization had a different idea—what if they looked at more detailed stats of players where they currently played rather than trying to guess their potential? Some very committed baseball fans had been collecting and analyzing nontraditional stats for two decades, but the A's were the first official baseball organization to take this approach seriously.

This new approach, which came to be called Moneyball after the book about the approach by Michael Lewis, was embraced by several decision-makers in the organization, including the manager, Billy Beane. They started digging into more obscure statistics like on-base percentage and slugging percentage and looking at which stats led to real payoffs in the sport. This enabled them to identify undervalued and underutilized quality players already in the league as well as stronger draft picks, whom they were able to bring onto the team with very little investment. Beane then used further data analysis to inform other decisions, such as the best order for the batter lineup, and subsequently created a strong, winning team again for pennies on the dollar in the philanthropic era. As an example, the budget team won 20 consecutive games in 2002, breaking an American League record.

The rest of the baseball world took note, and soon other teams that still had a large budget were able to build up their rosters. Moneyball is widely considered to have completely changed the sport, and it's still an arms race situation, with every team trying to figure out a way to use data in new ways to get another small advantage. Now, it's considered a risk to not study player stats in this way.

Example 2: Stopping Cholera in London in 1854

We return to London for one of the most famous examples of data analysis solving a real problem, the work of John Snow during a cholera epidemic that gripped London in 1854. At that time, people did not understand how illness was spread, as the concept of germs did not exist yet. Instead, it was believed that "bad air" was the general cause of diseases spreading from person to person. Snow was ahead of his time in many ways, even revolutionizing medicine by popularizing the use of chloroform in surgery as the first anesthetic. But his major contribution to data analysis occurred when he made a map of London during the 1854 epidemic and added a bar representing each death from cholera at each home location. It was clear from this view where the cases were concentrated, and based on this, he identified a particular water pump on Broad Street as the source, because most of the victims were getting their water from this particular pump. You can see a zoomed-in view of the map Snow used, focusing on the area around the Broad Street pump in Figure 5-2. Rather amazingly, he convinced authorities to block access to that water pump immediately, and the epidemic soon petered out.

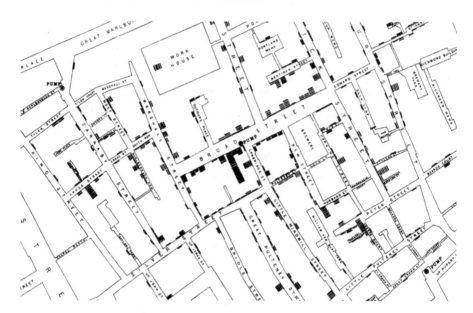

Figure 5-2. *The map John Snow created showing deaths from the 1854 cholera epidemic. Source: A cropped area from "File:Snow-cholera-map-1.jpg,"* `https://commons.wikimedia.org/wiki/` `File:Snow-cholera-map-1.jpg`

On the map, all of the darker black marks are small bars indicating victims. Some are spread out, but it's pretty obvious from looking at this map that something about the location close to Broad Street was exposing people to the disease. There's a huge stack of deaths on Broad Street right next to a dot labeled PUMP. People living near there were getting their water from that pump. It was something we call domain knowledge, basically knowledge about the world the data comes from, that allowed Snow to conclude it was the water pump, instead of a cloud of "bad air" hovering over the area for some reason. Snow was a doctor and scientist, and even though germ theory wouldn't be established for several more years, there were doubts among experts about the bad air theory. Snow thought about the day-to-day lives of people living in this area and realized that they would all be getting water, so that was a potential source. When they shut down the Broad Street water pump, nobody was certain that the epidemic would be stopped. But it was, and that realization that water could harbor disease added to scientists' understanding of disease and the ways it could be spread.

Fundamental Skills for Data Analysts

I've talked a bit about what data analysis involves. Data analysts must have certain skills and attributes to be successful. Although some people have a natural aptitude for data analysis, it is a field most people can learn to be good at if they have an open mind, develop a data mindset, and work on the skills I'll talk about in this section. A range of skills—functional, technical, and soft—are necessary for a good data analyst. Additionally, domain knowledge is critical to doing good data analysis. We will look at the main skills below.

Functional Skills

Functional skills are pretty high level and are the ones that help you decide how to go about solving a problem. They involve both natural attributes and high-level ideas and skills learned in courses like science and math. The most obvious functional skill is logical and systematic thinking. You need to be able to work your way through logical steps to come to important conclusions and be able to back them up. Part of this is being aware of your biases so they don't impact your work. Although this is partially a soft skill (discussed below), you need to be able to listen to other people and understand their perspective even if it's different from your own. Related to this are organization and the ability to follow a process. All data analysts need to have a good foundation in math, and some roles require statistics knowledge as well. A good data analyst will also have natural creativity. You hear a lot about the value of out-of-the-box thinking or intellectual curiosity, and it is absolutely true that the most exciting insights often come from someone looking at things in a new way. One of the attributes people who work in data analysis and data science will often talk about is a *data mindset*—this is basically having an open mind, not jumping to conclusions, understanding that data is a representation of something, and respecting the data in whatever form it takes.

Technical Skills

While the primary technical skills depend on the specific role, all data analysts will need to have a general comfort level with computers that will enable them to learn any particular software their role requires. At a minimum, a comfort level with Microsoft Excel and Word (or the Google equivalents) will be needed, and most data analysts are expected to be quite experienced in Excel, including with many of the more advanced functionalities like pivot tables and v-lookups. Many analysts will need to have an understanding of database systems that they will be working with, and they are often expected to use Structured Query Language (SQL) to interact with those databases. Although it is not as common for data analysts as data scientists, many analysts will also do computer programming, usually in Python or R (this trend is on the rise, too). Sometimes data analysts will even code in VBA (Visual Basic for Applications, a programming language embedded in Microsoft Office products) in their Excel work.

Soft Skills

Unless you are working on an entirely solo project you intend to never share with other people, soft skills are also important. The term "soft skills" generally refers to the set of different abilities needed for interacting with other people in order to get work done. Most of these skills are required for virtually any job, but there are some that are specific to the technical work that data analysts and data scientists do.

Some of these have to do more with a mindset or attitude than specific skills. Being generally adaptable is crucial, as things often change during a data analysis project. Having a growth mindset, or an understanding that you always need to be developing your skills in data analysis, is important. The field is constantly changing, and you have to keep up to date with your knowledge and skills. Additionally, keeping in mind the impact that your work can have will help you make sure you are acting ethically.

Time management covers an important set of skills. You will usually be working on more than one thing at a time. You must be able to understand what tasks are most important and have a decent idea how long each thing will take, and when something is really due, so you can understand what you should be doing at any given time. The ability to prioritize is necessary in most jobs. One aspect of doing data analysis work is that sometimes you can't work on something because you are either waiting on someone else or you are stuck at a step that has to complete before you can do the next steps (for instance, you may be running code that takes an hour to finish, and you can't proceed on that project without the results). So you need to be able to work on multiple projects at once. You need to be able to understand how each of your tasks works so you can effectively prioritize and allocate the right amount of time to each in order to meet deadlines.

Another set of soft skills involves communication with a variety of different types of people. Usually, everyone who's expected to be users or beneficiaries of a project are called *stakeholders*. This includes technical people, like your deeply technical data analyst peers, and nontechnical people whose brains will shut down as soon as you mention the term "p-value"—and everything in between. Frequently, the work you are doing is ultimately for people who are not technical, so you need to be able to explain your work and conclusions in ways they can understand. You will often find that people don't trust you or your findings unless they feel like they can understand them. They generally do realize they aren't going to understand all the "fancy math," as we often say, but they still want to get the basic concepts behind the information you present. This is one of the reasons data analysts and scientists often prefer the simpler—and more explainable—solution over the more complicated/fancier one. This strategy will be discussed later in the book.

One of the common ways you interact with others is through presentations, so being able to create an easily followed presentation is valuable. Communication skills are not one-way, either—it is important to be able to listen and receive feedback, which can come both from your customers and your leadership. Unfortunately, conflict with customers and leadership sometimes arises, and you need to be able to stay calm and negotiate or simply listen to understand what is necessary, even in cases where you may disagree. There are times when standing up for what you believe is right is important, but other times it is more prudent to stay quiet in the moment (and sometimes indefinitely). In some environments, rocking the boat can get you in trouble and even retaliated against, so you should always consider your circumstances when going against the grain.

Companies sometimes have particular cultures that define communication styles—for instance, some may want to avoid conflict and prioritize people's feelings by avoiding direct speech, whereas others prefer to keep everything clear and in the open, so direct communication is favored. Knowing how to avoid hurting people's feelings while still getting your message across is valuable in the first case, and being able to be direct and clear without being mean is valuable in the latter case. It's actually been found that organizations that favor direct communication and don't shy from conflict, but still respect people's feelings while finding constructive ways to address the conflict, are more successful, but the indirect style is found more often than not.

Domain Knowledge

Another important area in data analysis is called domain knowledge, which basically just means expertise in the type of data you are working with. We'll talk about it in more depth in Chapter 10, but for now know that domain knowledge is considered one of the three pillars of data science and is also extremely important in data analysis. It's basically specialized knowledge and practical understanding of a particular "domain," an area such as financial data, gameplay data, website usage data, medical data, retail data, and so on. Usually, it implies deep knowledge of even more specific types of data within those areas. If you do not have a decent understanding of the data you are working on, you cannot do good data analysis—or good data science.

"FREE" DOMAIN KNOWLEDGE

Most of us wouldn't know where to start if we were handed a bunch of data about windmills in Europe. But a lot of the time we know about something just from having been involved in that world. For instance, college students will all be comfortable with data about students, classes, and instructors. We would know that there can be a class, like ENG 301 American Literature 1, that really is more of an idea, existing as a course description in the college catalog, for instance. A more tangible class would be one that made it on the schedule of classes for a given term. Section 002 of American Literature 1 is scheduled in the fall semester of the 2022–2023 school year, at 3:30 p.m. Tuesday and Thursday, with Alex Thorne as the instructor and eventually with a list of specific students enrolled. Generally, both students and instructors will be associated with multiple classes, and these have to fit in a schedule to avoid overlapping times and so on. These things seem obvious to most of us, but that is the nature of domain knowledge. When you have it, it seems so natural and obvious that you often don't even realize that the things you know are actual knowledge. This can get you far on the road to good data analysis in the right kind of data.

Domain knowledge has to come from somewhere. In the class example, it clearly comes from living the experience. But a lot of the time, you aren't going to already have it and will need to develop it. This is especially true in the business world. Sometimes you can find material online or in books to learn what you need to, but often the only way to learn what you need is to find someone who already has the knowledge and can help you learn it. Such people are often called *subject matter experts*, or SMEs. It is true that you can sometimes figure things out purely by looking into the data, but you have to be very careful with this and must always validate your assumptions with an expert.

One of the challenges of working with a subject matter expert is that they often don't know what they know. If they were to describe a process or a system, they would likely leave steps or assumptions out because they are second nature or instinctive for them, so they don't think they're worth mentioning. For instance, in the class example above, imagine if they forgot to mention that instructors teach multiple classes. If you were completely unfamiliar with how class scheduling works, you might make the mistake of assuming each instructor only has one class or that an instructor teaches all sections of a given class. It seems impossible that anyone could think that, but someone who has no experience with secondary or higher education might not know. These kinds of incorrect assumptions can be devastating in data analysis, leading you down completely wrong paths.

The nice thing is that domain knowledge is usually transferable to some degree. If you work in the retail space for one company, you might go to another retailer and find that you aren't starting from zero again.

CRISP-DM: The Data Analysis Process

Although data analysis projects are all a little different, there is a general process to manage such projects. We often call this the data analysis lifecycle. Different data analysts may use slightly different processes, but many teams use a process called *CRISP-DM* (CRoss Industry Standard Process for Data Mining) for both data analysis and data science with only slight differences between them. It has the following six steps:

1. Business understanding

2. Data understanding

3. Data preparation

4. Exploratory analysis and modeling

5. Validation and evaluation

6. Visualization and presentation

Although these steps appear linear, there is a lot of iteration within the process. We don't just finish one step and move on to the next and never look back. Figure 5-3 shows how iterative the process can be.

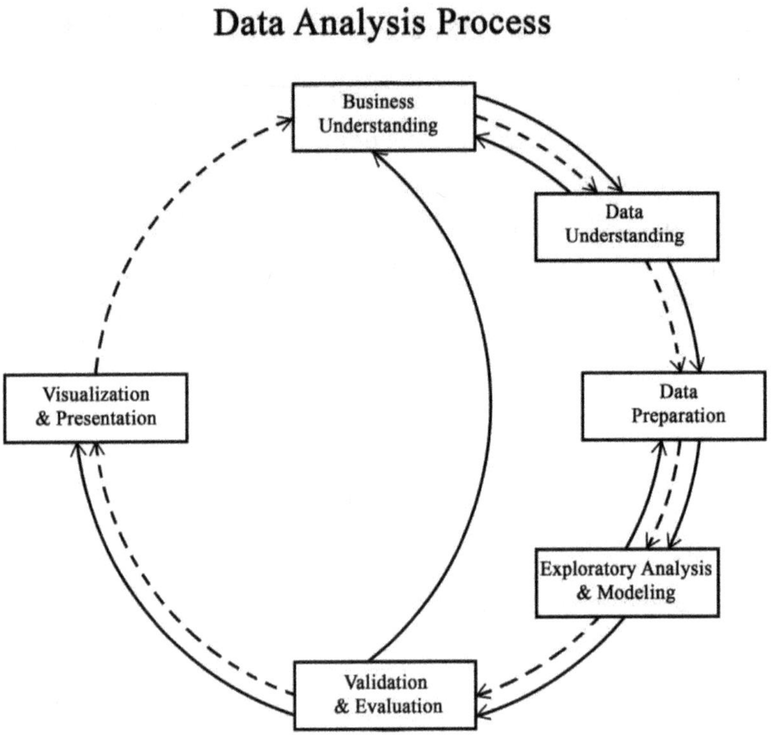

Data Analysis Process

Figure 5-3. *CRISP-DM data analysis process*

We will go over the basics of each step here, but see Chapter 21 for more details on the process.

Business Understanding

We always have to start with business understanding, which means understanding what your customer is trying to learn and defining your research questions. In other words, what business problem are they trying to solve and how will you investigate that? For instance, a football team might want to know which of their defensive players to trade. This would imply you'd want to look at some performance metrics on all the players. You'll have to work with the business to identify and define those, and you'll also have to determine if you have data for those metrics. This step often occurs in conjunction with the next, data understanding, and you may find yourself going back and forth between them a lot in the beginning. Talking to customers to try to generate research questions is called requirements gathering, something done in a lot of different disciplines. But this will involve more than quickly asking them what they want. You will have to begin developing domain knowledge during this process, if you don't already have it.

Data Understanding

After you have a good understanding of what the customer is looking for, the next step is to understand the data that is available. This also implicitly includes finding the data in the first place. Most data analysis work is done by teams that have access to established data sources, so you will likely already have access to these common sources. Before you identify the specific data sources to use, you need to know—in a general sense—what data you need to answer the research questions you came up with in the previous step. Imagine in our example that you and the business have identified four things to track among the defensive players: tackles, sacks, interceptions, and fumbles.

Once you know what data you need, you look for data sources that contain that kind of information and then start investigating the specific data source(s). If you're lucky, you'll have a data engineering team that can give you exactly what you need, or there will be documentation like a data dictionary for your source(s), but often you won't be so fortunate. It's also not always clear what exact fields are important, so this can be an iterative process. It should also be mentioned that once you have obtained the data you want, you may find that it doesn't quite allow you to answer the research questions you came up with. You might tweak the questions, but you will likely need to check with the business to make sure your new versions still will give them what they want.

In the football example, imagine that we found sources that have each players' tackles, interceptions, and fumbles, but nothing that specifically contains sacks (a specific type of tackle, on a quarterback). However, you've found another table that has timestamped tackles with a record of what position was tackled but not the player who carried out the tackle. You look again at your original tackle data and see it has some times recorded. You conclude that it might be possible to line these two sources up, but you're not sure it can be done or how much time it will take. This is when you would discuss it with your business stakeholders—how important is the sacks figure, and is it worth the time investment to extract? You wouldn't necessarily be able to answer this before going to the next two stages, but it's appropriate to talk to the stakeholders first. They might just tell you sacks aren't that important for this question.

Note that like with the example, you'll be working with the data some at this stage, but you won't really dig into it until you're working on the data preparation.

Data Preparation

Once you are comfortable with your understanding of the business needs and believe you have the right data, the next step is data preparation, which can take up a huge proportion of the process. There will be a later chapter dedicated to this topic, but data prep involves many things. It's everything from cleaning up messy text fields that have trailing white space, making sure your data in numeric fields is all in the expected range, identifying and dealing with missing or null values, and much more. We'll go more in depth in Chapters 13 and 14.

If you're at a company with data engineers who have prepared nice, clean data for you in advance, lucky you. In that case you may be able to skip some of this stage, but more than likely you'll still have some work to do here.

If the project will involve any modeling, you'll also need to do feature engineering, where new features are created based on existing data source(s) in order to improve modeling or analysis. Knowing what to create usually relies on what's discovered in the next step, so there is some back-and-forth with these two, as well. Feature engineering is almost always necessary in data science, even when we have clean data from a data engineering team, but it's not needed as often in data analysis. Determining what features are needed is its own process and is based on both how the data looks and what techniques they're going to be used in.

It's worth mentioning that there can be a lot of back-and-forth between this step and the next, exploratory analysis and modeling, because you may find things in the next step that require you to do additional data prep or change some of what's been done.

In our football example, some of the data prep might be creating a new table with all of the basic stats with player names and replacing null values with 0 based on your stakeholders confirming that is the right business logic. You would also make sure to save all the numeric fields as numeric data types. You may still be wondering about the possibility of deriving sacks from the two data sources. You likely wouldn't have enough knowledge to do that yet and will rely on what you find in the next step to determine if it's possible and necessary.

Exploratory Analysis and Modeling

After you've got the data ready, it's time to start doing what you've been wanting to do the whole time: dig into the data. Exploratory analysis and modeling is often another big step. How big depends on how deep you go. But as mentioned, you will probably be going back to the data preparation step a few times after finding problems or little gotchas with the data that you missed when you did the cursory investigation in that step. In other cases, you may even have to go all the way back to data understanding. You may discover that you need to find entirely new data. This is totally normal and does not mean you've done something wrong.

Exploratory data analysis (EDA) is always the first major step in "looking at the data." This isn't a haphazard process of opening a data file in Excel and casually looking at it. Exploratory data analysis provides a framework for investigating data with an open mind. It isn't a rigid set of steps that must all be followed, but it helps guide you through your early analysis. The statistician John Tukey named this process in the 1970s, and data analysts have been following his guidelines since. EDA can be thought of as a mentality or attitude, too. We want to approach looking at the data with a curious mind cleared of assumptions and expectations, because we never know what we'll find.

So we know that EDA is important and relatively simplistic, but what exactly does it involve? Basically, EDA is descriptive statistics, as we covered in Chapter 2. We generate summary statistics and basic charts. Summary statistics tells us about "location" and "spread" and involves metrics like mean, median, mode, and standard deviation/variance. Typical charts include scatterplots, bar charts, histograms, box plots, and pie charts.

The initial EDA that we do often focuses on individual fields. The summary statistics will be done on individual fields, but it's not uncommon to make plots with multiple fields, especially with line charts and scatterplots. A variety of fairly simple charts and graphs are useful when doing EDA. These are often simply ways of seeing the measures of location and spread visually, which can be easier to interpret. Often things that are difficult to see in purely numeric data can be glaringly obvious when visualized. We might make a line chart of values in one field broken down by values in another field (with different lines for the breakdown field) or look at the distribution of values of a field in a histogram. Outliers may jump out at you.

One thing worth mentioning is that when you are creating charts as part of your EDA, you might not be super picky about including axis and chart labels. This is okay if they are for only you to see, but any time you are going to share a chart with someone else, even another colleague, it's best to label the axes at a minimum, and a chart title is always a good idea, too. If more than one type of data is included, a legend should be added. It might seem obvious to you what the axes and chart components represent when you are creating them, but it won't necessarily be obvious to other people—or even to you a few months down the road.

EDA is critical, and depending on the kind of project we're working on, there may be deeper analysis necessary like with a data analysis project, or we might proceed toward modeling directly like with a data science project. Or, even more likely, we might need to go back to the data preparation step before embarking on some more EDA. Learning about your data is usually an iterative process.

With the football data, you'd look at the distribution of the metrics we do have (tackles, interceptions, and fumbles), looking for outliers or other anomalies we didn't see during data prep. But we still have the question about deriving the sacks from the two sources. You would look into this closely, seeing if you can figure out how to line up the different times in the two files. You might have to do quite a bit of work to figure this out, which is why you want to make sure it's worth it to your stakeholders. But if you figure out how to do it, you'd move back to the data preparation step and add this derived feature to your table.

You may have bigger questions to answer after you've done your EDA, and by this point you should have a good sense for what variables are available to you in the data. You will mostly understand which features are reliable and accurate and what their limitations are. So you will have a sense for how far you can take your data in answering your questions, especially as they get more complex. Looking back over your questions and trying to create more is a good first step after EDA is done.

The exploratory analysis steps you will follow after doing EDA will depend entirely on the goals of your project. You will likely have research questions that haven't been answered yet, or you may have new requests from customers, both of which will guide you. The next steps would usually involve more complex ways of slicing and dicing the data to focus more on how different fields interact with each other and drilling down to understand the data that pertains to the specific research questions. Some of the work at this stage may be looking toward the visualization and presentation step, but not usually creating final deliverables yet.

While all of the steps discussed above are common in most data analysis projects, there are important advanced techniques that some data analysts will use. Some of these will overlap significantly with what people think of as data science's main techniques, which is why they're considered advanced here. We are not going to cover them since they will all be addressed in later chapters.

The most common advanced techniques that data analysts use are testing for correlation, basic hypothesis testing, significance testing, and other tests like t-tests and chi-squared tests, all of which are a part of statistics. Being able to understand these requires a basic comprehension of probability, as well as distributions, starting with the one most people have heard of—the normal distribution, otherwise known as the bell curve. You've seen all of this in Chapters 3 and 4.

Another common technique is linear regression, also based in statistics. This is visually a lot like drawing a line of best fit through dots in a scatterplot, but it is of course more complicated to create. It usually has more than two variables, which means it's difficult to visualize, but it's still easier to understand than a lot of other techniques. It is considered quite valuable despite being a relatively simple approach, because it is easy to explain and rather intuitive to most people.

You might be able to do a linear regression (or its sibling, logistic regression) if your stakeholders have given you a bit more info. Imagine they've rated some of the players between 1 and 10. You could create a linear regression model trained on the metrics you've been working with to calculate a rating for the remaining players. They might be able to use that to select a threshold, where anyone with a rating lower would be traded.

Any data analysis project (and most data science projects) will have exploratory analysis steps and some deeper analysis, but only a few have modeling. Modeling is most common in data science projects, but some more advanced data analysis projects may involve modeling, especially something like linear regression. Sometimes it may even involve modeling with machine learning. This overlap between data analysis and data science is just the nature of the field—the expectations and work vary tremendously job to job and project to project.

Validation and Evaluation

Once you've finished the analysis, you need to validate your work and evaluate your models if you have created some. This is where you "check your work," something you should never skimp on. You have to make sure that you haven't made any computational errors or used the data in any way that will lead to a misunderstanding of the real world. This is not trivial to do and often can take a lot of time.

While validation is about ensuring that the work that was attempted was done accurately, evaluation has us considering if the work we did is the right work and whether it answers the questions the stakeholders had. There are different techniques for evaluation, and peer reviewers will also be evaluating your approaches during review. In *peer reviews*, coworkers look at each other's work to identify problems before they become permanent. Peer reviewers will look at the steps you took to produce the analysis and any code you wrote to help you identify any shady assumptions, bad logic, or coding errors.

Although sometimes validation and evaluation aren't done until after the work is "done" (it's not truly done until we know it's right), it's also common to check things along the way, and many teams will review each other's work while it's still in progress. It's often a good idea to seek input from colleagues to make sure you're on the right track. Even the best data analysts and data scientists make mistakes along the way—often small, but not always. Peer reviewing can be a great way to avoid going down a rabbit hole. Data science in particular is a very collaborative field.

You'd need to validate all the transformations you did on the football data in the data preparation step and any of the charts you created or tests you ran in the previous step.

Visualization and Presentation

Finally, we are at the last step. This is all the way over on its own on the left side of the diagram because it's quite different from most of the other steps. This is where you compile all your results and present them to your customers. Visualization is usually a huge part of a data analyst's job. It's standard to prepare a lot of charts during the exploratory analysis and modeling step, but this step is about presenting to your stakeholders. Visualizations are important, but often there won't be too many because you don't want to overwhelm the stakeholders with too much information.

Visualization is an important part of presentation, but usually there's more to it. How this is done can vary widely at different companies and also with whom we're presenting to, as well as the formality of the project. If this is a one-time project, you might be able to get away with showing some of the visualizations you created in the exploratory step. Some companies want everything in slide decks, so you might create a presentation that highlights your findings, including charts you created. Others might prefer Word documents. A lot of people are hung up on Excel and like to see the raw data, so you might create an Excel sheet with all your data and charts in it. (Excel is the bane of many data professionals' existence.) More technical companies may use Python notebooks or create a web app. There is a chapter on visualization and presentation later in the book that will discuss these topics more.

If this work was foundational for providing ongoing information to stakeholders, it might involve creating a dashboard that's automatically refreshed with updated data on some regular cadence. Some data analysts do create dashboards in tools like Tableau or Power BI and set up automatic data refreshes, but at a lot of companies there would be another team that would do that work, based on requirements you give them.

With your football results, a logical thing to show would be a table containing player ratings and their other metrics, ranked high to low. Other interesting trends or views could easily have been found during exploration.

Key Takeaways and Next Up

This chapter introduced data analysis, both a field in its own right and a subset of the work most data scientists do. We covered the major types of skills data analysis requires, including functional and technical skills like ways of thinking and specific tools or techniques, soft skills like communication and other people skills, and domain knowledge, which is having a breadth of knowledge about the world that given data represents. Good data analysis is systematic and generally follows a process model called CRISP-DM, which outlines six major steps: business understanding, data understanding, data preparation, exploratory analysis and modeling, validation and evaluation, and visualization and presentation. The major tasks in each of these steps were covered, although we will dig into each more deeply in later chapters.

Coming next is the chapter on data science, where we'll talk about how it developed as a relatively new field, address the values that define data science, talk about how organizations see data science, and finally talk about data scientist qualities and roles.

PRACTITIONER PROFILE: SANDIP THANKI

Name: Sandip Thanki

Job Title: Data scientist

Industry: Academia

Years of Experience: 13

Education:

- PhD Physics and Astronomy

- MS Physics and Astronomy

- BS Physics and Astronomy

The opinions expressed here are Sandip's and not any of his employers', past or present.

Background

Sandip Thanki is a physicist focusing on astronomy by training, earning a series of degrees culminating in a PhD in Physics and Astronomy. He studied stars and began his career as a professor. After becoming a department chair, one of his colleagues was a dean who worked with data a lot. He often pulled data for her, and he ended working with it on his own as he became curious. He moved into another role that required even more data work, and he loved it.

Work

Although Sandip started out working with student data as an academic administrator, he considers himself a true data scientist. He jokes that it's not that different from when he was working in astronomy—the job is the same as back then, saying that it's just the table attributes that have changed—now rows are students and columns are GPA, course credits, and financial aid status, while before rows were stars with columns of brightness and size. But this job is also different because he can actually do work that can change students' lives for the better, which he finds incredibly rewarding.

His job involves a lot of different activities, and he does a lot of reporting and answering ad hoc queries from a huge variety of stakeholders. Although it might not be obvious, his work often has high stakes because hundreds of thousands—even millions—of dollars can be on the line through grant proposals and other reports that have legal ramifications. He's always careful and methodical, but for those projects he triple-checks his work.

Sound Bites

Favorite Parts of the Job: Sandip loves the fact that his work can make a positive difference in people's lives. He also loves that every day is different and he never knows what interesting questions people will ask him.

Least Favorite Parts of the Job: He finds some of the queries he gets boring and basically meaningless when they're just being used to check mundane boxes. These are things like generating percentages of different ethnicities to get dumped into some report, rather than to drive efforts to improve opportunities for disadvantaged students.

Favorite Project: Sandip has lots of relatively small efforts that he's proud of. One involved digging into data to identify students who had stopped attending college even though they had been successful while attending and were close to completion. The simple SQL query he ran will lead to people's lives being changed after outreach efforts can help pull some of those students back into school and help them graduate.

How Education Ties to the Real World: The astronomy data he worked with wasn't totally clean, but next to data on people it was pristine. With students, there are so many ways to break things down and much more room for error in the data.

Skills Used Most: People skills are hugely important. The data science department is often intimidating to people, so being helpful—especially prompt and transparent—really helps build trust. Often stakeholders don't know what they need. He has learned to suss out the real requirements when people ask him for information—he knows how to ask the right questions so they can, in turn, ask the right questions of him. The last critical skill is methodical thinking and behaviors like taking good notes, documenting your work, and organizing everything. The ability to refer back and even reuse prior work is a huge time-saver.

Primary Tools Used Currently: SQL daily, Tableau weekly, and R and Python occasionally

Future of Data Analysis and Data Science: Sandip thinks that data is currently underutilized and more will be used to benefit people in the future once we've figured out better ways of anonymizing it. For instance, maybe it would be possible to warn drivers that an erratic driver is approaching them.

What Makes a Good Data Analyst: Patience, keeping the goal in mind, and being highly organized.

His Tip for Prospective Data Analysts and Data Scientists: His top two are (1) keep your stakeholders in mind at all times and (2) value simplicity over complexity. Only put a few charts on a dashboard and share spreadsheets with no more than a couple sheets. People are often turned off by complexity.

Sandip is a data scientist working in academia to help faculty and staff know how to best help students succeed.

CHAPTER 6

Bringing It into the Twenty-First Century: Data Science

Introduction

Data science has famously been called the "sexiest job" in recent years. This is because it's shiny and new and because it has the ability to offer deep insights into data that companies hold that could help them figure out how to be more successful. There's no doubt that its potential is powerful and real, but the hype is a bit overstated.

Data science does have a lot to offer, but it's not the easy, magic bullet so many people think it is. The amorphous terms "data science," "machine learning," and "AI" get freely thrown around and used interchangeably in the media and business world, with very little understanding. We'll address these terms below, but when I think of "data science," it's basically anything that a data scientist does in the search for insights and forecasting. This may include activities that can fall under other

© Kelly P. Vincent 2025
K. P. Vincent, *A Friendly Guide to Data Science*, Friendly Guides to Technology,
https://doi.org/10.1007/979-8-8688-1169-2_6

labels, including data analysis, as discussed in the previous chapter. What separates it from data analysis is that sometimes the techniques do pull from machine learning, so are different from data analysis.

It's clear that data science is a combination of different techniques to solve complicated problems and yield illuminating insights. But it isn't a crystal ball—data scientists are always trying to teach customers and other people that the maxim "garbage in/garbage out" is real. We'll talk about all the aspects of data science and how they relate to data analysis and other named disciplines in this chapter.

This chapter will cover the emergence of data science as a discipline and then go into a couple examples of data science in the real world. We'll define the field and its values and then talk about the three main ways organizations use data science. Finally, we'll talk about the qualities good data scientists have and the types of roles out there.

The Emergence of Data Science as a Discipline

Data science emerged out of a long history of different disciplines, but its real foundation is data analysis, with bits and pieces that fall under the umbrella term AI added in. Some definitions and histories of related fields will help us understand what data science really is and where it came from.

Artificial Intelligence and Machine Learning

Artificial intelligence (AI) is a large umbrella term, covering many disciplines and concepts. Historically, AI has been centered on the idea of artificial creatures—and later, machines—with human intelligence. Ideas around this have been around for millennia, with more recent examples including novels like *Frankenstein* and *R.U.R. (Rossum's Universal Robots)*.

The foundations of modern AI began in the 1940s with philosophers and some advanced mathematicians and proto-computer scientists like Alan Turing, but it didn't really get going until computers became more affordable. Enthusiasm and optimism in AI flourished in the 1960s and into the 1970s, when everyone, including practitioners, was convinced that we'd have machines equaling or exceeding human intelligence by the end of the century. Most of the early work focused on creating formal rules representing how the human brain worked, so a lot of it was rule-based systems rather than modern machine learning. These rules were often called *heuristics* because they are usually identified through experimentation and trial and error. Researchers also worked on early neural nets, language processing, automata, and more.

The US and UK governments had been pumping money into AI research, but that dried up in the mid-1970s after skepticism descended when nothing had really come out of the research. That period is called the first AI winter, but there was a small resurgence in the 1980s with the advent of expert systems and further development of neural nets. However, another AI winter started in the final years of the decade when outcomes still weren't impressive enough.

The 1990s was when what we think of as modern AI really appeared in industry. People found ways to use some of the ideas, but they'd tempered their expectations and weren't really trying to bring a mechanical Frankenstein's monster to life anymore. In 1997, IBM's Deep Blue chess-playing system beat the world's chess champion, which triggered more excitement in the field from both practitioners and the public.

When we look back from now, successes kept mounting from the 1990s into the 2000s, but it was an interesting time because a lot of it wasn't called "AI" back then. Instead, the wins were in specific named fields that are *now* considered part of AI, like decision support systems, but back then they avoided calling it AI because it had such a bad reputation of being overhyped. Probability and statistics began playing a bigger role, with what's now called machine learning starting to take off. *Machine learning* basically refers to any computer system that processes information without following a set of human-created rules. It adapts depending on inputs. We'll talk more about it in a later chapter, but it and other approaches were used in a variety of fields, including robotics, speech recognition, natural language processing, Internet search, and more areas. These were often considered under the umbrella of other disciplines like computer science, electrical engineering, and informatics.

The emergence of big data and deep learning led to the rehabilitation of the term AI, and now we consider many disciplines to fall under the umbrella term. Big data made more advanced techniques necessary because of the amount of processing required, and deep learning is just a term for more advanced (and bigger) neural nets. And of course, we also have data science nowadays, which has possibly lost its position as the sexiest field to large language models and generative AI, the vanguard of modern AI. It's worth noting that in a lot of ways, AI still is a term for big-dreamers, with most of the public thinking it's basically magic and can do almost anything. In reality, it covers a range of disciplines and techniques for processing data that can be used to solve a wide variety of problems and depends mightily on the quality of the data and the skill of the practitioners.

Data Science: Data Analysis's More Sophisticated Younger Sibling

In a lot of ways, data science is a cobbled-together field, and there isn't a true universal definition of it. As I've said, in my view, *data science* is basically whatever work people who call themselves data scientists do in the quest for informative and valuable solutions to aid decision-makers. Data scientists usually work with the kind of data people associate with businesses. That includes things like sales records from a retailer, information about members of organizations or rewards programs, performance numbers for computer systems, information about patients in the medical world, banking records, and much more. Data scientists would look into this data to draw conclusions about expectations of future sales, information about the behavior of members, identification of computer system problems, information about disease patterns in medical patients, potentially fraudulent bank transactions, and so on.

A lot of the disciplines that fall under AI would not typically be considered a part of data science, like robotics, speech recognition, and generative AI, but that doesn't mean a data scientist would never touch those fields. Another area that data scientists draw less on is natural language processing, but that is changing with the increase of text data that's of interest to business, especially from social media. Additionally, it's becoming increasingly common for data scientists—and everyone else—to *use* AI tools like ChatGPT in their work. But that doesn't constitute *doing* generative AI.

The most common aspects that data scientists use to do their data science work that are also generally considered to fall under the AI label include machine learning and some statistical techniques. In the past, the term *data mining* was also often used as a type of AI that involved working with data in databases, but you don't hear it as much nowadays (it was generally not as sophisticated as current techniques).

AI isn't the only area that data science draws upon—as I've previously said, I regard data science as data analysis made significantly more sophisticated. The goals of data analysis and data science are usually very similar—to better understand aspects of a business or area based on looking at data from that business or area in order to help decision-makers make better decisions and identify risks. Like data analysis, it also is under the general umbrella of data analytics, as mentioned in Chapter 5 and also to be discussed more in the next chapter, but data science generally falls primarily under the predictive and prescriptive types of analytics, where data analysis focuses more on the first two types, descriptive and diagnostic.

More on Terminology

In the business world, everybody's excited about "AI," "machine learning," and "data science," terms that get thrown around a lot. It's interesting to look at how all of these terms are actually being used in the real world. Google has a tool called the Ngram Viewer that will display frequency percentages of any search terms found in their Google Books corpus, which has books published through 2022. Let's look at how popular the terms are, even though missing the last couple years is not ideal. We can use Google's Ngram Viewer to see the popularity in published books up through 2022, starting in 1960, shown in Figure 6-1.

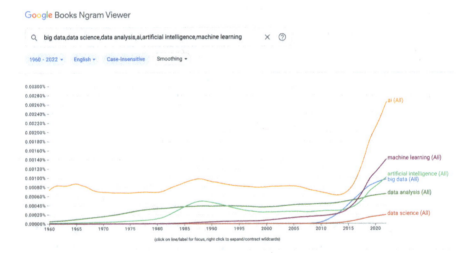

Figure 6-1. *Google Ngram Viewer for the words "ai," "big data," "machine learning," "artificial intelligence," "data analysis," and "data science" from 1960 to 2022. Source: Accessed January 17, 2025, at https://books.google.com/ngrams/graph?content=big+data%2 Cdata+science%2Cdata+analysis%2Cai%2Cartificial+ intelligence%2Cmachine+learning&year_start=1960&year_end= 2022&case_insensitive=true&corpus=en&smoothing=3*

Notice that the term "AI" is used way more than any of the other terms, even "artificial intelligence" spelled out. This is because it's about 90% buzzword—people use it without knowing what it truly means, ordering data workers to figure out how to use it, rather than turning to it as a solution to a real problem. Most companies need data science, not other types of AI. The term "big data" started taking off a little before "machine learning," but it's slowed down while "machine learning" has continued growing. This is because they're also very buzzy, as many of the slightly more informed leaders calling for AI know it's machine learning that drives modern AI, and everyone has been excited about big data as the foundation for most AI. The chart makes it clear how new data science is, while data analysis has been around steadily since the 1970s.

Examples of Data Science in the Real World

The term data science may be relatively new, but many smart companies and industries have followed a data-driven approach for decades. In some industries, it's actually a matter of survival, though when used well, any company can improve their lot with data. Here I'll talk about Coca-Cola as a data-driven company and fraud detection in insurance.

Project 1: Coca-Cola Is a Data-Driven Company

Coca-Cola has always been a technologically innovative company, with unique product design and aggressive and creative branding and advertising. But more interesting here is that they've been working with data and statistics for over a hundred years—their statistical department was created in 1923. In the early days, they collected and analyzed traffic patterns to pick the best billboards, identified the most lucrative retailers and instructed their sales reps to visit those twice as often, and used collected data on drugstore shopping habits to help train soda fountain operators to sell even more Coke (and to convince customers to buy other things—sweetening the pot for participating operators). They also took a stab at forecasting sales and profits, all before computers. Even more interestingly, the company's statisticians warned of the impending financial crash in 1928, unfortunately being proven correct not long after by the Wall Street Crash the next year.

Since the company has had sophisticated statisticians for a very long time, it's not surprising that they jumped feet first into big data and modern data science. Nowadays, data work drives almost all strategic decisions. It's a massive company operating in over 200 countries and collects data on everything it can, everywhere.

One of the innovations has yielded a trove of useful information on customer flavor preferences better than any target group study ever could because of the scale. Coca-Cola's Freestyle dispensers were launched in 2009 and allow customers to add a large variety of flavors to their Coke, Sprite, and other Coca-Cola products, like vanilla, lemon, cherry, and more, all from a snazzy machine and touch screen. Over 10 million drinks are dispensed in these machines every day, and it's enabled the company to create new flavors, like Cherry Sprite.

While the Freestyle dispensers are interesting, crunching those numbers doesn't require a lot of sophistication. But the company is doing other things that draw from AI more directly. One fancier initiative is in looking at pricing and promotional elasticity, which uses statistical or machine learning techniques to identify the impact of changes in pricing (the "regular" price) and the addition of promotions (including the type of promotion). The global reach of the company means that they can do more than look at historical data on pricing and promotions, but can actually run experiments to see the impacts changes have.

It also won't surprise anyone that Coca-Cola is using advanced techniques in its advertising, all while capitalizing on new platforms like social media and mobile phones. The ease of A/B testing and digital advertising in general has meant that they've shifted their advertising focus away from TV and into social media and other online platforms. This provides them with endless data, which their data scientists can then analyze in so many ways. Although it can be difficult to find information about the exact approaches they use—articles tend to focus on the buzzwords—it's clear that Coca-Cola has really embraced data and data science to drive their ongoing business success.

Project 2: Fraud Detection

Fraud detection by banks, credit card companies, and insurance companies was important even in the early 1990s, especially as more and more transactions were being done virtually. But it was simplistic, relying simply on transactional history without much other information. As the decade wore on, companies began to get more data and use more sophisticated detection methods.

Most companies never divulge their detection secrets, but a researcher worked with an insurance company in the mid-1990s, and we can see how they used what was a cutting-edge technique at the time.[1] The kind of home insurance fraud that the company dealt with was not generally through professional criminals, but rather regular people filing fraudulent claims for a variety of reasons.

The company wanted to automatically flag potentially fraudulent claims for further investigation. They also wanted an explanation for why it was flagged. They used a fuzzy logic system to accomplish this. This system scored each claim with a degree of likelihood between 0 (non-fraudulent) and 1 (likely to be fraudulent). The company established a threshold, and claims that scored below this threshold were automatically paid, whereas

---

[1] This example comes from *Fuzzy Logic and Neurofuzzy Applications in Business and Finance* by Constantin von Altrock, Pearson, 1996

others would be put forward for claims auditing. The information given to the claims auditor included the reason the claim had a high score, setting them up to investigate manually.

The system was not very complicated, only having seven inputs, including number of previous claims within a year, length of time as a customer, income of the customer, and an average monthly bank balance. Different variables flowed into three subsystems covering a customer's history with their insurance, banking history, and changes in personal circumstances. These three were fed into the two separate components: the scoring one that applied fuzzy logic to generate a number and the reason one that would give the claims auditor a starting point. One key benefit of the approach is that there wasn't one single factor that could trigger a flag—it's only the combinations of multiple that can generate a high score. For instance, someone who's had a lot of claims in one year might not be flagged if the other factors weren't sketchy—maybe they were just having a bad year.

One important aspect of this system was how the company established the threshold that would trigger the audit. This was done by having claims auditors manually evaluate and score over a thousand claims and having the system score those same claims. They could then adjust the threshold appropriately to get the flagged claims to line up as much as possible with the manually flagged ones. Interestingly, the system actually identified a handful of cases that were truly fraudulent that the human auditors did not identify. They then tested different thresholds. Obviously, there's a balance to be found here—if the threshold is too low, there will be too many claims needing manual investigation (which is expensive in terms of people-hours), but if it's too high, they'll miss genuinely fraudulent claims (which is expensive in terms of payouts). This balance is still important in all such systems today, even though today's fraud detection systems are way more complex.

Okay, So What Is Data Science?

Data science is the new alchemy. Wave your hands over the crystal ball and business insights and value will come bubbling out and turn into fat bars of gold. The only difference is that alchemy cannot produce anything real, while data science can. It just takes more than a quick wave of hands.

Companies throw a lot of other buzzwords around like "digital transformation" and "data-driven," but unless there's a real commitment underneath all the fluff to actually make changes to establish a real, usable data infrastructure, nothing is going to change. Companies often hire data scientists who come on board only to find that there's no data available except for a few spreadsheets floating around.

If it's not magic, what is it? *Data science* is fundamentally using computationally intensive techniques to find meaning in data. The point is that it almost always involves techniques that would take too long for a human to do by hand either because of the amount of data or the sheer number and complexity of computations. I've also said that data science often involves doing data analysis, but these are the parts that data analysts usually don't do.

One other thing that makes data science stand out from data analysis is the word "science" in the term. This might be there more because it sounds fancier, and data science is generally considered to have more cachet than data analysis. Science sounds more important than analysis. The truth is that there are some aspects of data science that might be science-y, as that's true of most technical fields. But honestly, a huge chunk of it is art.

And just like with data analysis, most of data science is finding, understanding, and preparing the data. This is where domain knowledge—business knowledge about the data you're working with—comes in as being critical. There is a chapter dedicated to domain knowledge later in the book. It's only after doing a great deal of work with the data (data understanding, data preparation, and exploratory data analysis) that modeling can be done and lead to valuable insights. Leaders are impatient

215

and can have trouble understanding this, so often data scientists have to explain it repeatedly. So another huge part of data science is dealing with leadership and other people, explaining things and planning details.

However, once the data is ready and the exploratory and other data analysis is done, data scientists can finally get to the part that gives their job its reputation: the modeling. Usually this involves machine learning and other types of techniques that fall under the AI label. There are later chapters dedicated to machine learning and natural language processing that will provide solid overviews of these areas, but for now know that this is the phase that often jokingly gets called "fancy math." There are lots of ways this work gets done, and it may include coding, using a GUI that allows drag-and-drop workflow building, or something in between. A later chapter will talk about the most common tools. Once the data scientists have done these steps and believe they've found interesting things worth sharing, or created a model that will generate forecasts, it's time to figure out how to present it to stakeholders. Another chapter will talk about visualization and presentation.

If this process sounds familiar, it's because it's basically the same as the process for doing data analysis that we saw in the last chapter. The modeling is done in the exploratory analysis and modeling step in the CRISP-DM process, which we saw in Chapter 5, but it's shown again here in Figure 6-2.

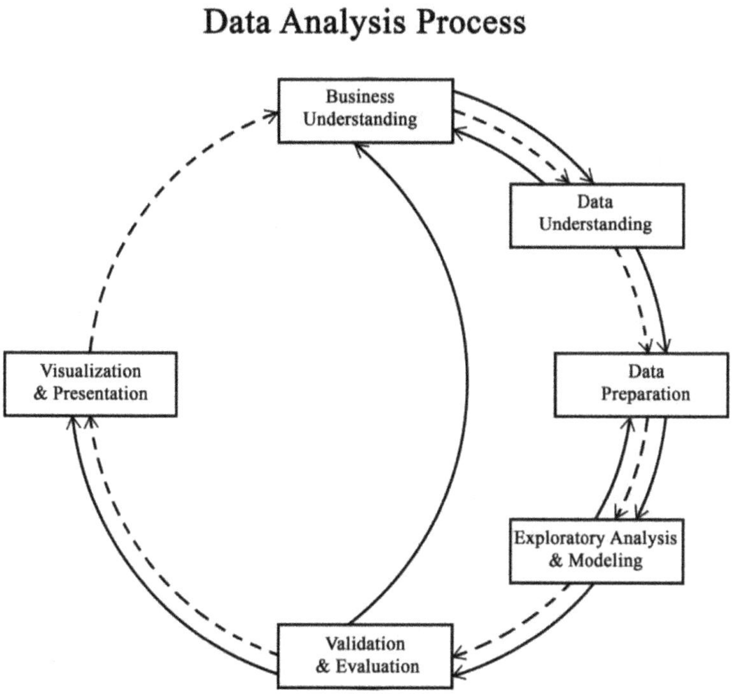

Data Analysis Process

Figure 6-2. *The CRISP-DM lifecycle*

One difference in the process between data analysis and data science is in the exploratory analysis and modeling step, because data scientists usually spend more time here because they are doing more complicated work. Additionally, the work that's done in the validation and evaluation step is also likely to be different from that in a data analysis project, with more emphasis and time spent on code review, but it's not restricted to that. Similarly, the visualization and presentation step may involve different types of outputs, depending on the project. But keeping this process in mind will help you understand and plan any data science project you work on.

For a final look at what makes data science what it is, see Figure 6-3, a Venn diagram based on one by a data scientist named Drew Conway showing how the three major areas of data science intersect to be true data science.

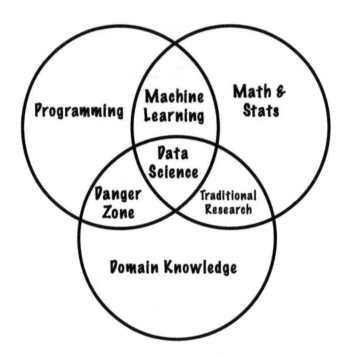

Figure 6-3. *Drew Conway's data science Venn diagram*

This diagram is a little tongue-in-cheek, but it really does capture the components of data science. "Programming" captures any sort of tool that can be used to work with data and apply math- and statistics-based techniques. This does include non-code-based tools like Excel, so doesn't always involve literal programming. "Math & stats" is self-explanatory, except that it does imply knowledge of specific types of math and stats, not necessarily everything. I've talked about the importance of domain knowledge, which basically means an understanding of a business or another area (without necessarily having technical knowledge).

If you look at the intersections of the circles, you can see some interesting things. First, the intersection of math & stats and domain knowledge without coding skills is the classic academic research of psychologists, social scientists, and other scientists, with studies set up and traditional statistical testing. This is a legitimate profession, but the work can tip into data science if these researchers can develop and use real coding skills. Coders who have good math and stats knowledge can do machine learning, but it's not data science unless they understand their data at a deep level. It's common for people who've learned some coding and how to do machine learning to start trying to find interesting problems, but without having a deep knowledge of what the data represents, the work has limited value. Finally, someone who's an expert in some domain and learns to code may think they know how to find things out, but without the rigor of math and stats, their results will be all over the place, which is why we label it the "danger zone."

Although so many people still think data science is magic and easy, people who are more aware know the secret: data science is hard, and for two main reasons—first, data is messy, and second, defining the exact problem to be solved is difficult. Data is both data scientists' bread and butter and the bane of their existences.

Data Science Values

Data science isn't one monolith, and every project is a little different from others, but there are still a lot of commonalities that require certain ways of working to be most effective. There are six core values that we see in good data science: simplicity, explainability, reproducibility, trustworthiness, understandability, and shareability.

Simplicity comes into play when you're looking for a solution—you start with the simplest that might work and only go with more complicated choices when the simpler ones aren't working. Simplicity sometimes gets a bad rap, but it's actually valued in almost all science and engineering fields. If you remember that it's the opposite of complexity, you might see the advantage. Complex machines and systems cost more and are harder to maintain. The same is true for any software. And as it turns out, the same is also true for any data science solution. The goal is generally to create a solution that is as simple as possible without compromising quality.

Explainability—how easily you'll be able to explain your solution to your stakeholders—is critical in a lot of situations. Being able to explain things to your stakeholders can be hugely important in some cases, especially when they're new to data science. It's crucial for building trust. If stakeholders can't understand what you did, they may not trust your results. Simplicity factors in here, because in general simple solutions are easier to explain than complex ones. You'll often hear people talk about black box vs. white box solutions. A *black box* solution is one whose choices fundamentally cannot be explained. The complexity is too high or calculations are masked. A classic example of a black box is a neural net. We'll go into more detail on neural nets in Chapter 15, but the lack of explainability is one reason it's not used as much as it might be in the business world. A *white box* solution, on the other hand, is easily explained. Two classic examples of white box solutions are linear regression and decision trees. We'll look at those here to see how they are explainable, but we'll talk more about them in Chapter 15. Explainability is not always paramount, so if black box methods achieve higher accuracy (which they often do), the unexplainable black box approach might be the right one to use.

Ordinary least squares *linear regression* (we usually drop the first part when talking about it) is a classic approach that's been used by statisticians for decades, which basically draws a straight line through data by minimizing the error (the difference) between each point and the line for all points. In a linear regression, the model is literally just a formula where you multiply each variable by a coefficient that has been optimized by the linear regression technique and add all the terms together. You can share this formula with your stakeholders, which makes it clear which variables are the most impactful. Figure 6-4 shows an example of a formula that represents a linear regression model that predicts rating in the video game data based on the other numeric features in the table.

```
Rating = 3.44 + 0.52 * ('Number of Ratings')
            + -0.32 * ('Plays')
            + -0.18 * ('Playing')
            + 0.36 * ('Backlogs')
            + 0.86 * ('Wishlist')
```

Figure 6-4. *A linear regression formula for predicting rating from several features*

This formula is interesting—Wishlist is the most important, even more than Number of Ratings. Plays and Playing both reduce the score, which seems counterintuitive, but Backlogs is more important than either and may counteract them. Models can sometimes have surprising values. It's also possible this just isn't a very accurate model.

Another easily explainable model is the decision tree, where each decision point can be visually represented, so a path can be followed. At the highest level, the *decision tree* approach basically splits the data into two sides by identifying the most effective split (where one side matches mostly one label or result better than other splits) and then repeats the process for each result until it reaches a result that is all or mostly one label (or a max depth is reached). An example of this would be: take the right branch if Number of Plays is greater than 100; otherwise, take the left. If most of the data greater than a hundred has a rating of 4 or greater, and most of the data less than 100 has mostly ratings less than 4, this would be a good split. Figure 6-5 shows a very simple tree on the same video game data, also predicting rating.

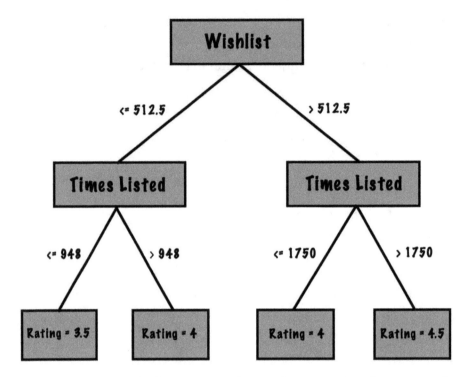

Figure 6-5. *A simple decision tree for predicting rating from several features*

This is a stupid-simple model that uses wishlist and times listed to determine the rating, only forecasting 3.5, 4.0, or 4.5 ratings. But it's easy to trace a path from the top node down based on feature values for a given point. This model only looks at two features, Wishlist (counts of users who've added it to their wishlist) and Times Listed (times users have listed the book), both simple counts. Decision trees have *nodes* (decision points on a feature), *leaves* (the final decision at the bottom), and *branches* (lines connecting nodes or nodes and leaves representing a particular decision range). If you want to predict the rating of a game based on its current values, with a Wishlist of 750 and a Times Listed of 975, you'd start at the top and check the cutoff for Wishlist. Our value of 750 is greater than 512.5, so we'd take the right branch. Then we look at Times Listed and see that our value of 975 is less than 1,750, so we'd take the left branch, which takes us to the leaf node with a predicted Rating of 4.

Although both the linear regression and decision tree models are a little technical and some stakeholders will struggle to understand them, they are fully transparent, and most people will be able to learn how to read them. Additionally, you could easily display the tree in a less technical way, even hand drawing one on a whiteboard to make it less intimidating. Transparency is the cornerstone of explainability.

Another scientific aspect of data science is the need for *reproducibility*, which means that it should be easy for you—or someone else—to run your process again and get the same basic results. So that means it's important to document your process in some way and make sure that any data will be consistently available to anyone attempting to repeat your process. One concept worth mentioning here is determinism—a *deterministic* process means that if you run it, you'll get the exact same result every time, where a *nondeterministic* process may have different results (sometimes only a little, sometimes a lot). Some techniques are deterministic and some aren't—I'll talk more about these in Chapter 15. Usually, one important factor in reproducibility is the random value generators available in computer languages, which aren't truly random. We refer to it as pseudorandom, and one way to ensure that your results are reproducible is to manually set the random seed in the code before running any algorithm that isn't deterministic, because that guarantees the same starting point.

One concept related to reproducibility is more about direct code *reuse*, which may be writing functions that someone else can call in their code or sometimes copying and pasting. The advantages of this are that everyone on the team is doing something the same way and that if you come back to look at work you did a year earlier, you may be able to rerun the code with only a few tweaks.

Trustworthiness is the data science value that encapsulates ethics, legality, privacy, and security, which often interact. Basically, this is about ensuring that the work you do stays within ethical and legal boundaries and does not compromise individuals' or companies' privacy and security. All of these will be addressed more in depth later in the book, but the important things to remember are that data scientists need to be responsible and understand the law as it relates to the data and processes they use and their companies' rules around privacy and security and have their own moral compass that keeps ethics in their minds at all times.

Understandability and *shareability* are another couple of related data science values and, respectively, refer to how easy to understand and easy to share your results are. Both of them apply to your stakeholders and your peers. Shareability with your stakeholders basically refers to the final step in the CRISP-DM lifecycle, which we first saw in Chapter 5 (see below in Figure 6-3 for another look). That final step, visualization and presentation, is all about sharing your results with your stakeholders. How you share will depend on your company and stakeholders, but this might be a slide deck or a web page on the intranet. But making your results easy to access isn't enough—they also need to be understandable (especially explainable, as we saw above). We've talked about the importance of understanding where your stakeholders are coming from so you can provide them with information that isn't intimidating and that they can understand.

Shareability can also harken back to reuse and reproducibility because it's common to share results, processes, and code with other data scientists, and understandability factors in there, as well. Most data scientists like to see what their peers are working on, but they also are going to be interested in different things than your stakeholders are. So you may be sharing different information, including code. Your colleagues might want to reuse something you've written or follow a process you've created. In this way, understandability also means making your code and process user-friendly to your colleagues—with clearly spelled-out steps, well-commented code, and other things that make it easy for someone else to understand (and repeat or reuse) it.

How Organizations Use Data Science

As I mentioned above, a lot of organizations talk about data science for attention while not actually being capable of doing it because of lack of data infrastructure and tools. But some organizations do successfully use it, sometimes effectively, sometimes not. There are three basic ways that organizations do use data science: understanding, planning, and automation.

Understanding

One of the most common uses of data science is to better understand the people or objects that the organization works with and how well the organization is doing. This generally involves looking at historical data. Data science often takes prior data analysis work a step further, but doesn't generally replace any data analysis already being done.

A pizza restaurant might want to learn more about their customers, and they might start a profiling investigation. They could use a machine learning technique called clustering to group their different customers based on attributes (if they're lucky enough to have data on their customers). They might find that there are four major types of customers: weekday lunchtime rushed customers, weekday evening diners, late-night drunken visitors, and weekend daytime dilly-dalliers. This could be really helpful in serving each type of customer with different specials and marketing.

The same restaurant might want to evaluate a marketing campaign they ran that was based on targeting these types of customers. They could look at sales data over many weeks before and after the campaign to see if there are differences. This can help them assess the campaign and its impact on the behavior of the four customer types. They might find that late-night specials on beer brought in more of the late-night visitors and increased profits significantly because of how much more profitable beer is than pizza.

Planning

Data science is also used a great deal for planning purposes. This can be both from looking at the success of different efforts and deciding which to continue with, but this is also where actual forecasting can come into play—so it can focus on either historical data or predicting the future, or both. Any prediction will rely on historical data to be generated.

The pizza restaurant might have been very happy with the success of their campaign, which they studied to identify the particular parts of the campaign that were most successful. This could be valuable information in planning which types of specials to continue offering and helping to figure out ways to change the less successful efforts.

But they might go even further and create a model that would forecast future sales based on the four customer types. This would be helpful in knowing how much to order of ingredients and cooking supplies.

Automation

A final way that organizations use data science is to automate things, which can happen in a variety of ways. This usually—but not always— implies an element of predicting the future, even though historical data is critical as with any forecasts. It could also involve automatically assigning new customers to sales reps in a customer relationship management (CRM) system, automatically creating an estimate of likelihood to complete a degree in a college advising system, or automatically matching people up in a dating app. Often companies are already doing these things, but they can bring in machine learning techniques to do them better (or more quickly).

For the pizza restaurant, this might involve automatically assigning labels to new customers, determining optimized routes for pizza delivery drivers, or having a table that's automatically updated every Sunday night with the forecasts for the next two weeks in order to interface with an ordering system and automatically order ingredients and supplies.

Data Scientist Qualities and Roles

There isn't just one type of data scientist. Different roles have different focuses. Some data scientists might be borderline data analysts because most of their work doesn't involve the more rigorous machine learning that others might do. Some might spend most of their time heads down in Python machine learning code and rarely come up for air. Still others might find themselves dealing with stakeholders and doing high-level planning work more than working with data tools. There's a chapter later that discusses more about different types of data science roles and how to find one that is best for you, which also includes some more career advice.

The functional, technical, and soft skills outlined in the previous chapter hold true for data scientists, even though different types of roles will emphasize some skills more than others. But some of the most important skills for data scientists include curiosity, patience, being methodical, and a growth mindset. Curiosity drives you to find the interesting questions to ask, which leads to the most valuable findings. Patience is required for a lot of reasons, but mostly because you can't jump into modeling even though you want to—you have to wait until the data is ready. Being methodical and having great attention to detail help you avoid mistakes and make processes that are logical and easy to repeat.

And finally, a growth mindset is necessary in this and many other technical fields—there are always new tools and techniques coming out, and you never know when one might be exactly what your project needs.

Key Takeaways and Next Up

Data science does have a lot to offer, but it's not magic, despite what many people in leadership believe. Data science has emerged as a more sophisticated variant of data analysis and has a lot to offer organizations that understand what it can and can't do. With good data, data scientists can help organizations understand their data, plan and make decisions, and automate time-consuming or inaccurate processes. There are several values that comprise data science, including simplicity, explainability, reproducibility, trustworthiness, understandability, and shareability. Simplicity says that data science solutions should be as simple as possible while still being useful. Explainability requires that the results solutions come up with can be explained (favoring white box over black box solutions). Reproducibility has data scientists following rigorous code practices and processes so the work can be reproduced (so we know the results are legitimate). Trustworthiness is about getting stakeholders and others to trust that your solutions are valid, ethical, and appropriately secure and private. Understandability is partially related to explainability, but it is more about the overall work being understandable, not just the results (which shares some goals with reproducibility). Finally, shareability addresses how easy it is for you to share your results both with your stakeholders and your colleagues (and leaders), which requires focusing on several things and relates to understandability. Note that these aren't always equally important—explainability in particular can be sacrificed if it's not required and higher accuracy is more important.

In the next chapter, we'll be talking about the idea of modern "data analytics," another term that is used in different ways. But here, it means an initiative that involves all aspects of working with data to extract insights, from getting the data in the first place to making it ready to use and to performing business intelligence (primarily basic reporting), data analysis, and data science on it. I'll talk about each of those areas and how they all fit together (or not) at different organizations.

PRACTITIONER PROFILE: LAUREN JENSEN

Name: Lauren Jensen

Current Title: Data scientist

Current Industry: Retail

Years of Experience: 10 in analytics

Education:

- Advanced Machine Learning post-baccalaureate certificate

- Statistical Analysis with R post-baccalaureate certificate

- BA in Business and Political Science

The opinions expressed here are Lauren's and not any of her employers', past or present.

Background

Lauren Jensen was always interested in politics and for a long time intended to be a political speech writer. She pursued that in college with a BA in Business and Political Science. In college, she managed to get a competitive internship at the Democratic National Committee during Barack Obama's campaign. Through that campaign, she was exposed to the way the campaign used analytics. When she saw that their analytics revealed that they were ahead in the battleground state of Ohio, Obama's rival's home state, they reallocated resources to other battleground states and still won Ohio. This was remarkable and only possible because of analytics, and it had her hooked.

Work

After graduating college and a brief stint working in a retail store, Lauren landed a job in marketing analytics at a retail company. The team wasn't doing much advanced analytics, but they were wanting to move in that direction.

Lauren ended up doing some grunt work working on some marketing mailers and coupon books that didn't initially involve any data science, but eventually there were some opportunities to do more analytics and some more sophisticated reporting, and she found that she loved it. She wanted to get better at it so she could move into a role that would involve a lot more data science, so she enrolled in a certificate program in R-based statistical analysis and later one in advanced machine learning. She managed to make an internal move at her company to a team dedicated to advanced analytics, so she kept developing her skills while doing interesting work that had real impact on the business.

After several years in that role, Lauren decided she wanted a new challenge, and she moved into consulting, which exposed her to natural language processing and generative AI, both of which were new and fascinating to her. She had more leadership responsibilities in that position, which she also enjoyed. She moved back into retail for a new role, where her wide experience was beneficial to a team that's still developing relationships with the business since analytics is fairly new to the company.

Sound Bites

Favorite Parts of the Job: Lauren loves working with different types of people. She also loves how the projects vary a lot in her work, where there are always new puzzles to solve—and usually more than one way to solve them, so figuring it out the best solution is part of the fun.

Least Favorite Parts of the Job: People have very unrealistic expectations about what data science can do and often decide what they want without discussing it with people who do understand what's possible. Often something might be theoretically possible, but not feasible at that time at that organization, for instance, because of data or infrastructure limitations. Lauren saw this in consulting especially, with strategy people making impossible promises.

Favorite Project: Lauren did some cohort analysis while a consultant. She investigated whether social media could be used to offset the loss of data because third-party cookie data is no longer available. She found that, yes, it could be, if you utilize the right platforms and create effective advertising. The most interesting thing about the project was that she found that when done correctly, it could be even more valuable than third-party cookies were.

How Education Ties to the Real World: Lauren has taught students and worked a lot with interns at her jobs, and she's found that so many of them expect everything to be easy and are surprised when it's not. Education doesn't focus enough on flexibility and problem-solving. She's had students expect her to give them a step-by-step plan for every problem and then be frustrated when she explained that it's not cookie-cutter—each problem is a little different. Students also aren't prepared for real-world data. It doesn't always make sense to replace missing values with the mean, for instance.

Skills Used Most: The biggest skill Lauren uses regularly is problem-solving. This is problem-solving of all types. Some recent examples include figuring out what she doesn't know that she should know, how to work with cutting-edge tools that don't have good documentation, and how to fix things when a project goes south. These involve critical thinking, and sometimes it can be especially hard when problems aren't always reproducible. Another set of important skills are those that involve working with people. This includes communicating, managing, and storytelling (basically, bringing them along with you).

Primary Tools Used Currently: Python and SQL have been critical throughout her career, and currently she uses JupyterHub and Snowflake for most of her work, relying on Google Slides for her presentations.

Future of Data Science: We're at a crossroads and need to decide where we should go next. Even though data science and AI are often overhyped, there still are going to be big winners—companies and people who adapt and adopt tew tech early and wisely—and the losers who stay still. The losers won't make it. This will mean job loss, among other problems. It's also important to realize that tech is outpacing regulation, and we haven't dealt with all the political ramifications yet, which we will have to do.

What Makes a Good Data Scientist: Problem-solving and critical thinking are crucial. If you have a role on the more technical part of the data science spectrum, you might be able to get away with weaker people skills, but most data scientists really do need to have good communication skills.

Her Tip for Prospective Data Scientists: Get familiar with cleaning data, because that's what you'll spend most of your time doing.

Lauren is a data scientist with experience crunching numbers and leading data science teams in marketing, retail, and other industries.

CHAPTER 7

A Fresh Perspective: The New Data Analytics

Introduction

We've already seen that data science is associated with a lot of other terms—data analysis, AI, machine learning, and so on. Here's one more: analytics.

Data analytics as a concept has been around for a while, but the term is starting to be used more as a catch-all term to describe a comprehensive data-driven approach that astute companies are turning to. A successful analytics initiative involves many disciplines, but the most important are business intelligence, data analysis, data science, database administration, data tool administration, and data engineering (which increasingly has a specialization called analytics engineering). All of these contribute to an analytics program that will help the company accomplish its goals. Other areas that are often part of an analytics program are machine learning engineering and productionizing support.

© Kelly P. Vincent 2025
K. P. Vincent, *A Friendly Guide to Data Science*, Friendly Guides to Technology,
https://doi.org/10.1007/979-8-8688-1169-2_7

I've already talked about data analysis and data science in previous chapters, and I'll talk more about business intelligence and machine learning in later chapters. For now, I'll quickly explain the terms we haven't discussed yet. *Business intelligence* (BI) is basically business reporting—charts and dashboards—that helps people make informed decisions. *Database administration* refers specifically to the management of the databases and some other data stores. *Data tool administration* is managing the various tools that data scientists and other data workers use. *Data engineering* involves capturing and preparing data for *data workers* (people who use the data, including BI, data analysts, and data scientists), and *analytics engineering* is a subset of data engineering where the engineers have more analytics expertise and can better prepare the data for easier use by data workers. *Machine learning engineering* involves implementing the machine learning models data scientists have created in order to make them more efficient and ready for implementation in production. *Production* refers to software running in the "real world"—where that software relies on the real data, where actual customers use it, where it automatically runs to create new data for a dashboard, etc. Putting something like a forecasting model into production is often the whole point of developing it, but there is also a lot of code that data scientists will write that is only used to generate results to be shared directly with stakeholders. Data scientists do sometimes have to do their own production deployments, so other *production or deployment support or tools* help with those tasks. Some data scientists will find themselves doing many—or all—of these tasks, but in bigger and more mature organizations, they are split out.

As an example of how all of these roles may interact at a mature organization that has different people in the roles, consider a company that hosts customer image files and automatically tags and classifies them, allows users to follow each other and share images with each other, and makes recommendations for new users to follow. There would be databases that store the image files, tags, labels, associated metadata, and other data, which would be managed by the database administrators (DBAs). Data engineers would be responsible for collecting additional data, possibly on customer usage and behavior, additional image metadata, and system behavior. They and analytics engineers would also be responsible for preparing that for consumption by data workers like the BI engineers and data scientists. BI engineers would create dashboards that leadership and others use to understand ongoing trends and behaviors of aspects of the site that they're interested in, like the number of users, the number of stored images, and the number of shares. Data scientists would look at the data and investigate it to answer business stakeholder questions, but they would also be responsible for determining if the recommendation engine needs to be modified. They might also prepare some data that is used internally in reporting. Machine learning engineers would be responsible for implementing any of the data scientists' processes that are going to production, like changes to the recommendation engine or automating the data scientists' data intended for reporting, and they'd work with data or analytics engineers to make this available to BI engineers.

This maturity concept—the idea that different organizations are at different levels of analytics maturity—will also be addressed in the chapter, along with the four classes of analytics, which generally require different levels of maturity to achieve.

Analytical Word Salad

We already looked at the usage of many of the terms used around data science in Google's Ngram Viewer in the last chapter, but it's interesting to see where "analytics" slots in. In Figure 7-1, you can see it falls after "AI" and "machine learning" but before any of the others. Remember that this is simply counting occurrences of the individual terms across writing through 2022.

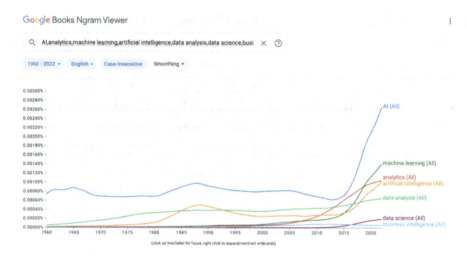

Figure 7-1. *Google Ngram Viewer for "AI," "analytics," "machine learning," "artificial intelligence," "data analysis," "data science," and "business intelligence" from 1960 to 2022. Source: Accessed January 17, 2025, at* $https://books.google.com/ngrams/graph?content=A$ $I\%2Canalytics\%2Cmachine+learning\%2Cartificial+intelligenc$ $e\%2Cdata+analysis\%2Cdata+science\%2Cbusiness+intelligence\&y$ $ear\_start=1960\&year\_end=2022\&case\_insensitive=true\&corpus=$ $en\&smoothing=3$

It's actually a positive sign that the term analytics was being used so much up to the end of the 2010s, because organizations are going to do best if they take the high-level analytics view rather than focus on the amorphous AI or the other specific disciplines. It's unfortunate that "machine learning" is taking over in popularity. However, I tried the search again with "data foundation," "data engineering," and "analytics engineering" added in, and the first two came in lower than everything else, and "analytics engineering" barely registered at all. Although more recent data might improve those numbers, it's still problematic since they are all crucial to any analytics program. You can't do any analytics work without having a strong data foundation built by data and analytics engineers. So much of the discussion of analytics is surface level and without understanding of what's really involved.

Data science and AI still have the sheen of newness, and people have heard all about the deep insights they can offer, promising them the keys to the vault of financial success. But they aren't the magic bullet that so many people think they are. There's no doubt that these fields are powerful and real, but the hype is overstated unless an organization commits to a true analytics initiative rather than cobbled-together pieces.

Examples of Data Analytics in the Real World

We've seen several examples of analytics already in previous chapters, but it will be helpful to look at some from the high level of an analytics program, rather than specific projects. First, we'll look at a wholesaler's efforts to lower costs and improve relationships with both customers and suppliers and then at the way a professional basketball franchise used analytics to improve their engagement with fans.

Example 1: A Data Analytics Initiative to Reduce Costs and Improve Customer and Supplier Relationships

A UK-based wholesale business supply company used a CRM (customer relationship management) software suite that had built-in reporting, but they found it was too limiting.[1] They wanted to be able to understand

---

[1] "Increasing profitability using BI and data analytics for the wholesale and distribution business" on Data Nectar at https://www.data-nectar.com/case-study/distribution-landing-analytics/

customer needs better, improve customer retention, optimize their inventory processes, and reduce the time it took to get paid by customers. They worked with a consulting firm to implement a new system with the goal of increasing revenue by 20% over the eight months after launch. Fundamentally, they wanted to switch to a data-driven approach to strategy and decision-making.

The company was operating out of one warehouse and shipping only to other businesses, but they had 3,000 office-related products and over 1,000 customers. Some of the challenges they faced included keeping the right products in inventory, which involved anticipating customer needs, and managing customer payments efficiently to ensure on-time payments. They also struggled to find the right opportunities for cross-selling and upselling of more profitable products. The business knew that their data would give them useful information about all these challenges, so they hired a consulting company to help them achieve this.

It was clear that what they needed first was a business intelligence system that would provide good reporting and dashboards for leaders to look at. The consulting firm helped them identify data they could use from their CRM system and some other sources, move the data into a different system their custom reporting could access, and create a business intelligence infrastructure. This allowed them to operate more efficiently because the additional information at their fingertips could help them make better decisions.

There were several steps involved in preparing the data, including basic processing and more advanced transformations, before it was loaded into a SQL data warehouse. Then data cubes and data marts with a variety of dimensions were created for easy use. The data prep ensured that the data would be usable by a variety of people, from self-service users to BI report builders. Ensuring good data security was part of the data processing and the storage management.

Several Key Performance Indicators (KPIs) and reports were created once the data was ready. These included ones showing customer segmentation in several areas; accounts receivable analytics including a breakdown of customers and identification of the best and worst customers; supplier analysis into pricing, delivery time, credit time, and margins; and inventory replenishments that helped reduce the cost of inventory.

The project was considered a big success because it increased the revenue in high-margin products by 15% and led to a reduction in payment time of 18% and a 17% reduction in inventory expenses. It also led to better negotiations and terms with some of their suppliers. The company believed that none of these would have been possible without a concerted effort to implement a comprehensive analytics program focusing on data, advanced reporting, and KPIs.

Example 2: A Basketball Franchise Engages with Its Fans Through Data

A professional basketball franchise was saddled with legacy IT systems that didn't allow for the analytics they wanted to do and made it harder to communicate with fans in effective ways. A new data analytics initiative allowed them to increase revenue, improve marketing conversion, improve fan engagement, and increase fan attendance at games.[2]

The franchise had an outdated CRM system that allowed communication with fans, but without good customization options, meaning that they might receive messages irrelevant to their interests and purchase history. Much of the interaction with fans happened via their mobile fan app, and they knew the app had great potential to do more.

---

[2] "How digital transformation helped benefit fans and the bottom line," EY, https://www.ey.com/en_us/insights/consulting/how-digital-transformation-helped-benefit-fans-and-the-bottom-line

The franchise had two key but not directly related goals. First, they wanted to replace this CRM system with a modern one to improve fan engagement. Then they also wanted to add more sophisticated digital tools to motivate sales reps to higher levels of performance. Like the previous company, they didn't think they had the internal expertise, so they hired a consulting firm to help them reach these goals.

They created a new CRM and enhanced the mobile fan app so the two were fully integrated, providing a wealth of info about fans to the franchises and leading to an improved experience for fans. Specifically, the franchise was able to track aspects of fan behavior like attendance and purchase patterns (of food, drinks, and merchandise). The franchise was able to analyze these things and make appropriate changes, such as price adjustments to fill otherwise empty seats. The app provided valuable information for fans, including parking details and where to find the shortest concession and bathroom lines.

The franchise also created several dashboards for the sales teams like leaderboards for weekly, monthly, and quarterly goals, which encouraged friendly competition and raised ticket sales significantly. Some of the enhancements to the app streamlined the process of purchasing merchandise and season tickets. These changes increased the sales of both merchandise and season tickets in the app by 50%. Although this had been a huge undertaking, the franchise was very happy with the results.

The Umbrella of Analytics

A lot goes into creating a successful analytics program, and companies need to be organized and intentional for it to be successful. This section will address the transition from the old way of doing analytics and the different disciplines within an analytics program.

From "Old" Analytics to "New" Analytics

A lot of companies have been doing analytics for years or decades, but the "new analytics" involves looking at it as a comprehensive program rather than a little bit of BI here and some analysis there. A comprehensive program usually will add rigor by centralizing the work to a degree and instituting more standards and best practices for the work that makes it more like traditional software development.

The different data worker roles are often developing software (especially BI developers making reports and data scientists creating models that will run in real time), so software and other system development processes can apply. The process usually involves splitting things into at least two environments: development and production. A *development environment* is where all development—new code and changes to existing code—is done. These environments are kept completely separate from each other, but usually development is a mirror of production as much as possible. There can also be another environment between dev and prod, usually called test, which is a step toward production where more rigorous testing can be done of code.

One issue that affects data analysts and scientists is that sometimes the development environment does not have access to real, production data. This makes sense in the software world and even in BI development. The reports and dashboards can be developed on fake or incomplete data, and then when deployed, they can be switched to show the production data. Final testing still needs to be done to ensure something wonky didn't happen, but it's a relatively small effort.

In contrast, data analysts and data scientists absolutely need access to the real data from the beginning, which can be a bone of contention at companies that don't understand what data analysis and data science are or how they're done. The early stages of the process including exploratory data analysis require access to the real data. Work like creating features and selecting the best ones also requires real data. It is meaningless to

243

build a model based on incomplete or fake data. It would be like giving your doctor completely made-up symptoms and leaving out the real ones and expecting them to give you an accurate diagnosis.

All of these issues are understood and handled at organizations that have a true analytics initiative. In fact, some organizations are so organized that they follow a relatively new paradigm called DataOps (see the sidebar for more info on DataOps).

DATAOPS

In DataOps, the entire lifecycle of data products, from data preparation to reporting and presentation, is made as rigorous as possible with automation and other principles from three other paradigms used in the tech world, Agile, DevOps, and lean manufacturing. Agile is a software development methodology we'll talk more about in a later chapter. It doesn't really work on its own for data science project management, but some of its principles are useful. We'll talk more about DevOps later as well, but it's also used in software development to automate both testing and deployment (leading to what's called continuous integration and deployment (CI/CD)). Its principles factor into DataOps. Finally, lean manufacturing is another paradigm used in manufacturing that emphasizes statistical process control, which translates in DevOps to the data that's in place being constantly monitored, with notifications being sent out if anything appears out of the ordinary.

Data Tools and Infrastructure

Data engineers need tools to build the data foundation, and BI developers, data analysts, and data scientists need tools to carry out their work. There's a virtually unlimited number of options. These might include any of the cloud environments and their various data-related tools, "on-prem" solutions that might be stored within the company's infrastructure, tools that need to be installed on people's computers, and more. All of these tools will require a mechanism to access the company's data, wherever it's stored. Ideally, someone other than the data workers themselves should be managing these tools and infrastructure, but that isn't always the case. But BI developers, data analysts, and data scientists aren't going to be able to do as much productive work if they're spending time managing infrastructure.

Data Foundation

As mentioned in the introduction, BI, data analysis, and data science rely on there being a solid *data foundation*—cleaned, prepared, and ready-to-use data—as a starting point. There are lots of ways to store the data, and we'll talk in more detail about the various types of data storage in a later chapter. But for now, some of the terms that might be used to describe how the data is organized and stored include database (SQL or NoSQL), data warehouse, data lake, and data mart. There will be different ways of interacting with these sources.

The most important aspect of the data foundation is generally not how it's stored—instead, it's the quality and usefulness of the data. There should be data available at the grain that people want to see (see the sidebar for an explanation of grain), which often means storing it at the lowest grain so people can combine and summarize it as needed. There need to be useful ways of combining the data—for instance, people will want to know what purchases customers are making, so having tables with purchases and tables with customer info won't be useful unless there's a way to connect the two.

Although ideally the data users would not be developing this foundation themselves, in less mature organizations, they might have to. But if they don't have time for a comprehensive effort and can only prepare data on a project-by-project basis, this is not a true data foundation.

AGGREGATION AND GRAIN

When working with data, it's common to aggregate, usually summing or averaging across a column to get fewer rows because they are at a different grain than the original data. For instance, consider the following in Figure 7-2, which shows some student data with GPA and awards.

Student Name	Student ID	Gender	Advisor	Age	GPA	Awards
Beto Cruse	430	Male	Miller	18	3.70	4
Billie Jean Harris	651	Female	Patel	15	4.00	2
Cindy Rodriguez	453	Female	Jackson	17	2.80	3
Edna Darwin	410	Female	Namath	18	3.70	0
Elisa Garcia	101	Female	Miller	18	3.12	2
Guru Choudhury	291	Male	Patel	16	3.41	0
Henri Dubois	456	Male	Patel	16	2.99	2
Ian MacAlastair	322	Male	Jackson	17	3.65	2
James Raj	601	Male	Jackson	16	2.76	0
Lakshmi Kumar	275	Male	Patel	15	3.91	1
Lisa Galton	243	Female	Miller	17	4.00	1
Mabel Johnson	613	Female	Miller	17	3.25	0
Mackenzie Dobson	186	Nonbinary	Smith	17	3.75	1
Parker Thompson	561	Female	Namath	18	3.20	1
Ryan O'Connell	384	Male	Miller	16	3.80	1
Sarah Smith	119	Female	Miller	15	3.56	0

Figure 7-2. *Some example student data*

If we want to summarize this data to get an overall picture of different advisors' students, we could aggregate GPA and Awards at the Advisor grain, which would result in only one row per advisor containing the average of the GPA and the sum of the Awards for all the rows for that advisor. It would look like Figure 7-3.

Advisor	GPA Avg	Awards Total
Jackson	3.07	5
Miller	3.57	8
Namath	3.45	1
Patel	3.58	5
Smith	3.75	1

Figure 7-3. *The same data aggregated at the Advisor grain only*

In this table, the exact GPA and Awards values don't appear in the original table at all except for Smith, who has only one advisee. Jackson has three advisees, so the GPA in this table is the average of those three students' GPA, and Awards is the total number for all three students. This is interesting, but maybe we also want to consider the students' ages. We could summarize at the Advisor–Age grain instead, as you can see in Figure 7-4.

Advisor	Age	GPA Avg	Awards Total
Jackson	16	2.76	0
Jackson	17	3.23	5
Miller	15	3.56	0
Miller	16	3.80	1
Miller	17	3.63	1
Miller	18	3.41	6
Namath	18	3.45	1
Patel	15	3.96	3
Patel	16	3.20	2
Smith	17	3.75	1

Figure 7-4. *Example student data aggregated at the Age–Advisor grain, with average GPA and total number of awards*

This table has more rows because most advisors have students of different ages, all except Namath and Smith. Any time you are aggregating or viewing aggregated data, you have to consider the grain.

Dates are a very common thing to tweak in different grains because we often want to view values like sales summarized at the week or month grain, rather than looking at daily values.

The data foundation is ideally created by data engineers (or in some places, analytics engineers), but the reality is that sometimes other people end up having to build it. These can include anyone at the end of the line in some way, including data analysts, data scientists, or business intelligence engineers.

Business Intelligence

Business intelligence (BI) is really just reporting—creating a range of types of reports from everything from Excel sheets to summary documents, to dashboards with charts, tables, and other numbers. BI reports are usually ongoing things that get refreshed and looked at regularly, not one-offs. Leaders and workers use all these to understand the state of various aspects of the business, which helps them make decisions. Many companies have been doing this for a long time even if they are otherwise immature in terms of analytics. Usually, the business intelligence team is also responsible for preparing the data they will be using in the reports, either within a tool like Tableau or Power BI or in a database (either on their own or in conjunction with data engineers and/or database administrators).

The BI developers—report creators—are usually not doing any work that would be labeled "data analysis." The analysis they do is simply in understanding and preparing the data in service of reports to share with decision-makers. Regardless, a lot of the analysis they do to understand the data is similar to what data analysts and data scientists have to do to learn about the data, but they stop after an understanding of what's there is achieved. Data analysts and data scientists use that information to do further work.

248

The focus of BI is almost always on past and current business, not the future. If any future data is displayed, it is just numbers in a table someone else has created or results of a simple formula that has been shared with them, not anything the BI team itself has generated from any sort of algorithm.

Self-Service Reporting

Companies can struggle to keep up with all the BI needs, so many create self-service analytics programs that allow nontechnical businesspeople to create their own reports. People often like it because they don't have to wait for someone else to build it, and they can also figure out what they need by trial and error themselves, rather than waiting for a developer to check in with an in-progress dashboard. The success of a program like this hinges entirely on the quality of the data available to users. Fully automatic reports—those designed to run as needed, but with fields and values determined by pre-programmed formulas created by BI or data analysts—are safe, but a lot of self-service involves nontechnical people designing their own reports, picking fields and aggregations. The data has to be both clean and easy to use for this to work. Specifically, it is difficult for users to know how to combine things when all the data isn't already in one table—joining data sources together properly often takes technical know-how. If the data isn't ready, they may produce charts with inaccurate information, which can affect decisions and also create trust issues with the analytics teams. It's not uncommon for a data analyst to do some analysis only to find that the results contradict self-service reports generated by nontechnical people. It can take extra work to show that there isn't an error in the data analysis work, but rather in the self-service report. In summary, successful self-service programs require well-designed data, guardrails, and good training (the users need to understand what they can and shouldn't do).

Data Analysis

We've learned all about data analysis in Chapter 5, but to recap, data analysis involves any of the following: looking into data and finding insights by doing exploratory data analysis, slicing and dicing the data in different ways to expose different aspects of the data, making charts and other summaries to illustrate characteristics of the data, and doing statistical analysis to reveal even more about the data. The end goal is to understand the business better, which requires a depth of understanding of the data that business intelligence developers might not need.

Depending on the organization, some of a data analyst's work might involve doing business intelligence, but it's expected that they do more than generate dashboards. But it's also common for a data analyst to build a dashboard to summarize true data analysis work they've done, so some of the skills often overlap. More experienced data analysts may also dip into data science territory, especially with one of the more simpler machine learning techniques of linear regression.

Data analysts are also usually focused on the past and present, but some more advanced ones may also be involved in some basic forecasting.

Data Science and Advanced Analytics

Data scientists go a step further than data analysts, as we've seen in Chapter 6. They use more complicated techniques than data analysts typically do, including machine learning, natural language processing, and other statistically advanced techniques. However, it's very common for data scientists at some companies to do a lot of data analysis and even business intelligence, especially when the organization is smaller or less mature and doesn't have a strong analytics initiative. Data scientists work with data from any time frame—past, present, or future—based on the business needs and questions.

Advanced analytics in this context is simply another term for data science. However, sometimes the term is preferred by organizations with a holistic analytics approach, because it conveys that context better than data science, which implies separateness from the other areas.

Analytics Maturity and Classes of Analytics

If you look up maturity models of analytics, you will often find charts showing four different classes of analytics, which they present as a maturity model. These include, from least to most sophistication, descriptive, diagnostic, predictive, and prescriptive, all of which will be described below. These are useful classes to understand and we'll discuss them below, but I think it's more beneficial to judge an organization's maturity by their available data, tools, and data capabilities rather than the classes of analytics they are doing. Organizations do the analytics that they need, and not doing the most sophisticated doesn't mean that they aren't capable.

A Maturity Model

The analytics maturity model has five basic levels of analytics maturity at an organization, which depends on the selection of analytics it is capable of. As mentioned above, some believe that the level is determined by the type of analytics they're doing, with data science being the highest level. But it's not that simple. In the lowest level, maturity level 0, the organization has no organized analytics initiative at all. That doesn't mean that there aren't people working with data, but it's not in a comprehensive program that will lead to success.

Ideally, an organization works their way up through the hierarchy a step at a time. At maturity level 1, they have at least the beginnings of a data foundation—data that business intelligence, data analysts, and data scientists can use in their work—and the right tools for those data workers to use. Both of these really are critical, but both of them can also be in progress—the key requirement for this level of maturity is that the need for data and tools is fully understood by leadership and they've committed resources to creating this data foundation and suite of data tools.

An organization at maturity level 2 would have a solid business intelligence effort—usually a team or teams that produce multiple reports a year for different stakeholders, all based on using the data in the data foundation. If they then grow their analytics initiative to include data analysis, that would bring them to maturity level 3. Adding data science to their repertoire would bring them to maturity level 4. Usually, the first data science an organization does is relatively basic, with less of a division between data analysis and data science than highly mature organizations have. Maturity level 5 would apply to an organization that is capable of doing more advanced data science projects. A level 5 organization therefore has a fully realized, mature analytics program that involves projects of varying complexity happening in all three disciplines in response to different needs from the business.

The reality is that many organizations are wholly jumping the gun, getting caught up in the hype of AI and data science and forming data science teams despite having no other analytics or often only BI. Often these organizations don't even have data that is anywhere ready to use (sometimes it isn't even available at all). Trying to do data science without a data foundation or other types of analytics is often a fool's errand. Good data scientists are skilled at figuring things out and finding solutions to many challenges, so they may still manage to do some good work. But it will take much longer than necessary and still be limited in scope. The organization's maturity level with this scenario is still going to be 0, because they don't know what they're doing in a systematic way.

One other implicit requirement at each level that isn't always obvious is that not only is that stage of analytics being done, but the team(s) have sufficient tools to do their work. Doing BI requires BI tools and doing data science requires numerous tools with enough compute power and memory to deal with large amounts of data, for instance. Organizations also often struggle with this, leaving people without sufficient tools that make them waste time finding workarounds or having to rerun things because their last effort has timed out. If an organization seems to be at level 4, where they have teams doing BI, data analysis, and data science on a solid data foundation, but the data scientists have no tools outside of Excel, that's not a level 4 organization. Excel is not sufficient for data science. It would instead be level 3, assuming the BI developers and data analysts have the right tools.

So the maturity level is 0–5 and matches the level they've reached where they've also implemented each of the prior ones in their analytics initiative. Table 7-1 shows a summary, with an analogy in terms of movement included.

***Table 7-1.** The maturity model levels with implemented parts of an analytics initiative*

Maturity Level	Implemented Analytics Stages	Movement
0	No data foundation	Stand still
1	Data foundation + tools	
2	Data foundation and BI + all tools	Crawl
3	Data foundation, BI, and data analysis + all tools	Walk
4	Data foundation, BI, data analysis, and basic data science + all tools	Run
5	Data foundation, BI, data analysis, basic and advanced data science + all tools	Fly

There's a huge range in different companies' levels, although it is possible to generalize a little about what to expect from different types of companies (but obviously, there are many expectations to these generalizations). Most startups and tech companies are at least at level 3, many even 4, because they've been using that information to drive decisions from early on. They have understood the space and value of data for their entire existence. Larger nontech companies that have been around a while are usually at 2 or 3 because they've put the time and effort into trying to understand their business, which means looking at their data through BI or even data analysis. So their data foundation is usually in place, even if it may not be quite suitable for data science work yet. Smaller organizations usually are much lower on the scale because they haven't taken the time and effort to put together a solid data foundation, even if they're trying to extract information from the hodgepodge of data sources they do have (like lone spreadsheets floating around).

Note that this is not fully black or white scale—some data foundation is better than none if it's still being built, for instance. But a data analyst who makes lots of dashboards but doesn't have time to do anything else shouldn't really be considered to be fulfilling the data analysis level, because all they're doing is BI, even if they have a different job title.

Four Classes of Analytics

The four different classes of analytics are in a hierarchy because each type is increasingly challenging. The first is descriptive analytics, followed by diagnostic, predictive, and prescriptive. A maturity level 1 or 2 organization will be able to do the first—descriptive—but probably not the second. The other classes of analytics simply require more maturity to be possible. Data analysts will be required for the second class—diagnostic—and generally data scientists would be required for predictive and prescriptive.

Descriptive

Descriptive analytics is data work that's concerned with what has happened in the past and often also what is happening now (or at least, right before now). This is the primary domain of BI and data analysis. It could take the form of reports as part of BI and the exploratory and descriptive statistics of data analysis. For instance, a report created by BI that shows the daily sales of each of a video game company's games up through yesterday, with additional summaries by month, is an example of descriptive analytics. Work from data analysts might go a bit deeper, but still would be describing the past, like with breakdown of the click results of A/B testing on a new online marketing campaign for the company's most recent game release.

Diagnostic

Diagnostic analytics is work that does more than describe what has happened, aiming to also *explain* why these things have happened. It's important to understand what has happened, so descriptive analytics is a part of this, but it's not the end goal. This is generally the domain of data analysts and sometimes data scientists. As an example, the dashboard showing daily game sales might allow the user to hover over the sales figure for a day to see more info about that day, including information about marketing, advertising, promotions, holidays, and significant national events like Amazon Prime Day and election day. Another click would allow the information to be overlayed on the chart so trends across the full sales chart can be seen. For instance, a multi-line chart for one game's daily sales could contain the sales figures, the number of ads served, the price of the game (reflecting regular price changes or discounts), and vertical lines on days with a holiday or major event.

Predictive

Predictive analytics sounds just like what it sounds like—the focus is predicting future trends. This is primarily the domain of data scientists because forecasting generally requires machine learning approaches. One of the most common predictive analytics tasks is to forecast sales for a short period of time into the future. This might appear on the same chart as described in the "Diagnostic" section—users could see the history, the explanatory items overlaid, and the next 14 days' forecasts. This kind of chart is powerful because there's so much info in one place.

Prescriptive

The final class is *prescriptive analytics*, which aims to help tell decision-makers what they could—or even should—do. This is pretty advanced and the domain of data scientists, even though a lot of them may never do this kind of work, because rather than giving info for decision-makers to use, we are telling them what to do. It doesn't have to be quite as aggressive as that sounds, however. We could have a system that looks at the impact of price changes and promotions on sales and suggests when a promo should be run or a price should be changed. The information would generally also include an explanation, such as what features led to the recommendation. A decision-maker wouldn't have to implement this, but it's a suggestion based on previous information that's used in complex analysis to consider more information than one human brain could.

Analytics Maturity and Classes Together

An organization's analytics maturity relies on a lot of factors, but these define how successful it can be in improving its business through analytics. Figure 7-5 shows the relation between maturity, the types of analytics being done, and the four classes of analytics when it progresses one level at a time.

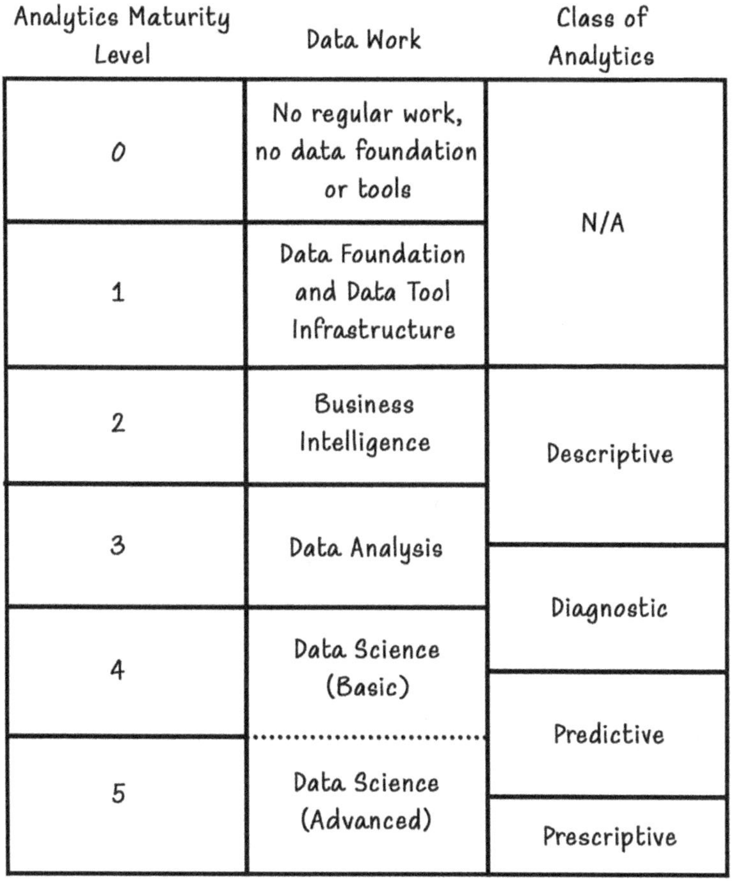

Analytics Maturity Level	Data Work	Class of Analytics
0	No regular work, no data foundation or tools	N/A
1	Data Foundation and Data Tool Infrastructure	
2	Business Intelligence	Descriptive
3	Data Analysis	Diagnostic
4	Data Science (Basic)	
		Predictive
5	Data Science (Advanced)	Prescriptive

Figure 7-5. *The analytics maturity level with corresponding data work and classes of analytics in the traditional trajectory of an organized analytics program*

As you can see, level 1 is simply setup and preparation for analytics to begin, but until level 2, no analytics is necessarily being done. At level 2, business intelligence is producing descriptive analytics. Diagnostic analytics becomes possible at level 3. Level 4 has added predictive analytics to the organization's repertoire. Finally, at level 5, prescriptive analytics is possible.

Building an Analytics Program

As I mentioned earlier, there is one way to develop an analytics program, which involves going through the steps of the maturity model from 1 to 5. But it's not something a single person or small team can really do on their own. It's critical that leadership understand the need and what it involves and are fully bought in. It's pretty clear that the goal of most analytics programs is going to be to drive competitive advantage or otherwise improve the organizations' performance. But leaders must understand the tremendous amount of work that must be done and how much time it takes before that value can be realized.

I've emphasized how important a data foundation is, and that is going to require people who can be data owners and govern the data. A data scientist who's been prematurely hired may have neither data nor tools, so they must convince leadership of the necessity. After that, they should be working with the people developing the data foundation to help them design a usable system. But other people will be necessary to this process.

Another maturity level 1 requirement is that data workers must have the tools to do their job, and this costs money. The money can be a challenge. Organizations often don't want to pay for compute resources (computer processors and memory, whether physical or in the cloud) and storage even though it's necessary to have enough computing power to do anything. If you don't have anything more than a laptop, you can't do more than basic data analysis and data science on smaller datasets.

A Caveat

Although there is a logical path to creating a solid data analytics program, many companies try to do one of the more advanced things without the others, which is always problematic. A common situation is starting a data science team without having data ready for them to use. This situation is challenging but can be remedied. But how? Usually, it will fall to the newly hired data scientists to fix things.

The most important thing is to ensure that a data foundation exists and the tools are there for data scientists to use (level 1). If they're lucky, these intrepid data scientists can explain this to leaders, but it's not a given that they'll listen. Some organizations simply aren't going to hear that they jumped the gun and will not understand that good data science can't be done without the earlier maturity levels attained, even the most basic one of having usable data. Sometimes all you can do is move on.

However, once a data foundation exists, the data scientists should try to convince leadership that business intelligence and data analysis teams should be put into place. Most likely, that won't be happening very soon, so the data scientists may simply have to fulfill those roles themselves. This can be very frustrating for such data scientists because it often means they aren't able to get to the real data science work they want to do.

Key Takeaways and Next Up

The idea of a comprehensive analytics program is relatively new in business, but it's part of becoming a digital company, which is the goal for most companies nowadays. Being able to use data to make decisions is critical to organizations staying relevant and getting ahead. But so many don't realize the amount of work that goes into having an analytics program that will allow insights to be found that can help leaders make decisions. In the chapter, we talked about the different components of a successful analytics initiative. We also talked about organizational analytics maturity levels and the different classes of analytics that can be done at the different levels.

In Chapter 8, I'll be diving into data security and privacy, topics that a lot of data scientists aren't particularly interested in. But it's important to understand some basic things, and the chapter will discuss those parts and why. I'll cover the different areas of data security, data privacy as a human right, and personally identifiable information (PII). Then I'll talk about the various types of security compromises that are out there and how to avoid them. I'll also address some data security and privacy laws.

PRACTITIONER PROFILE: TAYLOR HAGIN

Name: Taylor Hagin

Job Title: Business intelligence analyst/data consultant

Industry: Consulting

Years of Experience: 1 year as a data scientist, 15 in tech

Education:

- Galvanize Data Science Intensive Bootcamp

- BS in Psychology

- AS in Psychology

The opinions expressed here are Taylor's and not any of his employers', past or present.

Background

After getting his degrees, Taylor ended up in some jobs that weren't going to take him anywhere, so he researched a variety of options, including doing an MBA, another master's, and a bootcamp. At one job, he ended up doing some database work when the small company's database administrator left, and he enjoyed that work, but he realized he needed to learn to code. These interests helped him while he was identifying good options, starting with data science and other information jobs, all with good salaries. He analyzed all the options—he actually did an informal ROI study and decided the bootcamp was the best value. It was intensive and short, so he'd be earning real money sooner than with any grad degree. The bootcamp was truly intense, and it took him a couple of tries to get accepted, but once in he knew it was a good move. He wasn't a top performer, but he learned a ton and networked like crazy. He had several interviews that came out of his final presentation at the bootcamp.

Work

Taylor landed a job at an analytics consulting company and dove into the work. He worked with another data scientist, who was a great mentor, so he learned a lot from him. Although he enjoyed the data science work, he was eager to experience more in the tech world and ended moving into some different roles at the consultant company. He worked in project management and even sales there, which were both interesting and challenging, but eventually got burnt out with high-stress consulting. He stayed in tech sales for a bit and then moved into product management. These were still rather high-stress so he shifted into tech education, building documentation and learning models, and then got hit with a layoff. In his job search, he decided to focus on companies that are in the same niche as one of the companies he worked for and liked and soon landed the position he's currently in, which is great.

Sound Bites

Favorite Parts of the Job: Taylor loves coding and when you do something very complicated and are still able to explain it to less technical people. He also enjoyed the learning process when collaborating with someone more knowledgeable—it's the best way to learn. He also can't help but also enjoy the clout—people are impressed with you when you say you're a data scientist.

Least Favorite Parts of the Job: The worst thing about doing data science through a consulting company is that you usually don't get to see where your projects go long term. Additionally, although he does enjoy explaining complicated things to leaders, sometimes they're arrogant and not receptive— and often not nearly as knowledgeable as they think. Gathering requirements from people who don't have any idea what they want or need is always painful.

Favorite Project: One of the earliest data science projects he worked on with the consulting company was at a major retailer. He worked with another senior data scientist at the consulting company and learned a ton on a fascinating project that involved forecasting sales at specific stores for a large retailer. They created a fairly complicated ensemble machine learning model that relied on cutting-edge algorithms. He loved the way they had to put a solution together through a combination of knowledge and trial and error. This wasn't something where you could just Google and copy and paste some code.

How Education Ties to the Real World: Taylor's bootcamp was better than some master's degrees because he was working with real data and had live coding exercises twice a day, so he got used to working with messy data under a time crunch. A lot of master's programs don't even touch these things, which he learned after working with graduates from analytics programs. They often don't get the need for urgency and the importance of data prep.

Skills Used Most: Communication, taking quality notes, and asking a lot of questions. Technical skills matter, but they pale next to the soft skills.

Primary Tools Used Currently: GitHub, Python, SQL (SQL Server), Google, and ChatGPT

Future of Data Science: We're in a weird spot right now where people think generative AI will solve all DS problems. But it's way too unreliable and hallucinatory—Taylor thinks of it like inexperienced teenagers typing really fast. Sure, they generate stuff—but is it what we need? We'll probably revert to things that were reliable before. It's not by chance that banks are still using FORTRAN. It works.

What Makes a Good Data Scientist: The top three skills are communication, empathy, and curiosity. Empathy really is a part of communication because it enables you to understand people better, and curiosity drives discovery. But keeping ethics and security in mind and paying attention to best practices are all also very important. You have to always keep learning, which means risking making mistakes and admitting when you have less experience so you can learn from colleagues.

His Tip for Prospective Data Scientists: Network, network, network. Even if it's uncomfortable, look at how other people are doing it successfully and try to mimic that. If anyone is looking at a bootcamp, make sure to inspect them closely—ideally talk to someone who's attended it. Some of them aren't good. Taylor recommends the one he did.

Taylor is a business intelligence and data professional with a wide variety of experience in tech roles.

CHAPTER 8

Keeping Everyone Safe: Data Security and Privacy

Introduction

We've all heard the horror stories from major companies of data breaches, ransomware, or other incidents involving data being compromised. These things often happen because of human error somewhere along the way, but hackers and scammers are getting more and more sophisticated every day. Truthfully, most humans aren't that good at being careful, and it's more difficult than it should be to accomplish everything we need to do to keep data and our information secure and private. It's crucial that we—everyone who uses computers—do this. But it's especially important for data scientists and anyone who works with data to pay attention to security and privacy. Privacy is especially important for data scientists because they work with data that can be sensitive, and not all organizations have good practices around privacy, so it can fall to the data scientists to be personally responsible.

© Kelly P. Vincent 2025
K. P. Vincent, *A Friendly Guide to Data Science*, Friendly Guides to Technology,
https://doi.org/10.1007/979-8-8688-1169-2_8

Most of us have a good sense for what "security" and "privacy" mean in the general sense. Security's related to safety and basically means that we're safe from danger. Our house is secure if it has locks that should keep bad actors out of the house. Privacy is similar, but basically means the state of being free from observation by others. Obviously, both security and privacy are relative—regular home locks won't keep everyone out, and an intruder could always simply break a window to get in. Windows also provide a risk to privacy—if we don't cover them, people can see in.

Data security and privacy are simply these concepts applied to data. They're clearly interrelated, and the terms are occasionally used interchangeably, but they do have distinct meanings. *Data security* involves keeping data safe from nefarious actors who would use it to do harm to companies or people, which means storing it in a safe place, accessible to only the right people, and with protections to keep all others away. In some ways, *data privacy* is a subset of data security, because it refers specifically to keeping people's data secure, but it also comes with a perspective that people are entitled to control their data and control who sees it, even if there's no clear risk of harm. For instance, many people want to keep their shopping history private even if the clearest risk is seeing ads in their browser for the kinds of things they buy. Other people don't care and might even like that because they're more likely to find out about a sale on a product they're interested in if the ads they receive are targeted.

It's not just with browser ads—with both security and privacy, we tend to balance convenience with risk to determine the levels of security and privacy we perceive and that we're comfortable with. People don't all agree on what the right level is, and it's going to vary from person to person. Most of us don't want to board our windows up, even if that would make our homes more secure and more private. This mental balancing of convenience vs. risk that we do whenever data security and privacy are in play is one of the things that can be very dangerous if we aren't cognizant of our choices and the ramifications.

Humans are very bad at assessing risk in most areas, and security and privacy are definitely a couple of those areas. Things that are out of sight are hard for people to include in their calculation of risk. Some of us are instinctively suspicious when we receive a link in an email from someone we don't know, but for others, it doesn't raise any red flags. Links often lead to funny or interesting things, so if the trusting person is curious and doesn't have any alarm bells reminding them that they don't know anything about the entity that sent this link, including whether the sender is out to steal from them in some way, they might click it.

Obviously, this scenario is what the scammers count on. They don't need everyone to click—they just need a few right people to click. This human weakness—one person giving away credentials or clicking a link that installs software or otherwise gives hackers access—is almost always the basic avenue hackers and scammers use against both individuals and companies.

In this chapter, we'll look at a couple examples of security breaches. I'll talk about the various elements of data security and data privacy and how companies manage security and privacy. I'll go into the many types of data and privacy compromises and talk about what individual responsibility goes along with these efforts. Then we'll cover some of the laws related to data security and privacy.

Examples of Data Security and Privacy in the Real World

The nature of security and privacy is that the interesting stories about either are always about failures. If security is good, we'll never know what terrible things hackers almost did. The first example in this section shows a major security failing at a company that was irresponsible and should have done better, while the second is an example of what can happen when a single individual makes a "tiny mistake" by clicking a bad link.

Example 1: Massive Equifax Security Breach

In 2017, one of the biggest data breaches occurred when hackers broke into Equifax's systems and accessed them from May through July that year. Equifax didn't discover it until July 28, after the data of almost 165 million Americans, Brits, and Canadians had already been compromised. Some of the data that was stolen included full names, addresses, Social Security numbers, birth dates, driver's license numbers, credit card numbers, and dispute documents that had additional personally identifiable information.[1]

This breach is a clear example of something that could have easily been prevented. Equifax knew they had significant security weaknesses after an internal audit in 2015. They had many known security vulnerabilities that needed to be patched, and they were not meeting their own schedule for patching them. The patches were being applied haphazardly, with little oversight and no consideration of how important each patch was in terms of the specific systems or data it was intended to fix.

The particular vulnerability that caught them out was on software called Apache Struts, a web framework used with the Java programming language. In early May, Apache released a patch that they said needed to be immediately applied to all sites using it. Just days after the patch was released, security experts had already identified hackers trying to find

---

[1] "Credit reporting firm Equifax says data breach could potentially affect 143 million US consumers" by Todd Haselton on CNBC.com, September 8, 2017, https://www.cnbc.com/2017/09/07/credit-reporting-firm-equifax-says-cybersecurity-incident-could-potentially-affect-143-million-us-consumers.html

many companies' Struts-based websites that hadn't been updated. So it was clearly known as a significant vulnerability, but Equifax didn't apply the patch.[2]

Hackers broke into Equifax on May 12, 2017, via their credit dispute website that used Struts and hadn't been patched. Early in the breach, they accessed internal employee credentials, so they were able to search many additional internal systems, which allowed access to even more systems while appearing to be legitimate searches from internal users. This intrusion wasn't found by Equifax until late July.

Even that discovery was an accident and the result of further irresponsibility of Equifax. They had a network security certificate on an application that was nine months out of date, and they finally updated it on July 29. The updated certificate enabled detection of suspicious activity, and it picked up the hackers' work, so that's when Equifax employees were notified. They had blocked the breach by the next day, but didn't notify the public until September 7.

Afterward, the breach was investigated, and numerous weaknesses in Equifax's security profile were found to be at least partly to blame. Obviously, the most important was failing to update the Apache Struts site immediately as they were instructed, but the blasé attitude toward patching in general and insufficient process were the main reasons for that. Other things that contributed were a poorly designed and insecure internal network, a lack of good breach detection systems, and insufficient encryption on personally identifiable information.

---

[2] "Equifax failed to patch security vulnerability in March: former CEO" by David Shepardson in Reuters, October 2, 2017, `https://www.reuters.com/article/us-equifax-breach/equifax-failed-to-patch-security-vulnerability-in-march-former-ceo-idUSKCN1C71VY/`, and "DATA PROTECTION: Actions Taken by Equifax and Federal Agencies in Response to the 2017 Breach" by the US Government Accountability Office, August 2018, `https://www.warren.senate.gov/imo/media/doc/2018.09.06%20GAO%20Equifax%20report.pdf`

Equifax's overall handling of the breach was criticized widely, and the services they offered to consumers were substandard. They created two sites that would allow consumers to access information about their own info and whether it had been compromised. Both sites looked like phishing sites to security tools that detect such sites. Neither site was on the Equifax domain, so there was no way for consumers to know if they were on the real site or had stumbled onto a well-designed phishing site set up to look like the real site. Additionally, one of the sites didn't even work, returning random results when tested.

The criticism really damaged Equifax's reputation, and the company's stock price dropped significantly. The chief information officer and the chief security officer were both let go. Equifax also hired an external security company to investigate. They had to work with the FBI and the FTC. Various fines exceeded hundreds of millions in the United States and 11 million pounds in the United Kingdom. The company was also sued by individuals, and a class action suit was filed. An external national consumer credit protection organization began an investigation that was later dropped, but the Pennsylvania Attorney General led a coalition of 50 different US state attorney generals that reached a settlement with Equifax having to pay $600 million to individuals and states.[3] Although it was a win, that sounds like a lot more money than it is to Equifax, who's still chugging along.

---

[3] "AG Shapiro Secures $600 Million from Equifax in Largest Data Breach Settlement in History" on the Pennsylvania Attorney General site, July 22, 2019, https://www.attorneygeneral.gov/taking-action/ag-shapiro-secures-600-million-from-equifax-in-largest-data-breach-settlement-in-history/

Example 2: A Ransomware Attack Shuts Down Operations for Weeks

Ransomware always seemed a little far-fetched to me until I had a friend whose company got hit with an attack during the pandemic and she described the nightmare of trying to carry on with business at all, much less as usual.

My friend manages a team in the accounting department of a business that's part of a large multinational conglomerate based in Europe. She works in the Pacific Northwest of the United States for part of the business that cuts raw materials for use in manufacturing. There are around ten other locations in the same business across the United States and Canada, spread out from the west coast to the east coast.

The early January morning that everyone was coming back to their home offices after the winter holidays, my friend found she couldn't log into the remote desktop on her work laptop. She couldn't access her email or any of the other software her team used. Soon text messages were flying among her entire team—nobody could log in, and when they reached out beyond their team, they learned it wasn't only them. Something was obviously wrong, but it would be hours before they knew what.

Around noon, they finally got a message from IT: the company had been hit by a ransomware attack, and everyone was to just hang tight. The IT department had been scrambling since early that morning. Although some of the details wouldn't come out until later, an employee had clicked a bad link and inadvertently given hackers access to many of the company's systems. When IT had started receiving notifications from people unable to log in, they investigated and fairly quickly realized there was a huge problem, so they shut all other systems down. Even getting that noon message out to employees was difficult given that they didn't have immediate access to employee records and company email wasn't working. But the only good news was that because of the structure of the

271

conglomerate, there were some systems that weren't impacted by the attack, including human resources and payroll, which relieved a lot of the staff.

But on the ground, in the accounting department, that day was supposed to be the start of the month-end process, where all the financial records were finalized to close out December. They obviously couldn't do that, or anything, because they had no access to their invoicing system. They also couldn't pay bills to their vendors and couldn't reach out to them without access to the system or even their emails. On the second day, the department had everyone create outlook.com email addresses with their name and company name so they could at least communicate with each other, but they obviously couldn't email vendors from emails like jane_acmepress@outlook.com.

IT was working to get access going again one system at a time. But the priority was actually at the branches where the metal cutting was supposed to be happening, because they were completely blocked. Although the cutting systems weren't compromised, they couldn't access any of the orders so they didn't have any of the specifications for what they should be cutting. The longer they were backed up, the more the core of the business was going to be impacted.

On the third day, my friend's team regained access to SAP, the primary enterprise software that was hosted in another country so hadn't been compromised. They still couldn't access anything on their network drives, but were slowly finding some things they could do. Several systems they used were third-party or web-based, so they were safe. They were able to monitor the company's bank accounts so they could see money coming in, but couldn't tie that to any accounts or documentation internally. A few days in, my friend was asked to write a very carefully worded snail mail letter to send to their vendors (at least the ones whose mailing addresses they could round up) explaining that they were having some "technical problems" and would be paying invoices as possible.

And still, the metal cutting was stalled, with nobody able to do anything. They weren't able to contact customers about the backup, either. It wasn't until the second week that those orders were finally released and the cutting got going again. That same week, my friend's team was able to log into their remote desktops and access their email, but nothing else— and all their archived email was inaccessible. They wouldn't have access to the email archives until April. By the third week, things were mostly back to normal in the accounting department. They'd created a workaround for December closeout, and they were lucky to basically be able to handle January month-end as normal, where discrepancies due to the irregular December month-end were reported up the chain to headquarters.

Employees weren't told how much money was ultimately lost, but it had to be in the multiple millions. The company avoided paying the ransom, which is important because while they had to absorb the damage, the hackers did not benefit. The company did start mandating more and better training about security as well as put processes in place to avoid people falling for emotionally manipulative scams like the urgent-email-from-the-CEO-requiring-immediate-payment one.

Data Security

Data security is so important that almost all companies have a team usually called InfoSec (short for information security) that manages it. This team holds ultimate responsibility for data security at the company and all the policies and technological solutions in place to ensure it. Another important part of their responsibility is teaching individuals who work with their data about their own personal responsibility to protecting the company's data.

One of the reasons InfoSec has to teach people about their role in protecting the company's data is that most people have only a vague sense of how much danger is out there. It's the out of sight, out of mind problem.

People have often heard of the dark web, but don't necessarily know what it is. The *dark web* is a part of the Web that's available only via special tools that guarantee anonymity and privacy of users and is highly associated with illegal activity and places to access stolen data.

It's common to refer to the people who seek to steal data or harm companies or people by using their data in undesired ways *bad actors*. These are the hackers, the scammers, and even the spammers. *Hackers* are the people who gain unauthorized access to systems, whether *black-hat* (intending to do harm) or *white-hat* (those who work for companies to identify weaknesses in order to fix them). *Scammers* are people who use fraud or otherwise cheat people out of money or information in both the real world and the computer world. A *spammer* is someone who sends large amounts of emails or other unsolicited messages to people, usually with ill-intent but always to get something from people. It is shocking to a lot of people that these bad actors are not lonely young men in their mother's basements—there is a whole criminal network of these people, who often operate in certain countries out of mundane office buildings with company Christmas parties.

InfoSec is responsible for keeping all these people—who often operate via bots and other software—out of their companies' systems. They do this with a variety of policies and tools following some of the most important tenets of the industry.

Data security is usually considered to have three elements: confidentiality, integrity, and availability. These make up what's called the CIA Triad, which can be seen in Figure 8-1. *Confidentiality* means that only the right people have access to the data, which involves keeping outside hackers out, but also managing internal access. Not everyone needs access to human resources data or product ingredients, for instance. *Integrity* means that the data is in the condition it's supposed to be in—it hasn't been modified by someone, whether unintentionally or maliciously. This involves managing access wisely and is one of the reasons data scientists and other data users are often given only read access to data sources—

there's no reason for a data scientist to be changing data in source tables, and not having access keeps them from doing so accidentally. Finally, *availability* means the data is available to the right people when they need it.

Figure 8-1. *The CIA Security Triad*

Domains of Data Security

There are several major domains of security that InfoSec deals with: infrastructure, network, application, and cloud. Infrastructure and network security involve devices and software both physically and digitally, where companies are only responsible for digital security on applications and the cloud.

Infrastructure security involves protecting any hardware the company has that isn't specifically related to networking. In the old days, large companies would have huge server rooms that contained hundreds of computers, usually called data centers. These required a lot of work to keep running and safe, including expensive cooling and ventilation systems as well as physical security ranging from locking doors with keyed access and protections on the physical cabinets called racks that stored the flat, thin computers. These "on-prem" (on-premise) setups are becoming less and less common as companies turn to the cloud, partially because it means not having so much to do in terms of infrastructure security. But companies still have infrastructure devices, like employee laptops and audio–video equipment in meeting rooms. Infrastructure security is generally owned by InfoSec but relies on employees to handle their devices properly.

Network security involves securing physical access to network devices—such as with locking doors into rooms with the devices or boxes that prevent access to devices not contained in a room—as well as software and firmware protections like firewalls and VPNs (Virtual Private Networks). It's concerned with keeping bad actors from accessing the network in any possible way, whether directly from a device or through a computer's wi-fi connection. Protections that are installed directly on network devices are called *firmware*, but VPNs that are installed on users' laptops are considered *software*. This is also managed primarily by InfoSec, but also relies on individuals doing what they're supposed to do.

Application security is any kind of protections that exist in software running on people's computers or other servers that doesn't specifically relate to infrastructure or network security. It's usually focused on preventing access to data, either directly through databases or other data sources or through applications like Excel, customer relationship management software, or other tools that access company data. Application security is important to InfoSec, and they manage aspects of it like logins to company systems, but employees have a lot more impact on it than the other types of security. Databases are usually managed by a different team—database administration—and they're the ones largely responsible for database security and everything related to it, but they usually work in conjunction with InfoSec. People writing software— including data scientists—that runs on their laptops or other computers hold some responsibility for writing secure code, especially when they are writing software that relates to their company's website or APIs (Application Programming Interfaces, systems that allow other people to write code to download data from the company's data sources). Even software that exists entirely within the company's infrastructure needs to be written securely.

Cloud security is related to the other three types, but it's specifically for securing the company's cloud infrastructure, which involves virtual "devices" like virtual computers or databases, virtual communication tools similar in function to network devices, and software that's running on these devices or elsewhere in the ecosystem. Usually there is a designated team to manage the security and other aspects of the cloud, but like database administrators, they work with InfoSec.

Data Security Technologies and Approaches

There are several methods and technologies that are used for data security. The first relates to controlling who can access data and systems. *User authentication* involves creating accounts for each user so they can log into a system and the system tracks who's logged in, when, and some other info. *User authorization* is the next step, the process of assigning appropriate permissions to each user's account. Access can often be given at a very fine-grained level, like some users may have access to only certain tables in the database or even be restricted from seeing some specific columns or rows. Managing that level of access is onerous, so a lot of organizations manage things at a higher level.

Monitoring is also important and involves both real-time and scheduled monitoring of networks, database and software access and use, and physical site access logs, and anything out of the ordinary should be automatically flagged and raise an alert. Audits of everything, including plans, should be done regularly to make sure things aren't slipping through the cracks.

It's important to be able to recover in case something bad does happen. *Data backups* are copies of all critical data so that if data is lost because of a security failure or even just a storage failure, it can be recovered. These backups are usually taken on a daily basis and may even still be physical tapes that are deposited in a different location every day, so going back to the backups will involve a loss of data, but it's designed to not be catastrophic.

Planning and policies are also important. Policies and plans should be in place to define how the data is managed, backup schedules and plans, and how it's decided which users should have access to what. Risk assessments should be done and included in plans.

Finally, some of the most obvious methods of data security relate specifically to how the data is stored and moved around. *Encryption* is a method of modifying data with code so that it's not usable if accessed in place or intercepted in flight without the decryption key. Data can also be modified in some way to protect certain fields or rows.

A common data modification technique is *masking*, where the real values are replaced by different, meaningless values. The simplest approach is just to *null out* the real values—leave them blank, basically. Usually, the general form of the data would be maintained. The most basic is *replacement masking*, where each character or digit is replaced with the same value, so a Social Security number would show XXX-XX-XXXX in every row. There are a few other types of masking. With *scrambling*, the values are simply reordered, so a Social Security number 123-45-6789 might end up 831-96-2457. In *shuffling*, a real value from the same column but a different row in the table would be used. With *substitution*, a fake but realistic value would be used, where someone's first and last names could be replaced with random first and last names. Sometimes encryption is also used here for specific fields or rows that are particularly sensitive.

Note that masking and other modifications might happen in place—the data could be permanently changed—but normally a new table would be created that shows everything that's in the original with the modifications in place in the new table only. There would be very few people with access to the original table, and most people would only be able to see the new one. Other times the changes could be applied at the display point based on the viewer's access privileges, rather than actually physically in a table. For instance, one viewer with higher privileges might see what's in the column in the real underlying table when using the company's data viewer, but another user might see scrambled data in the same column when accessing the same table. Different database systems allow different types of control.

The particular type of masking is going to depend on the situation. For example, it's common for companies to have "prod" (production or real) data and "dev" (development) data that's like data in prod but is not complete or exactly the same. As mentioned in Chapter 7, dev data is crucial for report development, with the reports only being pushed to prod when they're complete. While they're developing, report builders generally don't need the full, accurate data, but they need the data to behave and look like data in prod. If the report is going to show and do operations on salaries, they need values in the salary column to behave properly in drop-down boxes or numeric filter boxes so they can test that those things are functioning correctly, but they don't care if the actual values are "right." It's typical for report builders to not even see the report with real values until they put it into production. Appropriate masking makes the process seamless.

In other cases, data scientists may be using similar data, but they will need core fields unmasked. But usually data scientists don't need personal information like people names, addresses, and so on, so those can easily be masked. Different users need different levels of access.

Data Privacy

As mentioned above, data privacy can be mostly considered a subset of data security, and most of the efforts to ensure data privacy mirror those used to ensure security of data in general.

Privacy as a Human Right

One thing that stands out about privacy is how human-focused it is. Most human-focused entities regard privacy as a human right. The United Nations includes the following as Article 12 of their Declaration of Human Rights: "No one shall be subjected to arbitrary interference with his privacy, family, home or correspondence, nor to attacks upon his honour and reputation. Everyone has the right to the protection of the law against such interference or attacks."[4]

This is in line with how most people feel. Generally, people think they should have control over who knows what about them, although social media has shifted things a lot. Younger people are more comfortable with more being known about them and aren't as protective over their personal information as older people. Still, when it comes down to their personal information being used in a way that harms them, even if it's only in a small way, people realize maybe they do want to know who has their data.

Historically, with their emphasis on profits, most companies have not shared this view as the data revolution has gotten underway. They're motivated to know as much as possible about their customers, and many have been collecting it unchecked for years. But people are becoming more aware of what their rights should be, and things are starting to change. Governments are responding to the shift in perspective with new

---

[4] "Universal Declaration of Human Rights" by the United Nations, Article 12, https://www.un.org/en/about-us/universal-declaration-of-human-rights#:~:text=Article%2012

laws, with the European Union's General Data Protection Regulation (GDPR), the California Consumer Privacy Act (CCPA), and the California Privacy Rights Act (CPRA) leading the way. These will be discussed more below.

These laws are becoming increasingly important because of the major problem of inaccuracy of data about us. Data aggregators buy data up from every company that doesn't promise not to sell your data, and they combine it all together to sell as a product to other companies. The problem is that it's a jumbled mess. Different people are conflated, their phone numbers, emails, other data getting all mixed up and switched around. Many of these aggregators think I share my parents' address even though I haven't lived there, or even in the same state, for decades. I get text messages about local elections intended for my mom. My parents have gotten multiple flyers addressed to me inviting me to sell their house since I'm nearing retirement (not even close). Another time they got a razor and an invitation to enlist in the army since I had just graduated high school (and was a man). Nope. Yet another time they got a sample of baby formula intended for me. These things are funny on the surface, but there have been countless issues where innocent people have been conflated with completely different people who have been convicted of crimes and subsequently been turned down for jobs or loans, with no recourse (which means it's just going to keep happening, every time they apply for something). They can't fix it, and most of the people on the other end of the job or loan application aren't going to let these people explain. It does seem like it should be a human right to be protected from being punished for other people's crimes and bad choices.

Types of Data Relevant to Privacy

Because privacy pertains mostly to people, we often refer to *personal data*, any kind of data that represents people in some way. It's generally considered the most potentially sensitive, across the board. This would include obvious things like name, age, gender, address, phone number, and email. But it also includes anything else related to a person, including their banking info, medical history, and purchase history.

There are many types of personal data. Table 8-1 summarizes many types. This is definitely not an exhaustive list; some types of data can be more than one type.

Table 8-1. *Several types of personal data*

Data Type	Examples
Physical	Height, weight, and hair color
Personal	Name and age
Location	Country and zip code
Medical	Diagnoses and lab results
Financial	Bank balances and credit score
Employment history	Companies worked for and job titles
Education history	College attended and degrees attempted
Criminal history	Convictions and traffic tickets
Identification	Passport number and family name
Account	Account usernames and library card numbers
Security	Passwords and your mother's maiden name
Political	Party registration and voting history
Subjective	Opinions and beliefs
Behavioral	Online browsing history and driving history

Sensitive data is any data that has the potential to be used to harm a person or an organization in some way, whether intended or not. Companies hold a lot of sensitive data that isn't related to people. Many things can be considered sensitive even when it's not obvious how it could be harmful. Obviously, trade secrets like ingredients and recipes of products are highly sensitive, as is most financial data. With personal data, almost everything is considered sensitive, even things that seem innocent like favorite color or hobbies, because it's impossible to know how someone might use that data if they got hold of it.

Sensitivity is always a factor when system administrators are giving people access to data. For instance, most companies are not transparent about salaries so only a small number of people would have access to those values—generally human resources has access and so would managers (at least for their direct reports) and higher-level leaders.

Another important thing to consider is that context can alter sensitivity even within a company. For example, gender is fairly mundane in a lot of cases, but people evaluating music school auditions do not need to know the gender of the auditioner (changing auditions so musicians' gender wasn't known several years ago revealed that there had been a huge bias against women in the music world). Similarly, using gender to determine credit limits on new credit cardholders is illegal, so it's best for the individuals (or systems) making those decisions to not even be able to see it. But for a kids' summer camp assigning cabins and bunks, gender's more important.

Personally Identifiable Information (PII)

Anyone who works in data that relates to people in any way will hear about PII all the time. *Personally identifiable information (PII)* is simply data that can be used to identify an individual in the real world. Many types of personal data qualify as PII, like name, Social Security number, and email.

On their own, things like account number and country aren't necessarily PII, but it is relative and contextual, too. For instance, if we think of zip code or post code in someone's address, there are many people who have each one, so instinctively it seems like it wouldn't reveal anything personal. But one of the most important aspects of PII is that you have to consider how it could be used, not just what's in an isolated dataset. If you're a small company selling medical supplies and medicine and you have a table with the total amount of insulin purchased per zip code, that seems not at all related to personal info. But if there are other tables that contain PII that include the zip code, people could join the different tables together. If there's only one person in a zip code, we know that person is buying insulin. There are plenty of people within the company who should have access to that data, but likely not everyone needs access to it, so the company should consider how to keep that info available only to the right people.

It is exactly these kinds of vulnerabilities that bad actors exploit and regular people never even think of as possible. Most of the time, these kinds of risks aren't really known, because most people using the data every day aren't sneaking around joining tables willy-nilly to see what sort of sensitive info they can uncover. This is why it's important to keep privacy in mind any time new data is being stored. The people responsible need to consider not only the new data but existing tables and how they relate to the new data.

There are some well-known examples of how somewhat innocent data and usage led to violations of privacy. The retailer Target once decided to send targeted advertising to pregnant women by identifying products pregnant people buy. One of the recipients was a teen girl whose mother knew about the pregnancy and was helping her daughter buy pregnancy-related products, but the girl's father did not know, and he was angry when his daughter received the ad mailer featuring pregnancy products. He reached out to Target to express his unhappiness. This is how the man found out about his daughter's pregnancy, a huge and potentially dangerous violation of privacy. Given how angry he was about her "mistakenly" receiving a flyer for pregnant people, it's not hard to imagine that he was even angrier about an actual pregnancy. People who are pregnant should be allowed to share that information as they see fit, not be outed by a retail store.

In another example, AOL released some search history logs in 2006 that were "anonymized." A *New York Times* journalist interpreted that as a challenge, and starting with the AOL data, he dug through a variety of sources so that he was able to find several specific people and tie them to their real search history. The article identified a specific Georgia woman who gave him permission to share her story, putting a real face to supposedly anonymized data. This led to a resignation and firings at AOL, but in some ways it was an innocent mistake in the early days of the data world. This is why privacy is so important, and it's critical to think about all the possible ways data can be tied together.

Security and Privacy Compromises

There are a ton of ways bad actors may try to harm people and organizations and a large number of reasons. Most of the time it's theft— they want money from you or, more often, data that they can sell for money. But in some cases, they want to harm the target by damaging their reputation. We'll look at the major ways they do it in this section.

Social Engineering

Social engineering is a technique hackers use to gain access to systems by tricking real people into giving them access in some way. This may be physical access, like an employee holding a door open for someone when everyone is supposed to scan in, or virtual, like manipulating someone via email to send money to a scammer. Most people find it awkward to let a door shut in somebody's face, so if there's a person right behind you and they look friendly, most of us find it difficult to not hold the door for them. If they add the extra sob story of having forgotten their card key and having an important meeting in three minutes, it's even harder. Scammers often use fabricated urgency or the threat of someone higher in the company hierarchy to convince people to do things without checking for validity.

Another type of social engineering also involves person-to-person interaction, but it involves convincing someone to give them virtual access, perhaps by giving them a password or even the name or other personal information about other employees.

The key point is that hackers are manipulative and rely on human nature and behavior to gain access to systems or places. Some of what they're doing can also be considered phishing, which we'll address next. Other types include pretexting (using a made-up story like being a tech support employee to trick someone) and baiting (leaving a physical device like an infected USB drive in a visible location and hoping someone will plug it into their computer).

Phishing

Phishing is an action by hackers that involves trying to get credentials or login details for systems, or to get information like PII, by convincing people to give it to them. It generally relies on social engineering. Phishing usually comes in the form of an email or a website, but there's another category of phishing called *vishing* (voice phishing) where people use phone calls to carry out the effort, and phishing via text messages is also getting more common.

Phishing emails usually impersonate a person or company and request sensitive information. They may be claiming to be coming from a superior who's claiming to be unable to access some information they need immediately, using social engineering to create a sense of urgency and exploit either the company hierarchy (people don't want to say no to people higher up than they are) or existing relationships. Bear in mind that hackers take advantage of the fact that companies usually have fairly standardized email addresses, like first name-dot-last name at the company domain, and the information can be publicly available on LinkedIn.

These emails sometimes direct users to click a website, which usually looks like a real website the user might recognize. But it's instead a mock website. If the user tries to log in, they will enter their username and password, and the hackers now have the user's details and can log into the real site with those details.

Sometimes phishing sites can show up in web searches or links from disreputable websites. During the Equifax breach, Equifax's own Twitter account inadvertently tweeted a link to a mimicked site. People need to be vigilant all the time and always check the URL bar to ensure they're on the real site. This was why Equifax's creation of the two domains not on their own (Equifax.com) was such a bad decision.

It's important to realize that phishing hackers are getting more sophisticated every day. It used to be that a phishing email was relatively easy to spot because of bad English (typos, grammar problems, unnatural language), the occasional use of random weird characters instead of regular letters, and low-quality graphics when present. But that's not the case anymore. Nowadays you need to stay alert and look at linked URLs in emails (are they going to a legitimate domain associated with the email, like amazon.com rather than amazonk.com) and think about whether it makes sense to be seeing a request like the one you've received (would Sally from accounting really ask you for those logins?). Additionally, they are getting more targeted. *Spear-phishing* is when a phishing email is targeted and customized to a specific person or organization, and an even more sophisticated type is called *whaling*, when the person targeted is a high-ranking employee like an executive.

Password Guessing

Password guessing is just what it sounds like—hackers try to get into an account by "brute force" guessing a password, trying a whole bunch of different ones. This can be done by an individual typing it in manually, like when people try to break into someone's computer in the movies, but more often it's automated by using a program that tries one password guess after another in extremely quick succession. In that case, once they've managed to get in, it tells the hackers what it is, which gives them information they might use for other systems for that user.

There are a couple of different strategies hackers might try. The first is just trying really common (and very insecure) passwords like "password" or "1234." So many people never change passwords from the default (especially on things like modems) or use simple and easy-to-remember passwords.

The second strategy involves utilizing knowledge of a person's life to try different things. They might try variants of someone's birthday, phone number, Social Security number, pet's names, children's names, and virtually anything else they've been able to find either on the black market or from people's social media.

Note that password guessing usually relies on having the username, although it is possible to try different ones out. But since most usernames are emails nowadays, they can often get that with relative ease. And as mentioned above, in the corporate world, there are common email structures that companies use, and often the username they use in other company resources is the same as their email before the at symbol.

Physical Theft

Basic theft is a tried-and-true way for people to get information they shouldn't have, and *physical theft* involves hackers stealing laptops, smart phones, external hard drives, flash drives, and other hardware. In some cases, this requires further work, such as figuring out a password or decrypting a device, but often hard drives and flash drives have no security measures, or they are easy to break.

Obviously physical theft involves hackers gaining access in the real world to these devices. A laptop or phone left on a car seat or on the table at a coffee shop while the user goes to the restroom is fair game for bad actors. They also can get into businesses or houses and steal information or devices that way. Sneaking into businesses can often be done via social engineering, like when someone lets a hacker through a secure door without scanning.

Malware and Attacks

Computer viruses and worms are two well-known and feared types of *malware*, software that exists to do harm in some way. Other malware includes spyware, wiper malware, and ransomware. All of these can wreak havoc in business and even individuals' systems, whether physical or virtual.

Most of these attacks rely on social engineering to some extent, at least in the beginning, because someone needs to fall for the trick to get the malware into a company network. People are always the weakest link in any security program.

Anti-malware software is critical in most computer systems.

Viruses and Worms

Computer viruses are software that gets installed on a computer or system and takes control of some of the software on the computer, or even the whole computer. The catch with viruses is that the user is tricked into installing the virus software themselves, because web traffic can't install software without permission. This unwanted software can do all sorts of things, like encrypting or corrupting data, files, or software. The user may have no idea it's there, although often it causes problems by slowing the system down.

Worms are related to viruses and do much of the same things, but they behave differently by replicating themselves and propagating to other computers. They're especially common in organizations because once a user brings a worm onto their computer, it can spread across the network with ease.

Spyware

Spyware is a type of malware that is intended to glean information after gaining access to a system. A common one is a keylogger, which is installed on someone's computer and makes a record of every keyboard press. This is common when people use public wi-fi networks. Another user can access the computer on the network and record everything the computer user types, including if they enter a username and password to log into a site. This is why using a VPN (Virtual Private Network) whenever on public wi-fi is always recommended. But keyloggers can collect more than usernames and passwords since they gather everything typed—including URLs and information typed into spreadsheets and other programs.

Trojan horses are another type of spyware, which involves a piece of software that's installed on a computer by a user who thinks it's something legitimate. It's common for the trojan horse to be hidden inside a file that is otherwise legitimate. It can then go do its thing with the user unaware that there is a problem.

Wiper Malware

Wiper malware is similar to viruses except that its primary purpose is to delete data and files or to shut down systems. This is about stopping operations rather than stealing information, so these kinds of attacks are usually done by countries (governments or individuals) attacking another country's infrastructure, military, and so on. Another type of group that may employ wiper malware is *hacktivists*, hackers who believe they are doing good by disrupting organizations doing things they don't agree with.

Ransomware

Ransomware is also similar to viruses and worms, but the first thing it does is encrypt data and files so no one can access them. Then the hackers demand a ransom, and if the company doesn't pay the requested amount by the deadline, they may delete or permanently corrupt the company's data and files or potentially release some of it on the dark web. The targets of ransomware attacks are often similar to wiper malware attack targets, but the goal is to get money. By attacking critical targets like the supply chain or hospitals, the hackers hope that victims will just give in and pay because the consequences of not being able to operate are so dire and far-reaching.

Other Cyberattacks

There are also some types of attacks against systems that don't require accessing the system directly. One common type is the *denial of service (DoS) attack*, where an attacker floods a system—like a website or API—with requests in order to overwhelm the system and prevent legitimate users from accessing it, as well as potentially bringing down the site or service. These attacks generally involve some networking trickery. A specific variant of the DoS attack is the distributed DoS attack, where multiple systems are attacking the same target. In these, attackers frequently hijack other systems and form a botnet, or even rent access to botnets other people have created, to send all the requests. Botnets can be expanded and can potentially grow exponentially. The use of botnets also makes it even more difficult to identify who's responsible because the computers making up botnets are themselves compromised and their owners may have no idea.

Although DoS attacks are never fully preventable, there are methods for detecting them. Having a plan for how to respond—and then following that plan immediately when an attack is detected—is important.

Scams

Scams are one other area that isn't necessarily technically related to security or privacy, but still makes for very bad experiences that can affect both individuals and businesses. There are many bad actors in the world who will take advantage of naïve people through various scams that generally rely on social engineering to manipulate people into doing things like withdrawing money from their banks and sending it to someone. This usually affects individuals more than organizations, but sometimes scammers will use their social engineering tricks to impersonate people in an organization to trick their "coworkers" into helping, either with company or personal money or resources.

Scammers are getting more sophisticated every day. It used to be easy to spot things like the infamous Nigerian prince email scams, but scammers have improved their grammar and manipulation skills. They've started impersonating organizations that people find intimidating like the police and IRS. Now they're even using voice deepfakes to trick people into thinking they're dealing with a relative or friend on the phone. Like with everything else, we need to be on alert and suspicious of anything that seems out of the ordinary. If someone is trying to convince you to send money, stop and ask yourself if this makes sense. Call or text the person who's calling you if it's supposedly someone you know on the phone. Remember that the IRS is not going to call you and ask you to send cash gift cards to some address.

Vigilance is key to keeping organizations, computers, and ourselves safe from the many people trying to get something from us for free.

Data Security and Privacy Laws

The problems we see related to data security and privacy are all relatively new, as computers and the Internet have allowed things that in the real world would be small, like someone finding a credit card, to blow up and happen on a large scale. Before the Internet, someone with a stolen credit card would have to physically go to different stores—or at least phone different stores—to use the card. Now they could sell the number quickly and thousands of people could use it until the credit card company detects the fraudulent behavior and locks the card. But in the early days, the fraud detection wasn't exactly top-notch. It's been getting better over the years.

Like credit card companies improving fraud detection, the law needs to catch up with this new world. There have been some important steps made in the right direction, particularly in the area of privacy. But it's also becoming increasingly common for governments to hold companies responsible when their negligence allows data breaches that affect citizens, primarily through fines. Similarly, the attacks that bad actors carry out are illegal, and if they can be identified, they can be fined and even imprisoned.

Most of the recent legal efforts that countries and states are working on relate to privacy rather than general data security, but some laws attempt to address both. The European Union is leading the way with its comprehensive privacy and security law that went fully into effect in 2018, the General Data Protection Regulation (GDPR).[5] It regulates how data on EU citizens is collected and managed, meaning companies operating in any part of the world have to follow it if they are dealing with data on EU citizens in any way. Companies can be given huge fines, up to €20 million, or 4% of global revenue for bigger companies.

---

[5] "What is GDPR, the EU's new data protection law?" at https://gdpr.eu/what-is-gdpr/ and "Complete guide to GDPR compliance" at https://gdpr.eu/

Consent and citizens' right to access their own data are huge parts of the GDPR. Organizations are required to gain consent only when it's "freely given, specific, informed and unambiguous" and explained in "clear and plain language." Figure 8-2 shows EU citizens' privacy rights as defined in the GDPR.

1. The right to be informed
2. The right of access
3. The right to rectification
4. The right to erasure
5. The right to restrict processing
6. The right to data portability
7. The right to object
8. Rights in relation to automated decision making and profiling.

Figure 8-2. *EU citizens' privacy rights in the GDPR. Source:*
`https://gdpr.eu/what-is-gdpr/`

Even with all those things handled, there are many requirements for data protection, which can be seen in Figure 8-3.

1. **Lawfulness, fairness and transparency** — Processing must be lawful, fair, and transparent to the data subject.
2. **Purpose limitation** — You must process data for the legitimate purposes specified explicitly to the data subject when you collected it.
3. **Data minimization** — You should collect and process only as much data as absolutely necessary for the purposes specified.
4. **Accuracy** — You must keep personal data accurate and up to date.
5. **Storage limitation** — You may only store personally identifying data for as long as necessary for the specified purpose.
6. **Integrity and confidentiality** — Processing must be done in such a way as to ensure appropriate security, integrity, and confidentiality (e.g. by using encryption).
7. **Accountability** — The data controller is responsible for being able to demonstrate GDPR compliance with all of these principles.

Figure 8-3. *Data protection requirements for organizations defined in the GDPR. Source:* `https://gdpr.eu/what-is-gdpr/`

Organizations must be able to demonstrate that they are following these data protection rules, as well as additional "appropriate technical and organizational measures" in data security. Larger organizations are required to designate a data protection officer (DPO), but most smaller companies can get away without one if they still make the effort to understand and implement GDPR requirements. Additionally, organizations are required to report data breaches to both government authorities and people whose data was compromised within three days.

Full compliance with the GDPR by all companies was required in 2018, and many of the laws in the United States that have come up since have looked to it for guidance. But it's a different story in the United States, where there's no comprehensive law at the federal level, and instead there's a patchwork of laws for specific contexts and from specific states.

The first modern law in the United States was the California Consumer Privacy Act (CCPA) in 2019, which dictated regulations for any company dealing with California citizens' data. In 2023, it was amended with a new law, the California Privacy Rights Act (CPRA), which replaced CCPA. Some of the most famous components of California's laws relate to consent and the rights of citizens to opt out of data collection and access of data about them. Like with the GDPR, companies must correct inaccurate information, and there is strong liability for data breaches, along with fines.

Several US states have followed suit and passed their own laws, including Virginia, Colorado, and almost 20 other states. Unfortunately, citizens' rights are a political issue in the United States; so much of the south and other areas do not have any privacy laws.

Conflict with InfoSec

One thing that's worth mentioning is that InfoSec is often one of the most disliked departments in a company. Even though everyone knows what they're doing is important, they are usually zealous in their efforts to maintain security and privacy, to the point of sometimes getting in the way of work by other departments. Getting access to data can require jumping through a long sequence of hoops, often involving extensive documentation and multiple approvals, especially high-level leadership approval. It can sometimes take weeks, which is frustrating when you're waiting to get started on a project and it doesn't make any sense that a senior vice president who has no idea how you do your work needs to sign off on your access. Often their rules seem draconian to everyone simply trying to do their jobs, and because InfoSec doesn't understand how other departments' jobs get done, they make things difficult or impossible. A particular scenario that has always driven me crazy is forced computer updates. They will often schedule these to happen overnight on individuals' computers, assuming that people aren't using them then. But data scientists often run models or data transformations that can take many hours to finish, so running them overnight is perfect, since you kick it off before you leave for the day and it's ready for you when you get in, in the morning. Unless your company forced a system update overnight and you lost everything and will have to wait a full day to get back to it ... Still, their hearts are in the right place, and it's always best to try to be friends with InfoSec since they're more likely to listen to you if they don't view you as the enemy.

Key Takeaways and Next Up

Data security and privacy are important, and it's everyone's responsibility to pay attention and follow good practices around them. There are four main domains of security: infrastructure, network, application, and cloud. There are several standard things that companies (and some individuals) do to protect their data and systems. Data privacy is also important and is increasingly being seen as a human right. Personally identifiable information (PII) is any data that can be used to identify someone, and sensitive data is personal data that has the potential to be used to harm a person or other data that may be used to harm any other entity. There are several security and privacy compromises that people need to be on the lookout for. These include social engineering, phishing, password guessing, physical theft, viruses, worms, spyware, ransomware, denial of service attacks, and scams. Finally, we looked at some of the laws coming in regarding security and privacy.

Chapter 9 will dive into ethics, which shares some elements with security and privacy, but operates differently. I'll define ethics and how it relates to people working for an organization and with other people and organizations' data. I'll address the ways we defer to computer-based systems, thinking them less biased than people. I'll cover the idea of data science ethics oaths and then talk about frameworks and guides to performing ethical data science. Finally, I'll talk about how individuals can be part of an ethical culture.

PRACTITIONER PROFILE: DARIUS DAVIS

Name: Darius J. Davis

Job Title: Senior security engineer, Blue Team, and founder/CTO of Southside CHI Solutions

Industry: Computer security

Years of Experience: 15 years in tech from system admin to security engineer

Education:

- Self-taught

The opinions expressed here are Darius's and not of any of his employers, past or present.

Background

As a kid, Darius was curious about how everything worked. Sitting in the back of the car and watching his parents drive, he started wondering how things worked—how did the car work under the hood? How did *everything* work? That curiosity stayed with him, and he first started learning about coding and technology after wanting to figure out how video games work. He started learning code and web design and, as a teenager, found a lot of freelance work writing HTML, building forms, and doing other web work through a site called Scriptlance. Building a portfolio through this work is what enabled him to break into the corporate world in tech.

Work

Darius's first corporate job in tech was a little intimidating because he felt out of place as a young Black man from Chicago, especially since he'd gotten in without a degree, but his social skills helped him learn and get more comfortable. His early job was in IT support, but he kept growing his programming and database skills and soon moved into a role doing software development and data administration for a small company. He had a couple of good managers who helped him get more adept at working in the corporate world, and he kept growing as a software engineer and eventually led a team in platform integrations. From there, he shifted into security, where he's been for a while. His background in software and administration gave him the wide view of tech while the position has also allowed him to learn more about network security and higher-level security aspects like data protection strategies and protecting sensitive information.

Sound Bites

Favorite Parts of the Job: Darius loves fixing and solving problems, especially when it involves applying knowledge he's gained through experience. As a leader, it's as satisfying to lead a team in solving problems, even when he's not the one specifically figuring out the answer. He also enjoys working (and adapting) in an industry that's constantly changing and evolving.

Least Favorite Parts of the Job: It can be hard to get people to listen, sometimes, and you have to really work to help them understand the why of things. Also, colleagues aren't always consistent—they might have done good work when you last collaborated with them, but the next time it might be a different story. The last thing is with governance, risk, and compliance, which defines some aspect of how people should do their work to minimize security risks, and often people don't like it. Related to that is the people often resent security as a blocker or just a "cost center" because the fruits of that work are often invisible, if it's working. People should remember that not hearing or seeing much from your security department is usually a good thing.

Favorite Project: One of Darius's favorite projects was building a data ingestion and reporting service at his first programming job. The company worked with CNC machines that drilled holes in aluminum plates that were sold as "vacuum tables." These sheets had to be made to an extremely high standard of precision because they were used to handle delicate materials that required a perfectly flat surface. Darius developed a system that took in data directly from the machine's serial port, analyzed it, and generated a comprehensive "flatness" report. This helped ensure customers about the precision and quality of the plates they were buying. He loved it because it was technically challenging but also incredibly impactful to the business. He had to both solve the technical challenges and think about the customer's side of things, and it opened his eyes to the value of supporting business goals and solving problems.

Skills Used Most: In order: problem-solving, communication, and technical skills. Communication is important because a lot of people in tech can be very siloed and can lose the bigger perspective. Social skills in general are key and can often be what separate successful people from others. Learning to speak professionally and diplomatically has been important to Darius's career.

Primary Tools Used Currently: Python for automating tasks; several security tools like CrowdStrike, Prowler, Kali Linux; plus intelligence from Discord and DarkWeb resources

Future of Data Science: The future of AI is great from a scientific perspective, but not necessarily for humanity. Now that we've put AI in the hands of regular people, we have to watch for the consequences. We have AI, and people are already inspired to go and do bad things with it. We need to pay attention and start focusing on training people, especially young people, to be responsible with it. Also, it's going to start feeding on itself, like AI training on AI-generated content. Companies and developers should be careful about what is released to the public.

What Makes a Good Tech Worker: Problem-solving and communication. Obviously technical skills are important, but those vary role to role, while problem-solving and communication are critical across the board.

His Tip for Any Prospective Tech Worker: Everyone needs a portfolio. This doesn't have to be GitHub with loads of code—it can be and that's great, but it's not a requirement. In your portfolio, make sure you're focusing on what you learned and how and also how you would apply that knowledge to the organization you're trying to work for. Although it is important to learn technologies relevant to your field, you don't want to focus *too* much on the tech itself—instead focus more on how you would use it to solve business problems. But when you do pick technologies to learn, do your best to make sure it or similar tools really are used in the area you want to go into, or it can be wasted effort.

Darius is a security engineer with a software engineering background and wide experience in tech, and he's also a cofounder of the Chicago-based security consulting company, Southside CHI Solutions.

CHAPTER 9

What's Fair and Right: Ethical Considerations

Introduction

Data science is an exciting field, and it's easy to get caught up in thinking about what you can find in the data and coming up with helpful predictions or an easy way to label something that used to be done by hand. But anything that can have a positive impact can also have unintended consequences, which may be negative, so it's important to anticipate those and make sure the work should still be done and, if so, how. Imagining and understanding unintended negative consequences can be difficult, however. It's important to have a systematic way of identifying potential impacts and either mitigating them or abandoning a problematic project.

This chapter will first cover a couple of examples of poorly implemented data science negatively affecting people's lives. But then we'll take a step back and talk about what the term "ethics" really means and why it matters. I'll address the challenges of balancing ethics and

© Kelly P. Vincent 2025
K. P. Vincent, *A Friendly Guide to Data Science*, Friendly Guides to Technology,
https://doi.org/10.1007/979-8-8688-1169-2_9

usefulness in data science models, because these often are inversely related. I'll introduce the concept of an oath that data scientists can take to promise to try to be ethical, and we'll look at a couple examples. I'll talk about some frameworks and guidelines that exist to help us do ethical data science. Finally, I'll talk about how data scientists can make sure they're part of an ethical culture.

Examples of Data Ethics in the Real World

A lot of data science—maybe even most of it—deals with people in some way. Much is relatively low stakes, such as finding patterns in people's shopping behavior to send them advertising more likely to lead them to purchase something. But some deals with things that are not low stakes, like people's jobs, health, or legal consequences. The first example we'll look at relates to an automated system that influences the future of defendants convicted of crime, and the second will look at a system that automatically rated teachers to determine whether to fire them or not. One of the major problems with both of these systems is the lack of transparency. It's not possible to definitively say why a person received the particular score they got in either system.

Example 1: Racist Recidivism Prediction

A 2016 ProPublica expose of a tool used in Florida to help judges determine sentencing revealed that the tool was heavily biased against Black people.[1] A for-profit company named Northpointe created the tool,

---

[1] "Machine Bias: There's software used across the country to predict future criminals. And it's biased against blacks" by Julia Angwin, Jeff Larson, Surya Mattu, and Lauren Kirchner, ProPublica, May 23, 2016, https://www.propublica.org/article/machine-bias-risk-assessments-in-criminal-sentencing

which specifically predicts the likelihood that a given defendant in a criminal case will reoffend within two years after the particular crime they are charged with. This information is given to judges, who use it along with recommendations from prosecutors to determine sentencing. ProPublica did a study of the results of the tool as used in Broward County, Florida, comparing them with actual reoffense rates and finding major problems. The main one can be summarized in Table 9-1, which shows that the algorithm was tougher on Black people than it should have been and easier on white people.

Table 9-1. *Rates showing racial bias from Northpointe's system*

		All	White	Black
Labeled Higher Risk	**Reoffended**	67.6%	76.5%	55.1%
	Did not reoffend	32.4%	23.5%	44.9%
Labeled Lower Risk	**Reoffended**	37.4%	47.7%	28.0%
	Did not reoffend	62.6%	52.3%	72.0%

The algorithm got it very wrong. Rows 1 and 4 are where it got things "right" (matching reoffenders with a high-risk score and non-reoffenders with a low-risk score), so percentages in the 60s for everyone together among those given higher risk scores and lower risk scores, and no glaring alarm bells are raised. However, that's still a pretty high error rate—it's getting it wrong about a third of the time. You have to wonder if we should really be using a tool that's wrong a third of the time to contribute to decisions about someone's liberty.

Setting that question aside, it's when you break the numbers down by race that you suddenly see a massive, discriminatory difference. Look closely at the middle rows. Among white people who were predicted to reoffend, less than a quarter did not reoffend, but among Black people, that number was close to 45%—barely different from random guessing.

Additionally, half the time, the white people it predicted would not reoffend end up reoffending, while that figure for Black people was only 28%. The model behaves like it's hesitant to label white people as higher risk, with no such qualms with Black people. This means that Black people are likely getting much harsher sentences—staying in prison longer—than they deserve, while many hardened criminals with the luck of having white skin are likely being released into the world quickly, only to reoffend. No responsible data scientist would consider error rates like this acceptable.

It is worth mentioning that the idea behind automated risk scoring of defendants isn't inherently a bad one—it actually has a lot of potential to remove bias because before systems like this, humans were making these decisions based on gut feelings and their own biases. If they're too lenient, dangerous criminals will be released too early and commit more crimes. If they're too harsh, people with relatively minor offenses will be sent to prison for longer than necessary, at unnecessary expense and risk of hardening these people into more dangerous criminals. Obviously, you want an automated system to be better than faulty humans, but it takes a great deal of careful work to ensure that.

We don't know exactly how Northpointe built their system, but they had a questionnaire with 137 questions that the defendants answered or that court staff filled out from their records. They would have had data on defendants, with information about their cases and subsequent reoffending (or not). We assume they used a subset of this data to train the model and then tested on parts of the data they hadn't used in training. We'll talk more about predictive modeling in Chapter 15, but this is how a predictive model is built—you train with some of the data and hold some out for testing to make sure you're getting a decent level of accuracy. Now that the model is trained, it can be used on other data. We don't know what kind of testing they did for bias (if any). Race itself wasn't included in the questionnaire—this is generally illegal—so it wasn't in the model. But we know that what are called proxy variables (ones that behave a lot like race does in a model) were present. Sometimes data scientists are so confident in the fairness of their models that they don't test for bias. But if you don't do this testing, you'll have no idea what sort of bias has been captured in your model. We don't know if they looked at the results by race, but if they did, they would have seen the significant differences between Black and white people's scores.

So if race wasn't in the model, how did it perform so differently on Black and white people? Even though the questionnaire didn't ask about race, it did contain many questions that social scientists know are proxies for race. A sample questionnaire is shared online.[2] Most of the questions seem reasonable, asking about criminal history—types of crimes, number of arrests, and whether the defendant was impaired by drugs or alcohol at the time of the current crime. But then there are the potentially damning and unfair questions. One asks the age the defendant was when their parents separated (if they did). Another asks how many of the defendant's friends or acquaintances have ever been arrested, and one more asks how often they see their family. Then there's one asking how many times the defendant has moved in the last year. It's hard not to notice that some of the questions address issues that would not be admissible in court cases because they are not relevant to the case at hand. And yet, this information is used to influence decisions about their futures.

There are actually many points in the use of Northpointe's tool that raise ethical concerns. The first is that several states, including Wisconsin, New York, and Florida, started using the tool without validating the results. They basically trusted the tool, probably assuming that automation is inherently less biased than humans. But as we've seen, Northpointe did not do a lot of their own validation. They did do some validation that found that their recidivism scoring was around 68% accurate. This is

---

[2] "Sample-COMPAS-Risk-Assessment-COMPAS-"CORE"" at https://www.documentcloud.org/documents/2702103-Sample-Risk-Assessment-COMPAS-CORE.html

closer to 50%—what random guessing would give us—than it is to 90%. In their study, they did not dig into the racial discrepancies any more than determining that it was slightly more accurate on white men than Black men (69% vs. 67%), even though it's not clear how they got those numbers.

In the end, in my view, the most unethical behavior was Northpointe's response to the ProPublica analysis. They simply claimed that ProPublica's work was flawed and there was no bias. They responded to the ProPublica article by saying, "Northpointe does not agree that the results of your analysis, or the claims being made based upon that analysis, are correct or that they accurately reflect the outcomes from the application of the model."[3] When your tool is being used to impact people's lives, it's your responsibility to make sure it's accurate.

If you do want to dig into how ProPublica did the analysis, you can. They have a page dedicated to explaining their analysis,[4] which includes a link to their GitHub repository that has the R code they used.

Example 2: An Opaque and Faulty Teacher Evaluation System

Like the first example, this one starts off with good intentions: to improve student performance at a Washington, D.C. school district. The district had some very bad results for quite some time, with only half of ninth graders making it to graduation, and low scores on standardized tests across the board. The district hired someone to fix the problem. The administration

---

[3] "Machine Bias: There's software used across the country to predict future criminals. And it's biased against blacks" by Julia Angwin, Jeff Larson, Surya Mattu, and Lauren Kirchner, ProPublica, May 23, 2016, https://www.propublica.org/article/machine-bias-risk-assessments-in-criminal-sentencing

[4] "How We Analyzed the COMPAS Recidivism Algorithm" by Jeff Larson, Surya Mattu, Lauren Kirchner, and Julia Angwin, ProPublica, May 23, 2016, https://www.propublica.org/article/how-we-analyzed-the-compas-recidivism-algorithm

assumed that it was low-quality teaching that was the problem, so they wanted to get rid of low-performing teachers. They created an automated teacher scoring system called IMPACT and planned to fire the teachers with the lowest scores. IMPACT was supposed to measure a teacher's effectiveness by looking at their students' scores on standardized tests in math and language arts, an approach they called a value-added model.

In Cathy O'Neil's book *Weapons of Math Destruction*, she talks through the example of one middle school teacher, Sarah Wysocki, who was relatively new but getting good reviews from her students' parents and the principal. She received a terrible score from IMPACT that was low enough that even when her other, positive reviews were brought into consideration, she was still below the cutoff threshold and was fired. She tried to find out why her score was so low, but the system was a "black box," and no one could tell her how it had calculated her score.

There are several problems with the approach taken with IMPACT and the district. As we learned in the chapters on statistics, sample size is hugely important. The more complex the data is—the more variables affecting it—the bigger the sample should be, in general. There are so many factors that affect an individual student's performance in a given academic year—family trouble, poverty, illness, bullying—that laying the responsibility for a student's performance entirely on the teacher doesn't really make sense. But the district seemed to think that measuring that for all 25–30 students of a given teacher should average out. But it doesn't. A sample size of only 30 is woefully small for such a measure, and it was irresponsible for the developers of IMPACT to ignore that.

Another problem with IMPACT is it was a one-shot system—they trained it once and didn't take error into consideration to tweak the model. Feedback is crucial to ensure a system isn't blindly spitting out junk results. Without knowing if the dozens of teachers who were fired were really bad teachers, there's no way to know if the system is accurate. O'Neil sums it up succinctly in the book: "Washington school district's value-added model ... [defines] their own reality and use it to justify their results."

As she says, this self-perpetuating type of system is extremely destructive to humans who deserve to be treated fairly and are instead being arbitrarily targeted for punishment based on a made-up and never-validated system.

BLACK BOX SYSTEM

A black box system operates without revealing how it works. As mentioned previously, some machine learning approaches give forecasts or labels that can be fully explained—the garden department's sales is forecasted to be this amount because it's summer, Friday, three days before a holiday, and there's a sale on lawn furniture. Those are four of the key values that the model uses to compute a forecast. Common algorithms that are used when this kind of visibility is important are linear regression and decision trees. But there are also a lot of algorithms that don't give you this kind of transparency, so we wouldn't be able to tell you why this department's forecast was $5,500 and this other one was $2,100. Any system that generates opaque results is called a black box.

What Is Ethics?

Most people know what someone means when they say the word "ethics," but for the purposes of this chapter, I'm going to define it because it doesn't always mean exactly the same to everyone. I'm also going to define some other terms that come up in the discussion of ethics, bias and fairness.

Definitions

I'll start with the most general definitions, from the dictionary. Merriam-Webster defines "ethics" as "a set of moral principles : a theory or system of moral values."[5] Since that definition leans heavily on the word "morals," it's helpful to see that Merriam-Webster defines "moral" as "of or relating to principles of right and wrong in behavior."[6] We're still fairly abstract with the reference to principles, but clearly, ethics is about establishing an understanding of right and wrong and preferring what's right.

When considering ethics as it relates to data science, we are talking about understanding whether the choices and actions we make result in consequences that are right or wrong, where wrong things lead to harm of people, animals, or even some organizations, whether directly or not. That harm can take many forms, but the goal of thinking about ethics in data science is about avoiding that harm altogether.

Human Bias

"Bias" is another word that is used a lot when discussing ethics in many fields, including data science, but it also has an additional technical usage in data science related to evaluating a model that we'll address in Chapter 15. Keeping with the ethics-related meaning, Merriam-Webster says "bias" is "an inclination of temperament or outlook," especially "a personal or sometimes unreasoned judgment."[7] Both parts of this definition are worth understanding—the first reminds us that everyone has a viewpoint (this is human nature), and the second reminds us that our perspectives aren't always fair to everyone. Human bias isn't a huge problem when we are fully aware of it because we can compensate for

---

[5] https://www.merriam-webster.com/dictionary/ethics
[6] https://www.merriam-webster.com/dictionary/moral
[7] https://www.merriam-webster.com/dictionary/bias

it to strive for fairness. But when we aren't aware, dangerous things can be embedded in data science systems and solutions that we build. Unfortunately, *unconscious bias* is incredibly common, when we have an internalized bias but do not think we are biased.

Fairness

Fairness was mentioned just above when I was discussing bias because one of the fundamental problems with bias is that it results in different groups being treated differently, some better than others, which is clearly unfair. The Merriam-Webster definition of "fairness" is clear on this: "fair or impartial treatment : lack of favoritism toward one side or another."[8] This is an intuitive definition. Unfair data science benefits one or more groups while doing nothing for—or even harming—another group or groups.

Why Ethics Matters

The short answer to why ethics matters in data science is people. The meanings I discuss above are important. In data science, we should always strive to avoid bias to achieve fairness to all people impacted by our systems and solutions, which would mean we are behaving ethically.

But why do we really need to behave ethically? One obvious reason is that we want to do the right thing—most people believe that behaving ethically in general is something good people do.

But most data science in the world is done by people who work for organizations. They may be less inclined to do the right thing for the sake of their employers. Employees often feel absolved from personal

---

[8] https://www.merriam-webster.com/dictionary/fairness

responsibility. If an employee's personal beliefs aren't enough to make them want to keep ethics top of mind, there are also more tangible reasons. Behavior by an organization that is deemed unethical by the public can have huge reputational and even financial impacts on them, as often things that start as reputational problems swiftly turn into financial problems. These kinds of things can be very impactful at large companies, but missteps by startups and young companies can be devastating. And don't forget that it can also have personal impacts to the irresponsible parties—remember the AOL people who lost their jobs after the insufficiently anonymized Internet search data was released.

At this point it sounds like the main reason to do data science ethically is because not doing it that way can get you in trouble. But the truth is that doing ethical data science can actually lead to better results, which can lead to greater business value. This may seem counterintuitive, but if data scientists are biased, that usually translates to them making invalid assumptions or forgetting certain things about the real world that they're representing in their data.

Imagine data scientists working on a diagnostic heart health system. They have data from many sources that they've incorporated into the system, including data from decades of research studies. Sounds good, right? But it's well-known that research on heart health was historically done only on men, for decades, until fairly recently. So the data scientists' system would be much better at diagnosing heart problems in men because it has so much more information about heart health in men, especially because women's heart attacks tend to manifest and play out differently from men's.

But they're stuck, right? It's not their fault there's not as much research on women. They could say their hands are tied and leave the world with a system that is biased because of decades of human bias embedded in the data—or they could make the more ethical decision to see if they can find additional data on women that they hadn't originally considered. Behaving ethically often involves thinking outside the box, and it requires paying attention.

The Balance Between Convenience, Utility, and Ethical Concerns

If there are so many good reasons to do ethical data science, why is it even a topic of discussion—why isn't everybody doing it? It's because it's often hard to do, it can be difficult to know what is ethical, and also there is a tradeoff both between convenience and ethics and between utility and ethics. In other words, sometimes being ethical means having models that take more work or even appear less accurate than when ethics is ignored.

The issue in the previous example of a heart health diagnostic system is that it's inconvenient, time-consuming, and potentially expensive to find the additional data that they hadn't originally counted on needing. On the other hand, not attempting to improve a model that we know underperforms on a subset of users is easy, but unethical.

Apparent utility is at issue in other cases. A lot of data can be used in models, but much of the data related to people is considered sensitive. But frequently this sensitive data actually appears to increase the performance of a model when it's included. In some areas like banking, it's illegal to include some of this info (gender and race in particular), but when there are no laws dictating it, it's up to the data scientists to decide what to include. They need to consider the downsides of including sensitive data and see if it causes any ethical problems.

The ethics of data science work can rarely be considered in isolation. There are cases where using gender would likely be relatively harmless— think in a retail store's recommendation system. There are certain products that more women buy than men and vice versa. And yet, remember the mailer Target sent to women they thought were pregnant, unintentionally outing a pregnant teen girl to her angry father, obviously not the right thing for a corporation to be doing. Now, picture a country with a government carrying out ethnic cleansing, which can also demand access to any company's data within its borders. That company may proactively consider whether they should collect customers' ethnic identities along with other data. Perhaps they know it would help their model, but if they don't collect it, the government can never demand they share it. Things are complex and the big picture is important.

False Utility, Perpetuating Bias, Technochauvinism

Many people continue to believe that data and machine learning can't be unfair, denying the role that human and systemic biases have because they're built into the data and the systems we use. This attitude has been labeled *technochauvinism* by Meredith Broussard,[9] encapsulating a belief that a technological solution will always be the best. We have to remember that in actuality, any data science system is only as good as the data put into it—something that is done by humans.

This means that just because a data science system "performs well" doesn't mean it is without bias. When you train on past data, your "well-performing" system is simply going to reproduce the same decisions that have always been made, and if those are biased, the system will be, as well. In the recidivism example above, some of the questions on the company's questionnaire were known to be proxies for race, but they publicly acknowledged that the model was more accurate with them. We don't know exactly how they tested their model, but we know that in the US justice system, Black people have always been given harsher sentences than white people for the same crimes, on average. And we also know that harsh sentences can "harden criminals," or turn less dangerous people into career criminals. This happens across all races. If the data the company

---

[9] She discusses this extensively in her book *More Than a Glitch: Confronting Race, Gender, and Ability Bias in Tech*, The MIT Press, 2023

was using for training had this consistently harsher sentencing for Black people, then the bias exists in the data itself. The model being accurate was just recreating the biases of human judges.

So it can be the case that an apparently less performant model can be a fairer model, even if that seems counterintuitive. In another example, some companies made early attempts to apply data science techniques to inform hiring decisions. These were often found to have a racial bias, which often came with proxy variables even when race was excluded. An uninformed person might ask, What's wrong with including race if it helps make better hiring decisions? The reason is because it doesn't make better hiring decisions—it makes the same hiring decisions as before. It's making discriminatory decisions, just like those made by humans in the past. It's known that hiring has had a racial bias for a long time, so using a model trained on the past simply repeats the past. This kind of thing is complex and difficult. But it's important to remember that models only have limited information. Imagine a scenario with a model that penalizes a Black person because Black people don't stay at the particular company as long as white people—but this is because the company has a hostile, discriminatory culture that Black people leave for their own mental self-preservation, not because the Black employees are less accomplished or committed than their white counterparts. These Black candidates are just as good as the white candidates, yet they are being penalized for their race by a system that is inherently unfair.

In a similar real-world case, Amazon created an applicant resume screening program in 2014 that was discovered to be highly sexist. Gender was not included in the model, but the model used natural language processing and again picked up proxy features, some of which were patently obvious, like resumes referring to women-specific things, like women's colleges or text like "women's flag football." It lowered scores for resumes with these markers. It also favored candidates who used strong verbs like "executed" and "captured," which were more common on men's resumes. Why was it penalizing women candidates? Because there's a huge gender bias in tech, where around 75% of employees are male. The only information the screening program was given was candidate resumes and a score from 1 to 5 given by a variety of (almost all male) Amazon employees, who clearly had a strong bias against women candidates. The model was not provided with information on the actual quality of the candidate, just a score given by biased men. So they have no idea if the candidates they hired were actually the best, and they have absolutely no way to know if the candidates they didn't hire were actually not worth hiring. So it was screening out women candidates because Amazon recruiting had always done that, not because the women were lower-quality applicants.

We have to remember that while computers can be used to reduce bias, you cannot fix a biased system simply by automating that same system. Training on biased data results in a biased system. These systems are largely self-fulfilling prophecies, despite those convinced that technochauvinism is a genuinely valid viewpoint.

Data Science Codes of Ethics and Oaths

Some of the early efforts at formalizing data science ethics involved the concept of an oath, inspired by the Hippocratic Oath that all doctors commit to following. Recently, several data science (and other) organizations have created oaths they recommend data scientists follow.

Like the Hippocratic Oath, an ethical data scientist oath would define the things that are important for all data scientists to do to ensure the work they are doing is ethical. It would amount to a "do no harm" promise.

An Oath from the Association of Data Scientists

Some organizations are trying to push for licensing of data scientists, but it's obviously voluntary at this point. I've personally never seen a job description mentioning licensing in any way. Still, this self-improvement effort could pay off. The Association of Data Scientists has a Chartered Data Scientist accreditation, and they require those with it to follow a code of ethics and standards that they define. They divide the code into five standards that chartered data scientists have duties toward: their own profession, their employer, their clients, their industry, and as a charter of the Association of Data Scientists. It's solid and worth looking at.[10] One disadvantage of it is that it is focused on aspects of a data scientist's career—their discipline, whom they work for, and their industry as a whole—with less focus on the data they may be working with and the people who may be represented in that data and consequently impacted by their choices. They do address the needs of the subjects of any data they work with under their obligations to the charter they hold, but the angle they come from is about protecting the integrity of the charter.

An Oath Inspired by the Hippocratic Oath

Another perspective some people have taken is to start with the Hippocratic Oath and transform it for data scientists. A consortium of different organizations called the Committee on Envisioning the Data Science Discipline: The Undergraduate Perspective created this one. It's in

---

[10] "Ethics & Standards for Chartered Data Scientists (CDS)," The Association of Data Scientists, https://adasci.org/cds/ethical-standards-for-chartered-data-scientists-cds/

a book the National Academies Press published in 2018.[11] They match each paragraph in the Hippocratic Oath and make it work for data science. You can see the entire oath in the sidebar (numbers added).

This oath also addresses obligations toward different areas, like the Association of Data Scientists' one does, but there is a much stronger focus on obligations to people in general, rather than on duties as an employee. #1 requires data scientists to respect their profession and the work of those who came before. #2 is about both the integrity of the profession and protecting people by using data appropriately. #3 is also about integrity and calls for consistency, candor, and compassion in the work and ignoring outside influence. #4 is again about integrity, having a growth mindset, and recognizing that all data scientists can learn from others, as no one knows everything.

DATA SCIENCE OATH

From the Committee on Envisioning the Data Science Discipline: The Undergraduate Perspective:

I swear to fulfill, to the best of my ability and judgment, this covenant:

1. *I will respect the hard-won scientific gains of those data scientists in whose steps I walk and gladly share such knowledge as is mine with those who follow.*

2. *I will apply, for the benefit of society, all measures which are required, avoiding misrepresentations of data and analysis results.*

---

[11] *Data Science for Undergraduates: Opportunities and Options*, from the National Academies Press, 2018, which can be read online or downloaded for free at https://nap.nationalacademies.org/catalog/25104/data-science-for-undergraduates-opportunities-and-options

3.   *I will remember that there is art to data science as well as science and that consistency, candor, and compassion should outweigh the algorithm's precision or the interventionist's influence.*

4.   *I will not be ashamed to say, "I know not," nor will I fail to call in my colleagues when the skills of another are needed for solving a problem.*

5.   *I will respect the privacy of my data subjects, for their data are not disclosed to me that the world may know, so I will tread with care in matters of privacy and security. If it is given to me to do good with my analyses, all thanks. But it may also be within my power to do harm, and this responsibility must be faced with humbleness and awareness of my own limitations.*

6.   *I will remember that my data are not just numbers without meaning or context, but represent real people and situations, and that my work may lead to unintended societal consequences, such as inequality, poverty, and disparities due to algorithmic bias. My responsibility must consider potential consequences of my extraction of meaning from data and ensure my analyses help make better decisions.*

7.   *I will perform personalization where appropriate, but I will always look for a path to fair treatment and nondiscrimination.*

8.   *I will remember that I remain a member of society, with special obligations to all my fellow human beings, those who need help and those who don't.*

If I do not violate this oath, may I enjoy vitality and virtuosity, respected for my contributions and remembered for my leadership thereafter. May I always act to preserve the finest traditions of my calling and may I long experience the joy of helping those who can benefit from my work.

—The Data Science Oath based on the Hippocratic Oath, in the book *Data Science for Undergraduates: Opportunities and Options* from the National Academies Press (2018),[12] numbers added

Numbers 5–8 get into protecting people. #5 addresses the importance of protecting people who are represented in the data that data scientists are using, by requiring them to be careful to ensure privacy and security, and also recognizing the great responsibility and obligations that come with handling data about people. #6 reiterates the importance of treading carefully with people's data by understanding that there can be real-world consequences to the work and also that there can be biases inherent in the data. It requires data scientists to think about the consequences, whether they are obvious or not. #7 says that there's nothing inherently wrong with using individual people's data, but that data scientists are required to ensure that the use is fair and not discriminatory. And #8 reminds data scientists that they are part of a larger society of humankind and therefore have obligations to their fellow humans, who all have different needs.

The final paragraph gets back to the importance of having integrity in the profession. Good data scientists will follow these rules and therefore do no harm and instead potentially help humanity. This is a general enough oath that any data scientist could commit to following it, regardless of the particular industry they work in.

Compliance

If we were to define an oath that data scientists should follow, it seems like there would be no harm in data scientists taking it. Personally, I think everyone should. Many of the data scientists I interviewed for the profiles

---

[12] Can be read online or downloaded for free at https://nap.nationalacademies.org/catalog/25104/data-science-for-undergraduates-opportunities-and-options

in this book specifically said they thought data scientists needed to take an oath or even be licensed, without me bringing it up beyond asking how ethics was involved in their day-to-day work. But it's hard to imagine this always being effective, for a variety of reasons. There's no real obligation to abide by the oath. It relies on the honor system. An oath is generally only taken once, and then the data scientist would go on doing data science, potentially for years, never thinking about what they agreed to again. When you're working, it's easy to get tunnel vision, and following an oath requires a big-picture view and regular reflection. Additionally, oaths are somewhat abstract and not directly tied to practice and also rather general since they're not very long content-wise. Ethical quandaries often arise out of very specific situations during the work, and having taken a general oath will not necessarily help the data scientist handle such a quandary.

There's also a counterintuitive danger with oaths—they could actually help companies and people mask unethical work, intentionally or not. Imagine a company that requires all their data scientists to take this oath. People could then think, "We can't be unethical, because we all endorsed this oath." If data scientists don't take the oath very seriously, thinking about it every time they face something new in their work, they will lose sight of its intentions.

This is actually a problem with a lot of performative initiatives companies take. They'll require employees to commit to being inclusive and diverse, avoiding sexual harassment, avoiding retaliation, and generally behaving ethically, but then turn around and ignore all instances of the disallowed behavior unless they think they're at risk of being sued. It's usually every employee for themselves, which means that for oaths to work, each data scientist has to take personal responsibility for taking it seriously.

Frameworks and Guides

If oaths aren't the only answer, what else can we do? Data science is really not a field where theory plays a huge role—it's a field of application and practice. So taking a more applied approach can be a lot more effective since it's more specific where oaths aren't. The best way to make sure the data science we're doing is ethical is to use a framework based on ethical principles to generate guidelines and steps to follow during the various stages of our projects. There are a variety of frameworks for data science practice in the world, but very few contain elements to deal specifically with ethics. We're going to go over a framework, the Responsible Data Science (RDS) Framework, that does in this section. It's described in the book *Responsible Data Science* by Grant Fleming and Peter Bruce.[13]

Principles

The RDS Framework is based on five principles that are crucial to ethical data science: nonmaleficence, fairness, transparency, accountability, and privacy. To be truly ethical, a data scientist should follow all five of these. However, it has to be acknowledged that following them isn't easy, and it may not always be possible. That does not mean we should not strive to follow them.

The mouthful *nonmaleficence* simply means the practice of avoiding harm through our choices and actions. Not every bad thing is preventable, but a data scientist needs to do their best to not cause harm. Although this one is obviously critical, it can be one of the most difficult of the principles to follow perfectly in the real world, depending on the scenario in which it's being used. Imagine a system that determines which people

---

[13] *Responsible Data Science: Transparency and Fairness in Algorithms* by Grant Fleming and Peter Bruce, John Wiley & Sons, 2021

in a refugee center will have the opportunity to apply for asylum by trying to determine greatest need through a machine learning algorithm. The people who aren't selected by the system could claim they were harmed. Sometimes we're dealing with no-win situations.

Fairness simply means what we already know—the state of equality in representation, leaving people with their dignity, avoiding discrimination, and justice. This applies to the whole data science process, but most importantly to outcomes. There are a lot of factors that can harm people if they aren't considered for fairness, including ethnicity, gender, social class, and disability status (among many others). Because there can be so many dimensions to fairness, it can be hard to ensure everything has been considered. Additionally, some people don't think certain factors deserve consideration in fairness, like sexual orientation or gender identity nowadays. Additionally, fairness doesn't have to be about characteristics of people (although it usually is). For instance, we might want to select final exam questions from a list, and we'd want to cover a range of topics, not have several questions related to one topic. Like shown in that case, fairness can be thought about as a means of achieving balance, and it's still important to always try for it, despite the challenges. Frequently, testing for fairness is done after outcomes are generated. The results are broken down in different groupings and compared (for instance, results for women vs. results for men). This should be done even when those factors (gender in this case) aren't used in the modeling itself. This is often how proxy variables for protected classes are found before they're released into the world like Northpointe's recidivism risk scoring system. It can't be overstated how important it is to look for surprises here. There's a phenomenon in statistics called Simpson's paradox, where when you look at the data all together, one trend is seen, but when you break it into subgroups, the complete opposite trend is seen for the subgroups. You cannot know in advance how things will break down.

Transparency is another critical principle, one we've talked about. *Transparency* in data science generally means that everything in the data science process can be explained, especially particular outcomes from a model. In both the teacher performance scoring and the recidivism risk scoring, the models were not transparent at all to those receiving the scores. Model transparency generally requires an algorithm that provides the features used to generate an outcome, although there are also some tools called *model explainers* that can be used to determine the most impactful features for a given outcome when the algorithm itself doesn't provide the details (but these don't work in all cases). As mentioned, it's not only the outcomes that should be transparent. Other elements of the data science process should also be clear, which is why documentation is important for the different steps during a project.

Accountability is the fourth critical principle, and it has a variety of implications related to data scientists following proper practices and standing behind their work. Part of it relates to compliance with laws and other applicable rules—a data scientist lying about not following them is not being accountable. It also relates to behaving with the integrity expected of a data science practitioner. Another area of accountability relates to people impacted by the outcome. Data scientists should be able to respond to user questions, and users should have options when receiving a harmful or negative outcome. At the moment, this is a fairly lofty principle, and there are few if any instances of accountability other than to follow laws and rules. A lot of people have experienced this problem with social media, being falsely flagged by an algorithm for a violation of terms of service and finding how difficult it is to get it remedied. Pre-Musk Twitter once claimed I violated their terms of service for posting a link to an article that I'd written about how harmful zoom-bombing is, which obviously doesn't violate their terms of service. Even their supposedly manual appeal process didn't fix the problem, as they responded saying that I had violated them, and that was it. There was nothing I could do except to delete the tweet to have my account reinstated. Things like loan or job application rejections are even less likely to have any recourse.

The final RDS Framework principle is privacy, which we've already discussed extensively in the previous chapter. But here privacy basically means we've not collected personal data we don't need, we've deleted any personal data we needed only temporarily, we've stored what we do have with appropriate security, we've made sure that sensitive data is removed before model training, and the data is appropriately protected so sensitive data can't be inferred after the fact.

Stages

The RDS Framework outlines how the above five principles can be put into practice in real data science projects. The authors describe it as a "best practices framework" with five core stages, and it can be seen in Figure 9-1. Note that most stages have some documentation that needs to be done, usually called deliverables in business.

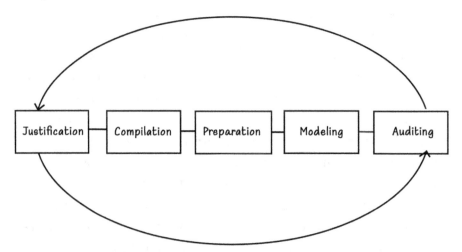

Figure 9-1. *The flow of the RDS Framework*

The framework starts with justification. *Justification* here is basically requirements gathering and defining the problem in the business understanding step in CRISP-DM. Considerations of likely technical solutions are also done here, and data sources would start being identified. It's important to consider ethical ramifications here. The authors of the framework recommend generating an impact statement at this point, which would formally consider and document the potential ethical concerns or consequences, with outside help engaged if the team members don't have sufficient knowledge. This is also when other similar projects might be looked at, which is one more reason documentation is so valuable.

The second stage is *compilation*, which is when all the materials necessary for the completion of the project are gathered, from the data itself to the tools that will be used in the solution. It can also involve other pieces that might not be obvious, like the metrics that will be used to measure performance, the algorithms, the data dictionaries, and even a project development approach and a project task management tool. There are a couple deliverables that should come out of this stage. The first is a document addressing the data itself more deeply than the data dictionary, including information about how it was gathered, how it's been processed, and how it will be used. This is generally targeted to nontechnical people and should address issues and especially ethical concerns. Note that in practice it can be hard to know some of this, especially how it was collected and processed, but the idea is to make the best effort possible. A second document that's for technical audiences is audit reports that go deep into the issues raised during the compilation step in the impact statement. Focus should be on ensuring fairness (and how that will be done).

The third stage is preparation, and this lines up with the data preparation step in the data science process. In *preparation*, we do any necessary data cleaning, feature engineering, and feature selection. It also includes the exploratory data analysis step in the data science process because in order to prep and create and select features, we have to have an understanding of the data. Rather than producing any brand-new documentation in this step, we should update prior datasheets as necessary (to log the data cleaning and any new features created and address how it impacts ethical considerations).

The next stage is the modeling itself. It can be useful to predefine performance metric values here as a baseline (lowest acceptable) and an optimistic high value. The final decision about what kind of algorithms will be used happens here, when the transparency of the ones being looked at is considered. It's also common to run each algorithm multiple ways with different hyperparameter values (this will be covered more in Chapter 15), and it's recommended that the performance of the algorithms with various combinations is logged and saved. Other things that should be documented include the rationale for selecting the final algorithm used and a discussion of the balance between performance and explainability (if a black box algorithm is chosen because of higher performance or vice versa).

The final stage is auditing. *Auditing* is when we take a step back and look at everything we did along the way. Note that if anything concerning is found, we may have to go back to the beginning and work through all the stages again. This is where checks would be done for bias on protected groups and other special groupings (like whether the model performs similarly on men and women or on white people and people of color). It's also important to make sure that suitable explanations of the model can be provided to stakeholders and that if this isn't the first audit, all previous issues raised during the other audit(s) have been solved. All of these things should be documented in each audit.

Being Part of an Ethical Culture

The most important factor in having a culture of ethical practice among data scientists is for the individual data scientists to care about ethics. Some could mandate the RDS Framework be followed, but if people don't take the various steps seriously—they might only give a cursory thought to most ethical concerns and create an insufficient impact statement in the justification stage—nothing is gained by following it. The datasheets and audit reports could also be superficial and insufficient. An outsider reviewing this documentation might not know that there's a lot left unconsidered. So, for the framework to help, people have to follow it in earnest.

It's unfortunately human nature to minimize the amount of work we have to do, and searching for ethical issues is work. How can we persuade people that it matters, that it's worth doing? I think the most convincing things are ethics horror stories. Personal stories always engage and convince people more than abstract ideas, especially when they can connect with those at the heart of the story. A similar approach is to have an open conversation among data scientists about the ethics issues they're finding and mitigating in their own projects. People learn from each other really well.

There are many things practitioners can do to minimize ethical problems, including focus on using algorithms that are explainable either because of inherent transparency or via model explainers that can be applied after the fact. On a more individual level, even one person can start a shift toward ethical data science. Even if the team isn't ready to start following the RDS Framework, one person can make a habit of bringing up ethical considerations at different stages of the process, and eventually some other people will start following suit. One thing worth mentioning is that truly following the RDS Framework will slow projects down a little, so it's unlikely any single person can follow the framework alone, as it will take them longer than other people to complete projects, which likely won't be acceptable to the team leadership. But then again, if the leaders can be convinced, it might be possible to get the whole team following it.

Key Takeaways and Next Up

This chapter dove into ethics in data science. I first looked at a couple of disturbing examples of the harm that can be done when data science systems aren't created ethically. Ethics, bias, and fairness were defined, and then we looked at why ethics matters, even though it is rarely considered during most data science projects. Usefulness, or utility, of data science systems is often seen in opposition to ethical data science. There is some truth that sometimes being cautious on ethical concerns can result in a less useful—or less performant—system, but often a model that appears better performing is simply doing a better job of perpetuating bias from the real world. I discussed the idea of data scientists taking an oath for ethical data science based on the Hippocratic Oath and the weaknesses of that as the primary effort for ethical data science. Instead, a more comprehensive framework called the RDS Framework was presented to ensure more ethical data science through integration into the entire data science process. Finally, I talked about what it takes to be part of and grow an ethical data science culture.

This is the final chapter in Part I of the book, where I've laid all the fundamentals needed to practice data science. Now you're ready to dive into the most important topics for practicing data science. We'll be starting by looking at the importance of domain knowledge.

PRACTITIONER PROFILE: HARVEY SCHINKAL

Name: Harvey Schinkal

Job Title: Senior data scientist

Industry: Consulting in various industries

Years of Experience: 3 as a data scientist, prior technical work

Education:

- Doctor of Technology (current student)

- MBA in Leadership

- BAS Medical Administration

- AAS Global Health

The opinions expressed here are Harvey's and not any of his employers', past or present.

Background

Harvey Schinkal started off working retail but knew he didn't want to do that forever. While an assistant manager at a home improvement store, he learned about business and reporting, which led him to try for something new. He worked at a couple different IT roles and realized he needed more formal education to accomplish his goal of working at a large company. He entered the military with the intent of using the GI Bill when he was done. After the military, he was interested in healthcare. He earned an associate's in Global Health and a bachelor's in Medical Administration, but then changed courses again after needing to find a job as the GI Bill started running out. He ended up back in tech and then really wanted to pursue data science. He found

Purdue University's Doctor of Technology program. It required a master's first, so he got an MBA and then started the doctorate. He's focusing on ethics and machine learning for the degree.

Work

Harvey has continued consulting in data science in various industries while working on the doctorate. As mentioned above, his work experience has been very varied, but most of it has been in IT in one form or another. After leaving the military, he got a consulting gig doing data visualization and learned that the tools had changed since he'd joined. He worked in Tableau and Power BI and learned a lot about visualization. He also worked in some other tech roles before being on some data science projects, and that really caught his interest. He read a few books about AI and data science, including Pedro Domingos's *The Master Algorithm*, which inspired him to go deeper in the field and led to where he is today.

Sound Bites

Favorite Parts of the Job: Harvey loves working with people in general, because they make the job interesting and rewarding, often even more than the tech work. He especially loves working with people he can learn from. He's found that it's not always someone more senior whom you can learn from.

Least Favorite Parts of the Job: Gathering requirements and working with nontechnical stakeholders can be difficult because sometimes people have unreasonable expectations about what's possible. Sometimes they aren't really willing to listen to the reality of what's possible.

Favorite Project: One memorable project was a customer retention system. When a customer said they were leaving this company, the customer service representatives used a system Harvey worked on that indicated whether that customer was retainable or not and helped determine which reward to offer them to keep them from leaving. There were actually two dozen models running at the end, and he'd created several of them from scratch and refreshed the others that were running already. This generated tens of millions of dollars from customers being retained.

Skills Used Most: Database programming, general programming, and visualization are the obvious ones. But also, good data instincts are important. For instance, you should be able to recognize when your model is overfitting or understand that some findings aren't actually that helpful (like that more women buy feminine hygiene products, for instance). Data detective skills are important and basically amount to asking good why questions. Domain knowledge and understanding how data is collected are also important. Finally, communication is huge.

Primary Tools Used Currently: SQL, R, Python, Tableau, Power BI

Future of Data Science: Harvey sees that there's still a growing need for data scientists, but with AI tools, the need isn't as high as it might be otherwise. AI is helping us increase productivity in data science by 10–15%, but it can't fully replace data scientists or software engineers.

What Makes a Good Data Scientist: Intellectual curiosity, strong interpersonal skills, technical aptitude, willingness to learn new tech, and flexibility are the most important specific skills, but soft skills in general are huge. It's also important to understand how algorithms work, even if you don't know the exact math behind them. This last point is relevant to the push at a lot of companies to provide self-service analytics tools to less technical people, because these people often don't have even the basic understanding of how these tools work, and data scientists might need to coach them.

His Tip for Prospective Data Scientists: Consulting and small companies are a great way to get some experience. Consulting can let you get a taste for different industries, so it can be valuable even if you don't think you want to do it forever. Also, internships can be worth doing before or even after you graduate, just to get some real-world experience. It's also good to read more general nonfiction books about AI and data science. Some classics include

Pedro Domingos's *The Master Algorithm*, Brian Christian's *The Alignment Problem*, Nate Silver's *The Signal and the Noise*, and Nassim Taleb's *The Black Swan*. These give you a good sense for the wider world of data science—who's using it, how it's being used, and what it can do.

Harvey is a data scientist with experience in a variety of tech roles and industries.

PART II

Doing Data Science

CHAPTER 10

Grasping the Big Picture: Domain Knowledge

Introduction

I've already mentioned domain knowledge several times in this book because it is critically important in data science. As a reminder, *domain knowledge* is knowledge that's specialized in some way, the opposite of general knowledge. People often claim domain knowledge of an industry or a field, but someone could be a domain expert in almost anything—healthcare, the Spanish language, video game playing, seventeenth-century horsemanship, the cozy fantasy genre, farm animal management, RC aircraft, the Pokemon card game, and pretty much anything you can imagine, whether it's a profession, a hobby, personal interests, life experience, or something else. You almost definitely have domain knowledge in several things, even if it might not be at expert level.

Okay, so you have domain knowledge. What's the point? In life, sometimes it's just fun to know about something you find interesting or enjoyable. But there are many times when such knowledge can be

© Kelly P. Vincent 2025
K. P. Vincent, *A Friendly Guide to Data Science*, Friendly Guides to Technology,
https://doi.org/10.1007/979-8-8688-1169-2_10

invaluable, even when it comes from only a hobby or passion. For a video game company wanting to learn more about their products and competitors, having an understanding of how video game players behave around games is helpful. A pizza restaurant would be better able to look at their data if they understand customer behavior and restaurant sales patterns. Not understanding that there are busier times of the day and differences between weekdays and weekends would make it harder to know how to break things down.

Examples of Domain Knowledge in the Real World

Almost every job involves domain knowledge to some degree, and it's especially important in the tech world, where technical people are building systems (of all types, not just data science ones) for nontechnical people in different domains. It can be especially important in business understanding and requirements gathering, and while you can pick up a lot of it on the job, not already having it can have negative consequences. We'll first look at an example where unique domain knowledge enabled a new business and then at another where insufficient domain knowledge contributed to a project failure.

Example 1: Clothing the Masses

The company Stitch Fix sells customized fashion to shoppers looking to get clothing suited to them with minimal effort. They sell boxes with five clothing or accessory items that are selected for each customer by an automated recommendation system and then manually tweaked by company stylists. It's an unusual idea, and the company's early success hinged entirely on the founder and CEO Katrina Lake's understanding of women shopping for fashion—which is domain knowledge. She points out that most online fashion companies try to differentiate themselves with faster shipping or lower prices, where she went for personalization.[1] Early on, she had trouble raising capital for the company because most venture capitalists are men who don't understand the clothing shopping world the way Lake does. The key differentiator was the personal element combined with the ability of an algorithm to consider many more options than an individual human could. The data science is based on the results of a questionnaire each customer fills out when they sign up plus feedback they give after each box. The algorithm creates a unique combination of five items, but then a stylist evaluates it, modifies it if necessary, and gives it the stamp of approval. What the data science does is make the entire enterprise scalable—it would be impossible for stylists to build a box from scratch, but if all they're doing is tweaking pre-selected items, it's far less time-consuming. The success of the concept was based entirely on Lake knowing how important the human element is in clothing shopping.

Example 2: Chickening Out

I worked on a project that wasn't successful largely because we didn't really have enough domain knowledge to design the right solution. It was also kind of a requirements gathering failure, but had we had more domain knowledge, we would have realized the solution we came up with wasn't really what the stakeholder needed. We were tasked with forecasting sales of Rotisserie chickens at several stores. We took the project over from an outside vendor who'd worked on it for several months and was forecasting at the daily level. When we took over, we started by looking at the data and also visited one of the delis where the chickens were cooked and displayed, and that gave us a lot of information about what that part of the business was like. Because we were able to leverage code from another

---

[1] "Stitch Fix's CEO on Selling Personal Style to the Mass Market" by Katrina Lake, in HBR's *10 Must Reads on AI, Analytics, and the New Machine Age*, 2019

project, we had a good model giving highly accurate forecasts within three weeks. But the project still failed, partially because they had unreasonable expectations for accuracy (even though we outperformed both their manual forecasts and industry standard accuracies, they wanted higher accuracy), but also because we didn't understand something important about what they needed (and they didn't understand what we could do well enough to ask for it).

Despite the visit, we still didn't fully appreciate the timing of everything. Chickens had to first be cooked, and they had a limited shelf life, so after a period of time, those that didn't sell had to be thrown out. Obviously, they couldn't just cook them all in the morning and leave them out all day. What they really needed was an hourly forecast. But because we had taken the project over from a third party who had been doing daily forecasts rather than hourly, we just proceeded with that approach. We were pleased with our quick results and had a nice visualization that showed the ongoing forecasts, accuracy, and a view of the specific features that went into each daily forecast. But in the end, the customer lost interest. I still think this is partially because we didn't actually give them what they needed—which was hourly forecasts. It cannot be emphasized enough how important it is to understand customers and their needs, and understanding their domain is the best way to do this.

What Is a Domain

I said above that a domain can be virtually anything. It can be a technical field—like data science, generative AI, software development—or a subject like healthcare or video games. Obviously, data scientists should have expertise in the data science field, but experience in other fields can also be valuable. I've especially found that my background in software development has been useful in data science, as a lot of data scientists learn less rigorous programming and have knowledge restricted to the code libraries used primarily in data science, and wider knowledge can help when it comes to more general programming, design, and best practices. This is especially useful with productionization.

It's often most useful to think of domains as being subjects rather than fields. In the business world, some of the most common domains that data scientists work in include retail, insurance, banking, and healthcare. A lot of the time, job listings will ask for experience in the company's domain, which is simply because there are little things that you start to understand when you work in one of these domains. Different areas have radically different data. Healthcare has a ton of data on patients, which is obviously going to be highly sensitive. Depending on what part of healthcare the role is in, there could be other types of data, including related to insurance, diagnosis, and high-level data related to public health. Retail will likely have customer data, but most of the data will relate to sales and inventory, so there's not as much PII involved.

Being familiar with laws and regulations also factors into domain knowledge. For instance, there is a lot of regulation in banking and insurance, and it's helpful for people to already understand this and know some of the specifics when they start a job in that industry. Similarly, it's helpful if people are already familiar with HIPAA when they start a job in the healthcare field (HIPAA dictates US laws around healthcare patient data). Being familiar with FERPA for jobs in higher education is also useful (FERPA laws cover privacy around student data in US colleges and universities).

But like I mentioned earlier, domains can also be much more specific. The video game industry is huge and employs millions of people, and often they prefer to hire people who are players themselves, so they understand everything that gets talked about in the company. If there's a project that's looking at different genres, it would be helpful for people to understand what many of the video game genres are and the ones that are likely to have crossover appeal. Someone who's never played any kind of game won't know that at first, and they also won't know what motivates someone to play a game in the first place.

Similarly, domains can also be specific to the company. I once worked on a project creating a system that processed Spanish text to assign a reading difficulty level, which was based on the company's existing system that did the same thing with English text. My colleague had worked on the English one and had great knowledge about how it worked. I hadn't worked on the English one, but it turned out that my intermediate knowledge of Spanish was very useful as we worked on the Spanish one. His domain knowledge on the English analyzer and my domain knowledge of Spanish meant we were the perfect team.

Domain Experts and Subject Matter Experts

People who understand a domain really well are called *domain experts*. Being a domain expert does not require someone to know literally everything about an area, but it does mean they have a lot of knowledge and are consistently gaining more as they work in that area. Even a little bit of domain knowledge can be very valuable, like my knowledge of Spanish was when I worked on a Spanish text analyzer.

For data scientists, domain knowledge is also valuable because it means they are familiar with the types of data science work that is done in those domains. Forecasting sales in a retail store is not the same as forecasting fraud on credit cards, for instance. But even more important is that there are certain types of approaches in different areas that are done that wouldn't make sense at all in others. For instance, in retail, there are some techniques for quantifying pricing and promotion elasticity, where you determine numbers that describe what happens when a price is changed (for instance, saying that if you increase the price by 10%, sales will drop by 4%, or if you temporarily drop the price by 15%, sales will jump 25%). The specific approach doesn't translate directly to other fields. Similarly, there's a technique for calculating what's called incremental sales when a new product is launched, which considers the sales of other products we would have expected to see (forecasting them), and then calculating how much of the total sales of the new product are new as opposed to sales that were taken away from the original products (losing sales of one of your own products to another of your products is called cannibalization).

There's another important term you will hear in this context, subject matter expert, which is usually abbreviated "SME" and pronounced "smee." A *subject matter expert* is someone with specialized knowledge in a specific area, just like a domain expert. The terms are largely analogous, but you tend to hear them in different contexts. The term SME is used a lot more and generally refers to someone who knows the data and business processes really well, often in the context of a specific company. This is basically just having domain knowledge of the company's processes and data. There might be a SME who knows the company's inventory management system inside and out, including the data that is stored and the processes used to create and manage it. SMEs often live on the business side of a company rather than the technical side, and it's incredibly common for data scientists to seek them out when they're working on a new project. Some SMEs may also have some technical depth related to the data—for instance, they might know source table names in the database and column names—but it's also common for them to only have higher-level knowledge of the data. Both scenarios are valuable, but data scientists might need to find someone else to bridge the gap between the high-level and database-level knowledge.

You tend to hear "SME" used less in relation to data scientists themselves who have become experts in that same area. We'd be more likely to call them domain experts (or even just not call them anything, but instead just recognize their expertise). A data scientist who's a domain expert at a company would most likely have a deep understanding of the data itself as well as the type of data science done in that domain at that company, but might have less understanding of all the exact business processes that a SME would know well. That's why a data science domain expert and SME work so well together, because they cover everything in that domain.

Why Domain Knowledge Matters

You first saw the data science Venn diagram in Chapter 6, but if you look at it again (Figure 10-1) with that lower circle highlighted, you can see how critical domain knowledge is to data science. All three areas are important, but domain knowledge is a core component of good data science. But how exactly is it valuable?

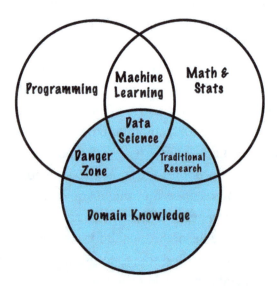

Figure 10-1. *The data science Venn diagram with domain knowledge highlighted*

By this point, you probably agree that domain knowledge is critical in data science. I talked in earlier chapters about how important it is to understand what the data we have represents in the real world, and domain knowledge ensures that you have that understanding. Let's look at several specific areas where it's most helpful.

Understanding Stakeholders

Being a domain expert helps a data scientist understand the world the stakeholders live in, which makes it easier to understand the problems they face and come up with better solutions, mostly because collaboration is smoother because of better communication (which comes from mutual understanding). Even having partial domain knowledge can be incredibly helpful. Any domain knowledge is especially valuable when you're dealing with stakeholders who haven't worked with data scientists before, because they don't always know what to expect or what information is important to share, and your domain knowledge can help you ask the right questions. Remember that stakeholders often don't know what they know, but your knowledge can prevent problems from that. Figure 10-2 is a funny reminder of how people don't realize what they know, whether in science or business.

Figure 10-2. *Average Familiarity, an xkcd comic. Source: Average Familiarity, xkcd.com, used under the Creative Commons Attribution-NonCommercial 2.5 License*

One other area that domain knowledge can matter in is credibility. This is actually true in all technical fields, where stakeholders often don't trust people building any kind of system for them. The reason for this is they've often had bad experiences with technical teams delivering solutions that miss the mark. A friend of mine works in restaurants installing point-of-sale systems (the registers where servers enter customer orders), and as part of her job, she has to train the staff on using the system. She learned early on to mention her years of experience as a server and restaurant manager at the beginning of training because it makes the people she's training trust her. Her knowledge also comes in handy when she's installing and setting up the systems, because she fully understands how these systems are used. Most of her colleagues have only technical experience and often don't understand things as well.

Similarly, at one of my data science retail jobs, the company culture held that nobody could really understand the business unless they had started their career in one of the stores, ideally pushing shopping carts. Bizarrely, they didn't count similar experience in other companies. There's not always much you can do in situations like that. The workaround my team had when we had a stakeholder with this mindset was to include one of our colleagues who *had* started his career in a store (our colleague was also a great data worker, so it was no sacrifice on our part).

Data Prep and Understanding

Domain knowledge also is important when it comes to working with the data itself. Understanding what the data really means is invaluable, and that sometimes cannot happen without domain knowledge. It is true that data can seem understandable without specific knowledge, but you'll also find that you've made invalid assumptions or misunderstood a column if you're not experienced with it.

For example, I worked for a retail company where almost all purchases were made by customers who had company memberships, so every sales transaction had a membership number stored with it. But there were occasional transactions that were done by non-members, and those were all assigned a dummy membership number (always the same value). Because there were loads of stores all over the world, if you broke down purchases by member number, that member number's total order count was many orders of magnitude larger than even the most spendy real member. If you didn't understand what this special membership number meant, you might make the mistake of including those transactions in your analysis, which would be very misleading if you were working at the member level, like calculating the average member spend.

At the same company, I worked on a project forecasting sales of a particular deli product in order to help the deli manager make decisions about how many of the product to make throughout the day. When we first started, we were seeing huge orders on some mornings, but not all. We didn't know what to make of it, but were able to learn (from SMEs) that these were orders placed in advance and picked up before opening, so it made sense for us to exclude them from our modeling. The deli managers would fully know about those orders in advance. Without that knowledge, we might have included them, and it would have reduced the accuracy (and value) of our forecasts.

Domain knowledge is also important in data preprocessing and feature engineering. We'll talk in later chapters about data prep, but one of the big things you have to deal with is missing data, and without domain expertise, it's almost impossible to know the right way to handle missing data. Domain knowledge informs both feature creation and feature selection, both of which we'll deal with in depth later. But it can help identify situations where multiple features really represent the same thing and shouldn't all be included in a model. As an example, if you were working on a system for a college that would optimize class assignments for instructors every semester, there would be fundamental things that would be assumed, and getting these wrong could be problematic. For instance, not realizing that instructors could teach multiple classes and even multiple sections of the same class would mess up the results. You'd need to consider many parameters, including their area of expertise, total number of classes an instructor can have, and whether or not they should be scheduled with back-to-back classes.

Developing Good Solutions

Having domain knowledge makes it easier for data scientists to come up with meaningful solutions, ensuring that the results will be valuable and useful to stakeholders. If you understand how stakeholders do business and how they use the tools they have, it's easier to understand where a data science solution might fit into their workflow. This can help you avoid giving them info that they don't need or that might even confuse things. I've found that sometimes including information that seems incredibly valuable to data scientists may not be as valued by stakeholders. This next example isn't strictly domain knowledge as it's more about understanding your customer's perspective (but sometimes understanding certain kinds of customers works like domain knowledge). We had done some forecasts on a project and had been rigorous, calculating statistical significance of each forecast. We thought it would be helpful for the stakeholder to see how confident we were in each forecast, but it actually had the opposite effect—they were intimidated by the term "p-value" and weren't used to the idea that results could be different levels of quality. They were used to dealing with third-party vendors who tended to just give them numbers for everything regardless of the significance of them.

Domain knowledge will also help us identify and mitigate potential problems and risks even before the project really gets started. It will also be helpful in solving unexpected problems that crop up along the way. One of the challenges working with retail data recently has been the impacts of the COVID pandemic. There were some pretty weird sales patterns in 2020, especially early on, but then there were significant supply chain problems that affected almost everyone. Usually when you're forecasting sales, you need a few years of sales, which still includes pandemic years. Understanding the fact that 2020 is not representative of typical years and the inventory problems that most places had means you can better deal with those anomalies, rather than ignoring them and having models that behave wonkily.

Finally, domain knowledge can help tremendously in measuring performance of the model. There are many standard metrics for measuring model performance, but choosing the best ones for your potential problem often isn't trivial. In many cases, particular domains use certain ones. It can also be important to know if false positives are preferred over false negatives or vice versa, which can inform which metric you choose. In medicine, flagging someone as possibly having a disease when they don't actually have it (a false positive) is vastly preferable to missing someone who really does have the disease (a false negative). Accounting for that sort of thing is easier with domain knowledge.

It's also important to realize that domain knowledge can be more important in domains that deal with humans. Domains like healthcare, customer service, and marketing can often be more difficult to model because of the complexity and (often) sparsity of the data, leading to lower accuracies, so any domain knowledge you can include can really make a difference. In domains like manufacturing and ecommerce, things are less complex and easier to define without special knowledge.

Security, Privacy, and Ethical Concerns

Another valuable thing that domain experts can offer is an awareness of possible security or privacy concerns as well as ethical issues that might arise—both potential and real issues can be anticipated in advance. Some of this can come from understanding general laws and regulations related to the security or privacy of the type of data involved and sometimes understanding company-specific rules and standards. It also can bring an awareness of how a given domain can impact people, whether employees, customers, or the general public. This means ethical difficulties can be avoided or mitigated, rather than cropping up unexpectedly and creating havoc (which could be anything from a major project delay to a public relations nightmare).

Becoming a Domain Expert

Unsurprisingly, there are two main ways to gain domain expertise: experience and studying. In general, both are going to be necessary to become a true domain expert. If you get a job in an area you've never worked in, you will definitely start learning things from your colleagues and stakeholders. Asking questions is important, as is trying to avoid making assumptions. It's also not uncommon for data scientists to sit with stakeholders for a day or a few hours (often called shadowing) to see how they go about their day and do their job.

Although practical experience and observation is valuable, it's rare that you can learn as much as you should only from working in that area. You'd be wise to also find other resources to learn from. The Internet likely isn't going to have information specific to the data you're working with at your organization, but it will have information on the domain in general. Depending on the kind of learner you are, you can find books, blogs, videos, and articles explaining things. Of course, not all sources are created equal. As an example, Medium.com has a lot of good data science articles covering all sorts of domains. Not all of the articles are good, but if you pay attention, you can usually identify trustworthy sources.

If you are trying to find a job straight out of a degree program or in a different area from where your experience is, studying is still a completely valid way to gain that domain knowledge. Sometimes, it can be useful to seek certifications to prove that you have achieved a level of knowledge, but this usually isn't necessary.

It's also really important to mention that once you have domain knowledge, even at the expert level, you will always have to keep learning. Fields are constantly changing, and you have to keep up.

Is Domain Knowledge Always the Best?

One important point to mention that feels very counterintuitive is that there are cases where domain knowledge can hurt the performance of a machine learning model. One of these cases is when it brings human bias into the picture when the expertise involves biased views. An example of this scenario is with the famously difficult game Go. Google DeepMind's system AlphaGo was able to beat a human champion. It was trained utilizing the domain knowledge of many Go experts. This was a big deal at the time. But then a new version of the system that was trained only on data and the game rules (without domain knowledge) was released—and it trounced the original version 100-0.[2]

There are also cases where what appears to be domain knowledge is simply wrong. I worked on a sales forecasting project where the store managers insisted that certain factors were critically important to their sales numbers, and our model did not pick those factors up because they weren't important. This could have been a teachable moment for the store managers, but instead they disregarded the information and didn't trust the forecasts. This really is just bias again, even if no one's getting hurt. It could be considered harmful to the managers and their ability to manage the stores effectively, but this is just how the business world works sometimes.

Remember that domain knowledge is just a tool—the vast majority of the time, it's valuable, but if you're remembering to always think about the ethical perspective, it's worth considering whether it might be wrong or there might be bias embedded in the domain knowledge itself.

---

[2] "1—The Importance of Domain Knowledge" by Haixing Yin, Fan Fan, Jiazhi Zhang, Hanyang Li, and Ting Fung Lau, August 30, 2020, from ML at Carnegie Mellon University blog, at https://blog.ml.cmu.edu/2020/08/31/1-domain-knowledge/

Key Takeaways and Next Up

This chapter looked at the importance of domain knowledge. A domain is anything somebody can know something about. Domain knowledge can be very valuable in understanding stakeholders and their needs. It can also significantly aid in data preparation and in creating good solutions. It's also important in identifying risks and security, privacy, or ethical compromises before they happen. Becoming a domain expert basically requires studying and getting experience if possible. Finally, we addressed the issue of domain knowledge not always being the most important thing—it can sometimes be biased or wrong.

In the next chapter, we'll be continuing to look at the important components of doing data science. One of the most important tools in the data scientist's toolbox is a programming language, usually R or (increasingly) Python, so we will talk about those languages at a high level.

PRACTITIONER PROFILE: MOMO JENG

Name: Monwhea (Momo) Jeng
Job Title: Data scientist
Industry: Software
Years of Experience: 10 as a data scientist, more as a software engineer
Education:

- PhD Physics

- BS Mathematics and Astronomy

The opinions expressed here are Monwhea's and not any of his employers', past or present.

Background

Momo Jeng started his career as a physics researcher and professor in academia. He did some programming as part of the physics work, especially running simulations. He wanted to move out of academia and applied for a software development engineer role and started a new software career at Microsoft. He did straight software development for a few years, but then missed research, so he started talking to people at work to learn about other roles he might be able to pursue. He learned about online experimentation and got on a team doing that, where he worked for five or so years. He enjoyed the work but wanted to try something different, so he moved over to a team doing natural language processing. He switched to another team doing online experimentation, where he still is.

Work

Most of Momo's data science work has been in online experimentation, where different things are being tested, like a layout on the Bing search engine or an ad design. This work has a bit more of a research feel than traditional software development, and he enjoys that. He works with large amounts of data, which means figuring out how to extract information from it. He's worked with data like search histories, where he's considered different ways of grouping it meaningfully, and tried to detect if people are satisfied with their search results. He loves the challenge of trying to figure clever ways to extract meaning. Also, in his role, experiment design is important, and often his team has to get creative to figure out ways to get at the information. This means designing the experiment exactly right so the data they get out of it will give them what they want. Sometimes the results are completely different from what was expected, and they have to dig in to see if there was something wrong with the study or if they really have found something unexpected. Sometimes the experiments can be challenging because you can't always control for everything. If you're looking at social media and divide users into control and experiment groups, you can't guarantee those groups won't interact with each other, which could compromise the study (depending on what is being studied).

Sound Bites

Favorite Parts of the Job: Momo loves digging into data, especially in the early phase, just trying to understand the data. He also loves the challenge of designing novel experiments and analyzing the results.

Least Favorite Parts of the Job: Meetings

Favorite Project: His team worked on what he called counterfactual experiments where they didn't run the typical A/B testing and instead predicted what the results would be. They validated this by running several fake experiments and then running some real ones, finding that their fake experiments were pretty accurate. To create this framework, they used some interesting statistical ideas like propensity weighting. This allowed them to get results for many counterfactual experiments quickly. Running real experiments on real data is time-consuming, meaning that there's a limit to how many models they can try out in a given amount of time on a finite number of users. The simulations allowed them to try many more models, so then they could use the best-performing ones in real experiments.

Skills Used Most: The most crucial is having good attention to detail. Ninety percent of data science is looking over data and making sure it makes sense, with cleaning, data validation, and more. He also uses a lot of traditional statistics. Corporate skills like communication, understanding hierarchy, generally working with people.

Primary Tools Used Currently: Python and an internal Hadoop-like map–reduce system (for working with the massive amounts of data they have)

Future of Data Science: He's hoping it doesn't, but it's not outside the realm of possibility that AI will be doing most data science in 20 years.

What Makes a Good Data Scientist: It's the attention to detail that most matters. He says it really helps to be the kind of person who's really bothered when something in the data doesn't seem right to where it needles you until you can figure it out.

Momo is a data scientist working in experimentation with a strong background in software engineering.

CHAPTER 11

Tools of the Trade: Python and R

Introduction

Learning a programming language is a required step for most data scientist positions. Data scientists generally don't do all types of software development—most won't be doing front-end work (like web development). Instead, data scientists usually work the most with scripts or notebooks that are single files. If you have to do your own productionizing, you'll probably also sometimes design multi-file software to carry out your modeling, and you may use other tools to take your code into production. Python and R are the two most common languages used by data scientists, and this chapter will introduce the basics of programming in common languages like these, as well as look a bit at how these two are different.

History of Data Science Programming

Programming languages and other computing tools have been used in data science for a long time, but the dominance of the primary languages used today—Python and R—has come about more recently. We talked in Chapter 4 about the development of statistical tools and how R specifically emerged

© Kelly P. Vincent 2025
K. P. Vincent, *A Friendly Guide to Data Science*, Friendly Guides to Technology,
https://doi.org/10.1007/979-8-8688-1169-2_11

from the proprietary language S as an open source alternative. Prior to that, statistical work was being done largely in SAS, a proprietary stats language, and in the FORTRAN language, a traditional language with good support for scientific and engineering programming (high numeric precision that allowed for accurate math and statistics). SAS was used more in finance and insurance with FORTRAN more widely. But FORTRAN was a little intimidating for less technical statisticians, which led to S being developed specifically for them. It caught on, but because it was proprietary, not everyone was able to use it. R was originally developed in the 1990s to help teach statistics, and it ran on top of S. Many programs written in R could run in S without modification. It is still a great language for learning statistics and programming, but it quickly spread like wildfire outside of teaching and into the general stats world. It's still heavily used by statisticians and now data scientists.

Python came into data science a little later, but it's starting to eclipse R in popularity among data scientists. Python started as a general-purpose language with particular strengths in text processing and efficient syntax (allowing for more in fewer lines). It attained a somewhat cult-like popularity, with the Python community calling themselves "Pythonistas," giving official best practices guide names like "The Zen of Python,"[1] and calling the original developer "The Benevolent Dictator for Life" (BDFL) because he maintained primary responsibility for it until 2018. It wasn't really used for data science until the Pandas package (an add-on for the core language) was released in 2009 and scikit-learn (the main machine learning package) in 2010. Pandas changed the style of programming in Python to be more R-like with the addition of data frames, the primary data structure in R. Data science itself was starting to take off back then, and programmers who were already using Python naturally started using Pandas and scikit-learn as they started doing data science, and its popularity also took off.

---

[1] You can find "The Zen of Python" and a reference to the "BDFL" at https://peps.python.org/pep-0020/

Although SAS is still around in some companies, the name of the game is R or Python, depending on your opinion. Unless you end up with a job that uses SAS, there's no reason to learn anything besides R or Python at the beginning. Both are perfectly legitimate languages for data science, both to learn and to develop. They do have different strengths, which might dictate which you would choose for a specific purpose, and most data scientists are pretty opinionated about which one they prefer. It's also common for teams to use one predominantly. On the other hand, there's no reason to not use both—plenty of teams do this, too. We'll talk more about how they're different and similar in the next sections. R is particularly good for statistics, and it's popular for visualization and in academia, while Python is good for machine learning and general programming, including building production systems.

Examples of Python and R in Data Science in the Real World

Our first example is a good example of how you don't have to choose between R and Python and can instead use them where their strengths lie. Then we'll look at how Python is used in the financial tech world.

Example 1: Hedging by Using Both

The hedge fund company Amadeus Investment Partners found success using a *tech stack* (the technical tools a team uses to get their work done) that includes both Python and R, plus SQL and other tools.[2] They have

---

[2] "Case Study: How To Build A High Performance Data Science Team" by Matt Dancho and Rafael Nicolas Fermin Cota on Business Science, available at https://www.business-science.io/business/2018/09/18/data-science-team.html

several different roles on their interdisciplinary team, and each role has specific tools it uses. They have technical SMEs who use R, Excel, and some other tools to explore the data. Their data engineers also use R, along with SQL and C++ (an older traditional programming language that's good for performance). Their data scientists use R for EDA and Python for machine learning and deep learning in particular. They also have user interface developers who also use both R and Python, along with other tools.

For Amadeus, they prefer R for most data analysis tasks, including exploring the data and visualization. There are some great packages in R for these tasks. But Python excels in the more computationally intensive tasks like machine learning, especially when they're working with the larger datasets that the company uses. It's also interesting that they seem to dictate exactly which tools people should use. A lot of companies allow data scientists to choose their tools, which they like, but it does lead to a disparate codebase that can be hard to maintain and reuse. By requiring data scientists to do their EDA in R, any future work can more easily build on earlier work.

One other point about this company that's interesting is that they do something quite unusual in data science: they hire top graduates from business schools and train them to be data scientists, rather than starting with people with technical backgrounds. They find that these graduates are able to pick up both Python and R reasonably quickly, which complements their more general soft skills and business knowledge from their degrees. This speaks to the relative ease of learning these two languages.

Example 2: Fast and Flexible FinTech

djangostars, a financial technology (FinTech) consultancy and custom software developer, prefers Python in all of its work.[3] They have many reasons why they find Python the best language for FinTech, especially for newer companies. One is that it's already preferred among mathematicians and economists, so any work that needs to be put into a system that comes from them is already halfway ready, even before "real coding" is getting started. While a lot of FinTech involves data science, there's much more to it than that, and Python's great as a multipurpose language. They say using Python combined with Django (a well-established Python-based web framework) is a great combination for getting products to market quickly. Additionally, that suite of tools is very flexible and easy to make changes to, something that's important in FinTech (as opposed to traditional banking). They also have found that because Python is relatively easy to learn, it makes collaboration and knowledge transfer to clients smoother because it's so easy for developers and data scientists to explain. Python's very well established, and there are free packages that do almost any imaginable thing.

The Basics of Computer Programming

Programming languages really are like human languages, just much, much simpler. I had a professor who got a computer science degree in the 1980s and convinced his college to count Pascal (another traditional programming language) as his foreign language requirement. That wouldn't happen today, but like human language, programming languages have syntax (the particular words, operators, and the order they're written)

---

[3] "Using Python for FinTech Projects: All You Should Know" by Artur Bachynskyi, October 28, 2024, `https://djangostars.com/blog/python-for-fintech-projects/`

and semantics (the meaning made by the specific text written following syntax rules). They also have idioms that develop over time, which are ways of accomplishing specific things that are conventional among programmers even if it's not the only way it can be written. A few languages also have fairly strict formatting rules that dictate style—Python is one, and Pythonistas fervently believe it's the only way to write nice-looking code.

Computer code is basically just a recipe written with the specific syntax of the particular programming language. Almost all programming languages use English words, and there are several words and operators that are used across most languages. Some of the most common operators include those in Table 11-1. These are mostly familiar items we know from basic math. They're largely used the same way in code. I'll talk a bit more about these below, as well.

Table 11-1. *Common operators in many languages, including R and Python*

Assignments and Control	Comparators	Mathematical Operators
=	>	+
:	<	-
	>=	/
	<=	*
	==	^
	!=	%

Some common keywords are in Figure 11-1. *Keywords* are words that always have a specific meaning in a language and can't (or shouldn't) be used anywhere in the code except in the defined way. All the keywords in Figure 11-1 are used to control the flow of programs, which we'll talk about below.

```
for
while
if
else
break
function
def
return
raise
```

Figure 11-1. *Common keywords in many languages, including R and/or Python*

Once you know the basics of syntax, you're on your way to learning how to do almost anything in code. Programming is very systematic, but once you get into it, you'll see how much creativity factors into finding good ways to solve problems.

There are a few different high-level ways to go about programming, called paradigms. The ones we will focus on here are procedural and functional, but another common one you'll hear about is *object-oriented programming*, which focuses on a certain type of modular design and can be done in both Python and R (it's easy in Python but a little trickier in R), although a lot of data scientists may never use it. *Procedural programming* involves a sequence of steps done one after the other. *Functional programming* involves the use of functions, which we'll define more further down, but basically involves having smaller units of code that can be defined and called repeatedly with different or no values for effective reuse.

Note that this chapter is talking about code, but you aren't expected to learn to write code in this chapter. There are some examples of Python and R code in the sections, but they're simply there to illustrate ideas and show the differences between the two languages. However, if you are a hands-on learner, there's nothing wrong with getting in a Python or R interpreter and trying things out. It's a great way to learn. Appendix A details how to install them and get started.

Traditional Programming

A computer program is simply a sequence of lines of code utilizing syntax and other text to tell the thing that compiles and/or runs the code what to do. Some languages are compiled with a compiler, which means they are processed and turned into a runnable file, or run in real time by an interpreter. R and Python both use an interpreter.

Code is formatted by using line breaks, white space (including indentation/tabs), and control structures such as curly braces ({ and }) to make it readable by both humans and the compiler or interpreter. Beyond this, there are many elements of writing code, which we'll cover in the subsections below.

I've included some example Python code in Figure 11-2 in case you've never seen code before. This doesn't do anything very interesting, but it shows some of the basics we'll talk about below. I'll also point to relevant parts in this code when we talk about it below. The left side is line numbers. Note that the code is color-coded—this is standard in most code editors.

```python
1    # variable declaration and assigment
2    num1 = 10            # int type
3    num2 = 99            # int type
4    num3 = 1.5           # float type
5    decision = True      # bool type
6    reason = "because"   # str type
7    numbers_list = [num1, num2, num3]     # list type
8    numbers_dict = {1: num1, 2: num2, 3: num3}  # dict type
9
10   # mathematical operators
11   two_nums_sum = num1 + num2
12   two_nums_diff = num2 - num1
13   two_nums_prod = num1 * num2
14   two_nums_quot = num2 / num1
15   two_nums_pow = num2 ^ num1
16   two_nums_mod = num2 % num1
17
18   # if-else
19   if num1 >= num2:
20       new_var = "bigger"
21   elif num1 == num2:
22       new_var = "same"
23   else:
24       new_var = "smaller"
25
26   # looping: for loop
27   running_sum = 0
28   for num in range(num1):
29       running_sum += num
30
31   # looping with a condition: while loop
32   counter = 1
33   while counter <= 10:
34       print("Hello time " + counter)
35       counter += 1
36
37   # function declaration and use
38   def lopsided_square(var):
39       var_square = var * var + 1
40       return var_square
41
42   lopsided_square(num1)
```

Figure 11-2. *Some example Python code*

Comments

All programming languages allow you to include comments in the code. The particular syntax for including comments varies per language, but Python and R both use the hash symbol, #. Any time you see that character on a line of code, everything after the # is a comment and won't be run by the interpreter, and everything before is normal code that will be run.

There are many opinions about how and how much to comment in code, but most people agree there should be some comments. Some people practically narrate everything that's happening, while others only include TODOs or mark things they think are very confusing. In practice, you'll find that you will forget what tricky parts of your code do when you come back to it later, so comments are highly recommended.

Variables

Variables are ways of storing data and values in code. You can assign a value to a *variable*, which is just a name to hold the value. Variables are for storing values like numbers, text, or other more complex data types. In some languages, including R and Python, almost anything can be stored in a variable, even a function. Variables are a way of having a placeholder that represents something else. It's commonly used when you're going to be doing something over and over and don't want to have to retype a value every time, especially if it might change later.

Python

Like almost all languages, Python uses the equal sign to assign variables. For instance, the code my_number = 7 stores the value of 7 in the variable my_number, so any time you see my_number in the code, you can mentally substitute the value 7. If you see my_number + 5, the result will be 12.

There are several rules when naming variables in Python. Only letters (uppercase or lowercase), numbers, and the underscore (_) can be used in variable names, and you cannot start the name with a number. You can start the variable name with either a letter or an underscore, but starting a variable with an underscore is used to indicate a particular scenario in Python, so for normal variables, you should start with a letter.

Lines 2–8 in the example Python code in Figure 11-2 are variable assignments, with a comment at the end of the line indicating what data type Python would currently assign to that variable.

R

R allows slightly different characters in variable names from Python. It allows all the same characters as Python but also allows the period (the period is an operator in Python, so it can't be used). It's actually quite common to use the period in R variable names. Names must start with a letter.

R is unique among languages in using a different operator for assignment instead of the equal sign: `<-`. The equal sign also works, but most R programmers prefer the other. It would look like this: `my.number <- 7`. It works the same as in Python, where that name means 7 any time you see it in the code.

Data Types

Data types are simply the allowed types of data in a given language, what you can store that the language will know what to do with. The most common basic ones (often called *primitive data types*) are *integer* (a whole number), *decimal* (a decimal number), *Boolean* (false/true or 0/1), *complex* (a value containing a complex number, the square root of –1), and *string* (text contained within markers, usually quote marks). Different languages will have different-sized versions of each of these.

There are more advanced types that can be available natively, be custom-defined, or be used through other packages (add-ons) in the language. These include the *array* or *list* (a sequence of values), the *map* (a sequence of pairs of a particular key pointing to a value), and the *set* (like an array or list without repeating values).

There are still more ways to store data, which are even more advanced and are almost never built into the language (so you have to use external packages). These are usually called *data structures*, a more abstract term indicating the behavior and relationship of parts based on the way the data is stored. These include *stacks* (last in first out, like a stack of plates at a restaurant), *queues* (first in first out, like a line at a concession stand), and *trees* (like a family tree). There are ways of working with the fancier data structures that make them very convenient in certain cases, even though those rarely come up in data science programming.

Finally, there are some big differences between different languages based on how they handle data types and variables. In statically typed languages (like Java and C++), a variable's type has to be defined at creation time and can never hold anything but that type of data. Python and R are both dynamically typed languages, which means you don't have to declare the type when you create a variable and a variable can hold different types of data at different times (it could hold a string and then later hold a number). Dynamic typing leads to less verbose code but means you have to be careful about changing variables.

In strongly typed languages, which most popular languages are, there are restrictions on how different data types can intermingle. Python and R are both strongly typed languages because when code is running, the interpreter is keeping track of all data types and forbids certain operations. As an example, one popular weakly typed language would allow you to "add" a string value (or variable) and a number value (or variable). "X" + 9 in JavaScript would give you "X9" (a string), but both Python and R would throw an error.

Note that both Python and R record a variable's data type based on the value assigned to it. So if you have assigned a variable the value of 7, Python and R will track it as an int until it's reassigned to something else.

Python

Python has all the basic data types plus two similar list types (called a list and a tuple, respectively) and the map (called a dict, short for dictionary). Lists and tuples are indexed starting at 0, so the first element is at position 0, the second at 1, and so on. Python only has one integer data type, int, in the current popular version (earlier versions had a larger integer called long). It also has only one decimal data type, float (it used to have a larger decimal type called double). The current int and float are as large as the previous long and double, which is why there's no need for them in modern code. The Boolean type is called bool and the complex type is called complex. Strings are stored in the str data type. You can find out the data type by calling the type() function and putting the variable name inside the parentheses.

R

R also has one of each of all the basic data types. It has two similar list types, vector and list. Both are indexed starting with 1, unlike Python. The integer data type is called integer and the decimal type is called numeric. The complex type is simply called complex. Strings are built with the character data type. Finally, the Boolean type is called logical. You can check the data type of any value by calling the class() function with the variable name inside the parentheses.

Operations

Operations are just the instructions on the action to take in programming languages. For instance, if you want to do mathematical operations in code, you'd use the operators seen in Table 11-2. These all work identically in Python and R, although the modulo operator is slightly different between the two. Lines 11–16 in the example Python code in Figure 11-2 show the six mathematical operators being used.

Table 11-2. *Mathematical operators in Python and R*

Purpose	Math Sign	Python	R	Example in Python and Result
Add two numbers.	$n + m$	+	+	6 + 2 gives 8.
Subtract one number from another.	$n - m$	-	-	6 - 2 gives 4.
Multiply two numbers.	$n \times m$	*	*	6 * 2 gives 12.
Dive one number into another.	$n \div m$	/	/	6 / 2 gives 3.
Raise a number to a power.	n^m	^	^	6 ^ 2 gives 36.
		**	**	6 ** 2 gives 36.
Calculate the modulo (remainder).	$n \% m$	%	%%	6 % 2 gives 0.
				7 % 2 gives 1.

One thing that's important in mathematical operations is the *order of operations*, which tells you which combinations of numbers and operators will be run first. This is the same as in regular math, and parentheses can be used to force some things to be done first. It's common when handwriting math to include parentheses for instant clarity, but in code, parentheses are usually only used when necessary because of the order of operations. For example, Table 11-3 shows some code, intermediate conceptual steps, and what the final result would be. These can be a little intimidating at first if you don't remember it from math, but after working with it for a while, it will become instinctive.

Table 11-3. *Order of operations examples*

Operation	Step 1	Step 2	Step 3	Result
6 + 2 * 3 - 1	6 + 2×3 − 1	6 + 6 − 1		11
(6 + 2) * 3	6 + 2 × 3	8 × 3		512
2 ^ 3 * 5 - 3	2^3 × 5 − 3	8 × 5 − 3	40 − 3	37
2 ^ (3 * 5 - 1)	$2^{(3 \times 5)-1}$	2^{15-1}	2^{14}	16384
2 ^ (3 * 5) - 1	$2^{(3 \times 5)}$ − 1	2^{15} − 1	32768 − 1	32767

Another type of operation that's used even more than math is comparison. We want to know if something is the same as something else or different or greater or less than. You can see these in Table 11-4. These always give you a Boolean value as a result. These work identically in Python and R.

Table 11-4. *Comparison operators in Python and R*

Purpose	Math Sign	Python	R	Example in Python and Result
Less than	n < m	<	<	6 > 2 gives True.
Less than or equal to	n ≤ m	<=	<=	6 >= 2 gives True.
Greater than	n > m	>	>	6 < 2 gives False.
Greater than or equal to	n ≥ m	>=	>=	6 <= 2 gives False.
Equal to	n == m	==	==	6 == 2 gives False.
Not equal to	n ≠ m	!=	!=	6 != 2 gives True.

There are a few other operators that are important with conditions: not, and, and or. not negates the value it's placed in front of. If the value is True, negated it's False, and vice versa. and and or are used to join conditions together to give a single Boolean value, where the value is True only if both parts are true with and and True if at least one is true for or. For instance, 6 > 2 and 2 > 6 would evaluate to False, but 6 > 2 or 2 > 6 to True.

These are the most important operators in Python and R. There are others that are used less frequently, including bitwise operators, but you're unlikely to need those in data science.

Program Control Flow and Iteration

One important facet of programming is making the code do what you want in the right order and, very often, repeat things. Order of operations is this on a micro level, but far more important is the control flow functionality. There are two common controls for making the program flow like you want: if–else conditions and looping.

If–Else Blocks

If–else constructions allow you to say if this, do one thing and if not, do something else. You can also nest these, so you could say, if this, do one thing; if this other thing, do a different thing; and if this last thing, do an even more different thing; otherwise, do something else. What this looks like if we spell out the logic is as seen in Table 11-5. This way of writing things out is called *pseudocode* (although normally you would have actual values like the block on the right), which is a code-like way of writing it that's not specific to any particular language, but makes what's happening fairly clear.

Table 11-5. *Framework for a three-condition if–else block*

Flow Concept	Pseudocode
if condition 1	if var1 > 0
then do thing A	then print "positive"
else if condition 2	else if var1 == 0
then do thing B	then print "zero"
else if condition 3	else if var1 < 0
then do thing C	then print "negative"
else do thing D	else print "something's wonky"

The code on the left in Table 11-5 checks condition 1 and will do thing A if condition 1 is true. Otherwise, it will check if condition 2 is true, and so on down. Note that the way these blocks work is that they "escape" at the first condition they meet. If condition 1 is satisfied, the block will do thing A and then go to the next thing after the entire block, not to condition 2. The only time it would do thing D would be if none of the conditions (1, 2, or 3) are true. The code on the right shows the flow if we're wanting to print the direction (positive or negative) of a numeric variable called var1.

Python uses three keywords related to if–else statements: `for`, `elif`, and `else`. To set off each condition, it uses a colon. R uses only `if` and `else`, using `else if` together as necessary, and needs curly braces to set off the different parts. See Table 11-6 for what the code in Table 11-5 would look like in Python and R. You can also see an if–else block in Figure 11-2 on lines 19–24.

***Table 11-6.** Example if–else code in Python and R*

Python	R

`if var1 > 0:`
`    print("positive")`
`elif var1 == 0:`
`    print("zero")`
`elif var1 < 0:`
`    print("negative")`
`else:`
`    print("something's wonky")`
	`if (var1 > 0) {`
	`    print("positive")`
	`}`
	`else if (var1 == 0) {`
	`    print("zero")`
	`}`
	`else if (var1 < 0) {`
	`    print("negative")`
	`}`
	`else {`
	`    print("something's wonky")`
	`}`

Looping

It's very common in code to iterate and do something over and over with slightly different values. These are usually more than basic, but conceptually it's fairly simple, and we can look at an example where we print some math results for several different values. Let's first look at some pseudocode, as in Figure 11-3.

```
for num in 0, 1, 2, 3, 4, 5, 6, 7, 8, 9
  sq_num = num * num
  print "The square of " num " is " sq_num
```

Figure 11-3. *Pseudocode for a simple for loop that prints the squares of the first ten numbers*

The type of loop shown here is the *for loop*, where the basic structure is for a value in a list of values, where the number of times the loop will run is the same as the number of elements in the list, with each run having the value be one of the values in the list. We always know how many times this kind of loop will run.

Another type of loop is the *while loop*, which will run over and over until a predefined condition is no longer met. This would look something like the pseudocode in Figure 11-4. In this case, we start with a variable called num and increase it each time the loop runs and only stop when the number gets to 10. As long as the condition after the while is true, the loop will continue. Additionally, you can also use the keyword break to exit a while loop at that point in the code.

```
num = 0
while num < 10
    sq_num = num * num
    print "The square of " num " is " sq_num
    increase num by 1
```

Figure 11-4. *Pseudocode for a while loop that prints the squares of the first ten numbers*

Note that most programmers prefer for loops over while loops because it's easy to have a bug in your while loop that means it runs forever (if you forget to increase num by 1 in the Figure 11-4 example, it would never exit). This is obviously an easy fix, but it still would be a hassle the first time you run it with the bug.

PSEUDOCODE AND CODE STEP-THROUGH

Pseudocode is basically the steps of an algorithm, basically code, written in human language. When you're first learning programming, it's often really beneficial to write things out this way first so you can understand what the logic and flow is.

It's also to write down what happens at every step of the pseudocode with actual values to see what happens. For example, a step-through of the code in Figure 11-4 would start off something like this:

num value	Result
0	The square of 0 is 0.
1	The square of 1 is 1.
2	The square of 2 is 4.
3	The square of 3 is 9.
4	The square of 4 is 16.
5	The square of 5 is 25.
6	The square of 6 is 36.
7	The square of 7 is 49.
8	The square of 8 is 64.
9	The square of 9 is 81.

This can help you see what's going on and find problems, especially in trickier code.

Python has the for loop and the while loop. The for loop uses the keywords for and in to set it up, and the while loop uses while. R has the same keywords but also has another loop similar to a while loop, which uses the keyword repeat. This one is just like a while loop, but it has no condition and only stops running when the keyword break is used inside the loop. So it always runs up to the first appearance of the keyword break. A while loop may never run at all if the initial condition isn't met. See Table 11-7 for examples in R and Python for a for loop running what's in Figure 11-3. In Python a way to get a list of every number from 0 to 9 is range(10) (not 1 to 10), so you'll see that in the code. You can get the same thing in R with the code 0:9.

Table 11-7. *Example for loop code in Python and R*

Python	```for num in range(10):```
	``` sq_num = num * num```
	``` print("The square of " + num + " is " + sq_num)```

R	```for (num in 0:9) {```
	``` sq_num <- num * num```
	``` print(paste("on the ", paste0(counter,```
	``` ("The square of ", num, " is ", sq_num, sep="")) ```
	```}```

Table 11-8 shows the code for accomplishing the same thing using a while loop. Note that in Python, there's a shorthand way of adding 1 to a number: num += 1.

Table 11-8. *Example while loop code in Python and R*

| **Python** | ```python
num = 0
while num < 10:
  sq_num = num * num
  print("The square of " + num + " is " + sq_num)
  num += 1
``` |
| **R** | ```r
num <- 1
while (num < 10) {
  sq_num <- num * num
  print(paste("on the ", paste0(counter,
    ("The square of ", num, " is ", sq_num, sep="")))
  num <- num + 1
}
``` |

You can see another for loop and while loop in the example Python code in Figure 11-2 on lines 27–29 and 32–35.

Functions

Everything we've talked about so far applies to procedural programming, but adding functions will take us into functional programming and can vastly improve code if we are performing tasks that are the same or similar over and over. A *function* is a block of code that does a specific task and may take variables to use and may return a value. It can have different names in different languages and contexts, including methods, procedures, and subroutines (these aren't all exactly synonymous, but are similar enough for this section). A function is considered callable, which means that you define it once and then run it later by including its name in the code.

A function has two high-level components: a function definition and the function body. These can look different in different languages, but the first line is always the function definition and contains the name of the function and input variables if they're needed. Input variables are called arguments or parameters. The function body starts indented on the next line or is contained within curly braces (or something else).

Functions can be small—even just a single line—or hundreds of lines long. In general, shorter is preferred, because it's considered best practice for a function to do one primary thing (even if it takes many, many steps to accomplish that one thing) rather than several different things. You'll use functions extensively in your code when you call functions that are native to Python or R or are in some of the packages you install. But you'll also often create your own, which is incredibly powerful.

Python

In Python, functions are called functions except when they are used in a certain context in object-oriented programming, but we won't worry about that. They are defined with the keyword def, followed by the name and opening and closing parentheses, with variables inside the parentheses if they are to be included in the function run. The function that appears in the Figure 11-2 code on lines 38–40 is just a silly function that returns the square of a number modified by 1. Line 38 is the definition, with the keyword def, the name of the function, and the opening parenthesis, one argument, a closing parenthesis, and the colon, which ends the definition and indicates the function body is coming next. In Python, you start it on the next line. The first line calculates the new value, and the final line returns it using the return keyword. We can call the function later by using the function name followed by opening and closing parentheses, with any variables to be sent included in the same order as in the function definition. The parentheses are required whether any variables are passed or not. Calling a function looks like line 42 in the sample code in

Figure 11-2, with the name and argument in parentheses. The argument can be a hard-coded value or a variable name.

Python has one additional type of function that's a little unusual. You can define what's considered an anonymous function with the keyword lambda. This creates a block of code that behaves like a function, but it's only called once, where it's defined, because it doesn't have a name. One of the most common uses of this is when you want to sort something a particular nonstandard way. Imagine that you have a list of lists of two numbers (a row number and a height), like [[4, 60], [8, 69], [3, 62], [1, 67], [2, 65], [9, 72]]. If you just call the Python sort() function on the overall list, it would sort of do what you might expect—sort all the two-item lists by the first number in each of them. But if we want it sorted by the second number, we could create a simple lambda function in the call to the sort function that tells it to sort by the second number.

R

Functions in R are similar to those in Python in how they're defined and used, but they look a little different. The same lopsided_square function seen in Python in Figure 11-2 above written in R is shown in Figure 11-5. The name comes first in the definition, followed by the assignment operator, the keyword function, opening and closing parentheses with the argument name inside, and finally an opening curly brace indicating the start of the function definition. The function definition follows on the next two lines, and it's closed with an ending curly brace. R also uses the same keyword as Python, return, to return a value.

```
lopsided_square <- function(var) {
    var.square <- var * var + 1
    return var.square
}
```

Figure 11-5. *An example function defined in R*

If we include `lopsided_square(3)` in the code, what we'll get back is 10.

Libraries and Packages

One thing that's important in coding in almost any language is that they're organized by purpose through libraries and packages. In Python the word "library" is sometimes used, but "package" is generally preferred. The base installation of the language includes many as default, but you can also install more. As an example, although basic mathematical operations are included in Python, there's a special package that includes more advanced math. There are also many third-party packages that are available and used in data science code. To use these kinds of packages in both Python and R, you have to import them first.

Python

In Python, you can import packages a few different ways. It's customary to do these at the top of your code. Packages contain classes or functions that you want to use, and how you import the packages determines how you refer to those items later. Table 11-9 shows some of the different ways to import a package called Pandas and how you would refer to a function later if you'd imported it that way.

Table 11-9. *Common ways of importing packages in Python*

Package Import Statement	Calling a Function from the Package
`import pandas`	`pandas.to_csv()`
`from pandas import to_csv`	`to_csv()`
`import pandas as pd`	`pd.to_csv()`

How you refer to imported packages relates to a concept called namespace, which defines what level variable and other item names apply to. The first way is the most basic, and the third way is the same except you've created an alias pd for the pandas package. When you want to call a function or refer to a class in a package imported either of these ways, you have to put the package name (or alias) in front of the function or item with a period between them, as shown in the second column. So sometimes the second method is convenient since you don't have to specify the name of the package every time you call a function in it, but it can have downsides. If you use the second method and write your own function and call it to_csv(), you will not be able to use both functions—the function that would be called would be whichever was the last one to be defined. This can be really confusing. Pandas is a critical package in data science, and the conventional way to import it is the third way, with the alias pd.

R

In R, there are a couple of key ways to import a package. The first is with the library() function, and the second is with the import package. There can be namespace challenges in R, too. By default, all functions in a package are usable with the function name only, but you can also include the package name followed by two colons to refer to a function in a specific package. See Table 11-10 for some example imports.

Table 11-10. *Common ways of importing packages in R*

Package Import Statement	Calling a Function from the Package
library(dplyr)	filter() or dplyr::filter()
import::from(dplyr, select, filter)	filter()
import::from(dplyr, select, dplyr_filter = filter)	dplyr_filter()

The first way is the most common, and you'd refer to the function by just its name or you could specify the package name as shown. The second way is similar, but it only imports the specified functions (in this case, select and filter are imported from the dplyr package). The third way allows you to rename a specific function, which can be important if different packages have the same function names. In the third case, select is imported as normal, but the package function filter is given the alias dplyr_filter. There are more ways to specify which functions are brought in, and you can see the online R documentation for more info on that.

Organization and Formatting

There are lots of styles and ways of organizing code, with opinions abounding. Some of the opinions relate to commenting. I once worked with someone who insisted that you don't need comments in well-written Python code because it's so easy to read. I do not agree with this, but it's true that you don't need a comment explaining every single thing you're doing. What your code looks like is important for a lot of programmers (especially Pythonistas). The goal is for it to be as readable as possible, which generally means a decent amount of white space and occasional comments. In R, curly braces are used to define different code blocks, and these generally start on the first line of the block and end alone on a final line at the end of the block. Python doesn't use code block enclosures like with curly braces and instead marks a block with a colon and relies on indentation to define the block that the code appears under. I'll talk below about some published style guides for both Python and R.

Error Handling

An important part of general programming is properly handling errors in code, especially when productionized or to be run by other people. The latter point is especially relevant when you're writing code that will be shared across your team or with other programmers. When you're just writing notebooks to run your own data science, you don't need to specifically handle errors in the code, because you yourself would be seeing the error and making appropriate corrections right then.

But imagine you've created a function that lives in a central code location people can use to do a particular transformation on a value. They would import your function and run it without knowing about the internal code inside it—but if you've done this right, users know exactly how to use the function because of your excellent documentation. Let's say it takes a couple of parameters and is supposed to return a transformed number. But if something goes wrong inside your function and it generates an error, the user doesn't have any way to troubleshoot the inside of the function—they may not even be able to see it. So you need to anticipate the kind of errors and capture them and provide a meaningful error message that is passed back to the user. Common scenarios that throw errors are dividing by 0 or providing the wrong data type. There are nice ways of including error handling in both R and Python, which you can learn about if you start working on code that requires it.

Security

One thing that programmers don't always think about is security. There aren't a lot of concerns around privacy or ethics that relate specifically to writing code, but code can be written that is less secure than it should be. This is generally less of concern for data scientists working exclusively with internal systems, but important if you're exposing your code to the Internet either by productionizing your system or creating an API (Application Programming Interface, a piece of software that runs on your network and allows other code to query it, so your code would generate the response and send it back).

Some of the areas of security that programmers should care about include authentication, encryption, error handling, input validation, output validation, and third-party package choice. We've talked about authentication and encryption in Chapter 8, but the point here is on ensuring you authenticate when necessary and set it up properly in your code and include encryption when necessary. Input validation and output validation are important for ensuring that they aren't going to cause any bad behavior. A common cyberattack is called SQL injection, which is where somebody is filling out a form and puts SQL code in a text box (like a field asking for a username or anything really). If they've written code that drops all tables in a database and no input validation is run on the text box, that SQL could run in the system's database. Input validation can block this. Output validation also ensures that no code or other problematic results would cause bad rendering in a web browser or display of an offensive image, for instance. I mentioned error handling above, which can be used in conjunction with input and output validation to handle detected problems. I also mentioned package choice because one of the beautiful things about open source languages like R and Python is that anybody can create a package and share it. It's not all sunshine and rainbows because packages that do bad things (whether intentional or not) can be released, but they'll get found out eventually. So it's recommended to stick with established packages instead of jumping on brand-new ones.

Data Science Programming

The defining thing in most data science and statistics programming is the data frame, the basic structure for storing tabular data both in R and with Pandas in Python. The data frame is actually a built-in data type in R, but in Python we have to use a package called Pandas, which gives us the R-like functionality.

Data frames store tabular data in a convenient and meaningful way. Table 11-11 shows a typical data frame.

Table 11-11. Example data frame

Pet_num	Species	Name	Sex	Birth_date	Breed	Color_1	Color_2	Date_died
1	cat	Chica	SF	2023-05-10	DLH	black		
2	cat	Zephyr	NM	2020-01-15	DSH	orange tabby		
3	cat	Darwin	NM	2014-06-11	Siamese	seal-point		
4	cat	Misha	NM	2012-05-06	DMH	brown tabby	white	
5	cat	Tank	NM	2012-08-25	DSH	orange tabby		2023-12-25
6	cat	Callie	SF	2011-04-30	DSH	torbie		2023-07-19
7	cat	Virginia	SF	2000-05-27	DSH	tabby		2014-03-15
8	cat	Maggie	SF	2006-09-18	DSH	Calico		2016-11-05
9	cat	Ash	NM	2010-3-27	DMH	Black		2011-03-20
10	cat	Newton	NM	2006-06-01	DSH	brown tabby		2018-08-10
11	cat	Vixen	SF	2007-04-17	DSH	brown tabby	white	2023-09-12
12	cat	Isolde	SF	1996-03-19	DLH	gray	white	2010-07-23
13	cat	Winston	NM	1995-05-13	DLH	black		2009-06-30
14	dog	Buddy	NM	1984-06-11	mutt	black		1996-09-04
15	dog	Barbie	SF	1983-10-04	Cocker Spaniel	Tan		1994-01-13
16	guinea pig	Rufus	M	1982-11-01		black	white	1984-12-08
17	hamster	Hammie	M	1981-07-18		tan		1982-10-02

Data frames are convenient because they allow us to do operations across rows and columns, and even the whole data frame, with single commands. Adding new columns and rows is easy. Grouping data and aggregations can be done in one step. For instance, adding a column Age based on the birth and death dates in the other columns can be done in a single line in both Python and R. In Python it would look something like this:

```
pets_df["Age"] = pets_df["Date_died"] - pets_df["Birth_date"]
```

There would be a little more involved depending on data types and what unit you wanted your Age column in, but this is all you have to do conceptually. It knows to take the whole column and subtract the other column with rows lined up and then adds it to the data frame as a new column at the end. This is much easier than the way it used to be when you worked with tabular data (usually this was going row by row in a loop and doing the computation on each pass of the loop, which takes a lot more time than using data frame operations).

Data scientists love data frames for working with lots of tabular data, but the other reasons they love Python and R so much are the statistics, visualization, and machine learning packages that are available in both languages. Table 11-12 shows some of the many packages data scientists use in Python and R. Some are built-in, but most will need to be installed (fortunately this is easy to do). It's beyond the scope of this book to talk about using these packages specifically, but we will talk about machine learning in a later chapter.

Table 11-12. *Data science packages used by data scientists. ML stands for machine learning and NLP for natural language processing*

Python Package	Use	R Package	Use
pandas	Data frame	data.table	More data frame manipulation
numpy	Numeric data manipulation	dplyr	Data manipulation
scipy	Builds on numpy	lubridate	Date manipulation
datetime	Date manipulation	ggplot2	Visualization
plotly	Visualization	stats	Statistics
matplotlib	Visualization	e1071	ML modeling
seaborn	Visualization	xgboost	ML modeling
statsmodels	Statistics	randomForest	ML modeling
scikit-learn	ML modeling and performance	tensorflow	ML modeling
tensorflow	ML modeling	naivebayes	ML modeling
torch	ML modeling	metrics	ML performance
nltk	NLP modeling	tidymodels	ML support
spacy	NLP modeling	shiny	Presentation

If you're coding in a cloud platform, you may have other packages that can give you big performance boosts. For instance, if you have access to Spark, you would use Spark data frames instead of Pandas, which would require using different machine learning libraries. Sometimes there are also platform-specific packages that are available. These can be very useful, but sometimes difficult when you're transitioning from one platform to another.

One other thing that's somewhat unique to data science programming is the use of notebooks, or systems that allow you to run individual lines of code rather than an entire file. In the old days, you'd have all your code in one file, and to run it, you'd run the entire file. All the steps would have to be run each time because between runs, you'd lose everything that was created during the last run. Notebooks function differently and allow you to run one cell at a time and maintain the variables and other objects you've created. This is a great innovation, especially in data science, because when you're working with large amounts of data, you don't want to have to repeat time-consuming steps, like reading and transforming a given dataset. A couple other advantages of these notebook-style options are that you can format notebooks for easy presentation and notebooks are also fantastic for sharing with colleagues.

Conventions and Habits

There are many conventions, standards, and best practices in programming, especially in software engineering. But a lot of data scientists don't learn these because they're learning with smaller pieces of code, working on solving a specific problem. This leads to overall messy code that is especially hard to reuse, a problem that compounds on teams with multiple people all writing messy code. It's highly recommended that you try to follow good practices because once you've learned them, they're instinctive and you don't have to think about it to write good code.

I talked a little about commenting above. This is definitely one area where developing good habits is worthwhile. Your future coworkers will appreciate you and possibly pick up the habit from you. But even more important is the fact that future you will appreciate yourself. It's incredibly common to write code and later not remember how it works or why you wrote it that way. A few comments can go a long way.

Most languages have published standards and best practices, and Python and R are no exception. Python's primary style guide is referred to as PEP 8,[4] with a secondary one PEP 257,[5] both of which are over 20 years old and still followed. There are also some ways to format your code automatically through packages or coding platforms. R doesn't have such a high-level style guide, but Google published one that a lot of people use.[6] It's relatively short. A lot of new programmers don't think style is that important, and it's true that you don't need to follow it to learn, but it's really good to develop the habit early on. Style guides exist to make code more maintainable (as in possible to read and understand later). Follow them consciously for a while, and you'll have it internalized in no time, so writing clear and highly readable code is second nature.

One thing you'll encounter in the programming world is how to name variables and table columns. The style guides do address this, but in general, variable names cannot include a space or begin with a digit, and while column names may technically allow both, you should stick with convention and avoid them. There's a bit of a war on about capitalizing and separating words in a multi-word variable name. Historically, it was common to use camelCase (all but the first word are capitalized), which is often confused with ProperCase (all words are capitalized), but a more recent entrant that's gaining steam in Python is called under_case (nothing or everything is capitalized with words separated by an underscore). It's fairly conventional to use under case in Python for functions and variables, but ProperCase for classes. In R, the period is usually used instead of the underscore even though underscores are allowed.

---

[4] "PEP 8—Style Guide for Python Code" available at https://peps.python.org/pep-0008/

[5] "PEP 257—Docstring Conventions" available at https://peps.python.org/pep-0257/

[6] "Google's R Style Guide" available at https://web.stanford.edu/class/cs109l/unrestricted/resources/google-style.html

There are many other conventions that can be followed, depending on the language and team preferences. In some languages it's common to start all of your variables with a single letter v, or with var, and sometimes to include the data type in the variable name. You can have your opinions on all of this, but generally you will go with the style of whatever team you land on.

Following all these rules can seem a drag at first, but it honestly pays off in producing code that's easier to read—even if someone else wrote it or it's months after you wrote it. Many data science teams will have their own standards, and you'll need to adapt to that and follow it, but if you land on a team without these, you're still better off having your own standards.

Picking Your Language(s)

If you are trying to decide which language to learn, there are many things that can influence your decision. If you're in a degree program, they probably have a language they use throughout. Obviously learning that one is crucial. If you already have some experience in one, it makes sense to stick with that. If you're studying on your own, you can try both out and see which you like better. If you expect to work mostly on the statistics side and probably won't be needing to deploy machine learning models, R might be the right choice. It is slightly stronger in pure statistics than Python. For the same reason, it might also be a good choice if you really want to learn statistics deeply. If you want to be competitive for jobs that might involve developing in other languages, definitely pick Python. Python shares more in common with other popular languages like Java and C# than R does. In general, Python's more popular in industry and R in academia. Finally, if you have no idea which to pick and nothing points you to one or the other, I'd recommend picking Python. It's starting to edge out R in popularity, and it's easier to go from Python to other programming languages than from R. You can also do a survey of data science jobs and see how many ask for R vs. Python.

You might be tempted to learn both, but I'd recommend getting at least to an upper intermediate skill level in one before starting to learn the other. There are a lot of similarities, but also some differences that can be gotchas. Additionally, if you are learning Python, I'd highly recommend also learning the parts of Python that data scientists don't always use. If you look for Python tutorials online, if they don't mention data science, they likely aren't going to teach Pandas and scikit-learn, but they are good for learning about the other areas. If you look for R tutorials, you'll be jumping right into data frames and statistics.

When you're looking for a data science job, even when teams prefer one language over the other, usually they don't care which one you already know, because it's considered relatively trivial to pick the other one up if you're already strong in the other. It's just like with human languages, where once you've learned one foreign language, it's easier to learn a second.

Key Takeaways and Next Up

This chapter started by looking at the history of programming in statistics and data science. The two most popular data science languages—R and Python—have both been around since the early 2000s, and they're both still used. We then dove into various topics related to traditional programming, seeing that there are a lot of similarities between R and Python, but some important differences. Commenting in code is important. Variables allow us to represent values with a name, making it easy to reuse. There are many different data types available in R and Python, including integers, floats (decimals), strings, Booleans, lists, and data frames. Many different operations can easily be done in code, including mathematical operations and comparators. A couple of the most valuable things coding allows are conditional behavior and repeating code, like running something over several examples of something. This sort of thing is handled with what we call control flow in programming—things like if-else blocks (which allow you to do one thing if a particular condition is met and something else if it's not) and loops (which run the same code repeatedly over a sequence of values). Another important way to make code efficient and reusable is with functions, which are blocks of code that can be called with specific parameters. Formatting and organizing code is also important, and paying attention to best practices and standards is invaluable. There are some data science–specific aspects to coding in Python and R, most notably the data structure called the data frame, a way of efficiently storing and working with tabular data. Finally, there are several things to think about when deciding which language to learn, including your career goals.

In Chapter 12, we'll be looking at data collection and storage. We'll start by looking at the different types of data beyond what we talked about in Chapter 1. There are two ways data can be collected, manually and automatically, but many possible scenarios within those options. We'll talk about the different types of risks we face in data collection and ways to mitigate them (or at least know about them). We'll talk about who owns the data and who manages it wherever it's stored. We'll get into databases and other ways of storing data, including spreadsheets and text files. We'll finally look at how data scientists tend to be involved with data collection and storage, which varies tremendously in the industry.

```
PRACTITIONER PROFILE: MAGGIE WOLFF
```

Name: Maggie Wolff

Job Title: Data scientist, product analytics

Industry: Travel tech

Years of Experience: 4 as a data scientist, 8 in analytics

Education:

- MS Data Science

- BA Communication

The opinions expressed here are Maggie's and not of any of her employers, past or present.

Background

Maggie always loved math as a kid, even competing in math competitions in high school. She also started learning to code in high school, specifically HTML and Pascal. She graduated during the dot.com boom and started college as a computer science major, intending to be a web developer. But the first class was really intimidating, and she thought it wasn't for her. She switched to communications, hoping it would lead to a more exciting career.

Maggie's first job out of college was working on marketing for accounting conferences. She soon switched to a role with a "scrappy marketing department" at a hospital where among other things, she managed websites a bit and, most importantly, learned web analytics. One thing she didn't like about marketing was how subjective it often was, but working with the data felt different. She loved working with it—no one else on her team was interested, but she kept digging deeper using Excel and other basic tools. The thought that she could be an analyst of some sort occurred to her, but at the time didn't seem achievable, so it fell to the back of her mind.

Work

Several years later, a VP saw her interest in data and analytics and moved her into a data analysis role, which she loved. She learned R and Power BI and was excited about what was possible with the right tools, so she started working on a master's degree in data science. After that, she was burnt out on marketing and moved into a product analytics role, where she's been for a few years.

Sound Bites

Favorite Parts of the Job: The need for logic and analytical thinking. This is so different from marketing, and now she feels she can be more impartial and focus on the data itself. It's a much more black-and-white world. She really likes working with product managers specifically because they have a better grasp of data than a lot of stakeholders. She also loves how there's always something new to learn and how nobody knows everything, but we can always learn something from each other.

Least Favorite Parts of the Job: All the extra work outside of doing analysis and data science, like installing programs, writing documentation. Having to figure out incredibly messy data is also a chore. And the "always learning" situation can have its downsides, too—sometimes you're doing a lot to keep up, which can be overwhelming (plus you're occasionally doing things no one else has done, which can take time to figure out).

Favorite Project: At one job Maggie created a new metric called User Effort Index that measured the amount of friction a user experienced trying to book a trip on the company's website. The need for this came out of talk about A/B testing stakeholders wanted to do to measure effort, but there wasn't a good metric to use to represent effort. So she started digging online, but there wasn't much out there. On her own, she identified several points that indicate a user is experiencing friction. She ran with it and used machine learning to predict what level of friction would cause them to forgo the booking. This was valuable and was adopted at the company.

Skills Used Most: Thinking quantitatively, having a mind that just naturally is comfortable with numbers. Curiosity is hugely important because you need to want to find out what's in the data, not look for what you think is there. Time-boxing is important, especially when doing EDA (it's easy to go down rabbit holes). Good communication skills are critical because you're dealing with complex ideas and your audience might not understand, but you need to translate for them to get their buy-in.

Primary Tools Used Currently: SQL, Snowflake, Python notebooks, Adobe Analytics, Tableau, some Power BI

Future of Data Science: Things seem uncertain, but Maggie thinks AI will make us more efficient but probably not replace us. She hopes that people continue to aim to be multidisciplinary and understand industry and context, which is something humans do much better than AI.

What Makes a Good Data Scientist: Quantitative instincts, a willingness to learn, solid communication skills. Another important skill is understanding that no one knows everything and it's okay to not be perfect. People are often afraid to start things because they don't want to fail, but as they say in a lot of circles, "fail fast" and get better faster.

Her Tip for Prospective Data Scientists: Try to find a mentor to help you learn about different types of industries and jobs. Sometimes once you start working, it can be hard to switch.

Maggie is a data scientist with experience working in marketing and product development and management.

CHAPTER 12

Trying Not to Make a Mess: Data Collection and Storage

Introduction

Two more aspects of data that are important, even if they're often not given much thought, are data collection and data storage. Who gets the data and how do they get it? And then where do they keep it so it's accessible? The ancient Egyptians and Sumerians sent officials out into their territory and recorded and stored their data on clay tablets. Medieval tax collectors did the same but used ink on parchment or paper.

If you're reading this, you obviously know that there's a lot more data collection happening now and at a much faster pace than someone riding a horse from house to house and writing some numbers down. Also, you know data must be on a computer somewhere for anyone who wants to do data science. But saying you're storing data "on the computer" is not specific enough—there are many ways to store data digitally. We'll look at some of the ways it's stored as well as how it's collected.

© Kelly P. Vincent 2025
K. P. Vincent, *A Friendly Guide to Data Science*, Friendly Guides to Technology,
https://doi.org/10.1007/979-8-8688-1169-2_12

Examples of Data Collection and Storage in the Real World

Examples of the power of well-thought-out data collection and storage abound, often fitting in with larger infrastructure like a move to the cloud or networking hardware. But these two examples show the value of improving your collection and data storage approaches in the face of challenging and fundamentally messy information and the value of looking outside traditional database systems. As with many things, thinking outside the box can change the game.

Example 1: A Crisis with Puerto Rico Addresses

Puerto Rico was hit with back-to-back hurricanes in 2017 that knocked out power for weeks, decimated much of the territory's infrastructure, and stranded a lot of people. It was a massive disaster that the United States responded to through several agencies. People would apply for aid, and the agencies had to approve it and then get that aid to people.[1] To do this, they were using addresses that federal agencies had, including the US Census Bureau. These particular addresses were collected by census enumerators. But they quickly found that the addresses that applicants put on their applications were extremely difficult to match to the official addresses the agencies had on file. They needed to match the aid recipients with known records because they couldn't send it off to unknown people. Census enumerators and residents often have different ideas on how addresses should be written. If you've only ever lived American suburbia, it's hard to imagine this happening. But a lot of places have nonstandardized addresses, especially rural areas in many countries,

---

[1] "Agencies Mobilize to Improve Emergency Response in Puerto Rico through Better Data," July 26, 2019, Federal Data Strategy, available at https://resources.data.gov/resources/fdspp-pr-emergency-response/

including the United States. There are probably still some places you can send Grandma Reynolds a letter with just her name, the town, and the state and it would get there. However, this inconsistency in Puerto Rico made it incredibly difficult for those on the ground to deliver aid.

The data was low quality for two primary reasons: the collection methods were likely insufficient to get addresses formatted the way residents would write their own addresses. However, there were also likely some genuine inconsistencies in the data. People sometimes write their own addresses in different ways at different times, possibly considering the particular purpose for sharing it in a given situation. This would be problematic regardless of the guardrails applied in data collection.

There was obviously nothing that could be done about the quality of the data at the time, and the agencies did their best to muddle through and deliver aid. They used locals as guides and aerial photographs to help, but it was still incredibly hard. The following year, several federal agencies got together to figure out a better way to store Puerto Rico addresses, ultimately creating a working group to tackle the problem. One of the challenges is that addresses on the island have a component called the *urbanización*, which can sometimes be the only thing that distinguishes two addresses, as there are a lot of repeated road names. This element is not present in other US addresses, so agencies don't normally have to deal with it. Figure 12-1 shows a couple of these (fictional but realistic) addresses, where the first line is the *urbanización*. The only difference in the addresses is what comes after the "Urb" on the first line. These aren't geographically close to each other—they just have the same street name.

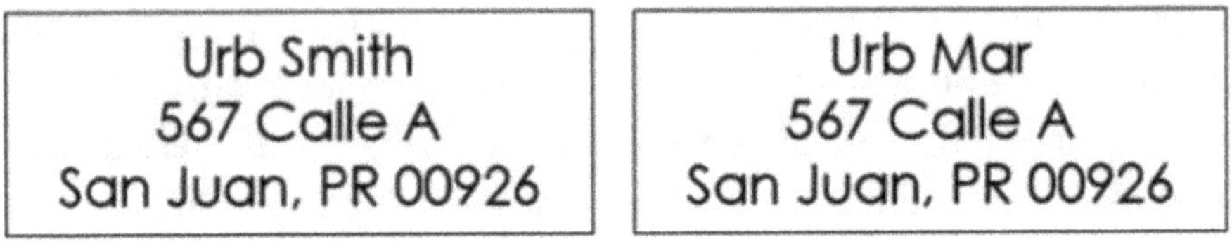

Urb Smith	Urb Mar
567 Calle A	567 Calle A
San Juan, PR 00926	San Juan, PR 00926

Figure 12-1. Two different but similar Puerto Rico addresses

The working group did manage to improve the situation with tweaks to data collection and storage, with some agencies doing on-the-ground work to verify addresses. Some also started using third-party, open source, or custom tools or data to validate and standardize and clean their data. All of this was necessary to overcome the limitations of data collection, both in terms of the challenges with humans collecting it and with the inherent nonstandard addresses.

Example 2: Shutterfly Says No to SQL

Shutterfly has a lot of data—billions of photos uploaded by millions of people who store, share, and buy products with the images printed on them. Their data storage infrastructure is massively important. Like most companies, they used relational databases in the beginning. I mentioned these in Chapter 1, but we'll talk about relational and other databases later in this chapter. Just know that the traditional and classic database system is relational and still widely used across organizations. The language these systems use (SQL) is a critical skill for data scientists. More recently, new types of database systems have been created, most of which are referred to by the umbrella label NoSQL because they're not relational.

Shutterfly found that with their relational databases, there were bottlenecks when people used their site.[2] These databases are also famously

---

[2] "Shutterfly brings scalability and user experience into focus with MongoDB Atlas on AWS," MongoDB, available at https://www.mongodb.com/solutions/customer-case-studies/shutterfly

expensive to scale up, which is usually done by adding storage, for instance. They decided to move out of the relational database world and ended up landing on MongoDB, a document database approach. We'll also talk a little about how this type of database works below, but the critical point here is they made a major shift in their database infrastructure and also how people had to work with the data. One of the benefits was that developers no longer had to design and follow rigid relational rules, but could instead store data "the way it appears in their heads," a huge change. The biggest benefit of MongoDB is that it's incredibly flexible, the polar opposite of relational databases. If the way they stored data in some of their MongoDB turns out to not be the best, they could easily change it. This could be done to improve the performance of querying or writing to the database, for instance. This makes scaling up to bigger and bigger systems trivial compared with scaling relational databases.

The results really were positive and the transition was not difficult. Shutterfly has multiple terabytes of data and used some migration tools to move their data from their relational databases to MongoDB on Amazon Web Services (AWS) in minutes and (importantly) without disrupting their website and services. The ability to scale up (or down) is especially important because Shutterfly sees peaks and troughs in usage around holidays and other key dates. The ability to scale down is itself somewhat unique, because usually companies need to make sure they have enough storage for their busiest times so they do have enough storage when they hit a peak. But with that approach, they would have more storage than they need all of the other times. Switching to MongoDB meant Shutterfly started saving money right away, with costs around 20% lower than their relational infrastructure. Figure 12-2 shows Shutterfly's front page, which wasn't impacted at all while they were migrating from a relational database to MongoDB.

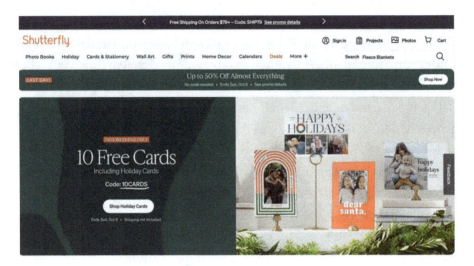

Figure 12-2. *Shutterfly powers their infrastructure with MongoDB. Source:* `https://www.shutterfly.com/`

Data Collection

In the bad old days, data collection was always manual. Someone went out and talked to people, or they observed something like factory activity, or they went outside and into nature and recorded things they saw. But nowadays, the vast, vast amount of data that's collected is obtained digitally. This may mean automatically (like websites that track your clicks or IoT (Internet of Things) devices collecting weather data) or manually via something like an online form. Sometimes manual is the only way, and often that data can be more valuable because it can get at info that is otherwise hard to get, like with researchers who interview individuals having specific experiences, such as a hospital stay or attending a particular event. We'll talk in the next two subsections about the different types of data collection.

Data collection is often a part of study design and Chapter 3 talks about that and sampling, so definitely look back at that if you're going to be collecting data as part of a study or experiment. But data is also frequently collected independently of any sort of study, so study design is not always relevant, and this chapter is focusing on the aspects of collection that must be considered after the fact and in the context of storage.

Manual Collection

There are many methods for collecting data by hand. All involve a person—the data collector—doing something to get data and recording it in some way. Many methods involve dealing with other people in the real world (or at least on the phone or virtually on a computer). These include interviews and focus groups, in-person surveys/questionnaires. These are commonly done in marketing and political assessment.

There are other methods where the data collector is still out in the world in some way but not dealing with people directly. They can still be recording data about people, as in the case of observation. For instance, a sociologist might watch how caretakers interact with each other while their children are playing at a park or count the number of teenage boys wearing shorts at a school on a cold winter day. Additionally, they may be observing something that has nothing to with people, like recording gorilla behavior or assessing the extent of fungal growth on trees.

Manual data collection can collect qualitative or quantitative data. Observation of people often involves some qualitative assessments of behavior, but usually even there the goal would be to get some numbers that can be crunched. Focus groups and interviews also can collect both, but there's often more qualitative data that the analyst will parse later to quantify some of it. Surveys and questionnaires usually produce more quantitative data, but they often include some sections where people can leave free text responses, which might also be later analyzed to quantify it. Figure 12-3 shows a typical questionnaire someone might fill out manually.

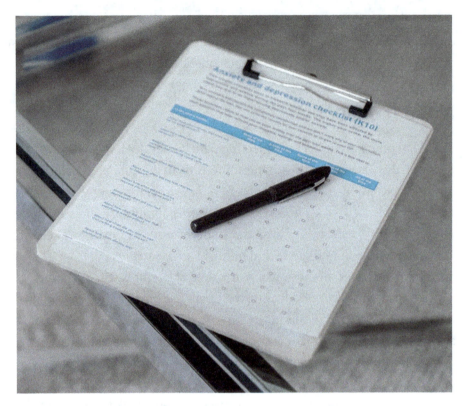

Figure 12-3. *A questionnaire for manual data collection.*
Source: Cropped version of a photo by Alex Green available at
`https://www.pexels.com/photo/crop-woman-in-office-with-`
`clipboard-5699437/` *and used under Pexels license (*`https://www.`
`pexels.com/license/`*)*

Automated and Semi-automated Collection

Automated data collection is the name of the game nowadays. It's hard
to do anything without something you're doing being tracked. Today I'm
sitting in Starbucks working, and I've been tracked since I left my house—
my car insurance tracks aspects of my driving through a phone app,
Google's tracked everywhere I've gone, there are records of my meal at the
Mexican restaurant I went to first because I paid with a credit card and

had the receipt texted to me, and the chai I bought at Starbucks is in both their transaction system and in my app since I used it to pay. If you don't want this kind of tracking done on you, you have to go to some lengths to avoid it.

So this category refers to both fully automated collection and semi-automated tracking. Fully automated is like what you see with weather and environmental sensors positioned around a large farm tracking environmental things and Google tracking everyone's location through their phone. The main point here is that neither side of the data collection has to do anything special for the data to be recorded, except initial setup (connecting the sensors to a network or someone having the right app on their phone). This would also include things like click behavior tracking on a website (the user is simply using the website as they normally would). See Figure 12-4 for an example of a weather sensor that automatically collects environmental readings.

Figure 12-4. *An example weather station that collects environmental readings. Source: From "File:2017-06-06 10 45 14 Present weather sensor on the Automated Surface Observing System (ASOS) at Ronald Reagan Washington National Airport in Arlington County, Virginia.jpg" by Famartin available at* https://commons.wikimedia. org/w/index.php?curid=63842646, *used under the Creative Commons Attribution-Share Alike 4.0 International License (*https:// creativecommons.org/licenses/by-sa/4.0/deed.en)

There are many tools that enable automated and semi-automated collection or at least make it easier, including barcode scanners (scanning UPCs at stores), RFID readers (getting info off a device with an RFID tag), optical character recognition software (scanning loan applications to extract fields), smart phones (phone GPS), and object recognition (license plate scanners). This is just a small sample, and new ways of collecting data emerge all the time.

Semi-automated data collection involves largely automated collection but where somebody has to do a little something special, like use a barcode scanner or key in values into a specific point-of-sale system. For instance, transactional tracking in retail is semi-automated because the cashier has to scan each of the customer's groceries (a manual process), but all the product details are retrievable from an existing database and from that point the transaction data goes into the database(s) automatically. Forms are another example of semi-automated, because someone has to create the form and someone has to fill it out, but everything else is automated. Figure 12-5 shows a typical multiple choice web form someone might fill out for a survey or study.

1. How many times have you visited [website A] in the last month?

- ○ None
- ○ Once
- ○ 2-5 times
- ○ More than 5 times

2. What was your primary reason to visit?

- ○ Make a purchse
- ○ Research before an in-store purchase
- ○ Comparison shop
- ○ Other

3. Were you able to complete your goal on the site?

- ○ Yes
- ○ No

Figure 12-5. *An example of a semi-automatic data collection form*

Quality in Data Collection

We've talked about how data is an abstraction of reality but how our aim is for it to meaningfully represent that reality. That means it needs to be accurate, which then means that the actual process of data collection—whether it's a person interviewing someone and writing down their answers or a bunch of weather sensors mounted on poles throughout an area—needs to be done well.

Fat Fingering, Typos, and Misspellings

These errors only occur during manual data collection. Just like with anything involving typing, data can come in with typos and other small errors. *Fat fingering* is what people call unintentional errors like typos. There can also be transcription errors, where someone working off a handwritten form might not be able to read what's written and will introduce an error on putting it into the computer, or recording errors, where the person writing down the original information spells something wrong or otherwise writes bad information (perhaps in the wrong field of the form).

One of the ways to deal with this is to have the data collector choose from existing options rather than always having free text fields. Also, depending on context, spell-checking can be done. Additionally, it's not uncommon to have a second person review any data entered by someone else.

Inconsistent Data

Another type of problem that occurs mostly manually happens when someone uses a field in different way on different records. A couple people collecting info on pets might put different info in the name field—one puts the owner's name in and the other puts the pet's name in. This especially comes up when there are different people collecting data, so training

410

people on what each field means is the best way to prevent this problem. Similarly, if subjects are filling out forms directly for their own record, clear instructions for each field ensure that this doesn't crop up as much.

Missing Data

Missing data is simply blank fields—basically, incomplete records. This can come about in both manual and automatic data collection, though it's more common in manual. It can happen when the data was truly never collected—someone didn't give the interviewer an answer or left a field blank in a form—or if it's unintentionally left out when transferring paper records into the computer, or a similar process of manually filling things out. Data can be missed during automatic collection as well, like if a user is surfing a website anonymously, we wouldn't have a username or if a particular weather sensor died, we would be missing data from that sensor. Missing data is always a hassle down the road, so we want things to be as complete as possible.

For the manual scenario, training the data collectors to understand how important complete data is will ensure they try to get data in all fields.

Misalignment Between Collection, Entry, and Storage

This is a problem generally in manual collection. Sometimes there are limitations on how data can be stored that can make it hard to save collected data. For instance, if the team will only have access to a relational database but the collected data isn't very tabular, it might be hard to enter it into the computer. If a system expects a single value in one field and two have been given, how would both be added or how would one be selected? If limitations like these are understood in advance, the data collection process can be designed to follow the restrictions from data storage and data entry.

Data Storage

Today, data is generally stored digitally in some way, but there are some things still stored on paper or even microfiche, a library mainstay pre-Internet (microfiche is a reel of transparent film with miniaturized versions of text and more printed on it). There are many more options for digital storage, including some old ones that aren't used for day-to-day stuff but are used for archiving. Some of the older types include magnetic tape (cassettes and video tapes), flash drives, and optical disks (CDs, DVDs, and Blu-Ray). All modern data that is intended for regular use ultimately lives on hard drives or solid-state drives, but these may be within an organization's own servers on premise or could be in the cloud, somewhere distant from where you're working on servers in a data center run by Amazon, Google, Microsoft, or even the organization itself. When magnetic tape or other types are used, it's for archiving (snapshots of data and systems for backup purposes, usually daily). I worked at a company where we made backup tapes every night and then took the physical tapes to another location once a week.

In the digital realm, there are several different ways to store data, but for the data that data scientists use, it's usually in databases, text files, or spreadsheets (mostly databases). There are many types or formats of all of these. I'll start by talking about spreadsheets and then text files, and then I'll move on to a much bigger discussion of database systems.

Spreadsheets

A spreadsheet is a common way to store data, and finance departments and many others have been using them for decades to store and work with data. They are pretty powerful, as the tabular structure is a natural way to store many datasets. When spreadsheets first came out, they revolutionized aspects of data analysis because of their charting and number crunching tools like pivot tables and v-lookups. While technology has moved on, many spreadsheet users have not. So data scientists will often find that they are pulling data from spreadsheets, even though they can be clunky to work with. Most data professionals view spreadsheets like Excel as the bane of their existences because of many limitations and poor use. Additionally, spreadsheets are useless with big data—the most popular spreadsheet software, Excel, can't handle more than a million rows.

Excel is the most common spreadsheet program out there, but there are some others. There are packages in both Python and R to read Excel sheets in. If a sheet is more than simple rows with a header across the top and data the rest of the way down (like if someone has put a title at the top and merged cells), it can still be read programmatically, but it's more difficult. Often if you end up with an irregular spreadsheet, it can be easier to copy and paste the relevant parts into a new file and save that separately. In this situation, you might even be better off saving it as a comma-separated value (CSV) file, which we'll talk about below. Figure 12-6 shows a spreadsheet with data on books that is difficult to work with. At first it looks just like a regular table, but if you look, you'll see that for books that have multiple authors, there's an extra line under that book only with data in the Contributor column and all others blank. This seems like a natural way to store data to a lot of spreadsheet users, but there are many issues with it.

ID	Title	Subtitle	Contributor	Contributor Role	Series Name	Series Number	Publication Date	ISBN	Publisher	Imprint
1	In Our Likeness	A Novel	Bryan VanDyke	Author			2024		Little A	
2	Competing Against Time	How Time-Based Competition Is Reshaping Global Markets	Gourge Stalk, Jr.	Author			1990	9780743253413	The Free Press	
			Thomas M. Hout	Author						
3	The Well of Saint Nobody		Neil Jordan	Author			2024	9781804549827	Apollo	
4	Kala	A Novel	Colin Walsh	Author			2023	9781838958626	Atlantic Books	Atlantic Books
5	Demon Copperfield	A Novel	Barbara Kingsolver	Author			2022	9780063251984	HarperCollins	Harper Perennial
6	Data Science for Undergraduates	Opportunities and Options					2018	9780309475631	National Academies Press	
7	Remember Me Tomorrow	A Novel	Farah Heron	Author			2024		Skyscape	
8	The Truth According to Ember	A Novel	Danica Nava	Author			2024	9780593642603	Penguin Random House	Berkley
9	Colton Gentry's Third Act	A Novel	Jeff Zentner	Author			2024	9781538756652	Grand Central Publishing	
10	The Loneliness Files	A Memoir in Essays	Athena Dixon	Author			2023	9781959030126	Tin House	
11	English as a Second Language and Other Poems		Jaswinder Bolina	Author			2023	9781556596575	Copper Canyon Press	
12	Autobiographies	Charles Darwin	Charles Darwin	Author			2002	9780140433906	Penguin Group	Penguin Books
			Michael Neve	Editor						
			Sharon Messenger	Editor						
			Michael Neve	Introduction Writer						
13	Reflections of Alan Turing	A Relative Story	Dermot Turing	Author			2021	9781803990125	The History Press	
14	Turing	Pioneer of the Information Age	Jack Copeland	Author			2012	9780198719182	Oxford University Press	
15	Butterfly Economics	A New General Theory of Social and Economic Behavior	Paul Ormerod	Author			2012		Pantheon	
16	A Deadly Education		Naomi Novak	Author	The Scholomance	1	2020		Del Rey	
17	Digital Transformation Success	Achieving Alignment Delivering Results with the Process Inventory Framework	Michael Schank	Author			2023	9781484298152	Apress	

Figure 12-6. *A sample spreadsheet that's difficult to work with because of "extra rows" in rows 15–17 that really go with row 14*

This makes this sheet hard to work with. Generally, you want one row per "thing" (whatever is being represented, in this case individual books). A better way to organize this information if it has to be in a single spreadsheet is shown in Figure 12-7. In this case, there are multiple Contributor columns, so each author can be added in a distinct column (some of the contributors extend beyond the image). This works here, but if there were 20 authors, it would be ungainly. These are considerations that also factor into database and table design, which we'll get into later.

ID	Title	Subtitle	Series Name	Series Number	Publication Date	ISBN	Publisher	Imprint	Contributor1	Contributor1 Role	Contributor2	Contributor2 Role
1	In Our Likeness	A Novel			2024		Little A		Bryan VanDyke	Author		
2	Competing Against Time	How Time-Based Competition Is Reshaping Global Markets			1990	9780743253413	The Free Press		George Stalk, Jr.	Author	Thomas M. Hout	Author
3	The Well of Saint Nobody				2024	9781804549827	Apollo		Neil Jordan	Author		
4	Kala	A Novel			2023	9781838958626	Atlantic Books	Atlantic Books	Colin Walsh	Author		
5	Demon Copperfield	A Novel			2022	9780063251984	HarperCollins	Harper Perennial	Barbara Kingsolver	Author		
6	Data Science for Undergraduates	Opportunities and Options			2018	9780309475631	National Academies Press					
7	Remember Me Tomorrow	A Novel			2024		Skyscape		Farah Heron	Author		
8	The Truth According to Ember	A Novel			2024	9780593642603	Penguin Random House	Berkley	Danica Nava	Author		
9	Colton Gentry's Third Act	A Novel			2024	9781538756652	Grand Central Publishing		Jeff Zentner	Author		
10	The Loneliness Files	A Memoir in Essays			2023	9781959030126	Tin House		Athena Dixon	Author		
11	Other Poems	English as a Second Language and			2023	9781556596575	Copper Canyon Press		Jaswinder Bolina	Author		
12	Autobiographies	Charles Darwin			2002	9780140433906	Penguin Group	Penguin Books	Charles Darwin	Author	Michael Neve	Editor
13	Reflections of Alan Turing	A Relative Story			2021	9781803890125	The History Press		Dermot Turing	Author		
14	Turing	Pioneer of the Information Age			2012	9780198719182	Oxford University Press		Jack Copeland	Author		
15	Butterfly Economics	A New General Theory of Social and Economic Behavior			2012		Pantheon		Paul Ormerod	Author		
16	A Deadly Education		The Scholomance	1	2020		Del Rey		Naomi Novak	Author		
17	Digital Transformation Success	Achieving Alignment Delivering Results with the Process Inventory Framework			2023	9781484298152	Apress		Michael Schank	Author		
18	Lesbians on the Loose	Crime Writers on the Lam			2015		Launch Point Press		Lori L. Lake	Editor	Jessie Chandler	Editor
19	A Natural History of Dragons	A Memoir by Lady Trent	A Natural History of Dragons	1	2013	9780765375070	Tor		Marie Brennan	Author		
20	His Majesty's Dragon		Temeraire	1	2006	9780593359549	Random House	Del Rey	Naomi Novik	Author		
21	Legends & Lattes	A Novel of High Fantasy and Low Stakes	Legends & Lattes	1	2022	9781250886088	Tor		Travis Baldree	Author		
22	Dirk Gently's Holistic Detective Agency		Dirk Gently	1	1987	9781478782997	Gallery Books		Douglas Adam	Author		
23	Nothing More to Tell				2022	9780593175934	Delacorte Press		Karen M. McManus	Author		

Figure 12-7. *The same data from Figure 12-6 with data organized better for a single spreadsheet*

There's another point on spreadsheets that's worth mentioning. Sometimes, stakeholders are so used to and dependent on spreadsheets that they expect their results to also be in spreadsheets, even if charts or other visualizations would be way better. This drives most data scientists crazy, and they usually try to gently introduce stakeholders to charts, which sometimes works. And…sometimes not.

Text Files

Plain text refers to files that can be opened in text editors like Microsoft's Notepad or Apple's TextEdit and still be readable. It's a common way to store data, with the comma-separated value (CSV) format used all the time in the data world.

A CSV file is literally just a text file that is formatted where a comma indicates a column separator. If the text being saved contains a comma, then double quotes are put around that particular field text. If the text contains double quotes, then an extra double quote is added in front of it. If the text contains double double quotes …well, you get the idea. There

415

is a way to handle anything. The nice thing is you don't generally have to worry about the details because you have programs that will read and write CSV files properly. If a nefarious or oblivious person gives you a CSV without proper handling of all these specific characters, it can be difficult to figure out, but it can be done.

Excel can open CSV files without difficulty, and you can generally edit them there safely and save them (assuming it's not more than a million rows). However, a CSV is just text, so it doesn't save any formatting you add in Excel, like bold column headers. If you used Excel to open a CSV version of the book data in that was in Figure 12-7, it will look the same without bold headers or different-sized columns. If you want to really see what a CSV truly looks like, you can open it in a text editor. The book one would look like Figure 12-8. A blank value is completely left out, so you can get a series of commas all together like at the end of most of the records shown in Figure 12-8, since most books don't have multiple authors.

Figure 12-8. *A view of the book CSV file in a text editor*

Serialized Files and Other File Formats

There are other ways that data can be stored, but using them in Python with Pandas and in R is generally simple because you just need to use the right library that's built to read such files, and then it's a line or two. Sometimes data or other objects are serialized and stored in what in

Python is called a *pickle*. Anything can be saved this way, and it's common to save machine learning models this way.

Another format that's becoming popular in cloud platforms is *parquet*, which is a good way to save a large amount of data because it can split across multiple files. Parquet files also work well with tools that split up time-consuming operations so the parts can be done simultaneously (this is called *distributed computing*, and it's common when working on cloud tools with large amounts of data).

There are more types of files for storing data emerging all the time, so it's good to stay on top of things.

Databases

Database systems are a daily part of almost all data scientists' work. Databases are generally better than spreadsheets and text files for a variety of reasons, including from a storage management perspective and because of the ease of working with them. Database management systems are software that handle the storing of all information in databases. Users interact with it through a few different ways, often with an interface hosted by the database that allows SQL to be written and run. Data scientists also use this interface, but much of the time, they interact with the system through Python or R code written in other environments.

In the tech world, the term *database* itself basically means a collection of related data that can be accessed through a database management system. You'll often see the systems themselves referred to with the abbreviation *DBMS* (database management system), which can be any type of database. Most DBMSs organize things at a high level into what they call databases, an object that can contain other objects like tables or other hierarchical objects. As a typical example, the Oracle DBMS has databases that can have many schemas, which have tables and other objects living inside them. A schema in this sense is simply an organizing object (for instance, permissions can be granted at the schema level

417

that will apply to all objects within it). In the Oracle setup, referring to a specific table requires the database name, the schema name, and the table name itself.

SCHEMA

The term schema is also used in the database world in a different way, to mean the overall logical structure of a database, often implying a visual representation of it. In most cases, it would include table names, table columns, and relationships between the tables. For non-relational databases, it would show different objects. It's very common for developers and data scientists to look at these to understand how to work with data in a particular database. Figure 12-9 shows the schema of some tables we'll be looking at below in the "Relational Database System" section.

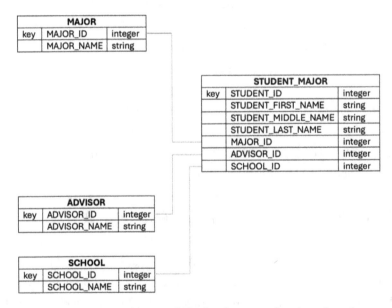

Figure 12-9. An example simplified schema of a tiny database

The diagram shows four tables. The STUDENT_MAJOR table stores the student name along with their major, advisor, and school. Three other tables hold only an ID and a value for major, advisor, and school. The lines show relationships between the tables. STUDENT_MAJOR stores a major ID that's linked to the MAJOR table and does the same thing for advisor and school.

This diagram is over-simplified as normally more information is conveyed, especially about the relationships between fields. Also, note that this kind of diagram is also called an *entity-relationship (ER) diagram* in the relational database world.

From a historical perspective, the majority of database systems used the relational organizing paradigm and a language called SQL to interact with it. More recently, non-relational database systems that use a paradigm called NoSQL have become more available because they are well-suited to certain types of data. As we saw, Shutterfly used a NoSQL database system in the example above. However, despite the availability of NoSQL, most data still lives in relational database systems. I'll talk about both types below.

Data scientists who work with data in databases primarily *query* the database by writing code that will bring back specific data from its tables, but they also create, modify, and manage tables. Every DBMS has some mechanism for interacting with the database in some way.

Relational Database Systems

Relational database management systems (usually abbreviated RDBMSs) are well-suited to structured data and rely on the relational data model (see the sidebar for more info). These and other database systems are ubiquitous in almost all organizations, as is SQL, the language that lets people interact with the data.

RELATIONAL DATA MODEL

The relational data model is a fairly intuitive way to handle tabular data. Columns are defined and individual rows are stored, with a value corresponding to each column. Typically, each table represents a specific thing (students, sales transactions, products, etc.), and they are linked together to establish relationships among the data. It describes the way many tables can be virtually tied together and defined with a language called SQL that enables people to easily do operations on the tables, such as retrieve specific data, add rows to a table, or combine multiple tables. It also revolutionized data storage in the early 1970s and was how data was stored for decades before alternative methods emerged.

SQL

SQL—short for Structured Query Language—is a simple language that allows you to interact with RDBMSs. Some of the most common RDBMSs include MySQL, PostgreSQL, Microsoft SQL Server, and Oracle. Each one has a slightly different flavor of SQL, but the majority of the syntax is consistent across systems.

Learning SQL best happens when working with data so you can try different things out. It's very different from most other languages, so learning Python or R doesn't directly help you, although some aspects of R and the Python library Pandas use SQL-like concepts to work with data.

SQL skills are crucial for almost all data scientists, but there's one other thing you'll be expected to take a stand on: the pronunciation of the language. Most people say it just like the word "sequel," but some people pronounce it "es-cue-el." Both are correct, but people get haughty about their preferred pronunciation, so you might want to see where you land so you're prepared when asked to pick a side.

RDBMSs allow several types of operations, but the ones data scientists use most are data definition language (DDL), data manipulation language (DML), and data query language (DQL). DQL and DML are used more than DDL by data scientists. DDL contains commands that allow users to define and describe the structure and relationships of data by creating, modifying, and deleting tables. Creating and modifying tables involves defining columns, including data types and unique identifiers. DDL also allows users to define relationships between tables, generally done with identifiers called keys. *Primary keys* are unique identifiers in a table where no two rows can have the same value in that key column (or columns). It's best practice to have a primary key in every table, usually just an integer value that increases by one for each new row, like a record ID. Foreign keys define relationships between tables. A *foreign key* is a column in one table that is connected to a key (usually a primary key) in another table, tying those tables together.

DML and DQL operations are more common for data scientists. DML encompasses the data-modifying commands, which basically means adding rows to a table with the INSERT command, modifying rows with UPDATE, or deleting existing rows with DELETE. There are some other commands, but these are the core ones. DQL only has one command, SELECT, which is used to query the database, or pull data out of it. This is by far the most common command that data scientists use.

One of the important concepts in the relational data model that relates to database design is *normalization*, which is the practice of breaking data into multiple related tables rather than storing it all in one big table. Storing everything in one big table usually results in lots of data redundancy, or repeated data, which vastly increases the physical size of the data being stored on a computer disk. The main goal of normalization is to reduce the size of the data stored. It does increase the complexity of working with the data, but the tradeoff is considered worth it. It helps to understand it when you are querying databases, too.

To understand database normalization, imagine a college that stores information on their students. They might store it in this simple way as shown in Table 12-1, often called a flat table because everything is in one place.

Table 12-1. *A small "flat table" with some basic student major data*

Student ID	First Name	Middle Name	Last Name	Major	Advisor	School
123	Sarah	Dawn	Morgan	History	Maria Nettles	Arts and Sciences
456	Vijay		Reddy	History	Maria Nettles	Arts and Sciences
789	Leroy	N.	Smith	Computer Science	Jack McElroy	Engineering
850	Veena	Swati	Srinivasan	Computer Science	Jack McElroy	Engineering
1000	Marvin		Garcia	Industrial Engineering	Linda Marlowe	Engineering
1111	Ted		Rowe	English	Maria Nettles	Arts and Sciences

But notice the redundant data: the values "History," "Computer Science," "Maria Nettles," "Jack McElroy," "Arts and Sciences," and "Engineering" all appear more than once. It's not always obvious, but text is always bigger than numbers in computer storage land. In a table this tiny, it wouldn't matter, but if there were thousands of students, the extra space required would start to add up. Normalization would call for splitting this table into multiple tables.

Considering what values are duplicated, a first step at normalization would be to pull majors, advisors, and schools into separate tables and assign a number to each value and store that number in this table. Those three main tables would look like Tables 12-2 to 12-4.

Table 12-2. *The normalized Major table*

Major ID	Major Name
1	History
2	Computer Science
3	Industrial Engineering
4	English

Table 12-3. *The normalized Advisor table*

Advisor ID	Advisor Name
1	Maria Nettles
2	Jack McElroy
3	Linda Marlowe

Table 12-4. *The normalized School table*

School ID	School Name
1	Arts & Sciences
2	Engineering

The new partially normalized table can be seen in Table 12-5. All of the majors, advisors, and schools are replaced with IDs that point to the value in the new tables. The IDs are still repeated, but numbers are smaller than text in databases so they take up less space.

Table 12-5. *The first pass at normalization of the student major table*

Student ID	First Name	Middle Name	Last Name	Major	Advisor	School
123	Sarah	Dawn	Morgan	1	1	1
456	Vijay		Reddy	1	1	1
789	Leroy	N.	Smith	2	2	2
850	Veena	Swati	Srinivasan	2	2	2
1000	Marvin		Garcia	3	3	2
1111	Ted		Rowe	4	1	1

If this stored thousands of students, it would be much smaller than the previous version. But it also requires foreign keys as mentioned above, which are what those numbers are under Major, Advisor, and School in the table. The fact that these numbers appear more than once in each column isn't very concerning because it's still taking up less space and it's still reduced redundancy-based storage space.

Those of you with an eagle eye will have noticed that the student major table is only partially normalized, because there is still some redundancy. Every row that has Major ID 1 will also have School ID 1, because the History major is always in the Arts and Sciences school. (Realistically, you'd have to confirm this, but this is generally how it works.) Additionally, if we assume that each major has only one specific advisor, we could pull all that out into another table and then only have a column for Major in the student table. We could modify the Major table to store all of this, using foreign keys to the Advisor and School tables. It would look like Table 12-6. Then the further normalized student major table can be seen in Table 12-7.

Table 12-6. *Further normalized Major table*

Major ID	Major Name	Advisor ID	School ID
1	History	1	1
2	Computer Science	2	2
3	Industrial Engineering	3	2
4	English	1	1

Table 12-7. *Further normalized student major table*

Student ID	First Name	Middle Name	Last Name	Major
123	Sarah	Dawn	Morgan	1
456	Vijay		Reddy	1
789	Leroy	N.	Smith	2
850	Veena	Swati	Srinivasan	2
1000	Marvin		Garcia	3
1111	Ted		Rowe	4

There are other advantages to a normalized structure. Imagine if the advisor for the History degree changed to a different person. That would only have to be updated in the Major table (to point to a different advisor in the Advisor table). Or if an advisor's name changed, it would only have to be updated in the Advisor table. In both of those cases, the change would be reflected everywhere. We could get to the advisor for a given student by looking at the Major ID in the student table, then looking at the Advisor ID in the Major table, and then looking at the row in the Advisor table with that last ID.

Note that the structure we went with in this table is not perfect. Storing the major in the student table might not be the best choice. What happens if a student is a double major? Any normal database system would not let you store two different IDs in the same column. So a logical next step would be to pull major out of the student table—and use the student table only for personal information about the student (perhaps with birth date, address, permanent address, and more). Then we would create a totally new table that stores just the Student ID and the Major ID—and if a student had more than one major, there would simply be two rows with that student's ID, with different Major IDs. An additional advantage of this approach would be the ability to track timestamps on a student's major— for instance, they were a chemical engineering major for one year before switching to history. We could track the start and end dates of each major, which would be impossible if everything was in one table. These are all options that can be considered when a data model is being designed.

One other thing is worth mentioning: notice that the advisor is stored as a full name. You almost never want to store names this way. It is generally considered best practice to store first, middle, and last names in separate columns, as we've done with the student names. This isn't always trivial to do after the fact, since it is not always clear how to split up a full name, as some people have two last names and it's the first of the two that is considered the primary last name, and there are other culture-specific naming conventions. In our example, we would ideally create a table with the first name, middle name, and last name. Then this table would store the advisor ID, like Table 12-8 shows.

Table 12-8. *A better Advisor table*

Advisor ID	First Name	Middle Name	Last Name
1	Maria		Nettles
2	Jack		McElroy
3	Linda		Marlowe

Just looking solely at this table, we have no idea what advisor or school is associated with a given student, but it's easy to run a query to join those tables together. Similarly, we could run a query to generate all the information in the original flat table by joining all the tables correctly. This is what data scientists do all the time—run queries that stitch different sources together to get the data we need to do the work.

This is an incredibly common way to work with data, so understanding the relational model is important—but it's definitely one of those things that gets easier and more intuitive with practice.

NoSQL Data

In more recent years, other ways of storing data have been increasing in popularity. The umbrella term *NoSQL* is often used to emphasize how they do not use the relational data model. These systems were developed in response to limitations of the relational database model, including with scalability, performance, and inflexibility. This is an area of data storage that is under active development. It's common to see NoSQL used in the cloud rather than "on-prem" (stored inside a company's own computer systems), but it can be set up anywhere.

One of the funny things about NoSQL databases is that many NoSQL DBMSs have built a SQL-like language on top of the system, despite it not being stored as a relational model. So working with a NoSQL database can often feel just like working with a relational database system, where all the differences are under the hood. This is because of how ubiquitous SQL is and the fact that NoSQL systems are new (most professionals know how to use SQL, but learning a new query language would take time).

There are several different main classes of NoSQL databases: document store, key–value store, wide-column store, and graph. *Document stores* hold semi-structured "documents," usually in JSON or XML format. They are very flexible and don't have to match each other, but they're not good for complex transactions. The database mentioned in the example, MongoDB, is a document store. Some JSON we saw earlier in the book could be stored in a document store, as can be seen in Figure 12-10, which shows three animal records in a document store.

```
{
   'animals': {
      'id': 1,
      'species': 'cat',
      'name': 'Marvin',
      'age': 14,
      'age_unit': 'years',
      'sex': 'neutered male',
      'breed': 'domestic short hair',
      'colors': {
         'color1': 'tabby',
         'color2': 'brown'
      },
      'deceased': True
   },
   {
      'id': 2,
      'species': 'cat',
      'name': 'Maddox',
      'age': 8,
      'age_unit': 'years',
      'sex': 'neutered male',
      'breed': 'Siamese',
      'colors': {
         'color1': seal-point',
      }
   },
   {
      'id': 3,
      'name': 'Pelusa'
   }
}
```

Figure 12-10. *An example of three records in a document store*

A *key-value store* holds keys that each appear only once in the database, and each key points to one value. The trick is that the value itself can hold key-value pairs. They are flexible and are suited to less structured data. At the top level, one key is stored and it points to a value, which can

429

hold many things. Figure 12-11 shows an example of three animal records in a key–value store, similar to the document store.

Key: animal1 Value: {"name": "Marvin", "species": "cat", "age": "14", "age_unit": "years", "sex": "neutered male", "breed": "domestic short hair", "deceased": "True"}
Key: animal2 Value: {"name": "Maddox", "species": "cat", "age": "8", "age_unit": "years", "sex": "neutered male", "breed": "Siamese"}
Key: animal3 Value: {"name": "Pelusa"}

Figure 12-11. *An example of data in a key–value store*

Wide-column stores are superficially similar to relational models because they also store data in tables and rows; however, columns aren't strictly defined at the table level and different rows can have different columns. This allows benefits of relational structure without the inflexibility. As you can see in Table 12-9, representing this basically just looks like a sparse relational table, but under the hood the apparently "null" values are not part of the row at all.

Table 12-9. *An example of three records in a table in a wide-column store*

ID	Name	Species	Age	Age Unit	Sex	Breed	Deceased
1	Marvin	Cat	14	Years	Neutered male	Domestic short hair	True
2	Maddox	Cat	8	Years	Neutered male	Siamese	
3	Pelusa						

The last type of NoSQL databases is *graph databases*, which store data in nodes and edges (the lines that connect nodes together). Nodes generally store nouns, like people, places, or things, and they can have additional properties. They're usually represented by ovals visually. Edges define relationships between the nodes (often using verbs). Figure 12-12 shows three animals and a house, all nodes, and three relationships (edges) defined as lines between the nodes.

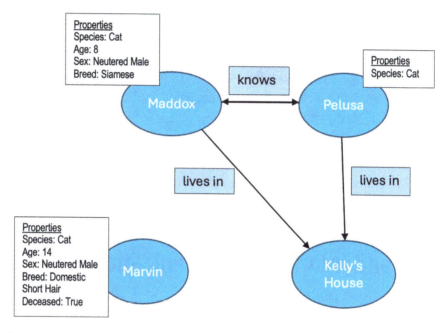

Figure 12-12. *An example of data in a graph database*

The NoSQL databases can be quite different from RDBMSs, and they often have specific use cases. You might never work with data outside an RDBMS, but being aware of other options may come in handy someday.

Data Care and Maintenance

I've already mentioned data cleaning and prepping, but I can't emphasize their importance enough. The next chapter will dive into data prep for data scientists, but there also is an element of data cleaning and preparation that is done more at the data storage level. This is clearer when there are different teams managing the data, like when analytics or data engineers are responsible for bringing the data into the standard locations that data scientists use. They often will do some basic transformations and apply simple business rules to get the data ready to use.

These data teams should ensure basic quality of the data, such as making sure that unique keys are truly unique in the data, no duplicates are present, and there are no empty fields where not allowed. Data scientists generally shouldn't have to concern themselves with this stuff, but the reality is that at a lot of companies, data engineers may not exist or may not do their due diligence. So it's important for data scientists to be ready to check the quality of the data they're using unless they're extremely confident in the data engineers' work. This confidence usually only comes about after data has been used a lot and all the kinks have been worked out or when the data engineering team is very mature.

Data and Information Organization

Although DBMSs ensure data is organized in certain ways, there are other ways of organizing data across multiple systems. There are a lot of terms floating around in this area, including data lakes, data warehouses, data silos, and data marts.

A *data silo* is a collection of data that is locked down so areas and systems other than the intended users cannot access it. This often isn't an intentional thing—often data silos grow slowly out of a particular area's needs. For instance, over time, HR has built up a lot of data that isn't shared outside of HR. That is probably appropriate for much of the data, but there may be parts of it that would be beneficial to other areas of the company without compromising privacy and compliance. Additionally, silos that have grown separately across different areas will have what is basically the same data stored in different silos, often in different forms. It may be exactly the same, which is inefficient and using unnecessary storage, or slightly different, which could lead to different results between the different areas. Generally, companies want to avoid data silos because they increase storage costs and the duplicated-but-different data can cause big discrepancies in analytics that use different silos' versions of the same data. Note that this isn't about access privileges—nobody wants the whole company to have access to the HR data, but parts of it should be available as appropriate to other areas and systems, and silos keep that from being possible.

A *data warehouse* is a centralized and highly structured system that brings together data from many areas into one to aid in reporting and analysis. Its target users are BI. The data warehouse organizing components are fact and dimension tables. Dimensions are generally the "things" and facts are the "actions." For instance, in a video game shop's data warehouse, there would be a dimension table of the store's products, which would have info like name, console, price, genre, and so on. There may also be a customer dimension table that has info on each customer, like name, membership number, and address. Then there would be a fact table representing all of the store's transactions—the purchases of different products that different customers make.

See Tables 12-10, 12-11, and 12-12 for such a small data warehouse. All the products are stored in a dimension table called Products (Table 12-10). This has general information about each one. Then there is a fact table Orders (Table 12-11) storing what are usually called order headers—one row per order even though there can be multiple items purchased. This information has several columns recording various things at that high level. The Orders table does not mention games at all, because games are sold at the line item level, which is what is stored in the Ordered Items fact table (Table 12-12). This one is associated with an order in the Orders table because it shows the information for each item in the order. It has two foreign keys in it—Order ID, which points to the Orders table, and Game ID, which points to the Products table.

Table 12-10. A tiny data warehouse dimension (Products)

Product ID	Title	Console	List Price	Genres	Release Date	Average Rating
1	Starbrew	Wii	39.99	Adventure;Puzzle	2018-05-16	3.45
2	Breed for Speed	Wii	49.99	Racing;Sport	2019-02-01	4.01
3	The Last Tribunal the First	Xbox	39.99	RPG;Turn Based Strategy	2020-07-14	2.98
4	Pineview Estates	PlayStation	49.99	Adventure;RPG	2021-10-31	3.90
5	Pineview Estates: The Backlot	PlayStation	49.99	Adventure;RPG	2022-01-08	3.99

Table 12-11. A tiny data warehouse fact table (Orders)

Order ID	Cash Register	Subtotal	Local Tax	State Tax	Order Total	Total Discount	Shipping Charges	Total Charged	Shipped Date	Transaction Start Timestamp	Transaction End Timestamp
1	1	29.99	1.05	1.95	32.99	10.00	5.99	28.98	2024-01-17	2024-01-15 15:03:17	2024-01-15 15:03:45
2	4	99.98	3.20	6.75	109.93	0.00	7.99	117.92	2024-01-19	2024-01-18 02:14:31	2024-01-18 02:14:42
3	1	109.97	3.29	6.30	119.56	20.00	7.99	107.55	2024-01-19	2024-01-18 17:58:19	2024-01-18 17:58:41
4	2	99.98	3.74	6.50	110.22	0.00	7.99	118.21	2024-01-19	2024-01-18 21:47:30	2024-01-18 21:47:48
5	1	49.99	1.17	3.50	54.66	0.00	5.99	60.65	2024-01-20	2024-01-19 01:39:42	2024-01-19 01:39:51
6	3	39.99	1.24	2.20	43.43	0.00	5.99	49.42	2024-01-20	2024-01-19 09:00:34	2024-01-19 09:01:00

Table 12-12. *A tiny data warehouse fact table (Ordered Items)*

Order Item ID	Order ID	Game ID	Subtotal	Local Tax	State Tax	Discount	Transaction Timestamp
1	1	3	29.99	1.05	1.95	10.00	2024-01-15 15:03:45
2	2	4	49.99	1.60	3.37	0.00	2024-01-18 02:14:42
3	2	5	49.99	1.60	3.37	0.00	2024-01-18 02:14:42
4	3	1	39.99	1.20	2.29	0.00	2024-01-18 17:58:41
5	3	3	19.99	0.60	1.15	20.00	2024-01-18 17:58:41
6	3	2	49.99	1.49	2.86	0.00	2024-01-18 17:58:41
7	4	4	49.99	1.87	3.25	0.00	2024-01-18 21:47:48
8	4	4	49.99	1.87	3.25	0.00	2024-01-18 21:47:48
9	5	2	49.99	1.17	3.50	0.00	2024-01-19 01:39:51
10	6	1	39.99	1.24	2.20	0.00	2024-01-19 09:01:00

There are different approaches to data warehouse design, but data scientists usually aren't involved in creating them. Data warehouse tables are usually "flattened" (less normalized), which is why they're convenient for BI (the users don't have to join tables together to create reports). It's also common to have aggregated tables in data warehouses. They're somewhat focused in purpose, but they can still grow very large. On the other hand, a *data mart* is simply a smaller, more-focused data warehouse, usually restricted to a particular area.

A *data lake* is a centralized and highly accessible repository of different types of data, basically the polar opposite of a data silo. Its contents can be sourced from other databases, data warehouses, and any other type of data. Data lakes are supposed to make data easy to work with so there are a lot of benefits to data scientists. One catch is that data lakes bring data into a centralized place without doing many (if any) transformations. The idea is that it's been made available, and users can figure out how it needs to be transformed based on their specific needs. Data lakes do require careful management (especially in data integrity and security), or they can become "data swamps."

Consequences of Storage Choices

All of the different storage choices come with pros and cons, often dependent on the intended use of the data. For instance, highly normalized databases are very efficient from a storage standpoint, but it can be costly to query them because many tables need to be joined together to get at all the information. This is one of the reasons data warehouses are used for BI—there are fewer tables to join together to get at one piece of information. BI reports query the database every time someone opens the report, and it's considered bad form if the viewer has to wait more than a few seconds for the report to fully load.

Security is also easier or harder to manage depending on storage choices. If the entire company's data goes into one database, the security team will have to manage all of the objects within that giant database. That's one reason it's more common for many databases to exist, each with certain purposes. Teams can be granted access to different databases, making access management a little simpler.

Data Ownership and Management

I've talked rather abstractly about the "other people" who manage the data, other than the data scientists themselves. Things vary at different companies, but common roles related to data collection and storage are data engineers, analytics engineers, and database administrators. *Data engineers* are the ones generally responsible for moving data around—doing transformations, cleaning, and so on—before depositing data into a database or another predetermined location. *Analytics engineers* are specialized data engineers, with more expertise in the needs of analytics, with the idea that they can make good transformation choices to make things easier for data scientists and other analytics users. Finally, *database administrators* (DBAs) are the ones who manage the database infrastructure itself, including (but not limited to) the database software, sometimes the machines the software is installed on, performance and optimization, and security and access rights.

I've also mentioned the importance of data quality and data governance. This is an area that is important, but often not done, or not done properly. While the DBAs manage the data where it physically lives and the data engineers handle moving it around, who really owns it and is responsible for how good it is? This is always a difficult question to answer. Data governance tries to answer this and also control other aspects of data management.

Data governance is a data management function that aims to ensure that the data is of high quality, usable, secure, and has high integrity. Some of this is high level and deals with issues like data regulations, but we're more concerned with how it relates to the work data scientists do. Usually, data governance people would work with security and DBAs, but not with data scientists. One of the things they do is try to identify data owners. Data owners can sometimes be expected to be very technical, but there are also cases where a business data owner is sought—someone with the business expertise on the data, as in domain expertise. Someone in HR should own

HR data, for instance. It shouldn't be a data scientist, data engineer, or DBA. But getting people in the business to own data is difficult because people don't really want the responsibility since it seems like a lot. The concept of data owners is also somewhat new, so people aren't used to the idea. The business data owner may not know all the field names and so on, so there also need to be other, more technical people who bridge the gap between the business and data scientists, BI, and so on (the ones who have to actually query the tables themselves). Realistically, sometimes the data scientists have to fill that role themselves.

The Data Scientist's Role in Data Collection and Storage

Data scientists have varying roles when it comes to both collection and storage of data. Most won't deal with collection very often. Those in marketing and other areas may run A/B testing or other experiments that can involve planning for data collection. Data scientists also don't often have to make storage decisions about core data. However, they will find themselves creating their own data sources to share with other data scientists or to use repeatedly within a project. Being aware of all the options can be important for that. Usually, it's as simple as storing a table in a team's database, but sometimes it's more complicated. If you just bear in mind that there are always tradeoffs with storage choices, you'll be ahead of the game.

Key Takeaways and Next Up

In this chapter we looked at both data collection and data storage. I talked about manual and automatic collection, where manual would be something like filling out a form and someone having to manually enter that data into a computer and automatic is where there's no direct human involvement to get data into a database. Most things nowadays are somewhere in between, like when you fill out a digital form—which you do manually, but when you click submit, the data goes into a database automatically from that point. There are a lot of ways errors can enter data, including typos, data that is inconsistent between records, and missing data.

I talked about the three primary ways data is currently stored, including spreadsheets, text files, and various databases. We dug into relational database systems and talked about normalization, a database design style. Relational databases have been the mainstay in data storage for decades, but I also talked about several different types of NoSQL databases, the alternative to relational databases.

In the next chapter, we'll start our discussion of data preparation by looking at data preprocessing. Data preprocessing is focused primarily on cleaning the data up and making sure it accurately reflects what it's supposed to represent. It includes things like dealing with missing values, duplicates rows, inconsistently formatted data, and outliers.

PRACTITIONER PROFILE: SHAURYA MEHTA

Name: Shaurya Mehta

Current Title: BI engineer

Current Industry: Retail

Years of Experience: 5 years in data analysis, data science, and BI

Education:

- Master of Professional Studies in Analytics

- BTech in IT

The opinions expressed here are Shaurya's and not any of his employers, past or present.

Background

Shaurya was interested in computers, so he went to university for IT. He started working in a software testing role where he was testing software for function and performance, a lot of it manual, so he learned about working with data. He especially enjoyed working on software that dealt with large volumes of data, and after more research into what sort of jobs were out there, he realized he wanted a career in data. He started learning on his own, but eventually decided he wanted a degree in it. He came to the United States for his master's.

Work

While working on his degree, Shaurya got an internship doing data science at a large retailer and loved it. He learned a lot and, after graduating, worked at the same company. His skills were developing in several areas. During the pandemic, supply chain problems cropped up at his company, and he got interested enough in supply chain management that he took a new job working in that arena. After a while, he transitioned into another interesting area, doing data analysis and BI work on data from the retailer's website. He enjoys the work because it's interesting to see changes over time and to help the website team make improvements to it.

Sound Bites

Favorite Parts of the Job: Shaurya finds digging into data fun because he always finds interesting things, but this also means learning about the data, which means he can do more with it later. All of this contributes to his ability to solve stakeholder problems, which is also really rewarding when he helps the company save money or operate more efficiently.

Least Favorite Parts of the Job: Dependencies on other teams can be frustrating, because often they don't have time to help you or do the thing you need because they have other priorities and you can be blocked, which leads to project delays.

Favorite Project: Shaurya has a couple of projects he really enjoyed. In one, deliveries were running late, and the company wanted to understand why. He worked on understanding the entire lifecycle of a product being shipped, including the time in the warehouse and when it was shipped out. Then he dug deeper into the shipping step by looking at data on vendors and the delivery services themselves (UPS, USPS, etc.). He actually was able to figure out where the delays were, and the company was able to use that to improve things. In another project, he looked into the processing of items in a warehouse and the number of people working on it. Looking at the data, he found that when there were more items being processed, there were actually few people and vice versa, which surprised everyone. This led to changes and major financial savings.

Skills Used Most: Having a good eye for data and the curiosity to want to know more about it. It can be time-consuming, but you need to go through your data and see what's in it and how it connects to the real world. Another important skill is patience when working with data and also people. It's also really important to find the story in the data. If you can't find the story, you need to ask different questions.

Primary Tools Used Currently: SQL, Databricks, SSMS for Azure, Excel/Google Sheets, Power BI

Future of Data Science: There is still more coming in the AI space. But we still need people because only they can truly have business knowledge. The ideal situation is that people will use AI to speed up their work. It's not just one or the other.

What Makes a Good Data Scientist: The most important thing is having an eye for data. Knowing the algorithms is not even close to enough—it's maybe 10% of what you need to understand. You need to be independent and willing to take the initiative and seize opportunities. Having a strong computer science background is also a huge benefit because it helps you write more efficient code. Also, it's really important to understand your company's business and of course the data, because only then can you come up with meaningful features for your ML models.

His Tip for Prospective Data Scientists: Choose a niche and develop your skills there even before you're looking for a job. Spend time researching that domain and learning about it as much as possible. But also be patient— sometimes things take time. But always have a growth mindset for skills in both coding and analysis.

Shaurya is an experienced retail data professional with experience in data analysis, data science, and business intelligence.

CHAPTER 13

For the Preppers: Data Gathering and Preprocessing

Introduction

By now, you probably know why we need to get the right data and prepare it properly—a forecast or analysis is only as good as the data that went into making it. The garbage in/garbage out maxim really is true in the data science world, and the chance that data in its rawest form will be good in a machine learning model without processing is very slim. That doesn't mean that the data doesn't contain massively useful info, but it's not inherently ready for prime time. Simone Biles didn't turn into a world-renowned gymnast overnight—it took years of prep, finding and honing particular strengths, and filling in gaps meaningfully.

Gathering data seems pretty self-explanatory, but it is not always straightforward to choose the right data for machine learning or analysis. As we saw in Chapter 10 on domain knowledge, you need to know a lot of the problem you're working on and the kind of data you need and have available. If you are collecting your own data, you'll want to pay attention

K. P. Vincent, *A Friendly Guide to Data Science*, Friendly Guides to Technology, https://doi.org/10.1007/979-8-8688-1169-2_13

to the previous chapter and Chapter 3's discussion of experiment design. More than likely, you'll be using data that already exists. This means you have to find it and then prep it for use.

What exactly does "data prep" mean in the real world? It's not just one specific set of steps. The prep you'll need to do depends on what you're trying to accomplish, the type of analysis or machine learning you're planning to do, and the type of data you have. Sometimes when data scientists talk about data prep, they're referring to all aspects of preparing any data to feed into a machine learning model, including both data preprocessing and feature engineering. In this chapter I'm focusing on the first phase, preprocessing, and will leave feature engineering for the next chapter. There isn't a hard-and-fast rule separating data preprocessing tasks from feature engineering tasks, but I'll focus on the basics, including cleaning, formatting, combining data, and handling "bad" values.

Examples of Data Prep in the Real World

Having the right data in the right condition is so important that it can make or break a project. As mentioned above, feature engineering is really critical to most machine learning, but you can't do good feature engineering if the data isn't ready for it. That's where data gathering and preprocessing comes in. In the first example below, a hospital had a system that was supposed to predict hospital readmissions of pneumonia patients in order to better decide treatment plans, but they realized far into the project that the data they were gathering was missing a crucial aspect of the information needed to make a decision, and their predictions were skewed because of it. They hadn't gathered the right data. In the next example, we look at how getting good results on one of the most popular machine learning problems that burgeoning data scientists try out, predicting which passengers would survive the Titanic, relies on good data prep.

Example 1: Missing Life or Death Information

This is a project aimed to help decide hospital readmissions of pneumonia patients in order to aid care decisions (and reduce costs). We know that different cases of pneumonia will have different levels of severity, and appropriate treatment depends on that assessed level, especially whether the patient needs to be admitted to the hospital or can convalesce at home. At this particular hospital, asthmatic pneumonia patients are always admitted to the hospital, into intensive care, regardless of the apparent level of their pneumonia's severity.[1] Because of this fail-safe for asthmatics, they ended up having a higher survival rate than the general population in the study. This effectively skewed the data because it made it look like having asthma was not a risk factor at all, which led to the model disregarding the presence of asthma in a patient in its model.

The way this happened was that the data on asthmatic patients was missing from the overall dataset, because they weren't evaluated based on their level. They were never fully considered in the process that the data scientists were trying to automate. It's always critical to make sure you have the right data to go into model development.

Although this was an honest mistake, if those on the project hadn't realized the problem and had released this system into the world, imagine the damage it would have had on asthmatic patients at hospitals that couldn't automatically admit them and instead had to decide using this tool as a guide. This would have been a huge violation of ethical healthcare.

---

[1] "Intelligible Models for HealthCare: Predicting Pneumonia Risk and Hospital 30-day Readmission" by Rich Caruana, Yin Lou, Johannes Gehrke, Paul Koch, Marc Sturm, and Noémie Elhadad, available at `https://www.microsoft.com/en-us/research/wp-content/uploads/2017/06/KDD2015FinalDraft IntelligibleModels4HealthCare_igt143e-caruanaA.pdf`

Example 2: Everybody's Favorite Machine Learning Problem

Kaggle.com is a popular site that hosts competitions and data for use by data scientists. It's a great place to cut your teeth on some real data science, and you can even use their platform to run code rather than install everything on your computer. People create notebooks with solutions and analysis and share it on the site, so you can learn from others and grow a reputation. Many of the competitions do have real monetary prizes, but the site is mostly beloved as a community for growing data scientists. So what's the example related to data prep? The most entered, rolling competition on the site is one for predicting surviving the Titanic disaster.[2] At the time of writing, almost 15,000 participants submitted over a million entries.

What has emerged out of this problem being solved over and over again is that the single most important part of a good solution is appropriate data preparation. When I checked the leaderboard, the top 335 entries had perfect scores (100% perfect prediction of survivors vs. non-survivors) on the test set Kaggle uses, which is not provided. Honestly, this is actually indicative of a type of overfitting (a problem with model training where the model doesn't generalize well and is too aligned with the specific data it was trained on), which we'll talk about later. We know that luck played some role for everyone who survived, so it shouldn't be entirely predictable, but we do also know that there were some features that were very important factors in predicting, including gender and class ticket.

If you look through some of the notebooks that contributors have shared with their high scores, you'll see that they've all done solid EDA, looking at each field for patterns and missing values.[3] They've looked at correlations to identify collinear features (ones that seem to be giving the

---

[2] https://www.kaggle.com/competitions/titanic

[3] For example, see https://www.kaggle.com/code/computervisi/titanic-eda

same information so we shouldn't include both). They used what they found in their EDA to choose the features to focus on, clean the data, and especially fill missing values. With so many people solving it, we know without a doubt that data preprocessing (and feature engineering) are crucial.

Data Gathering

The absolute first step in data gathering and preparation is to understand the business problem. What is it you're trying to accomplish by using the data? This influences the datasets you'll look for and use.

The actual process of gathering the data can sometimes be arduous. Unless it's data you already have access to, it will involve asking around to different people just to find the data and then possibly requesting permission to access it. Once you have it, you'll have to do EDA on it to be able to know whether you can actually use it, what fields you're likely to use, and what needs to be done to prepare it.

As part of gathering the data, you will also want to find out anything you can about how it was collected and especially if there's any known likelihood (or certainty) of there being errors in the data. This knowledge can be invaluable when dealing with problems.

The next part of the preparation, data preprocessing, is intended to make the data as usable as possible and ready for any significant feature engineering. It's also very common that you will discover in the data preprocessing phase (and also in the feature engineering phase) that you don't have all the data you want, which means you start back at the beginning and try to identify important sources you could get your hands on. This iteration is totally normal and may involve talking to stakeholders repeatedly.

Combining Data Sources

It's incredibly common for data scientists to need to combine data from different sources, also called *integration*. Most of the time, when considering relational tables, this means adding columns rather than adding rows. For instance, imagine a pizza delivery restaurant that has sales data in one table, where each sale includes a coupon ID if a coupon was used. They want to know how popular each of their coupons is and also calculate how much sales were "lost" in the discount. They have another table that has the coupon ID and other information, including the discount amount. If they combine those tables together—*join* them, in relational data language—they can see those patterns more easily in one table. Imagine they also want to learn what kinds of customers use each of the different coupons so they can have more targeted coupon sheets left at people's doors. Fortunately, they also have a table with some basic customer info. They could also join the customer data to the combined table already created, because each order is by a single customer and that customer ID is in the sales table. Once all the data is in place, they can do various manipulations—aggregations and filtering—to see what they can learn. For instance, they might aggregate sales at the weekly level to see if the weeks that they release new coupons are busier than other weeks.

Data Preprocessing Techniques

Some aspects of data prep might be done by data and analytics engineers, but data scientists will almost always still have to do some of their own prep. We often use rather imprecise language when talking about data preparation. You'll hear a variety of terms—cleaning, cleansing, wrangling, munging—all of which can refer to the same basic ideas. Data is almost never truly ready for analytics, as it always has *noise*—small to big errors, inconsistencies, and other surprises that can wreak havoc on analysis and data science work. One goal of EDA is to identify those values, and the goal of preprocessing is to figure out what to do with them and to make the appropriate changes.

Note that preprocessing can help us deal with many aspects of the data, even when it's not technically "wrong." We might decide to change the format of a date field to be stored in a certain order, for instance.

Duplicate Rows

Duplicate rows often hide in plain sight if you forget to look for them. But they can seriously bias models, especially if there's a particular pattern to why they're duplicated. Sometimes a duplicate row may be a literal duplicate of another row, with every column having the same value. These are the easiest to detect. Other times, only some of the rows need to be duplicated for the record to be considered a duplicate. If you had a table of people and two rows had the same Social Security number, name, and birth date, but different places of birth, something would definitely be fishy and you know one of those rows is bad. How to deal with duplicates depends on context and business knowledge. You always have to think about how a record got duplicated, which can inform what you do to fix it. Even in the scenario where every value is duplicated between two rows, there might be cases where they aren't truly duplicate entries—they could both be valid. For instance, imagine a table storing people's names, birth dates, and state of birth, but no other info. It's unlikely but not impossible for two people with the same names to be born on the same day even in the same state.

Formatting and Standardization

When data is stored properly, there isn't always a lot of formatting that needs to be done. But we aren't always so lucky. We want all the data in a particular column to be consistent and formatted the same, so sometimes cleanup needs to be done to make that happen. You might have a field that stores a weight in pounds sometimes and kilograms other times, indicated with another field (this is common in companies that operate globally, where things are stored in local units, like measurements and currency). It might make sense to create a new version of the table with two new columns that explicitly store weight in pounds and then in kilograms. This would involve converting the original field based on the second field that indicates which unit it is.

Formatting can be especially important with text values, where things like email addresses, physical addresses, and phone numbers can require standardization. Usually, this type of data is split into different columns and/or stored in raw form to avoid the need to format it in the database, but if you're dealing with data that hasn't been built this way, you may need to do some formatting. For instance, you'd prefer that phone numbers store just the numbers without punctuation, simply like 9435551234. But if some of the records store it with dashes only and others use parentheses for the area code, you will want to deal with that if you intend to use the number in any way.

A particularly common type of data that can be a nightmare because of formatting is dates. Major DBMSs have date, time, and date–time data types that can properly store dates and times, but it's not uncommon for them to be stored as text even in databases. And if you're working with text files or even native spreadsheets, you'll be working with text dates. Although humans are used to working with short forms like 1/4/24, that is not 100% interpretable without some extra information. Is it the month first, American-style, or is the month second? Is it 2024 or an earlier century? A lot of people who work with computers prefer the style 2024-01-04, which is always the four-digit year followed by the month and then day, both with leading zeroes. These are trivial for code to process, and they sort properly even if they're stored as text. But if the people who came before you weren't nice and stored text dates in another format—or, even worse, in different formats—you'll have your work cut out for you to standardize that. As the end of the twentieth century approached, there was mild worldwide panic over the "Y2K" bug, which was the fact that years had been stored as the last two digits only. It ended up not being that big a deal because the critical systems were fixed before the clock ticked over to 2000, but there was a lot of uncertainty over it.

Handling "Bad" Values

There are a lot of values in data that can be considered bad in some way, but the two most obvious are outliers and missing values. A lot of new data scientists instinctively want to just throw outliers out and reassign missing values the average of that column's other values. Both of these choices are almost always the wrong ones—but not in every case.

Outliers

First of all, outliers aren't necessarily bad data, because they might actually be correct. Sometimes things really are out of the ordinary. Other times, outliers are simply wrong. The mom of one of my friends had a driver's license that mis-listed her weight as 934 pounds—this was pretty obviously outside the realm of possibility for someone driving a car. If you had a dataset with this value in it, you would need to figure out what to do with it.

Outliers can also indicate that something else is going on. At a job I had at a retailer requiring membership, all of the orders were associated with a membership number. But there were occasional orders that were recorded with a dummy value of 99 because of some extenuating circumstances where the real membership number was not known. If we summed sales by member, member 99 would be leaps and bounds bigger than all other individual members. It didn't actually represent a member, so if we were looking at things as they relate to members, then we should throw those records out. But if we're not concerned about membership and only care about total sales at individual stores, we shouldn't throw those records out. Context and business knowledge matters. This is another reminder to think about what the values in the data actually mean.

In terms of identifying "outliers," there are fairly standard criteria, such as those that are below or above 1.5 times the interquartile range. But how you want to define them depends on your project and context.

Imputation

Missing values present another quandary that requires a decision, again based on context and what you know about what the data represents. Sometimes it makes sense to fill a missing value with another value, a process called *imputation*. Some of the common ways to impute are to fill empty values with 0 or the mean or median of the entire numeric column or the mode (most common value) of a categorical column. But what if those empty values indicate something completely different from the value you just dumped in there? In a numeric column storing credit card balance, imagine a bug in a system that couldn't handle a negative value (indicating a credit owed to the cardholder) and instead left the value empty in those cases. If you filled those with the average of all credit card balances, it would be hugely inaccurate. In this case, imputing with a 0 might be a good solution—but if you don't know why a value is missing, how would you know that that's the right approach for every single missing value in that column?

You always have to consider the data and context. Imagine a column called death_date. If that's empty, it might mean that the person isn't deceased, or it might mean the date is simply not known. If your data is of people who lived in the 1800s, you know it's the latter, but if it's people born in the 1900s, it could be either. Depending on what you are trying to do with the data, you probably don't want to fill it with some arbitrary value because anything you add will simply be wrong. Alternatively, you might put in an obviously unrealistic date like 12-31-9999, which you know is made up, but would allow some computations to be done on the field.

Note that this process is one of the ones that might make more sense to save for the feature engineering process, and look at it in the context of other features. For instance, if you know that some of the null credit card balances are there because of the problem mentioned above, you could explore the data a bit and see, if over the past few months, the cardholder has been charged no interest, you might be willing to infer that those particular cardholders have a credit and can safely be assigned a 0 balance.

A common technique to handle imputed values is to add a new Boolean column that indicates if the other column was imputed or not. That can go into a model, which can improve the performance. It can effectively use the value if it was not imputed and ignore it if it was.

Key Takeaways and Next Up

This chapter looked at the data preprocessing part of data preparation. We looked at a couple of examples where having the right data mattered and where data preparation was critical for the performance of a machine learning model down the road. We addressed the process of gathering data and combining different data sources. Then, we addressed some common preprocessing steps, including dealing with duplicate rows, formatting, and handling "bad" values. All of these steps can fall under different names, most commonly cleaning.

In the next chapter, we'll finish the data preparation part by covering feature engineering. This will address working with individual features to transform, combine, or split existing ones and create new ones based on those. There are also techniques that consider all the features as a single set and work on optimizing that. This includes feature selection and feature projection (or dimension reduction). It's generally a good idea to have fewer features, but only because having too many can be a problem. The most important thing is not to just have a few features, but to have the *right* features.

PRACTITIONER PROFILE: SETH DOBSON

Name: Seth Dobson

Current Title: Analytics engineer

Current Industry: Consumer packaged goods

Years of Experience: 10

Education:

- PhD Anthropology

- MA Anthropology

- BS Anthropology/Zoology

The opinions expressed here are Seth's and not any of his employers, past or present.

Background

Seth's background in college and through to his PhD was anthropology and zoology, where he ultimately specialized in studying facial expressions in non-human primates. He spent seven years in academia as a professor. The research was fascinating, and he enjoyed the data and statistical analysis part much more than most of his peers, so he developed more expertise in both. After a few years, he decided he didn't want to stay in academia. He started looking at ways he could use his data analysis skills and statistical chops in industry. He learned about the data analysis and data science fields and started looking for jobs there. At first it was difficult because his background was rather unusual, but ultimately a recruiter looking for someone with a background in behavioral science and statistics reached out to him. He'd landed his first job in industry.

Work

Seth found himself working in finance, which at first seemed odd for an anthropologist, but much of the work still related to human behavior and insights that could be gleaned from that. His specific work was not very deep into data science, but the role was positioned inside an advanced analytics center of excellence, so he was exposed to a wide variety of data science and analytics. He already knew R from his academic work, but he developed SQL and SAS skills there. He also found data science Twitter and started learning more about the field, also becoming increasingly excited by it. In his roles, he learned more about data (and analytics) engineering, and when he landed in a new data science role where there was insufficient data to do real data science, he pivoted into a new role, leading an effort to establish just such a data foundation. This has proved invaluable to data scientists, BI developers, and other data workers at the company.

Sound Bites

Favorite Parts of the Job: Seth loves the remote work in the job he has now, which makes it easier for him to have a good work/life balance. He also loves the social aspect in terms of connection and social activity. Meeting smart and interesting people is also interesting.

Least Favorite Parts of the Job: He doesn't love commerce and corporations and would prefer to work somewhere that benefits society in a clearer way. But he's still thrived here in a way he didn't feel in academia.

Favorite Project: His first big project at the finance company was about employee coaching, and he found a way to improve that very effectively. The employees he was working with had jobs with high turnover because they were pretty difficult, and there was a lot of burnout because of a focus on metrics. The company had started a coaching program to try to help employees develop the skills to make their job less difficult, but they weren't having very good results. The one-on-one coaching involved one employee helping or mentoring another in order to improve their skills, and usually the two employees were at different levels of the company hierarchy. Seth went back to his primates and body language roots and had the company (video) record the sessions, which he then analyzed afterward. He tracked things like smiling and other gestures and was able to quantitatively show that smiling—even if it was forced—made a huge difference. It's a classic prosocial behavior that builds trust, which was really important for the kind of coaching they were doing (which was primarily to help the coachee identify their own roadblocks). He created some visualizations that made it clear why some of the outlier coaches (the ones who had always been smiling) had overperforming teams. This work was very well-received and was incorporated in onboarding and training. He also won an award for this work, which gave him confidence in knowing he could do well in industry.

Skills Used Most: People skills are the most important. Anybody can learn coding, but not everyone can communicate well and develop relationships. If you have the most advanced machine learning model but nobody likes you, they won't use it.

Primary Tools Used Currently: DBT Cloud, BigQuery, GitHub, SQL, and R. Everything is all cloud now, to the point where he doesn't use any software that's installed on his computer (except the web browser).

Future of Data Science: The future of everything is AI. From an augmentation standpoint, there's room for AI to enhance a data scientist's work and make it easier. This doesn't even have to involve LLMs, but they can definitely improve throughput and work/life balance.

What Makes a Good Data Scientist: Intellectual curiosity and being comfortable with a lot of uncertainty because they go hand in hand. Being a good storyteller is also very important, since that's what you're doing at the end of the day (the data doesn't speak for itself).

His Tip for Prospective Data Scientists: Try to get a breadth of experience outside of data. Having a well-rounded intellect can be invaluable.

Seth is an experienced researcher and data professional with expertise in human behavior, data analysis, data science, and data engineering.

CHAPTER 14

Ready for the Main Event: Feature Engineering, Selection, and Reduction

Introduction

Working with features is part of the overall necessary data preparation process before running a machine learning model, and it can be massively important in getting good results. In data analysis, we don't usually do much feature engineering. Feature engineering is generally associated with machine learning because the features you use can make or break a machine learning model.

© Kelly P. Vincent 2025
K. P. Vincent, *A Friendly Guide to Data Science*, Friendly Guides to Technology,
https://doi.org/10.1007/979-8-8688-1169-2_14

So what exactly is a feature? We've used the term throughout the book interchangeably with "variable" and sometimes with "column," which isn't wrong. A *feature* is simply a value you have in your data that represents a characteristic of what your data represents, and it's the preferred term in machine learning, although features used in training can also be called *independent variables*. The outcome—whatever we're trying to predict—is called the *target variable* most often, but can also be called the *dependent variable*. I'm going to primarily refer to features and the target variable in this chapter.

Features can be understandable and intuitive or almost completely opaque, depending on the context and domain. In data about video game players, features could be intuitive ones like age, gender, console(s) owned, games owned, and game ratings and reviews. Or for something less intuitive, you could have data on a bunch of images, each with a hundred features with names like SHAPEFEATURE_1, SHAPEFEATURE_2, and so on.

In this chapter I'll be talking about *feature engineering*, which focuses on working with individual features and creating new ones based on aspects of the existing data, and *feature set improvement*, which involves looking at all the features as a whole and picking the right ones or combining them to make the overall set the most effective. Feature set improvement includes both *feature selection*, picking specific features among the ones you have, and *feature projection*, or *dimension reduction*, which takes existing features and combines them using mathematical techniques into new features that can be fed into a machine learning model. Feature projection is most common when you have a high number of features, hundreds or even thousands.

Examples of Feature Engineering in the Real World

It's generally accepted that feature engineering is part of the data preparation necessary before machine learning, but there have been some researchers who have demonstrated this in a couple of different domains by comparing machine learning done with and without good feature engineering. In both cases, it's clear that doing good feature engineering improved the models.

Example 1: Expert Feature Engineering

A research group studied the effect of including data scientists during feature engineering in addition to domain experts in two medical outcome forecasting applications.[1] The first predicted falls and fractures in patients. The second predicted bone defect side effects from antiepileptic drugs. They had forecasting models for and compared performance with feature sets that were selected simply by domain experts (the standard) and ones that involved a data scientist working with domain experts to bring in more sophisticated features. They did all the standard machine learning tasks that are appropriate, including data prep, feature engineering, feature selection, and balancing the dataset. The study ran the feature engineering and selection two different ways, with and without the data scientists. And what they clearly found was that while the performance was okay with only domain experts, bringing in a data science expert improved the

---

[1] "Case study—Feature engineering inspired by domain experts on real world medical data" by Olof Björneld, Martin Carlsson, and Welf Löwe in *Intelligence-Based Medicine*, Volume 8, 2023 available at https://www.sciencedirect.com/science/article/pii/S2666521223000248

performance dramatically. We'll talk about performance measurement in a later chapter, but in the first project, adding a data scientist took the accuracy from 61% to 82% and, in the second, from 80% to 91%.

Example 2: Detecting Radiation More Effectively

Some other researchers did a similar study that showed that among many techniques that can be used to achieve more accurate machine learning results, quality feature engineering can bring the most improvement.[2] They also found that feature engineering could also lead to better explainability in the model itself. They were working with radiation detection in three conditions: with obstructions, with background, and with neither obstruction nor background. Their model predicted the angle for the direction the source radiation was located. They used a random forest model and also did the normal data prep and machine learning training like the researchers in the previous example. They ultimately found that feature engineering improved the results in all three conditions by a large factor.

Feature Engineering Techniques

After you've worked through your data preprocessing as discussed in the last chapter, you're now ready to do the even more important work of fine-tuning and improving your features. There are many techniques that you may do at this point, but the first focus area is working with individual features, whether transforming existing ones and making interaction or other features.

---

[2] "Feature Engineering: A Case Study For Radiation Source Localization In Complicated Environments" by Matthew Durbin, Ryan Matthew Sheatsley, Patrick McDaniel, and Azaree Lintereur, 2021, available at https://resources.inmm.org/annual-meeting-proceedings/feature-engineering-case-study-radiation-source-localization-complicated

You will often find that many of your features are perfectly solid on their own, without much modification (beyond cleanup and other preprocessing). There's nothing wrong with using features as is, and you don't need to get carried away doing fancy feature engineering just because you think you're supposed to. But sometimes it can make a huge difference.

During your preprocessing work, you will have gotten to know your data. If you have a smallish number of features, maybe 50 or fewer, you've probably learned a bit about many, or even all, of the features. You may already have a sense for ones that are strong, ones that need to be investigated a bit, ones that seem unnecessary or redundant, and ones that likely shouldn't be included. In this case it can be easy to know where to get started with your feature engineering. But if you have a lot of features or if they are not very intuitive to work with, you may not have a great sense for the quality of your individual features.

In either case, you can do some more work to know what steps to take next. Some of the techniques that are discussed in the "Feature Selection" section further down can be useful to help get started working with features. But in addition to knowing your particular dataset, feature engineering is always hugely informed by domain knowledge. Knowing what types of transformations to make or how to combine features is much easier when you know the domain.

Just keep in mind that the whole reason for doing feature engineering is to find the most meaningful ways to represent whatever you're modeling or investigating, whether it's pizza restaurant sales or patient diagnoses, because that's the information that will be most valuable in a machine learning model.

Note that usually when you create a feature based on another feature (or features) in some way, the original features are usually excluded going forward (or you might just do the transformation in place).

Transformed Features

There are a lot of different ways you can transform your existing features to create new features. These are usually done considering only an individual value, but some may involve using info from the entire column. Many involve mathematical operations or statistical techniques.

A common one is the *polynomial feature* that's created by raising one feature to a power of two or more in order to capture a nonlinear relationship to the target variable. A statistical one could involve calculating a quantile, which would involve looking at the entire column before assigning a value to each row.

But there are many other possibilities. It's common in retail to track things in units. Imagine a pizza restaurant that wants to understand overall sales of their pizza. Because pizzas come in different sizes, there's no clear way to combine units of different sizes without losing information. But they could create a new variable representing square inches of pizzas sold instead, and then all those values could be summed. See Table 14-1 for what this might look like.

Table 14-1. *A new transformed feature for a pizza restaurant, Sq. Inches Sold*

Order #	Size	Quantity Sold	Sales	Sq. Inches Sold
1	Personal	1	12.50	45
2	Large	1	21.00	154
3	Medium	2	34.00	226
4	Large	1	25.00	154
5	Personal	3	37.50	135
6	Large	2	47.50	308

Often when you create a transformed feature, you wouldn't use the other features, specifically in the same model. This is because sometimes it leads to including the same information more than once, which can skew results. (This is basically the collinearity issue that I'll address later.) Because the new pizza feature (Sq. Inches Sold) combines Quantity Sold and Size (via a specific multiplier), we wouldn't want all three in the same model. We could look at including either Quantity Sold and Size or Sq. Inches Sold.

Interaction Features

Interaction features are new features you create from at least two other features based on some kind of mathematical operation, frequently multiplication or addition, but others are possible. These are usually done across a row alone. You don't multiply a couple features for the heck of it—interaction features are intended to capture the relationship between different features and how that relationship specifically might affect the target variable. If you have several related features, it might make sense to take the min, max, or average among them as a starting point for an interaction calculation, rather than including them all.

An example of an interaction feature that a pizza restaurant might use based on the new feature we added above, the square inches sold, is the total sales per square inch. See Table 14-2 for that calculation.

Table 14-2. *An interaction feature for a pizza restaurant, Sales per Sq. In.*

Order #	Size	Quantity Sold	Sales	Sq. Inches Sold	Sales per Sq. In.
1	Personal	1	12.50	79	0.158
2	Large	1	21.00	154	0.136
3	Medium	2	34.00	226	0.150
4	Large	1	25.00	154	0.162
5	Personal	3	37.50	236	0.159
6	Large	2	47.50	308	0.154

Often when you create an interaction feature, you would not use the original features it involves in further work, similar to transformed features. The new pizza feature involves dividing Sales by Sq. Inches Sold, and we also know that Sales is related to both Size and Quantity Sold, so it starts to get convoluted. But sometimes you don't know if the original features or your new interaction feature will perform better in the model, so sometimes you will test them all. This is why feature selection is important. We'll talk more about it below.

Dummy Variables

Most of the techniques we've talked about have applied only to numeric data. Another common technique that's especially useful with categorical data is creating dummy variables. *Dummy variables* are derived from a single original feature that has several values. Each value is turned into a separate column representing that value and a binary value indicating if it's the value in the original column. For a given row the dummy variables are all 0 except the one that matches the original column's value.

See Table 14-3 for an example involving the type of pizza ordered. Here, each row represents the order of a specific pizza, so we record what type of pizza it is only in a single column corresponding to that type with a binary value.

Table 14-3. *Four dummy variables created from the Type field*

Order #	Type	Supreme	Vegetarian	Meat-Lover	2-Topping	3-Topping	Multi-Topping
1	Supreme	1	0	0	0	0	0
2	2-Topping	0	0	0	1	0	0
3	3-Topping	0	0	0	0	1	0
4	Multi-Topping	0	0	0	0	0	1
5	Vegetarian	0	1	0	0	0	0
6	Meat-Lovers	0	0	1	0	0	0

Like the other types of features, generally when dummy variables are created, the original feature is excluded from further work. So we would use the dummy features moving forward in our model and not the Type feature. One thing worth mentioning is that if we have n different types, we really only need $n - 1$ dummy variables. In the above example, the final dummy variable (Multi-Topping) being 1 is exactly the same as the other five all being 0. Sometimes the final one is dropped for that reason.

Combining and Splitting Features

You may also just want to literally combine features. Imagine data that has age stored in two different fields, years and days. Once you've confirmed that the days column holds the number of days beyond what's in the years column, you can simply combine them. You could create a new field in total days by converting the years field to total days and adding the days column value or by converting the days column into a decimal representing the proportion of the year that's complete and add that to the years column. Either way, you've got a new column that has combined two others.

It's also common to split a column into more than one. This is different from creating dummy values because it takes data in a single column and splits that into multiple columns with specific values from the original column, all on the same row. It's common to store sex and neutered status in a single column in data about pets, but you might want to have that in two separate columns, one storing only sex and the other only whether the animal is spayed or neutered or not. These might be more valuable as binary features rather than a single feature with four values.

Table 14-4 shows an example of splitting and combining in some pet data. If we want a field that includes both the owner name and pet name, we could easily combine the last name and pet name into one. Similarly, we could split the joint sex and neutered status into new separate fields.

Table 14-4. *New features from combining and splitting, with original data on the left and new on the right*

Pet Name	Owner Surname	Pet Sex/ Neuter Status	Coded Pet Name	Pet Sex	Is Neutered
Michael	Smith	Male/Intact	Smith, Michael	Male	No
Marlowe	Jessup	Male/Neutered	Jessup, Marlowe	Male	Yes
Luna	Vintner	Female/Spayed	Vintner, Luna	Female	Yes
Bella	Rodriguez	Female/Spayed	Rodriguez, Bella	Female	Yes

Like with other feature adjustments, we usually would not use the old fields (the ones that were used in a split or combination) with the new ones in future modeling.

Encoded, Frequency, and Aggregated Features

Most of the techniques we've talked about in this section so far have involved operating on a single example or row. There are also many valuable features that can be created based on what data the entire column has. There are numerous ways to derive quality features, but we'll talk about three main approaches here: encoding, frequency counts, and aggregation. All of these things can be used in conjunction with the other techniques in this section, such as to create interaction features.

Target Encoding

A common technique is *target encoding*, where you encode categorical features based on the proportion of each value that occurs in a particular target value. It's especially common in anomaly detection tasks (like fraud detection—when the thing you're looking for is very rare), but can be used

in other scenarios. For instance, assume we have patient data for people
that includes type II diabetes diagnosis (present or not). We know certain
characteristics are associated with having diabetes, such as BMI, age, and
activity level. If we bucket both BMI and age and have activity level as a
categorical feature, we could see how many of each bucket have diabetes
present. For instance, maybe 18% of people aged 60–70 have a diabetes
diagnosis, so we could store the value 0.18 in the new target-encoded
age feature. Every example that is in the 60–70 age bucket would have the
same target-encoded age, but these can be very valuable in conjunction
with other features in a model. We could do the same with BMI and
activity level.

See Table 14-5 for an example of what this might look like. In this data,
we bucketed age, BMI, and activity level and calculated the proportion of
each that had a positive target value (a diabetes II diagnosis) and added
that value to the original data without keeping the buckets in the table.

Table 14-5. *Medical data with new target-encoded features*

ID	Age	BMI	Activity Level	Target-Enc-Age	Target-Enc-BMI	Target-Enc-Activity
1	56	29	Low	0.09	0.19	0.26
2	63	26	Medium	0.18	0.19	0.15
3	21	20.5	High	0.01	0.04	0.02
4	37	26.5	Low	0.04	0.19	0.26

Frequency Encoding

Another type of encoding is *frequency encoding*, which is similar to
target encoding, except you encode the count of the value of a particular
categorical feature. In the diabetes data example above, we could add a

new feature that stores the total count of the particular age bucket across the whole dataset. If there are 3,450 people in the age bucket 60–70, every example with that age bucket would have 3,450 as the value in the new feature. Table 14-6 shows the data from Table 14-5 frequency-encoded.

Table 14-6. *The same medical data as in Table 14-5 with frequency-encoded features*

ID	Age	BMI	Activity Level	Freq-Enc-Age	Freq-Enc-BMI	Freq-Enc-Activity
1	56	29	Low	73	131	192
2	63	26	Medium	121	131	113
3	21	20.5	High	11	26	19
4	37	26.5	Low	27	131	192

Scaling, Normalizing, and Binarizing

Scaling, normalizing, and binarizing all involve looking at a single column across all rows and transforming each row's value according to how it relates to the values in the column overall. Scaling sets a range with a minimum and maximum, like 0 and 1 in what's dubbed *min–max scaling*, which results in each row's value proportionally transformed to fit that range. We also sometimes want to get several, or even all, features on a similar scale, which we do through a process called *normalization* that scales many features with the same defined range. Note that there are some types of machine learning algorithms that require normalization, so it's an important one to understand. You could also *binarize* your data and convert all values to 0 or 1 depending on a threshold. Table 14-7 shows some student data before scaling, normalizing, and binarizing.

Table 14-7. *Some student data pre-transformation*

Student #	Age	Scholarship Value	# of Clubs	English Grade	Math Grade
1	18	0	2	81	89
2	20	1,500	0	57	60
3	21	25,000	3	80	92
4	21	17,000	0	93	94
5	18	9,000	4	90	78
6	22	0	1	49	81
7	19	12,000	2	96	91
8	21	750	1	75	81
9	20	21,500	3	86	79
10	18	15,000	0	62	58

If we scale Age, Scholarship Value, and # of Clubs with min–max scaling (a range of 0–1), we will have also normalized those because we're using the same scale for all three. We can also binarize the two grades on a pass–fail scale (60+ is passing). Table 14-8 shows what this would look like.

Table 14-8. *Some student data after scaling, normalizing, and binarizing*

Student #	Age-Scaled	Scholarship-Scaled	Clubs-Scaled	English-Grade-Bin	Math-Grade-Bin
1	0.00	0.00	0.50	1	1
2	0.50	0.06	0.00	0	1
3	0.75	1.00	0.75	1	1
4	0.75	0.68	0.00	1	1
5	0.00	0.36	1.00	1	1
6	1.00	0.00	0.25	0	1
7	0.25	0.48	0.50	1	1
8	0.50	0.03	0.25	1	1
9	0.25	0.86	0.75	1	1
10	0.00	0.60	0.00	1	0

Aggregation Encoding

We can also use group statistics based on aggregations. For instance, if we're working on credit card fraud detection, different types of cardholders may have different behaviors that are "normal" and not likely fraudulent. We can use group statistics to get powerful features. For instance, assume we have a huge table of individual transactions by thousands of cardholders. If we group cardholders into typical monthly spending ranges, with the lowest being up to $1,000 and one of the higher ones $20,000–30,000, a transaction for $900 might be out of the ordinary for the lowest spender, but a drop in the bucket for the higher spender. We might do something with several steps to create a new feature based on this idea.

We could take the average of all transactions by cardholders in each group and then take the difference between the average for the cardholder's group and the individual transaction value, so we'll get a new number that we can store as a new feature.

As an example, let's say we want to flag transactions that may need further investigation for fraud. We just want a simple screen that will only flag some, but we're not worried about false positives as these transactions will just be sent for further automated investigation. See Table 14-9 for an example of some transactions for three card members. Table 14-10 shows the average monthly spend per member through the previous month, and Table 14-11 shows a running total of the previous 30 days' spend for each member through a given date.

Table 14-9. *Some credit card transactions before transformation*

Trans. #	Transaction Date-Time	Member #	Transaction Amount
1	2024-05-02 13:04:53	1	450.23
2	2024-05-02 18:09:11	2	567.33
3	2024-05-03 00:34:30	3	75.65
4	2024-05-03 12:04:55	1	11,200.30
5	2024-05-03 14:21:39	1	101.92
6	2024-05-04 07:11:45	3	34.96
7	2024-05-04 10:34:22	2	120.55
8	2024-05-04 17:43:56	3	1,356.45
9	2024-05-04 23:54:23	1	1,204.67
10	2024-05-05 11:08:23	2	156.96

Table 14-10. *Average monthly spending per member through last month*

Member #	Member Average Monthly Spending
1	9,500
2	700
3	1,100

Table 14-11. *Running total spend over the last 29 days by member*

Member #	Date Through	Previous 29-Day Spend
1	2024-05-01	7,589.64
2	2024-05-01	2,587.45
3	2024-05-01	441.67
1	2024-05-02	8,039.87
2	2024-05-02	3,154.78
3	2024-05-02	441.67
1	2024-05-03	13,145.79
2	2024-05-03	567.33
3	2024-05-03	117.32
1	2024-05-04	12,545.61
2	2024-05-04	367.82
3	2024-05-04	1473.77

To determine if we should flag a transaction, we calculate a column called Remaining Spend by taking the member's average monthly spend from Table 14-10, subtracting the previous day's 29-day spend from Table 14-11, and finally subtracting all prior transactions for the current day. If this value is negative, we set the Inspect Flag column to 1. This can be seen in Table 14-12.

Table 14-12. *Flagging card members who are spending more than their normal amount*

Trans. #	Transaction Date Time	Member #	Transaction Amount	Remaining Spend	Inspect Flag
1	2024-05-02 13:04:53	1	450.23	1,460.13	0
2	2024-05-02 18:09:11	2	567.33	−2,454.78	1
3	2024-05-03 00:34:30	3	75.65	582.59	0
4	2024-05-03 12:04:55	1	11,200.30	−9,740.17	1
5	2024-05-03 14:21:39	1	101.92	−9,842.09	1
6	2024-05-04 07:11:45	3	34.96	947.72	0
7	2024-05-04 10:34:22	2	120.55	12.12	0
8	2024-05-04 17:43:56	3	1,356.45	−408.73	1
9	2024-05-04 23:54:23	1	1,204.67	−4,850.46	1
10	2024-05-05 11:08:23	2	156.96	175.22	0

Given how many of these transactions are flagged, this is a hypersensitive check and should probably be modified. But an approach like this can be powerful in a lot of situations.

Imputation

You may have already done some imputing—replacing missing values in your features—during your data preprocessing, but creating new features can sometimes involve new missing values. So you should check for this situation in your new features and impute as necessary.

Reshaping the Data

Another type of transformation you might do involves reshaping the data, most commonly through pivoting. *Pivoting* is a transformation that takes many rows and reduces the rows by creating new columns from the values in some of the original columns. It also involves aggregation, such as summing or averaging. For instance, take some data based on what we saw in Table 14-1, with some additional records, as you can see in Table 14-13.

Table 14-13. *Some sales data by pizza size*

Order #	Date	Size	Quantity Sold	Sales
1	2024-03-01	Personal	1	12.50
2	2024-03-01	Large	1	21.00
3	2024-03-01	Medium	2	34.00
4	2024-03-01	Large	1	25.00
5	2024-03-01	Personal	3	37.50
6	2024-03-01	Large	2	47.50
6	2024-03-01	Medium	1	18.00
7	2024-03-02	Personal	2	24.00
7	2024-03-02	Medium	1	16.50
8	2024-03-02	Large	3	59.00
8	2024-03-02	Personal	2	20.00
9	2024-03-02	Large	1	21.00

If we want to see the data per order without having multiple rows for a given order, we could do that. We could also aggregate at the date level. We want to see both quantity and sales summed up for each row. Table 14-14 shows the data at the order level and Table 14-15 at the date level.

Table 14-14. *Pivoted pizza sales data by order number*

Order #	Date	Qty-Personal	Qty-Medium	Qty-Large	Sales-Personal	Sales-Medium	Sales-Large
1	2024-03-01	1	0	0	12.50	0.00	0.00
2	2024-03-01	0	0	1	0.00	0.00	21.00
3	2024-03-01	0	2	0	0.00	34.00	0.00
4	2024-03-01	0	0	1	0.00	0.00	25.00
5	2024-03-01	3	0	0	37.50	0.00	0.00
6	2024-03-01	0	1	2	0.00	18.00	47.50
7	2024-03-02	2	1	0	24.00	16.50	0.00
8	2024-03-02	2	0	3	20.00	0.00	59.00
9	2024-03-02	0	0	1	0.00	0.00	21.00

Table 14-15. *Pivoted pizza sales data by date*

Date	Qty-Personal	Qty-Medium	Qty-Large	Sales-Personal	Sales-Medium	Sales-Large
2024-03-01	4	3	4	50.00	52.00	93.50
2024-03-02	4	1	4	44.00	16.50	80.00

There are other ways to reshape data, including unpivoting, which basically just reverses the process of pivoting. If we took Table 14-15 and did a quick unpivot on it, we'd end up with six rows per day, each one with a column with the values "Qty-Personal," Qty-Medium," etc. and a second column with the corresponding quantity or sales total (4, 3, etc.), like you can see in Table 14-16.

Table 14-16. *An unpivoted version of Table 14-15*

Date	Type	Quantity
2024-03-01	Qty-Personal	4
2024-03-01	Qty-Medium	3
2024-03-01	Qty-Large	4
2024-03-01	Sales-Personal	50.00
2024-03-01	Sales-Medium	52.00
2024-03-01	Sales-Large	93.50
2024-03-02	Qty-Personal	4
2024-03-02	Qty-Medium	1
2024-03-02	Qty-Large	4
2024-03-02	Sales-Personal	44.00
2024-03-02	Sales-Medium	16.50
2024-03-02	Sales-Large	80.00

Realistically, this is pretty ugly, and more work could be done to make it better, like having separate columns for Quantity and Sales. Sometimes transforming data takes several steps.

A Balanced Approach

As I mentioned above, most of the time you know something about your data and the domain it's in, so you know transformations and interactions that might be the most meaningful. It's always wise to do a little research to find out what transformations are common with the kind of data you have, because there are often conventional transformations or interactions.

Sometimes you can learn things or create a powerful model with just the right feature. So it can be tempting to just try every possible transformation or interaction and rely on the next step—feature selection—to find the valuable ones. This is a bad idea, however. Feature selection techniques aren't perfect, and the more junk (and let's be real—most of these features would be junk) that gets thrown at it (or a machine learning algorithm), the more likely you are to confound it. These approaches can sometimes find the needle in the haystack, but just as often, they'll find a bunch of features that appear to work well just from coincidence. You should stick to transformations and interactions that make sense. This is why data science is science and not just hocus-pocus.

FEATURE EXPLOSION AND THE CURSE OF DIMENSIONALITY

The name for the problem of too many features being created is *feature explosion*, when you have so many features that a machine learning model cannot estimate or optimize effectively. Feature selection is the most common task used to lower the number of features, but techniques called regularization and feature projection are also used. Both will be discussed below.

The *curse of dimensionality* is a set of phenomena that happen in high-dimensional data. It comes about because of how the number of features explodes combinatorially. The challenge is that you need a lot of records to meaningfully represent the many combinations of features. But high-dimensional data is normally sparse, and the examples are usually relatively "far" from each other. This means there's not good representation of the different combinations of features so there isn't enough info to train an accurate model.

So there are limits on how many features you should have based on the number of examples you have. One rule of thumb is to have at least five training examples for every single feature in your feature set. Although this works as a rule of thumb, there is always a balance in the number of features and training examples. Other factors can influence the number of examples needed, including the complexity of the model and how much feature redundancy there is.

But the reality is that sometimes you just have to try and see, as sometimes you can have a relatively high number of features to training examples and still get good results. I worked on a project that forecasted sales at stores, and we created a model for each individual store using five years of training data aggregated at the daily level. So that's only around 5 * 365, or a bit over 1,800, training examples per store. We had about 450 features, and at first we thought that was too many. But early exploration was promising, so we went ahead and deployed a model that worked well. It ended up running for over a year and got accuracies consistently in the low nineties most days (on legitimate unseen data). We also built a model explainer into our dashboard, which told us the features that had been most important in each day's forecast. We saw features that made sense and differed day to day, the sign of a healthy model. There are very few hard rules in data science.

Improving the Feature Set

Identifying and even making good features is hugely important in data science, but you still have to figure out which ones to actually use in your *feature set*. Regularization and feature selection are common approaches used to pick features and will be discussed below. There are additional techniques used to improve the feature set in an approach called feature projection (or dimension reduction) that combines features together so that there are fewer in total but the most important information is retained.

This part of data preparation can especially help with overfitting and underfitting. *Overfitting* is when your machine learning model is trained to perform really well on the particular data you have trained with, but it's too specific to that data and won't generalize to different data. Some of the patterns it's found in training are only present in that data. The whole point of a predictive model is to get forecasts on future, never-before-seen data, and an overfitted model will perform badly there. It's often considered akin to memorization, like when someone memorizes a bunch of facts but doesn't really understand any of them so they can't apply that "knowledge." We'll talk more about this—including how to detect it—in the next chapter.

Underfitting is the opposite—the model is too simple and doesn't capture the important patterns in the training data. This happens when you don't have the right features that can characterize the patterns.

Feature Selection

Although lasso regularization can effectively get rid of features, there are several other important ways to select features. They're generally grouped into three groups: wrapper methods, filter methods, and embedded methods.

Wrapper Methods

Wrapper feature selection takes the general approach of picking some features and seeing how they perform (based on an error rate) on a predictive model trained on some of the data and tested on another part of it, repeating this on a newly picked feature set, and so on. This can seem kind of backwards since you're trying to find the features to use in your own model and it's powering ahead and trying its own. These approaches tend to require a lot of computation power because they run through so many feature subsets, but they also can be highly accurate. The high computation cost can make it impractical (or even impossible) when you're working with a lot of features. Additionally, the resulting feature sets are sometimes good with only machine learning approaches similar to the predictive model used in selecting them, but this is not always the case.

One popular class of techniques is stepwise, which can be forward, backward, or bidirectional. *Forward selection* starts with no features and adds each feature one at a time, running a model to determine performance after each feature change. *Backward elimination* is the opposite, starting with all features and removing one at a time and testing that. Bidirectional moves both forward and backward until there are no more combinations. These approaches don't have to try every single combination, but can use statistical techniques to decide which to remove each round.

Filter Methods

Filter methods use some kind of calculation on features and then select ones that meet some criteria on that calculation (usually a threshold cutoff). The *variance threshold* approach can be used on numeric features and involves calculating the variance across a feature and dropping ones below a certain threshold.

Pearson's correlation is used in another couple filter techniques. First, correlation between features can be calculated. When features are correlated with each other, this is called *collinearity* (sometimes *multicollinearity*). When you have two or more variables that are highly correlated, you don't want to include more than one of them, because it overemphasizes that particular bit of information and can have wonky effects on your results. As mentioned above, when you have derived features from other features, you will often see collinearity when looking at all of them.

It's also common to check that there's correlation between a feature and the target variable, but this is never definitive because features can interact with each other to impact the outcome but not necessarily be strongly correlated with the target on their own. This is the primary reason machine learning is valuable—math can consider a lot more information at once than humans can, so it can find these relationships even when they're subtle or only important in certain cases. You can drop features below a particular correlation level.

Although correlation and variance are common filter metrics, there are other filter approaches. One involves using a calculation called mutual information that can compare pairs of features to help rule some out (similar to correlation, but useful for nonlinear relationships).

Like many things, filter methods require some nuance. Imagine a dataset of middle school students that includes age in both years and months. These two features would obviously be highly correlated, although the months feature contains a bit more info than the years feature. You wouldn't want to keep both. With adults it probably wouldn't matter which one you pick, but because these are fairly young people, months could make a difference if you're looking at something like maturity. There's a bigger difference between someone who's 146 months old and 156 months old than there is between someone who's 912 months old and someone who's 922. Like many things in data science, picking features can be an art that you learn about as you develop in your career.

Embedded Methods

Embedded methods combine the good parts of both filter and wrapper methods. Like wrapper methods, these methods test features on the performance of a predictive model run on a train and a test set of the data. The difference is that instead of only looking at an evaluation metric over several iterations, embedded methods actually pick features during the training process.

Regularization is a popular set of embedded techniques that reduces the impact of certain features to be included in a machine learning model in order to prevent overfitting and increase the generalizability of the model. Sometimes it might lower the performance on a training set slightly, but it is far more robust against data that's different from the training data (and would do better on the testing set).

There are two main types of regularization, ridge and lasso, and a third method that combines the two called elastic net. All of these involve complicated mathematical operations but can be done with a few commands in code. Both ridge and lasso penalize complex feature sets by coming up with weights (penalties) to be applied to individual features (a multiplier for the feature value). Lasso allows these weights to be calculated to be 0, which means it effectively removes those features from the set. Elastic net combines ridge and lasso linearly and can do a better job of picking the penalties.

The other main type of embedded methods are tree-based methods, including random forest and gradient boosting. Feature importance is revealed by the way that the trees are built during training. These techniques will output feature importance scores, and you can select the top *n* scoring features.

Feature Projection and Dimension Reduction

The techniques we've talked about so far have all maintained features in the original form before selection. Features might be removed, but they are intact. There's another group of techniques that combine features based on mathematical techniques to take us from high-dimensional feature space to low-dimensional. With these, the goal is to reduce the number of features, but there is no need to maintain the original features in any way. These methods can be great when you want high accuracy in your models but don't need to explain the forecasts in any way. As I've mentioned, transparency and model explainability can be really important in business, when stakeholders need to understand how you're getting the forecasts. But if that isn't your situation, there are many good techniques for making your feature space smaller. Note that when we talk about high-dimensional space, we're talking about a high number of features, as each feature represents a dimension of the data.

Principal Component Analysis

Principal component analysis (PCA) is the classic dimension reduction technique, first introduced in the early twentieth century, which does a linear mapping and maximizes the variance in the lower-dimensional space. This involves some linear algebra and statistics, specifically the covariance matrix and eigenvectors, but can fortunately be done in code. The way this works is by first standardizing the features so they're on the same scale. These new values are then used to compute the covariance matrix and do a linear algebra transformation. This creates multiple unit vectors (these are the "principal components"), and it's possible to then measure the proportion each component explains of all of the variance in the feature space. Normally, you do this and select the top components (how many depends a lot on your particular problem). These can be powerful, but the original features are completely obscured. Note that this does assume the data has linear relationships and may not work on nonlinear data.

Linear Discriminant Analysis

Linear discriminant analysis (LDA) is another commonly used method, which generalizes the statistical method Fisher's linear discriminant to find linear combinations of features that separate two or more. It's actually similar to ANOVA and linear regression and works with continuous data. Like PCA, it creates new dimensions that mathematically combine several original features. Unlike PCA, LDA uses class labels and aims to maximize the separation of the different classes. Yet it still results in a lower-dimensional space like PCA, with features obscured from the original.

Other Techniques

Both PCA and LDA work with linear data. There are variants of both that use methods to allow the handling of nonlinear data. Autoencoder and T-distributed stochastic neighbor embedding (t-SNE) are a couple of techniques for nonlinear data. There's another technique called maximally informative dimensions that keeps as much "information" (a concept that has a specific meaning in information theory) that's in the original data but still reduces the feature space.

Key Takeaways and Next Up

We covered four main approaches to improving your data before putting it into a machine learning model and sometimes before doing your data analysis. We consider both individual features and the feature set or feature space as a whole. We started with techniques that transform or combine individual features to get higher-quality features. Then we talked about improving the feature set as a whole, rather than individual features, which is done either by selecting the best features to create a stronger feature set, or techniques that mathematically represent the feature space and transform it, coming up with entirely new features that represent the ones that used to be there, but not in a one-to-one relation with the original features.

Coming up is the chapter that dives into machine learning. We've been building up to it, and we're going to talk about the two most common types of machine learning—supervised and unsupervised—and then a bit about some of the other types, including semi-supervised, reinforcement, and ensemble. We'll talk about the most popular algorithms of the different types and some of the considerations about what kind of problems they are commonly used for.

PRACTITIONER PROFILE: TYREE GILES

Name: Tyree Giles

Current Title: AI/ML engineer

Current Industry: Healthcare

Years of Experience: 10

Education:

- MS/PhD in Computer Science (current student)

- BS Mechanical Engineering

- BFA Painting

- BA Philosophy

The opinions expressed here are Tyree's and not any of his employers, past or present.

Background

Tyree loved understanding the world as a kid, but he wasn't a "math kid" and was even discouraged from going to college. He went anyway and studied philosophy and art. But when Malaysian Air flight MH370 crashed in the Indian Ocean in 2014, he was fascinated by the mystery—how could we lose an entire plane? He wanted to understand, and this led him toward engineering. He worked on a mechanical engineering degree and delayed taking a famously difficult weeder computer science class, but when he finally took it, he loved it and it was like talking to an old friend. He became a TA in that class and learned so much more and then did go through Andrew Ng's ML class. After graduating, he wanted to go into the ML world, but had to find a job that would hire him with a mechanical engineering degree.

Work

Tyree found the perfect job in telecommunications, where he worked on a proof-of-concept model to predict satellite failures. This was highly accurate, but the company realized they didn't have enough of the data they needed, so the project got cancelled. He learned about the extremely high failure rate of ML projects (estimated at 90%) and decided to learn more about DevOps and MLOps because the end-to-end ML project pipeline is often where things fail. When his company migrated to the cloud, he had to learn a whole new set of tools, and he realized that tech was an ever-changing world and data science and domain knowledge aren't enough anymore, so he started an MS program in computer science that will lead to a PhD. His current role is in healthcare, where he uses NLP on doctor's notes and other text sources to search for root causes of issues, identify outbreaks, and other things, all in service of public health.

Sound Bites

Favorite Parts of the Job: Tyree likes the people he works with on a tight-knit team and loves the problems he gets to solve. Some data science problems aren't that interesting (who cares if this green sells more widgets than that green on some website), but at his current company, they rotate assignments fairly quickly so it never gets old.

Least Favorite Parts of the Job: Sometimes fellow data scientists don't have good engineering practices, like following coding standards and best practices or testing code (ML code can be tested to some degree).

Favorite Project: One of Tyree's favorite projects was a mechanical engineering one he did as his senior project. Buildings experience thermodynamics—gaining and losing heat. The normal way this is measured is to shut a building down, fill it with smoke, and use sensors to determine where the smoke leaks out. But this is highly impractical, and he thought of an easier way based on experience working in construction. He used thermal cameras to take photos from outside and also placed temperature sensors inside and outside the building. He classified materials using semantic segmentation (glass, brick, etc.) and combined all that data to compute the energy being lost within 70% accuracy of previous smoke tests.

Skills Used Most: Critical thinking, inference process, being good at applying knowledge (not just having it), taking ownership and being accountable, and finally being a good communicator.

Primary Tools Used Currently: Python, PySpark, Databricks, PyTorch on Azure or AWS

Future of Data Science: There's a lot of talk about GenAI and LLMs, but he doesn't think they will dramatically change things in data science, even though he does use them alongside Google searches. Among people he works with, it seems like education is important and you need more than the basics. This doesn't have to be formal education, but you need to have some deeper technical knowledge to do well in data science now, where domain knowledge used to be the most important knowledge. This is especially true as neural nets become increasingly popular.

What Makes a Good Data Scientist: A solid grasp of the data science fundamentals. Being open to new ways of doing things and not limited to the algorithms you know or the ways things have been done before. Willingness to dig deeper technically, like learning more about how computers work, where you might learn different solutions to problems that other fields have dealt with that are new in data science.

His Tip for Prospective Data Scientists: Learn as much about computer science as you can, even taking courses like data structures, algorithms, operating systems, object-oriented programming. This can really change things.

Tyree is a data scientist with experience and expertise in telecommunications and healthcare.

CHAPTER 15

Not a Crystal Ball: Machine Learning

Introduction

Although GenAI (which relies on specific types of machine learning algorithms called neural networks that will be discussed below) is all the rage now, machine learning has been a part of our world for many years, in all sorts of fields. It's baked into fraud detection with banks and credit card companies. Amazon's been recommending products with it for decades. It's been helping us try to control the flow of spam emails for as long. It determines what we see on social media sites, and it's been used in healthcare for many years. Speech recognition has been keeping us away from human customer service representatives for longer than we'd like. And of course, businesses have long used it to forecast sales and almost anything else imaginable. Marketers, especially online marketers, embraced it early on. Map apps use it to warn you about unexpected traffic slowdowns. It's used in courtrooms to inform decisions about defendants. You get the picture.

Some of it is really good for people and some of it is really bad, with a lot in between. In this chapter, I'll cover the important aspects of machine learning as a discipline practiced by data scientists. We'll start with a couple of examples and then talk about the five types of machine learning. The key ones are supervised and unsupervised, but there

© Kelly P. Vincent 2025
K. P. Vincent, *A Friendly Guide to Data Science*, Friendly Guides to Technology,
https://doi.org/10.1007/979-8-8688-1169-2_15

are a few other approaches that can also be useful (semi-supervised, ensemble, and reinforcement). Then we'll dive into several supervised techniques, followed by unsupervised techniques, before moving on to some discussions of model explainers and machine learning coding and implementation. We'll finish the chapter by looking at some of the challenges data scientists face when carrying out machine learning, including the curse of dimensionality, data leakage, overfitting and underfitting, and the bias–variance tradeoff.

Examples of Machine Learning in the Real World

We know machine learning has been out in the wild for a while now, but our first example, the Netflix Prize, brought machine learning more into the mainstream conscience than anything had before, probably because of the $1 million award that was on the line. The second example is one use in the medical field where results have been making meaningful improvements in patients' lives.

Example 1: The Netflix Prize

In 2006, Netflix offered a prize open to anyone not associated with Netflix (with a few other exclusions) who could beat their in-house system that predicted user ratings by a certain amount (they used the performance metric root mean squared error, which we'll talk about in the next chapter). This would help them improve the recommendations they gave to users. If the system predicted a high rating for a movie by a particular user, it would make sense to recommend that movie to them if they hadn't watched it yet. And they'd want to avoid recommending movies that the user was predicted to give low ratings to.

Contest participants were given over 100 million four-field records for the train set, with user ID, movie ID, rating date, and rating. This isn't a lot of data, but within six days of launching the contest, a group had already beaten Netflix's in-house predictor. Within another week, two more had. This whole thing took the burgeoning data science field by storm, with several leading teams shifting placement on the leaderboard with each new submission. Over the next couple of years, there were two progress prizes to teams, many team mergers, and the final prize awarded to a team that was a combination of three original teams who'd done well early on. There was another team whose scores matched the winning team's, but the winners had submitted theirs 20 minutes before the other, so the first group of seven men walked away with $1 million (that's a very expensive 20 minutes).

The winning team was required to publish their approach, which was actually done in three separate papers. The three original teams each published a paper on their solution and the final solution that was a linear blend of those three models. These are the papers:

- "The BigChaos Solution to the Netflix Prize 2008"[1] by Andreas Töscher and Michael Jahrer

- "The BellKor 2008 Solution to the Netflix Prize"[2] by Robert M. Bell, Yehuda Koren, and Chris Volinsky

---

[1] "The BigChaos Solution to the Netflix Prize 2008" by Andreas Töscher and Michael Jahrer, November 25, 2008, available at `https://citeseerx.ist.psu.edu/document?repid=rep1&type=pdf&doi=f0b554683b425a0aad3720c8b0bd12 2eaa3c9b35`

[2] "The BellKor 2008 Solution to the Netflix Prize" by Robert M. Bell, Yehuda Koren, and Chris Volinsky, December 10, 2008, available at `http://www2.research.att.com/~volinsky/netflix/Bellkor2008.pdf`

- "The BigChaos Solution to the Netflix Grand Prize"[3] by Andreas Töscher and Michael Jahrer

Solving the problem involved some complicated modeling. Because the data was so simple and there was no obvious way to connect it to other data, there wasn't any data prep really necessary. So the teams were able to dive right into modeling. But with over 40,000 teams participating, it was obviously going to take some pretty sophisticated work to beat everyone else out. Many methods were tried along the way, but it was found that combining results from several different models (a technique in machine learning called ensemble learning) was necessary, but there was much tweaking of the ensemble solutions to reach the final one. The prize is considered to have advanced the machine learning field by having so many different techniques and approaches being compared directly on the same problem, a true apples-to-apples comparison.

One interesting thing about the whole situation is that Netflix had planned a second iteration of the prize, but instead canceled it because they got sued over privacy concerns. Although the lawsuit was later dropped, it serves as a big reminder about how it's almost impossible to truly anonymize any data that relates to people. Despite there being only four fields in the data with only IDs for the person and movie, someone was able to identify some of the people in the dataset by matching users and ratings on IMDB (the Internet Movie Database website).

Example 2: Machine Learning in Radiography

Analyzing medical images, scans of patients' bodies, is still an incredibly error-prone task for humans because it requires such intense attention to detail on top of expertise. Various solutions involving machine learning

---

[3] "The BigChaos Solution to the Netflix Grand Prize" by Andreas Töscher and Michael Jahrer, September 5, 2009, available at https://www.asc.ohio-state.edu/statistics/dmsl/GrandPrize2009_BPC_BigChaos.pdf

have vastly improved things by helping humans catch subtle things in the images indicating problems.

Breast cancer is the most common cancer worldwide. It can affect anyone but is most common in women over 40. It's well-established in healthcare that women should start having annual mammograms once they reach that age, meaning there are many opportunities to catch cancer even at an early stage. But it still relies on the radiologist detecting it. Every body is different, and some situations like dense breast tissue make it even more difficult to identify cancer.

The company Lunit has a tool called INSIGHT MMG[4] based on deep learning (neural networks, which we'll discuss below) that specializes in reading mammograms. They claim a 96% accuracy on cancer detection, which is obviously good, but one of the other benefits they provide is speeding up the process of detecting cancer from images. This tool in particular is good at detecting it even in dense breasts. With some retrospective studies (looking at earlier scans of breasts that were diagnosed later), they found that the system correctly detected 40% of those cases on the earlier scans—where humans had missed it.

This is exactly the perfect thing machine learning should be used for because it's only positive. It doesn't replace human expertise, but speeds things up significantly when there is no cancer, and also helps flag more difficult cases that doctors can then spend more time looking into, which improves the likelihood of survival by the patient.

---

[4] "Lunit INSIGHT MMG," available at https://www.lunit.io/en/products/mmg

Types of Machine Learning

Machine learning is fundamentally different from traditional programming because of what goes in and out of a program that's been written. In most traditional programming, we write code where we know what the output will be because that's what we're writing it to do. We can test every step of the way because we know the intermediate steps and what they should yield as well. Figure 15-1 shows the flow in traditional programming. The set of rules going in is the desired behavior that we write in the code, which the computer runs. Many programs will also rely on some data being inputted, anything from records to some simple configuration details. Think of the Computer box in the diagram as the software program we've written based on the rules. Running the program generates the output, the Computed Answers. In the cases with data input, we still know what the output should be for each data input or combination of data inputs, and any testing we do of the Computed Answers will be based on the known inputs and outputs. This type of software is also called *deterministic*, which means that every single time it runs with a specific set of inputs, the output will be exactly the same.

Traditional Programming

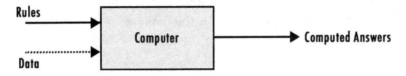

Figure 15-1. *Traditional programming flow. Rules are required but the need for data depends on the needs of the program*

An example of this kind of software is code that calculates aggregate sales values for a pizza restaurant. A programmer codes the rules for how the data should be calculated and enters the data of all the transactions and any other values to be included, and the computer generates the aggregations. If you run it with the exact same inputs, every time it runs, the output aggregations would be identical. As another example, back when there was panic over the Millennium bug in the summer of 1999, I wrote a Java applet Millennium countdown clock that ran on a company's intranet, counting down the number of days, hours, minutes, and seconds to midnight on December 31, 1999. The only data used in that program was the current system time.

There's one exception to the known output and determinism in traditional programming, and that is when we've coded anything using randomness. Programming languages can generate random numbers, which can be used to direct different behavior determined by which numbers were generated in the code. Imagine you want to randomly generate a Dungeons & Dragons character based on different numbers for their characteristics. A character is defined by a particular set of basic identities (like race and job) and numbers representing a variety of things like skills and abilities. For instance, if there are nine possible character races, you want it to randomly pick one, and you want it to randomly pick a number for the character's dexterity between the min and max possible. The output you want from this program is different and unknown every time you run it. However, you'd know the parameters of the output—all the ability and skill scores will be between defined mins and maxes, and you know what possible races, professions, and backgrounds they can have because it's one of a specific list. Another common use of randomness is in simulation, where scientists may want to define a starting point and see how things play out over time, which would be at least partially driven by random events. Chaos theory came out of weather simulations run this way.

Note that in programming languages, the randomness is considered pseudorandom because with a specific starting point, the random number generator will generate the same number in the same order, so in this case it's actually deterministic—the same every time—even though the program is still running with "random" values. This is sometimes the desired behavior, like when you're running an experiment and want it to be reproducible. This is controlled by setting a number that's called the *seed*—the starting point—when you create the random number generator in the code. If you want each run to be the same, you set the seed to the same number each time. In contrast, if you want nondeterministic behavior, you need to set the seed on something that varies, like by turning the current date and time into a number and setting the seed to that value. The output would be different every time.

Note that even many programs with randomness are still considered traditional programming. That's because inputs are known and outputs are based on the rules and the known inputs, and it ends there. Things are different in the machine learning world because you don't provide the rules in advance and never know what the output will be. The two predominant types of machine learning are supervised and unsupervised. I'll show diagrams of these two different types below in Figures 15-2 and 15-3 and another type, semi-supervised, in Figure 15-4. We'll talk about all of them below.

But first we need to address the two types of problems that machine learning is trying to solve: classification and regression. The goal of *classification* is to label data by predicting the target variable value—for instance, to determine what genre a video game is, whether someone has a disease or not, or what type of flower a given data point is. The number of possible labels can vary—I once worked on a project where we had personal accounts of people's experiences during World War II and we wanted to label them with the primary topic, which was a multi-class problem because there were seven possible topics. We ended up treating them as tags so each account could have multiple topics, so we looked at each label independently and had a binary model for each to determine whether it fit that topic or did not. This approach is not uncommon because it can be difficult to measure the performance of a multi-class model.

Regression is the other type of problem machine learning is used for and is what we call the forecasting of numeric data. Something like sales forecasting—predicting the next seven days of sales at a pizza restaurant—is a very common regression task. Others might be predicting the number of video game downloads or the number of patients in the emergency department each day.

There are different algorithms for each of these problems, but some techniques have variations that allow them to handle both. I'll talk about them below.

Supervised Learning

In *supervised machine learning*, we provide data to the computer so it can figure out patterns and rules within the data (this step is called training the model), and it outputs those computed rules in what we call a model. There is one minimum requirement of supervised ML—that we include the target value in the input data, which makes it *labeled data*. The fact that we are providing it with the target variable is why it's called "supervised" — we're providing it guidance.

There are two types of tasks supervised ML is generally used for, both of which require the trained model: prediction and inference. We'll talk in detail about each of these below. Supervised ML is really important in the business world and is also the most common ML done in organizations, with prediction the most common task.

Prediction

Once we have a trained model, we put it plus new data that wasn't included in the first step into the computer, and it outputs the predictions connected to the new data. Note that this new data needs to have the same features that the model was trained with. This can be a bit of a gotcha for new data scientists. I'm going to talk a bit more about what exactly has to go into a prediction-generating run at the end of this section.

It's not quite true that we don't provide the computer with any "rules" at all in the first step, for two reasons. The first is that we are providing specifically prepared data that has features we believe are important. This is why data preparation and feature engineering are so critical. The second reason is that the target value is provided. Figure 15-2 shows the two-step process of predictive supervised machine learning.

Supervised Machine Learning

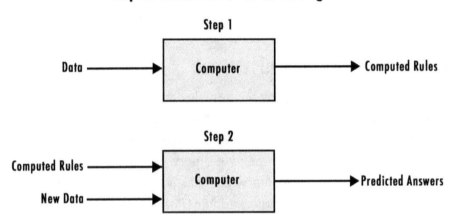

Figure 15-2. *Predictive supervised machine learning flow. The first step trains the model, and the second outputs the predictions*

Step 1 in the figure has a few specific parts. Overall, this is training the model, but in supervised ML, "training" involves testing before finalizing the model as well. The first step after preparing the data is to split the data into train and test sets, and also sometimes an additional validation set, in what's called the train–test split. See the sidebar for more info on the train–test split.

TRAIN–TEST SPLIT

There's a bit of an art to splitting data into train and test sets, plus validation if desired. Common splits are on percentages of the data, as 80–20 train–test or 70–20–10 train–validation–test. But how to pick the 80%? When you're not working with time-based data, there are many ways to pick your train and test sets. Usually this can be done randomly, using the same techniques discussed in sampling in Chapter 3. This approach means you will have good coverage and your test set will be testing as many of the patterns the model picked up in the train data.

If you're working with time-based data, especially with time series approaches, you will typically pick the latter 20% of your data for test and use the first 80% as train. One disadvantage with this is that your test data may not cover all the detected patterns in your model. For instance, if you have two years of data, you may only be able to test on five months of data, which is just over 20%. What if some parts of the year are easier to forecast than others? If those dates happen to fall in summer and early fall, the model may do really well. But when more time passes and you're running the model in late fall and winter, it might perform less well, and your original testing would not have picked up this fact. The vice versa could also be true.

Each of these train, test, and validation sets should be completely distinct from the others, with no data points appearing in more than one. Additionally, all this data must have the features we care about as well as the target value, whether that is a label like Has Diabetes or Does Not Have Diabetes or a number. Step 1 will take all that information and figure out generalized rules to get from a combination of features to a particular target value. That is the model, or the Computed Rules in Figure 15-2.

The high-level process is basically to train a model based on the train dataset (Step 1) and then test it on the test dataset to see how it performs (Step 2). We'd measure performance with a variety of metrics, which I'll be covering in the next chapter. We'd train many different models on the train set and test on the test set, trying different algorithms or different settings in each algorithm. Note that one of the parts of training the best models is not only trying different algorithms but also doing what's called hyperparameter tuning.

Most of the functions that you use to train an algorithm in R or Python have hyperparameters that need to be set when run, which are passed as arguments to the function when you call it in the code. These are things that define aspects of the machine learning algorithms that are being used. We'll talk about some of these when we talk about the algorithms below but an example is for a decision tree, when it would control how deep it could go. There are usually default values for most of these hyperparameters, but that doesn't mean those defaults are good for your particular problem, so this is a necessary step.

Hyperparameter tuning is the process of finding the best combination of hyperparameters that result in the best-performing model. There are different methods for doing this, but for now, just understand that this is what a validation set is used for. The tuning is done by picking hyperparameters and training on the train set, then testing it with the validation set, and repeating this until the best-performing model is found. The training set is then trained a final time with those hyperparameters and tested on the test set, and this gives the final performance of the model. There's also a method called cross-validation that involves dividing the training set into several different training and validation splits rather than using a dedicated validation set. This approach is especially valuable when trying to avoid overfitting, when the model is built too specifically to the training set and isn't generalizable. I'll talk more about methods for hyperparameter tuning and cross-validation below. There are convenient tools that allow you to do many different runs with a few lines of code.

After we've run through many iterations of Step 1 and Step 2 and identified and tested the best-performing model, the model is ready to use on new, unseen data to get the predictions for that data. That would mean rerunning Step 2 with the new data. A typical example of supervised ML is forecasting something like sales. A pizza restaurant may want to know how much dough to prepare in advance of the evening shift, so they could create a model based on inputs from their historical sales data and other data like weather, sporting events, holidays, and other date information (known data). Then they could provide that information to the model for the future dates (unseen data) and get forecasts of the total number of pizzas expected to be sold. This would help them know how much dough to prepare. The forecasts are the end goal of supervised machine learning.

What We Need to Make Predictions

Sometimes new data scientists struggle with understanding what we need to provide to a trained model to get predictions. As mentioned, we need to have values for all the features that the model was trained on. Let's say we have a model designed to predict daily sales at a store. It's February 1st and we have a model trained on the last three years of data, up through January 31st, and we want to forecast daily sales for the next seven days. We used a variety of features, including day of the week, week of the month, holidays, and several historical sales features. We have to have all of those features for the next seven days. Imagine a table with seven rows in it and a column for each feature mentioned above. Some people are surprised that we need to have "future data"—but the model is based on the values for each day, so they're required.

That means we need to be able to have those future values. Having features that are impossible to represent for the future is a mistake newbies often make. For instance, if you have a feature "yesterday's sales," you'd have no problem training a model with that. But you couldn't provide it when you're getting your predictions, because you have no idea what "yesterday's sales" would be for day 7, since we don't have the values of sales from day 6 yet. One way to deal with this is to use the forecast for day 6 as the "yesterday's sales" value in the day 7 prediction, but this isn't always the right solution because of the uncertainty.

Inference

The other use of supervised learning that's less common but still valuable is inference, which is where we generate a model in order to inspect it. The idea is to look at the features and values the model specified so we can understand what's important in the model. This could mean just throwing all the data into Step 1 from Figure 15-2 and seeing what comes out in terms of features and values, or it might involve running both Step 1 and Step 2 on multiple models to pick the one that performs best and then inspecting only that model. We would need to use a transparent algorithm for this to be useful, or a model explainer, in order to understand which features contribute to the output. In this case, we probably wouldn't even need to split the data into train and test sets. Instead, we'd train the model and inspect the model if it's an algorithm that will reveal what the features and values it used. Transparent approaches like linear regression and decision trees are perfect for this because they tell us exactly what features are important. But another tree method, random forest, will only output the most important features for the entire model, but not give any info on values.

Another option is to use a model explainer, a tool that will reveal the important features of a nontransparent algorithm (we'll talk more about these below). To use the model explainer, we actually run the train data through the model to generate "forecasts"—just the values the model would generate based on the provided features. We could then use a model explainer to get the most important features for each specific forecast and study that. In that case we might focus on the ones that are fairly accurate, but inaccurate ones could also be interesting.

One relatively new and developing area of machine learning is causal inference, which is specifically dedicated to finding the cause-and-effect relationships between variables. For instance, if you want to know what the impact of changing the price is on sales, you would normalize the other variables and look only at those two.

Unsupervised Learning

Unsupervised learning is similar to supervised in some ways, but the key difference is that we are not providing it with a target value, which is why it's termed "unsupervised." *Unsupervised machine learning* identifies patterns in the data without guidance (other than the feature engineering that was done during data prep), rather than figuring out how to get from particular features to a certain value as in supervised learning. We aren't actively teaching it.

In unsupervised learning, there isn't a requirement of a train–test split. It can be a one- or two-step process. The first step is basically running an algorithm on your prepared data so two things are generated: (1) the "answers" you asked the algorithm for and (2) a trained model you can run on new data with the rules the algorithm determined during its run. The most common unsupervised ML is clustering, which simply breaks the data into different groups, so the output is your original data with a group specified. This lets you see which data points are grouped together, which can let you dig in and find what characteristics this group has that are different from other groups. This may be the end of your immediate work, but you also have a trained model available to you. With optional Step 2, you could run new, unseen data through the model to get groups for that data as well. Figure 15-3 captures this flow.

Unsupervised Machine Learning

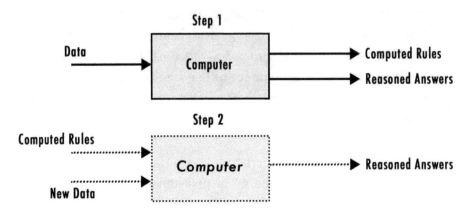

Figure 15-3. *Unsupervised machine learning flow. The first step trains the model and outputs the labels, such as groups. The optional second step allows the model to label new data*

Semi-supervised Learning

Semi-supervised machine learning is a combination of supervised and unsupervised because we provide both labeled and unlabeled data. Sometimes labeled data is expensive or very hard to come by, especially in the large quantities often necessary in machine learning. If unsupervised learning simply won't be enough for the problem being solved, there's a way to augment the existing data to get more "labeled" data that can be used in a supervised learning model. This can work well, but doesn't always. It's definitely a "your mileage may vary" approach that depends on your domain and data.

There are some general limitations to using this approach. One key one is that the unlabeled data needs to be similar to what's in the labeled data in terms of content. If we were trying to identify genres of songs and we had a set with labeled rock, pop, and reggae songs, adding in unlabeled songs that include country and rap in addition to the other three wouldn't lead to good results. This is pretty intuitive, but sometimes it's hard to know if your unlabeled data is similar to the labeled data or not. Some EDA usually can answer this question.

There are a few different techniques of semi-supervised learning, but I'm only going to talk about the self-training approach, a three-step process. Self-training semi-supervised learning starts with creating a *base model*, which is just a model trained only on the small amount of labeled data. Then we run the unlabeled data through the model to get our predictions—the labels for the unlabeled data, usually called *pseudo-labels*. One requirement of this second step is that there is a confidence generated with each prediction. You can then take all of the data that has pseudo-labels with a high confidence, often around 80%, and combine that data (the subset of the unlabeled data with highly confident pseudo-labels) with the original labeled data and train a new supervised model with that much larger dataset. Figure 15-4 illustrates the entire flow.

Self-Training Semi-supervised Machine Learning

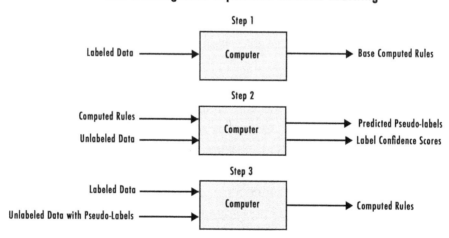

Figure 15-4. *The basic self-training semi-supervised machine learning flow*

You might have guessed that this is not usually a very robust technique and depends tremendously on the quality of the pseudo-labels. Because we're using so much more of the pseudo-labeled data than the real data, if anything goes wrong with that generation, the entire model will be wrong. This could happen if a feature is picked up in the original train set as being more important than it really should, when the feature would be overemphasized in the pseudo-labeled dataset. That doesn't mean you shouldn't try it, but you need to be careful and go through your results with a fine-toothed comb. Note that when it does work well, it's usually because points close to each other in the feature space usually have the same label, so pseudo-labels that appear close to others likely will have the right value.

Reinforcement Learning

Reinforcement learning is considered the third paradigm of machine learning, next to supervised and unsupervised. No training data is provided, but some rules of behavior are specified. Reinforcement learning is usually characterized as an "agent" exploring an unfamiliar environment and working toward a goal. Imagine a robot exploring a room trying to find a cabinet that has a door it needs to close (the goal). It's not going to get anywhere with just the goal—it also has to have a policy on how it makes decisions to pick something to do and a reward that it receives after taking desired actions. These goals, policies, and rewards are all specified in advance.

One of the fundamental challenges in reinforcement learning is finding the balance between exploration (like the robot exploring the room it's in) and exploitation (using the knowledge it has learned about the room). One of the interesting things about reinforcement learning approaches is that they don't always choose the locally optimal option (what would give the biggest reward at that particular moment) and instead can work with delayed gratification, making small sacrifices for a bigger payoff. Reinforcement learning's greatest strength is optimization— finding optimal paths, whether physical or virtual.

It's not used as much in data science as in other AI areas, but can be useful in marketing with recommendation systems. It was also used in AlphaGo, the system that beat the best Go player as discussed in Chapter 10.

Ensemble Learning

At its core, *ensemble machine learning* is not really a "type" of machine learning so much as a method of combining different ML models to get better performance than any of the models had on their own. Sometimes people refer to it as committee-based learning because of the way it combines results. The idea is fairly intuitive—it's known that when people are guessing the number of marbles in a large jar, for instance, many guesses averaged out will be closer to the right number than most individual people's guesses. The overestimates and underestimates tend to cancel each other out, each compensating for the others' inaccuracy. Even the best machine learning model has some limitations, like error, bias, and variance (we'll talk about these later), but when we combine several models, those problems are compensated for by the values in other models.

There are two primary methods to train and combine different models, either sequential or parallel. In the parallel approach, multiple models are trained independently on different parts of the training data, either with all models using the same algorithm (homogenous) or models trained with different algorithms (heterogenous). In the sequential approach, each model is trained and fed into the next model training. Then the various model outputs are combined at the end using a majority voting approach in classification (picking the prediction that appeared the most among the various models) and taking the average of all in regression problems.

There are three other techniques that can be used: bagging, stacking, and boosting. The word "*bagging*" is an amalgamation of "bootstrap bagging" and involves creating new datasets by resampling from the train set and training separate models with the same algorithm on each newly generated dataset. The results of the models are then combined to get the final predictions.

Stacking uses the heterogenous parallel approach by training several models and then training another one on the output of the first round of models. This approach of creating the last model is called *meta-learning*, because it's training on results that have come out of other models rather than the original data. The final model generates the final predictions.

Boosting is a sequential ensemble approach that improves poor predictions by focusing on running subsequent models with the incorrect predictions to improve them. A first model is trained on a train dataset, but it splits the results of that into correct and incorrect and then creates a new dataset including the incorrectly predicted records, but making some kind of adjustment based on those (like weighting them in the new dataset). This process repeats over several models, and they're combined at the end with computed weights.

Supervised Techniques

We're going to cover several popular supervised techniques in this section. There are more than will be mentioned, but a lot of the time, a data scientist's work involves sticking with the most common ones. And it's totally normal to hit up the Internet when your regular toolbox isn't enough.

Time Series Analysis

Time series analysis is an old statistical approach that's somewhat fallen out of favor because many other machine learning techniques do better, so I'm not going to say much about it. But it's good to know about it because there are times it can work. Time series analysis relies on time-based data, such as weekly downloads of a particular video game at a store. One limitation is that classic time series analysis does not allow features in the way that other ML algorithms do. We have the unit of time and the numeric value being tracked, such as sales or downloads. Figure 15-5 shows a typical chart we'd see in time series analysis.

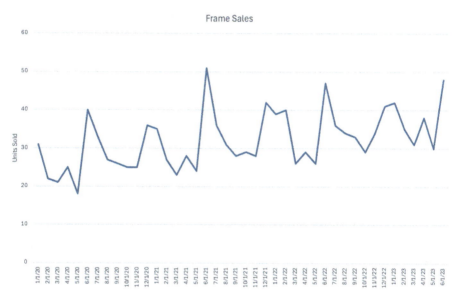

Figure 15-5. *An example chart suitable for time series showing monthly sales for a fictional small frame shop over a few years*

Figure 15-5 clearly shows an upward movement of the data. This is called the *trend* in time series analysis, a consistent general increase or decrease over a long period. We can also see *seasonality*, a pattern of regularly repeated increases or decreases, such as a boost in sales every June and December as we see in Figure 15-5.

These are pretty easy to spot visually, but there are tools for automatically identifying both trend and seasonality (or periodicity), as well as some other patterns. These tools let us *decompose* the original time series to see the individual charts showing the patterns like trend and seasonality. This is useful both because sometimes they're not as clear in the chart as we'd like and because we can use these characteristics in forecasting.

We can also use tools to generate forecasts based on the prior data in the time series. There are other algorithms based on time series that do a bit more. One popular one is algorithm *AutoRegressive Integrated Moving Average* (ARIMA), which looks at the time series from three different perspectives. The first is *autoregression* (AR), which focuses on the relationship between one data point and its prior (or lagged) points. The second is called *integrated* (I), which uses a difference method to effectively remove trend and seasonality. And the third is the *moving average* (MA), which considers the data point and the moving average of the past few observations. All of these are parameters that can be specified when running ARIMA. Note that there are some assumptions that are required to be able to use ARIMA, most importantly that the data is *stationary*, which means that the mean and variance don't change drastically over time.

Even though time series approaches don't always yield the best results, it's not too hard to try it out and see how it performs. But because we're usually only using one "feature"—the past values of the same variable we're trying to forecast—the patterns of the data must be confined to that variable alone. In other words, time series assumes that nothing external other than time features like month, day of the week, etc. determine the future values. Note that there are options that allow some external values to be included, notably a version of ARIMA called ARIMAX. It allows one additional variable, which can be a single feature or a combination of several (represented as one in the technique).

Generalized Linear Models

Linear regression is a very popular technique for both understanding data and doing predictions. It's also one of the first techniques done by statisticians and data analysts that falls outside of pure statistical testing.

It has the advantage of being quite intuitive and transparent, as well as easy to explain to stakeholders. Conceptually, *linear regression* specifically refers to taking a set of data—imagine a scatterplot with a bunch of points spread out on it—and drawing a line through it. There is a specific technique to draw the line, which we'll discuss below. That line represents the relationship between the two variables, or what the expected values would be within that plot—and if we look beyond the values already on the chart, we would see what the expected values would be outside that range. Forecasting, or looking outside of the range of the data we've built our model with, is called *extrapolation*.

Linear regression is a well-known specific technique, but it actually falls under the family of statistical regression techniques as *generalized linear models*. The name can be misleading because some of the techniques produce nonlinear models, like polynomial regression and Poisson regression, but the name holds. The two most common generalized linear models are linear regression and logistic regression. Linear is used for regression—to generate numeric outputs—and logistic generally for binary classification, generating non-numeric labels. Both will be discussed below.

Linear Regression

Linear regression is the most straightforward, intuitive, and simplest technique in the family of regression. The term "linear regression" technically involves modeling one variable (the target variable, or Y) based on only one other variable (the predictor of feature, X), and this isn't all that helpful most of the time, because it's rare that one single variable will predict another well.

One of the most common examples where using just two variables gives us a not-terrible model is weight vs. height, where height is X and weight is Y. Figure 15-5 shows a sample of 75 of the height and weight measurement from a larger dataset.

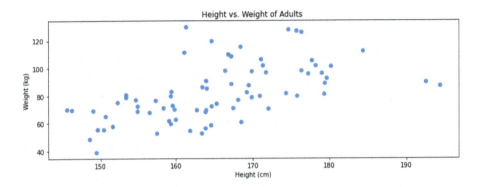

Figure 15-6. *Height and weight of adults*

There are several assumptions that must be made for a linear regression to be valid. The three most important are linearity, strong correlation, and X variable independence if there are multiple X values. We need to make sure that the data appears close to a clear straight line in a scatterplot, and there must be a reasonably strong correlation between the X and Y variables (usually greater than 0.5 or less than –0.5). People often don't bother to do these checks, but the regression won't give reliable results if they aren't met. Another important step when doing any regression is to understand the error that is inherent in the model (no model can be perfectly accurate).

Looking at Figure 15-6, we can see that the data is somewhat spread out and there isn't a line that all the measurements are clustered around, but there's a definite upward trend. Visually, this looks linear-ish. We can also calculate the correlation between height and weight. Using Python on this sample, the Pearson correlation came out to 0.565, which is over the 0.5 threshold. (A caveat: The original full dataset does not have a correlation over 0.5, but this particular random sample did, so I'm using it for illustrative purposes.) Now to the question of independence, it's pretty obvious that height and weight are not independent, because someone who's 6′6″ is not going to weigh 100 pounds unless something is very out of the ordinary. The same is true for a shorter person and a higher weight.

But we can set that aside for the ease of understanding linear regression. Honestly, there are very few sets of two variables that are not dependent that would yield a good regression, so we're stuck with this example.

The heart of linear regression is the method it uses for determining where the line should go through the points on the chart. We'll talk about that, but Figure 15-7 shows the above plot with the regression line through it. Intuitively, that feels right. If you'd drawn it yourself based only on Figure 15-6, it would probably look something like this. But how is it actually calculated?

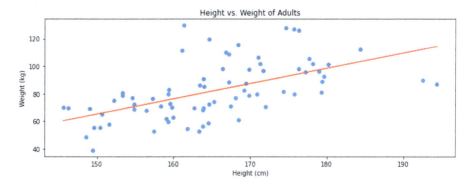

Figure 15-7. *Height and weight of adults with a regression line*

Linear regression calculates this line by minimizing the sum of all the squared errors for each point. The error is the vertical distance from the line to the point, but in regression, this is usually referred to as a *residual.* There are different ways to calculate the overall error, but the traditional linear regression, *ordinary least squares regression,* uses what's called the *sum of squared error* (SSE), which is calculated by squaring each residual and then summing all of those squares. Note that we square it because we don't want too-high values to cancel out too-low values, which would happen when summing. This approach is virtually impossible to do by hand because what it effectively does is try a bunch of different lines, calculate the sum of squared error, and then pick the line with the lowest sum. We can see how this works if we look at the residuals of the first 15 points in our dataset, as you can see in Figure 15-8. The black line connects the original point with the regression line, and the residual value is the length of the black line in the chart.

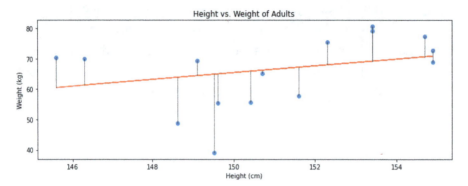

Figure 15-8. *First 15 data points in height and weight data with residuals shown*

That's how that line is created, but we still need to know what the actual model looks like. Fortunately, it's a simple equation that goes back to high school math. It's basically the $y = mx + b$ formula we learned for plotting a line. Remember that y is what's plotted on the Y-axis; m is the *slope*, or quantified angle, of the line multiplied by x, which is what's plotted on the X-axis; and b is the *intercept*, where the line crosses the Y-axis. This is the exact formula outputted in single-variable linear regression with specific values fit in. In linear regression, there is also presumed error (every model has error because we know a model cannot be perfect unless the variables are perfectly correlated), represented by an e in the formula, and the intercept is usually written first as a sort of "0" X value. So in regression we write it like this: $Y = X_0 + mX + e$. If you're curious, the formula for the line in Figure 15-8 is $Y_{weight} = -101 + 1.11 \times X_{height} + e$.

Multiple Linear Regression

Most of the time, when a data scientist mentions linear regression, they actually mean *multiple linear regression*, which is simply a linear regression where we are modeling one variable Y with several other X variables. In the height/weight example, we know that weight is determined by far more than height. There are a lot of different things that might affect weight, including gender, fitness level, age, and many more. The first three alone would vastly improve our model from above.

In order to run a multiple linear regression, the requirements mentioned above hold true for all the features to be included, which means that each must be correlated with the target and independent from the other features. We also still need to check for linearity between each feature and the target. We can still check for linearity visually if there are only a handful of features, by plotting scatterplots of each individual feature against the target and ensuring it looks linear. There are actually more systematic ways to check all of these than what we talked about above, but I'll leave it to you to explore them.

There are also some additional assumptions that ordinary least squares regression makes, which can be checked after the model has been generated. One is that the mean of the residuals is close to 0. The residuals should also be normally distributed, and there should be no autocorrelation in them, which would mean that there is a pattern of some sort in the data that wasn't captured in the model. You can also look these up.

One challenge with learning multiple regression is that because there are more than two variables involved, we can't make a chart to visualize the full regression (the three additional features mentioned above would require a five-dimensional chart). But the good news is that understanding it doesn't require visualization—we can just look at the formula to see what's happening. What we're doing with multiple regression is still the line that minimizes the residuals, the differences between the line and each Y value. The multiple linear regression simply adds more X features, so it looks like $Y = X_0 + m_1X_1 + m_2X_2 + \ldots + m_nX_n + e$.

This formula is what the regression outputs and what we can share with stakeholders. It's easy to explain. In our example of predicting weight with height, fitness level, and age, to share the formula with stakeholders, we'd rename the X values and end up with a formula that looks more like this: $Y_{weight} = X_0 + m_{height}X_{height} + m_{fitness}X_{fitness} + m_{age}X_{age} + e$. We'd likely make it less math-y before actually showing them the formula though. What the values of the features are would have to be determined, but let's assume height is in inches, fitness is a rating between 1 and 5, and age is in years (with fractions allowed). What this says is that we multiply each of those numbers by a specific multiplier that the regression has calculated and then add them all together to get Y_{weight} in pounds. To get a new predicted weight for a new person, it's simply a matter of plugging in their X_{height}, $X_{fitness}$, and X_{age} values.

The ordinary least squares approach works just as it does for single-feature regression, but the line it's building is just in multidimensional space. If you're having trouble wrapping your head around this, first try to imagine how it would work in three-dimensional space—the distance would no longer just be "vertical," so the residual lines would not all be going in the same direction as they are in one-feature regression. The jump from two-dimensional to three-dimensional adds complication, so the jump from three- to four-dimensional spaces adds even more that we can't visualize. But it's still a "straight line" through that space.

Weights in Regression

In practice, regression is sometimes overly simple, and we have to make adjustments to get it to work well. One of the common techniques is to add a numeric weight to some of the terms, so the weighted X values would be multiplied by a weight value before the regression is run.

For instance, imagine a model that has X values that might have been measured differently from each other. This means that, for instance, all the X_1 measurements are less precise than all the X_2 measurements. Maybe Javier measured the length of something with an old ruler and had to round measurements to a quarter-inch, but Sarah had a better ruler when she measured the height and hers are recorded to the 1/16th-inch level. In this case, Javier's measurements are less precise and could be given a lower weight to prevent that feature from having too much influence in the model.

Logistic Regression

Logistic regression uses a particular statistical function called the logit model as the foundation of the regression. You don't need to fully understand that to do logistic regression, but the basic way it works is that if you take a linear combination of features (basically the right side of the regression formula shown above, like $X_0 + m_1X_1 + m_2X_2 + \ldots + m_nX_n + e$), that value will equal the natural log of a particular fraction involving the target variable. I won't show the math, but that results in the regression generating a curve between two binary values.

Like other techniques, there are some requirements for and assumptions made by logistic regression. We're talking about binary logistic regression here, so the data must be binary. Another is that the dataset shouldn't be too small (this varies depending on your data, but usually a few hundred records with a handful of features is fine). The features need to be independent of each other.

Logistic regression operates somewhat like linear regression in that it determines a decision barrier between one target value or the other by minimizing a function with a particular technique. It uses some additional math to calculate the coefficients on the formula, $logit(Y) = X_0 + m_1X_1 + m_2X_2 + e$. The model assigns outputs probabilities that each record is of class 1 and of class 2 (which is just 1 minus the class 1 probability) and then uses a threshold (by default, 0.5) to apply the appropriate label. Above 0.5 goes with class 1.

If we look at another real example, we have another dataset with heights, this time of kids. We've got gender, height, and age and we want to predict gender based on the other two. Although it's true that boys are taller than girls, there's still a lot of overlap so our model won't be great with just these two features, but it should be better than chance. I created the model and then plotted the values of height only (I could only plot the target and one feature in a 2D plot) along with what's called the logistic curve, basically the decision boundary. See Figure 15-9 for what it looks like, with the curve in red. In datasets with more clearly delineated data, the curve is more S-like, flatter at the top and bottom and a sharper diagonal in the middle.

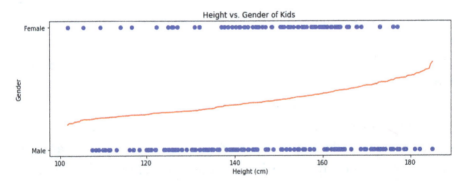

Figure 15-9. *The logistic curve for height and gender in the kids dataset*

Nonlinear Regression

Nonlinear regression is a regression that generates a nonlinear function that combines the model parameters in a nonlinear combination. Some data may not lie in a straight line, but it still fits a recognizable and quantifiable pattern. There are several types of nonlinear regression, as the fitted function could be almost anything, as long as it fits the data. However, not all models that have a nonlinear function are considered nonlinear regression, because if the function can be represented as a linear combination of the model parameters, it's still considered linear.

A common technique to linearize a nonlinear function is to take the natural log of both sides of a function if at least one predictor variable is a power of the number e. Don't worry about fully understanding this. Knowing that this is out there can help you find it when you need it, even though most of the regression you will use is likely to be linear.

Trees

Trees are another very popular, versatile, and useful class of supervised machine learning algorithms. They don't require linear data and can be used for both classification and regression. And even better, they can handle categorical features. Tree algorithms basically split the data repeatedly with if-then–else scenarios, starting at the top with all data, then splitting into multiple groups based on a rule, and then continuing down each branch, until it gets to a stopping point and gives the current batch of data a label (in classification) or assigns it a number (in regression). The logic that is generated is extremely intuitive in a visualization that looks like a family tree.

I'm first going to talk about the simplest tree algorithm, simply called the decision tree. Technically, there are a few different algorithms that are considered decision trees, so I'm just going to talk about the most common one. After discussing decision trees, I'll talk about random forest, which is an extension of the decision tree by making many of them and combining all the different results to come up with a single result.

Decision Trees

Decision trees are probably the most intuitive machine learning out there. Linear regression is easily explainable, but it still requires thinking about math, which will make some stakeholders mentally shut down. A decision tree can be drawn and be intuitively understood by anyone. A couple of other reasons it's so useful are that it doesn't require the data to be linear and can handle categorical features. I'm going to show a basic decision representing a simple spam filter (technically, a spam labeler) as a reference (Figure 15-10).

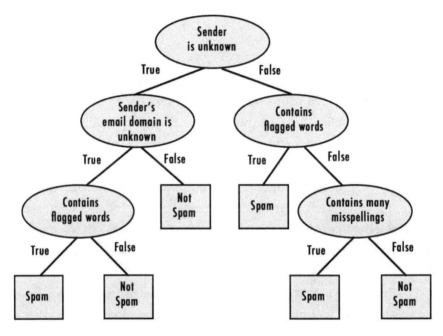

Figure 15-10. *A simple decision tree that labels email spam or not*

A decision tree is composed of nodes and leaves. A *node* represents a split being applied to a group of data, so we have a batch of data and we're going to apply an if-then–else question to the node and send the data down two (or more) paths, depending on the answer. It's often represented as an oval. In Figure 15-10, all the ovals are nodes, and each contains the question being asked at that point to split the data into two different groups. A *leaf* is the end point of any branch, which represents a group being assigned a particular label or numeric value. They're often displayed as rectangles. All the rectangles labeled "Spam" or "Not Spam" in the figure are leaves. The paths going from a node to another node or a node to a leaf are shown as a single line. The answer to each node's question is shown next to the relevant path line. In the case of the tree in the figure, all of the questions are binary, so they all say either "True" or "False." There is also a special node—the first one, at the top, called the *root node* (or just the root). The root node in the figure is the top one "Sender is unknown." We call the complete path from the root node to any leaf a *branch*. Note that the same "question" can appear in different branches of the tree, as "Contains flagged words" does in the figure.

This is an intuitive display. The way to understand how to use a constructed decision tree on a new record is to first ask and answer the question in the root node based on the values the record has and then go down the left path if the answer is True and the right path if it's False. Then you ask that new node's question and follow the correct path based on the answer. And so on, until you hit a leaf and can't go any further. That leaf is the label or value that will be assigned to that record.

That's how to use an existing tree, but how do you create it? As mentioned above, there are actually several different algorithms for training a decision tree. They have a lot in common, but I'll describe the general approach. The basic concept is that for a given batch of data, you decide what the most effective split will be. What that means is that the goal is to find a question that perfectly splits the data into one leaf where all the records have one target label and another leaf where records all have a target label different from that in the first leaf. This is what ideally happens at leaf formation, but along the way it has to split the data when that perfect question isn't there. So it looks at all the possible features it could question and calculates a metric based on how the data would be split. You don't really need to understand what these metrics are, but entropy (from information theory) and information gain are commonly used.

We'll look at how the tree in Figure 15-10 could have been built by looking at a small sample of a dataset of emails. See Table 15-1 for the data being used to train the decision tree. In reality, much more data would be needed to train this model, but we can still see how data and questions are used to build a tree. I'll go down one branch, looking at each step.

Table 15-1. *An example email dataset*

Email ID	Sender Unknown	Email Domain Unknown	Contains Flagged Words	Contains Misspellings	Is Spam
1	True	False	True	False	False
2	True	True	True	True	True
3	True	True	False	False	False
4	True	True	True	True	True
5	True	True	False	True	False
6	False	True	False	True	True
7	False	False	True	True	True
8	False	True	False	False	False
9	False	False	True	False	True
10	False	False	False	False	False

By first asking if the sender is unknown, the data is split into two sets shown in Tables 15-2 (Sender is unknown = True) and 15-3 (Sender is unknown = False). The algorithm would then calculate a metric score (whatever metric it's using) and, after looking at other splits, pick the one with the best score.

Table 15-2. *After the first node split: Sender is unknown = True*

Email ID	Sender Unknown	Email Domain Unknown	Contains Flagged Words	Contains Misspellings	Is Spam
1	True	False	True	False	False
2	True	True	True	True	True
3	True	True	False	False	False
4	True	True	True	True	True
5	True	True	False	True	False

Table 15-3. *After the first node split: Sender is unknown = False*

Email ID	Sender Unknown	Email Domain Unknown	Contains Flagged Words	Contains Misspellings	Is Spam
6	False	True	False	True	True
7	False	False	True	True	True
8	False	True	False	False	False
9	False	False	True	False	True
10	False	False	False	False	False

The algorithm now has two different sets of data to consider, with a mix of different labels in each. I'm only following the left branch (Table 15-2), so we now need to split the data in that table since not all the labels are the same. It tries different questions and calculates the metric on each, finding that asking about email domain has the best score. We split on "Sender's email domain is unknown." The data that flows down the True path is in Table 15-4 and the False path in Table 15-5.

Table 15-4. *After the second node split: Sender's email domain is unknown = True*

Email ID	Sender Unknown	Email Domain Unknown	Contains Flagged Words	Contains Misspellings	Is Spam
2	True	True	True	True	True
3	True	True	False	False	False
4	True	True	True	True	True
5	True	True	False	True	False

Table 15-5. *After the second node split: Sender's email domain is unknown = False*

Email ID	Sender Unknown	Email Domain Unknown	Contains Flagged Words	Contains Misspellings	Is Spam
1	True	False	True	False	False

At this point, as can be seen in Table 15-5, there's only one label represented in all the data on that branch, since it's only one row. This means there should be a leaf with that label Not Spam. The split on Sender's email domain being unknown still has different labels, so we need to split again. The algorithm again tries and scores different questions, eventually deciding on "Contains flagged words." That split can be seen in Tables 15-6 and 15-7.

Table 15-6. *After the third node split: Contains flagged words = True*

Email ID	Sender Unknown	Email Domain Unknown	Contains Flagged Words	Contains Misspellings	Is Spam
2	True	True	True	True	True
4	True	True	True	True	True

Table 15-7. *After the third node split: Contains flagged words = False*

Email ID	Sender Unknown	Email Domain Unknown	Contains Flagged Words	Contains Misspellings	Is Spam
3	True	True	False	False	False
5	True	True	False	True	False

Now, that split has yielded two sets that each have only one label present, which means these can be nodes with the appropriate label. And now if we repeat this process on the right branch of the root node we have a decision tree trained that can be used to classify new emails that come in.

When using real data to build a decision tree, the splits are never perfect, and most leaves will have some records with different labels. There are a lot of different ways to control the building of a tree, which is primarily why we can end up with misclassified records in training. In reality, we don't want the tree to get too deep, which is what would happen if we insisted on each leaf having only records with the same label. We'd also end up with leaves that have only one or a very small number of records. Those situations are basically overfitting (we'll talk more about it below). So we impose some limitations, like max depth and minimum number of records required in a node to split it (once we get to a node that splitting would lead to too-small nodes, we'd call it a leaf with the most common label).

Regression trees are built similarly, but it's not as straightforward without a single label. It would instead be trying to group values that are close to each other, rather than all being the same label.

Random Forest

Random forest is an approach that builds on the decision tree algorithm to create an even better-performing model (usually). It's actually an ensemble learning approach, falling under the bagging approach. Fundamentally, random forest creates a bunch of decision trees following the basic method described above, but there are some extra tricks that make it more effective. Because it's based on decision trees, it's just as flexible and powerful. It can do classification or regression and categorical data in addition to numeric.

Like most machine learning approaches, random forest does require a large amount of data, but that data can be fairly noisy as long as the noise is randomly distributed. The many decision trees created need to be independent from each other and balanced. Both of these are ensured in the standard approach, which I'll describe below.

The two key components of the random forest approach are bagging— taking random samples of the original data (with replacement)—and taking random subsets of the features for each sample. Then individual decision trees are trained for each of these combinations, usually in the hundreds. The bagging portion generates many different subsets of the data. Because these are random samples sampled with replacement, they're very representative of the data and have low variance. That's the basic bagging approach, but random forest throws in the trick of subsetting the features, too—usually the number of features included is the square root of the total number of features. This trick ensures that the many trees aren't too similar or correlated with each other. So each subset of the data also only has some of the features. This is powerful because it means that the result won't be dominated only by the most impactful features—instead, different features can show up as important in more trees, which allows for more subtle trends to be captured.

After all these trees are trained, the data is run through all of them, and the outputs are either determined by majority vote (in classification) or averaging (in regression).

The project I've mentioned where my team forecasted daily store sales is a good example of the difference between decision trees and random forest in practice. We had over 450 features on top of a couple thousand records, so it was a good candidate for random forest. Our features included a total of 18 features for each of seven major US holidays. These features were things like "10 days before Christmas," "9 days before Christmas," and so on, all the way through "Christmas day," and then we had "1 day after Christmas" for every day through "7 days after Christmas." Each holiday had different patterns in terms of the kind of impact it had—spending was huge during the lead-up to Christmas, but died off after. But because we were looking at year-round data, only 18 days a year could have any of these features (18 days where a Christmas feature could be True vs. 347 days with only False values, or less than 5% of the data). So, a single regular decision tree would be unlikely to pick up any Christmas features—for most of the data, other features would be far more important—unless it was very, very deep. But deep decision trees are a bad idea, so a decision tree approach didn't yield good results. Random forest *did* pick up on the Christmas and other holiday features and performed very well. In an earlier effort on this same problem, we'd had pretty good results with a time series approach, but random forest outperformed every algorithm we tested.

The main disadvantage to random forest is the weak explainability. Generally, the best thing to do is explain the decision tree approach so they can grasp that. Then you can explain that hundreds of these trees are trained on slightly different data, and the final label is determined by majority voting or averaging. In my experience, people have mostly been comfortable with that.

Another way to explain it is to talk about the features that are important. The random forest algorithm will output the most important features in the model overall, but that doesn't tell us a lot, especially about specific forecasts. I've mentioned model explainers in the past, and we used one on the daily sales forecasting and got really good info on each forecast. This is how we knew that the holiday-related features were coming up on the few days of the year where they could be relevant.

Naïve Bayes

Naïve Bayes is a method for classification only that came to prominence in early efforts of spam filtering, because it does really well identifying spam. It's not one specific algorithm, as there are several that can be used, but they follow the same basic steps. All rely on Bayes' Theorem and its conditional probability formula, where the terms can be flipped around. They also require all the features to be independent from each other and that each contributes equally to the outcome. These are the reasons it's been christened "naïve." An advantage of this approach is that it doesn't need a large set of data to train on.

The basic method followed here is to calculate the posterior probability (part of the Bayes' formula) for each feature. For instance, in a spam classifier that only considers the presence of individual words or phrases to determine spam status, the posterior probability (such as that the email is spam given that it has word w in it) is calculated based on the probability of anything being spam (the proportion of spam in the training set) and the prior (the probability that word w is present given that the email is spam). The posterior for each word for that email can then be combined to determine the overall label as spam or not (with a defined threshold value).

One disadvantage with Naïve Bayes over other methods is that we get no information on individual features (remember that the assumption is that they're all equally important). This is clearly different from regression and tree approaches. But it's simple and can work well in many domains.

Neural Networks

Neural networks, sometimes called artificial neural networks, are another class of mostly supervised learning algorithms (there are some variants for unsupervised problems) that have been around for a long time, waxing and waning in popularity. They've been big the last ten years or so as many new types have been developed and seen very successful usage. And, of course, Gen AI is based on advanced neural nets (usually called deep learning), so they've exploded in popularity. You'll often hear that neural nets were inspired by the human brain by mimicking the structure and function of neurons and synapses, which is true, but that doesn't mean that neural nets are any closer to human intelligence than other machine learning algorithms. A lot of practitioners don't like this comparison because of the inaccurate implications, but it is where they came from. Still, a simple network in the style of modern neural nets was used in the 1700s to predict planetary movement (long before we knew anything about the way animal brains work).

A *neural net* has the general structure of inputs, hidden (intermediate) layers, and outputs. See Figure 15-11 for the general structure of a neural net. There are huge numbers of connections, and as we add internal layers, the number of connections also explodes. The internal layers are opaque, and with so many of them, it's difficult to explain why a network assigned a particular output to any input. They can be incredibly powerful in many areas and are still used all the time. Note that neural nets with two or more hidden layers are called *deep neural nets*, which is where the term *deep learning* comes from. One of neural nets' strengths is learning from their mistakes, which makes them very adaptive.

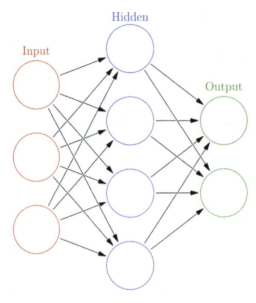

Figure 15-11. *A general view of a neural net. Source: From "File:Colored neural network.svg" available at https://commons. wikimedia.org/wiki/File:Colored_neural_network.svg, used under the Creative Commons Attribution-Share Alike 3.0 Unported License (https://creativecommons.org/licenses/by-sa/3.0/deed.en)*

As mentioned, there are many types of neural nets. We're not going to dig very deep into any of them, but we'll talk about the basics of each, and if you want to know more, you can go forth and Google. The earliest modern neural net was the single-layer perceptron, but it wasn't that useful, and the next that came in was the multilayer perceptron, also called the feedforward neural network. Feedforward means the data flows through the network in only one direction, from input to layer to the next layer and so on to the final output. This means that data will only hit one node in each layer. Each node has a weight and threshold that determine whether it sends the data further along, but these weights and thresholds are adjusted during training, having started with random values. The sheer number of nodes means that random starting points are the best option, and the adjustments are done until the network outputs data with each label together, which is how it knows what to adjust. A common technique for adjusting weights is *backpropagation*, which starts at the end and steps backward through the different layers for a specific input–output pair, adjusting a loss function using knowledge of how each weight affects the output. This process allows the weights to be improved.

Another important neural net is the *convolutional neural net* (CNN), a feedforward neural net that determines features by a particular type of optimization. It was the main innovation that helped improve image processing and computer vision, and it's still used, although there are some other methods coming in to replace it. The reason it's so valuable is that it uses some tricks to significantly reduce the number of connections between nodes in the network, which makes it much more efficient and less prone to overfitting that can happen in fully connected feedforward neural nets. But they still use a lot of computing resources.

Recurrent neural nets (RNNs) are a different type of neural net that work on sequential data (like time-based data and language data) where each input is not considered independent from the others—the previous elements in the current sequence are "remembered" with a hidden state that gets passed along, which influences subsequent weights. Unlike the other types we've talked about, the weights are not different on every node. Instead, the weights on nodes in one layer have the same value. Standard RNNs can develop problems with longer-term sequences, but do well on smaller things like words in a sentence or values in time series data. A variant of RNNs that have better and longer-term "memory" are called *long short-term memory* (LSTM) neural nets. They work well, but take a lot of training.

One more type of neural net is the *transformer*, a new one that has somewhat superseded RNNs and CNNs because it doesn't have the same limitations, especially in terms of resource usage. It's the primary tool used in GenAI (the GPT in ChatGPT is short for "generative pre-trained transformer"). These have some innovations in terms of how words are treated. At input, a sentence is converted to an embedding, which represents the semantics (meaning) of each word, and then each word is assigned a number representing its position in the sequence. There are some functions used that can discover relationships between the words. The network makes some adjustments to the layers and then works as a feedforward neural net. Transformers do still require a lot of training, but they're more efficient than RNNs and run faster on new data.

The final type of neural nets I'll mention is the *generative adversarial network* (GAN), which is actually two neural nets that interact with each other in a prescribed way to generate output. You can see the overall structure in Figure 15-12. One of the nets is called the *generator*, which generates data based on random data. The generator's main goal is to generate plausible output. This generated data is fed into the second net, the discriminator, as negative training examples. The *discriminator* is a classifier that also has real data coming in as input, so its task is to learn to tell what the real data is, as opposed to the fake data coming out of the generator. The outputs from the generator feed directly into the inputs of the discriminator.

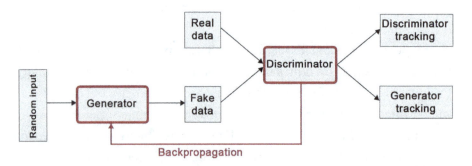

Figure 15-12. *The generative adversarial network structure and training behavior*

Backpropagation is used from the discriminator outputs all the way back to the generator and uses the calculation to adjust the generator weights (note that it's only the generator that's changed this way, for consistency of behavior). The goal of the overall system is what's called *convergence*—when the discriminator is guessing (basically, flipping a coin) whether its inputs are real or fake, which is the point when the generator is performing best. The training of the system is tricky because it's hard to know the exact point that convergence has been reached, and if the generator keeps sending to the discriminator after this point, then the discriminator can start acting wonky, where its (basically) random feedback through backpropagation negatively affects the generator.

Once the training of the system is complete (convergence has been reached and recognized), the generator can be used to generate data. You still have to send input into the generator, but random noise will generate images.

k-Nearest Neighbors

k-Nearest neighbors (*k*-NNs) is a supervised machine learning algorithm that looks at data in several dimensions and identifies each point's *k*-"nearest neighbors" (*k* can be any value, including 1) and can be used for both classification and regression. It can be used with a variety of methods of measuring distance to define "near," and higher-dimensional data (i.e., with lots of features) can require special methods. *k*-NNs can be used with nonlinear data and on both numeric and categorical data. One important thing to know with *k*-NNs is that the data must be standardized so that it is all on the same scale.

An interesting fact of *k*-NNs is that there really isn't a true "training" step like with the other supervised algorithms we've looked at. You still need a "train" dataset, but this works out to be basically just a reference set. The way the algorithm works means that very large datasets can take an unacceptable amount of time to process. Look at Figure 15-13, the data from the kids dataset we looked at earlier. It has 15 labeled data points with just two features, height and age, and with blue representing boys and red girls.

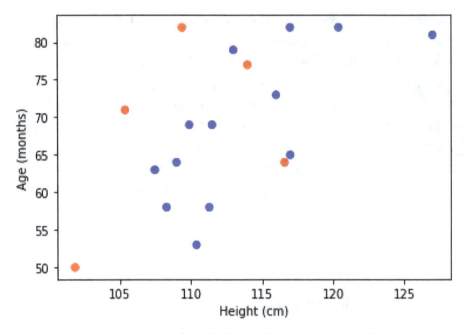

Figure 15-13. *A sample of the kids height dataset with blue representing boys and red girls given to k-NNs*

If we give this dataset to *k*-NNs as our "training set," this serves as our reference set. We set *k*, the number of neighbors to consider in classification, to 3 (a common value for this). Now, if we have two new points as seen in Figure 15-14, we can see how the classification works. The first new point is in purple in the upper left and the second in green in the upper middle. We can visually identify which three reference points are nearest to the purple point and green point, which are marked by black lines. In the case of the purple point, two of the three points are red and one is blue, so we would classify the purple point as red (female). The green point has four points fairly close, but although the nearest is red, the next two are both blue, so this one would be classified as blue (male).

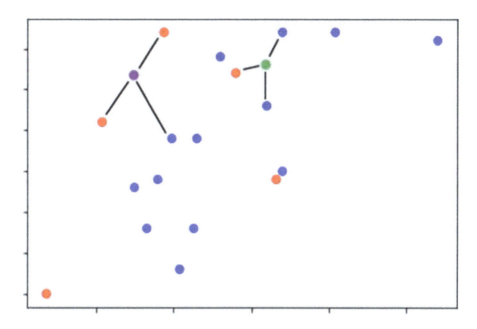

Figure 15-14. *Classifying two new points based on our "trained" k-NNs*

k-NNs is technically a classifier, but an average can be taken among the point's nearest neighbors to be used in regression. In the case of the new points above, if the labels had been numeric values instead, we would just take the average of the three nearest points. This is pretty intuitive, too, and simple to implement. Usually, there's some work involved in finding the right *k*, but it works well in a lot of cases, especially with more than two features. To find the best *k*, you generally create a bunch of *k*-NNs with many different numbers and run the test set on each to see which number does best.

k-NNs isn't used as much as it used to be, but it's still powerful and simple when there aren't too many features. It's still used in many areas, including in financial market forecasting, some recommendation systems, and image processing.

Support Vector Machines

Support vector machine (SVM) is another supervised machine learning algorithm primarily for classification, which finds a way to separate groups of data. It can handle both linear and nonlinear data, although there are different variants for the two purposes. In the linear variant, it finds a linear hyperplane in problems with many features.

HYPERPLANE

A hyperplane is the general name for a "line" through space that is one dimension less than the space. For instance, on a two-dimensional chart, you can draw a line within that space, which is just a straight line (a line is considered one dimension). Similarly, a two-dimensional plane can be placed in three-dimensional space. In four-dimensional space, it would be a three-dimensional object (a cuboid, like a cube but with different-sized sides). You can visualize the cuboid, but not the space. But from there it's completely abstract. Just think of a hyperplane as being like a line that divides hyperspace that's only two dimensions, and everything from there is an analogy.

The point of the hyperplane is to divide the space to separate different classes of things. Figure 15-15 is an example of what SVM is trying to do by creating a line that serves as a boundary between classes. In standard SVM, it is linear, so it has to be a straight line, but there are other variants of SVM that can produce non-straight hyperplanes. Note that in Figure 15-15, there are misclassified points, one blue mixed in with the reds and two reds mixed in with the blues. Sometimes this can be solved with a nonlinear hyperplane, but error is inevitable in machine learning.

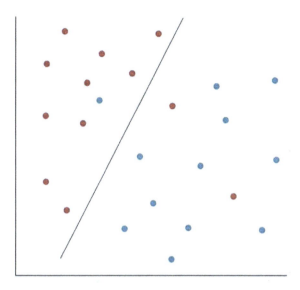

Figure 15-15. *An example of SVM's goal in separating classes in data with two features*

SVMs are very powerful, but like other algorithms, they can overfit when the number of features gets too large, because of the curse of dimensionality (which we'll talk about below). They are used in a lot of tasks, including text classification and image object detection. It's also worth mentioning that there are variants of SVMs that can do regression, although it's less common.

Unsupervised Techniques

With unsupervised learning, you just want to see what the data can say about itself. You aren't providing train data so that it can learn some rules to apply to new data. Instead, you provide your carefully curated and prepared features, tell it to break a leg, and let it do its thing. It will come back with information for you about what it found in the data.

There are two primary types of unsupervised learning that are used a lot in data science: clustering and association mining. In clustering, the methods identify groups of data that are similar to each other. With association mining, the techniques give you much smaller groups of things that are associated together. I'll talk about each of these below, touching on a few different algorithms for each.

Clustering

Clustering is a popular approach to identify similar points in your data. For instance, a retail store may want to identify different types of customers so they market to them differently. We could imagine a solution to this problem as a supervised problem if we already have some customers labeled with their type. We could then train a model on those and run the remaining, unlabeled customers through it. But that requires three things: knowledge of how many groups there are, what those groups are, and then data labeled with those groups. What if we don't have that info? Clustering comes to the rescue because it requires none of those things (technically, one approach does require you to specify the number of groups, but you can experiment to find the right number).

k-Means Clustering

k-Means is the first clustering method data scientists usually learn and use. *k*-Means clustering identifies a pre-specified number of clusters by putting each data point in the cluster that it's closest to, defined by the center of the cluster, which is the mean of all the points in the cluster. But how do they get the groups, since they're not labeled beforehand? *k*-Means takes a top-down approach, first looking at all the data and partitioning it into what are called Voronoi cells (see Figure 15-16 for an example of data broken up into Voronoi cells).

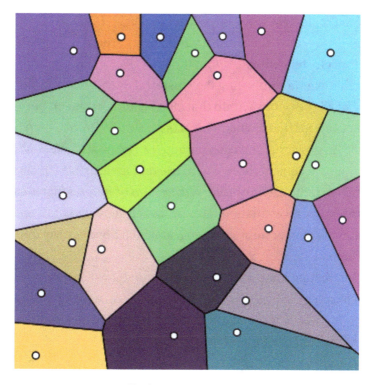

Figure 15-16. *Voronoi cells for a random dataset of 28 points*

The white points are our dataset, and the polygons are the Voronoi cells. Note that they aren't determined arbitrarily. To make them, you would draw a line between each point and its immediate neighbors and then draw a line perpendicular to that at the halfway spot between each two points. You extend that until it hits another line, and you'll end up with these sharp-cornered shapes.

This is a conceptual starting point. If you set k to 28 and each of these points were your cluster centers, any new points within this space would be assigned to the cluster corresponding to each Voronoi cell. The most common algorithm for calculating k-means is slow but intuitive and is called naïve k-means. You start by specifying k, and the algorithm first assigns k random cluster centers. Then every point in the dataset is assigned to the nearest of these k cluster centers. This is when the Voronoi cells factor in, because data points within any Voronoi cell are assigned to that random cluster center.

After this, we start adjusting the clusters. For each cluster, we take the mean of all the points within that cluster, and that mean becomes the adjusted center for that cluster. This is one round. This two-step process is repeated, first reassigning each point to the nearest cluster and then recalculating the center. There is a metric that measures each cluster center and its points called *within-cluster sum of squares* that is calculated after every center recalculation. This process is repeated until the within-cluster sum of squares stops changing, at which point we say the algorithm has converged.

One interesting thing about k-means is that it will always converge, although it isn't guaranteed to find the optimal clusters. This is because the starting points are random. It's normal to run it a few times and see if the clusters are similar or different (or radically different). Additionally, there are other ways of measuring "nearness" than the method described above, but they aren't guaranteed to converge like this simple approach is.

Note that there are some ways to assign the initial cluster centers other than fully random values (a couple are Random Partition and Forgy). You can Google to learn about these.

We've talked about how to run k-means, but not how to know what in the world your k should be. Sometimes you might have a sense for what is likely to be in the data, so you may have some numbers that seem worth trying. If you're trying to identify different types of customers of a store, a k of 50 is probably unreasonable—how would a company differentiate marketing or other services for 50 different types of customers? Additionally, the higher your k is, the more data you need for the algorithm.

Fortunately, you don't have to know the right k in advance, as there are some methods that allow you to systematically pick a good one. There are several, but I'll talk about the elbow and silhouette methods here. Google for more.

The elbow method is simple and it is commonly used, but it is not considered extremely reliable because it can be hard to define where the "elbow" is. With the *elbow method,* you determine the possible number of clusters, perhaps two to ten as an example. You run k-means on each of those and calculate the final within-cluster sum of squares value for each run, then plot that against the number of clusters. We are looking for the point where the within-cluster sum of squares goes from dropping quickly with each additional cluster to significantly slowing down or leveling off. See Figure 15-17 for an example of what a couple of these plots might look like.

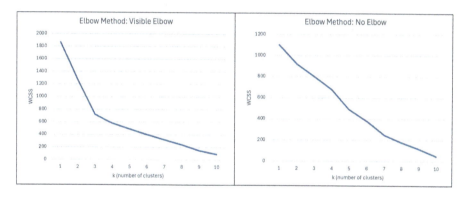

Figure 15-17. *Two elbow charts to determine the best k. The left has a detectable elbow where the right one doesn't*

In the Figure 15-17 charts, $k = 3$ on the left chart is what we're looking for—a change in how quickly the within-cluster sum of squares is dropping. But there's no point like that in the chart on the right. This is why this isn't a method people always use. It's considered subjective, and if you change the scale or size of the graphic, the elbow might look more or less "elbow-y." But you can always quickly generate this chart, and if you're lucky, you might see an obvious elbow.

Another common method of determining a good k is the silhouette method. With the *silhouette method,* you're measuring how well each data point has been clustered by looking at how much it has in common with the other points in its cluster and how different it is from points in neighboring clusters. It involves calculating a score per point and then over the entire result, with a value ranging between –1 and 1, where high numbers indicate good clustering and low bad. This works fine on datasets with a smaller number of features, but the curse of dimensionality rears its ugly head again when there are a lot of features.

The truth is that like a lot of machine learning algorithms, some trial and error is common. The need to provide the number of clusters makes this one take a bit of work, but it can be incredibly powerful in identifying meaningful groups.

Hierarchical Clustering

There are some other clustering methods that are considered hierarchical, with two primary types of hierarchical clustering algorithms, agglomerative and divisive. Agglomerative builds clusters bottom-up, with each point starting as its own cluster and gradually working up to combine clusters appropriately. Divisive is top-down, where we start with one cluster and start splitting that into clusters based on characteristics of each cluster. All hierarchical approaches generate a tree-like structure called a dendrogram that shows the hierarchy created going from every data point in its own cluster to a single cluster with everything (agglomerative) or vice versa (divisive).

Both of these approaches use a distance measure to inform splitting or joining decisions. The distance is measured between individual points of the dataset to measure what's called dissimilarity. Then another measure called cluster linkage looks at how different sets of the data are.

In the most common agglomerative method, clusters are joined in pairs in each step. An interesting thing about this is that it proceeds all the way to the point where everything is in one cluster. After running the algorithm to completion, you decide at what level in the tree you want to cut it, and you take the clusters as they are in that level. See Figure 15-18 for an example of this. It shows agglomerative clustering performed on the iris dataset, a famous publicly available dataset that a lot of people play with when first learning machine learning. It has data with labels for three types of irises.

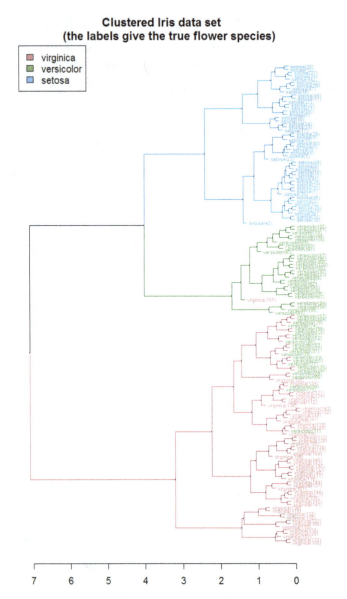

Figure 15-18. *Agglomerative clustering dendrogram on flower data. Source: From "File:Iris dendrogram.png" available at* https://commons. wikimedia.org/wiki/File:Iris_dendrogram.png, *used under the Creative Commons Attribution-Share Alike 4.0 International License* (https://creativecommons.org/licenses/by-sa/4.0/deed.en)

This algorithm starts on the right, with each individual flower data point in its own cluster. Its class is indicated by color. Then the algorithm starts grouping them, two at a time, and moving up to generate different levels of the tree. You can see where the color changes that many of the *versicolor* ones start getting grouped together and ultimately end up in a cluster with the *virginica* ones. But what you can do is pick where you want to cut the tree to see what the clusters are. Let's say you decide to cut it at the 3 on the X-axis. Imagine a vertical line going up from there, and it would cross four lines, each of which represents a distinct cluster. You can look at how the flowers are grouped at that point. Everything to the right under that line is in that cluster, so you can see there are two clusters splitting the *virginica* flowers, and one each for the versicolor and *setosa*, with some in the wrong clusters.

Divisive clustering works basically opposite to agglomerative, starting with one cluster and ending with every point in its own cluster. Visually, it looks similar in that it creates a dendrogram like in Figure 15-18; it's just built differently. You can Google for more info on how.

Association Rule Learning

Association rule learning is another class of unsupervised machine learning. You might have heard about the beer and diapers story, where men stop by the store after work Friday to pick up diapers and snag a case of beer on their way to the register. It sounds apocryphal, but apparently it's not—a group did association rule mining at a drugstore and really found that there was a strong coupling with beer and diapers, along with over 20 more pairs of products.[5]

---

[5] "Diapers, Beer, and Data Science in Retail" by Nate Watson, July 17, 2012, available at https://canworksmart.com/diapers-beer-retail-predictive-analytics/

Association rule learning (sometimes mining) looks at itemsets—groupings of items as pairs or bigger sets—to identify items that lead to a purchase of another item. The beer and diapers story is relevant because retail is the classic use case for it. It's even frequently called market basket analysis because we're looking at the items in somebody's shopping cart. The rules themselves contain directionality and basically can be thought of as if–then groupings. If they buy a pack of diapers, then they also frequently buy a case of beer.

The primary approach to association rule learning is the Apriori algorithm. We've heard this term before in other contexts (Bayesian stats, including Naïve Bayes), but here it lays out a specific approach to association rule learning that utilizes certain probabilities. It identifies the rules by first looking at baskets, or all the products in a single transaction. Then it creates itemsets, which are composed of groups of at least two items. For each itemset, a score called *support*, the percentage of all transactions that this potential itemset occurs in, is calculated, and any with scores over a certain chosen threshold are kept. The ones with high enough support are dubbed *frequent itemsets*. From that pared-down set, all possible rules are generated. Note that there will be a ton of these if there are more than a handful of frequent itemsets. The rules are written like {milk} ⇒ {bread} or {milk, bread} ⇒ {butter}, and we call the left side the *antecedent* and the right side the *consequent*.

The next step is to calculate a score called *confidence* on every possible rule as the percentage of transactions that have the antecedent also have the consequent. Confidence gives us a good indication of the significance of the pairing. Just like how we threw out itemsets with a support below a particular threshold, we throw out rules that have a confidence lower than another defined threshold.

This is good information, but sometimes confidence can be misleading on its own, so we calculate another score, called lift. *Lift* is based on support calculations and tells us the presence (and strength) of three possible relationships: (1) if the presence of the first item leads to the second item, (2) if there's no connection, or (3) if they have the opposite effect (the presence of the first item makes it less likely that the second will appear, implying substitution). The lift score tells us how much more likely the second item appearing with the first item is (a lift of 4 means it's four times more likely compared with the general population). One more score that gives us a further sense for the validity of the association, *conviction*, can also be calculated based on both support and confidence.

There are some other algorithms for calculating association rules, but Apriori is popular and reasonably efficient.

Model Explainers

One topic related to machine learning is the relatively new area, still heavily under development, model explainers. A *model explainer* is a tool that takes a trained model and predictions on unseen data and determines which features in the trained model contributed most to each individual prediction. This is invaluable in black box models, or just when model explainability is low. As we've talked about, decision trees are great for explainability, but they have some downsides that are mitigated when we switch to random forest. But random forest is largely not explainable. You can get feature importance scores for the overall model, but it tells you nothing about a specific forecast. My team had a visualization on a sales forecasting model where you could hover over a specific day's forecast and it would display the top ten features in terms of how much they contributed to that day's forecast. This was invaluable for the stakeholder, but it also made us more confident in our model, because it had picked up on meaningful features for each day, among the many features we had.

There are a few model explainers out there right now, with different benefits and downsides. Some only work on linear data, some will work on any model, and some on only specific models (algorithms). The one I've used with good results is called LIME (Local Interpretable Model-agnostic

Explanations), and libraries exist in both Python and R. You can Google the topic to find several to explore, starting with SHAP (SHapley Additive exPlanations).

Implementation and Coding Machine Learning

I've talked about a bunch of very important and valuable machine learning techniques and how they're trained or run. But when it comes to actually implementing any of them—writing the code—there are some other techniques that are very important. All of the methods discussed above are available in one library or another in R and Python, and that's great. But you can't just throw data at a machine learning function and expect to get good results.

I talked above about the importance of hyperparameter tuning and how that involves working with a validation set or using cross-validation. There are libraries that will help you do this. You provide the data, the model you are using, and all possible values you want to try for each parameter. A grid search will try every combination of all parameters and track the performance of each. If you are trying more than a few parameters with more than a few values, this can obviously take a lot of time, because the number of combinations of so many values will be very large. Fortunately, there are also libraries that will allow you to select a subset of all possible combinations of parameters and only run those. These tools run the algorithm with each combination, store the performance, and report back on the best-performing set of parameters.

Hyperparameter tuning through grid search on a validation set is great, but if you're just running everything on your full train and validation sets, you may not end up with the best model for handling new data. Since much of data science is about creating a model that can take in never-seen data and return something meaningful, this is important.

A technique called *cross-validation* automatically breaks your data into smaller train subsets and runs everything (including the hyperparameter tuning) on each of those subsets. This makes the model more robust and helps prevent overfitting and bias (both of which we'll talk about below). There are libraries for doing this in R and Python as well, making it relatively easy.

Like everything else in data science, there are several different cross-validation techniques to choose from. You can ask Google for more info on these and when they're best, but a common one is called *k-fold cross-validation*, where you split the train dataset into *k* (10 is common) subsets. You then run the model a total of *k* times, with a different subset used for testing on every run. This is done in lieu of working with a specific validation set.

One last thing worth mentioning is that there are times when the science part of data science is more prominent than others. This doesn't mean we literally follow the scientific method, but some data science teams are as systematic as possible. Others are messier. It's common to talk about running experiments when we're testing different types of algorithms. It's also common when we're working on a solution and we know we will be trying different algorithms, to first run one that we expect to perform less well than others (but still do something better than random guessing). We call this the baseline, and it's not uncommon for that to be a time series model in a lot of regression forecasting. Having a baseline gives us a reference point for all the other algorithms we try. This can be a rewarding way to work because there's a lot of clarity on what's going on.

Challenges in Modeling

Machine learning has a lot of moving parts, and it's full of the potential for errors or just bad luck. These can be mitigated by understanding some of the pitfalls and how to identify and deal with them or even avoid them altogether. I'll talk about overfitting, when your model is trained too closely to your specific training data and won't generalize well; underfitting, when your model has missed important patterns; imbalanced data, when you have many more of one label than the other(s); the curse of dimensionality, when you have too many dimensions to get good results; and data leakage, when you give your model an unfair advantage it can't sustain.

Overfitting and Underfitting

Overfitting and underfitting are both instances of a model not performing well, just in different ways. I'm going to talk about them and the bias–variance tradeoff, which is a way of balancing models to avoid both overfitting and underfitting.

Overfitting results when you have a model that has learned very specific traits of your train data (and test data) that may not be there in future data so it will not generalize well. It can also happen when your data is low quality, with a lot of noise. If the model "learns patterns" in the noise, it's overfitted (also, just wrong). Take a look at Figure 15-19, which is the same data as in Figure 15-15. In Figure 15-15 we have a simple straight line that separates the data in that figure into blue and red labels, with three data points (two reds and one blue) misclassified. Compare that with Figure 15-19, where we have a squiggly line representing the trained separator for the model that's so specific to the three likely outliers. This will create negative consequences when we're classifying future data.

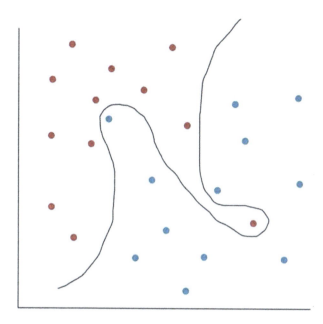

Figure 15-19. *Data with an overfitted model*

It's not hard to see that this model is weird and likely wrong. Let's say two new points come in to be classified, as shown in Figure 15-20. This shows two new data points with the correct class marked in the center of a green ring. But because these are just in the wrong spots within the circuitous decision boundary, they will be misclassified. The new red point that's near many other red points will be marked blue, and the new blue point that's right next to a correctly classified blue point will be marked red. It's pretty clear that this isn't what we want out of a model.

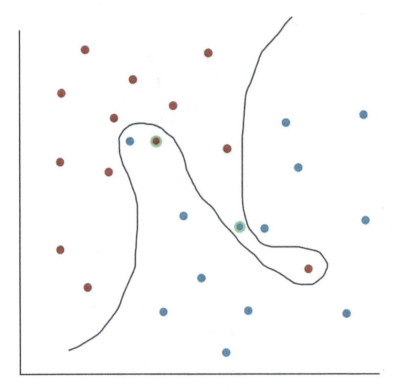

Figure 15-20. *Two new data points that need to be classified by this overfitted classifier with the correct class indicated in the center*

Overfitting comes about for a lot of different reasons, including the curse of dimensionality (discussed below), where too many features can cause it. Models can also overfit because of particular traits or tendencies of algorithms. For instance, if you let a decision tree get too deep, it can seriously overfit because it tries to account for every single quirk of the training data. Random forest is less prone to this because not every data point is in every tree that is built (among the hundreds). Fully connected feedforward neural nets and SVM are both prone to overfitting as the number of features increases. Often simply training too long can cause it (like with deep decision trees). Basically, it isn't hard to end up with overfitting if you don't understand how the algorithms work reasonably well.

There are ways to prevent overfitting. As mentioned, don't train a model too long—early stopping can be beneficial. Feature selection, as previously discussed in Chapter 14, can really help (both reducing the number of features and getting rid of features that don't add much and may confound the algorithm, leading to wrong conclusions about their importance). Regularization is another technique that helps, also discussed in Chapter 14. Other ensemble machine learning methods can also be valuable, where different models starting to go down feature rabbit holes can cancel each other's tendency to overfit.

Overfitting is a problem, but the opposite problem of underfitting also needs to be avoided. *Underfitting* is when a model misses patterns that are in the data. It can come about if a model doesn't train long enough, if there isn't enough data, or if the data isn't representative. The solution to the first problem is clearly to train longer (but not *too* long). The second problem can often be mitigated with semi-supervised techniques to increase the amount of data, or to create synthetic data through other methods (discussed below in the "Imbalanced Data" subsection). But if the data isn't representative because we don't have enough examples of different characteristics, these techniques can't help us. We simply need to find more data to get better representation.

You can see how overfitting and underfitting are at opposite ends of a spectrum. We actually refer to managing this spectrum as the *bias-variance tradeoff*. Underfitted models have *bias* because they make overly simple assumptions, which leads to decisions that perform poorly in both the train set and test set. Overfitted models have *variance* between the train set and test set, meaning that they do well on the train set and poorly on the test set (or on other unseen data) because they are too sensitive to the train data. The reason it's called a tradeoff is because some things that can be done to reduce bias can actually increase variance and vice versa. For instance, feature selection can lead to variance reduction because it makes models simpler, but it can also lead to increased bias. Regularization in some linear models lowers variance but increases bias. In decision trees, the deeper a tree goes, the higher the variance. There are some ways of calculating bias and variance, but they're pretty involved and not necessary for most day-to-day work.

Imbalanced Data

One of the other problems we can see in classification in particular is imbalanced data, where we have labels that don't all appear a similar number of times. For instance, in anomaly detection, the vast majority of the data is not anomalous (by definition), so we have only a relatively few examples with the anomalous label and many more with the non-anomalous label. This can be problematic when it comes to training a model to detect that.

The anomalous label scenario is an example of extreme imbalance, where one label represents less than 1% of the total data. But even less drastic imbalance can matter in binary classification. It's considered moderate when the label only appears up to 20% and mild up to 40%.

There are a few ways of dealing with imbalance. One is to downsample and upweight the majority class (the label that appears a lot). *Downsampling* means taking a sample of the majority class that reduces the discrepancy between the minority and majority classes. For example, if we had 1 anomaly vs. 100 non-anomaly, we could take 5% of the majority class, where we'd end up with only 5 non-anomalies to 1 anomaly, going from 0.9% to 17%. The next step in this approach is to *upweight* the downsampled majority, which means marking the example to be treated as more important during training.

Another common technique a lot of people use is called *SMOTE* (Synthetic Minority Oversampling Technique), which does the opposite of the above technique—it increases the number of minority class samples.

It brings the dataset into better balance by creating new (synthetic) examples with the anomalous label to be used in training. The technique uses k-NNs to identify the space around a given real example, then picks one neighbor, and creates a new synthetic data point between the original and the neighbor.

The Curse of Dimensionality

I mentioned this problem some above. The *curse of dimensionality* is what we call the situation that results in a whole host of problems because we have too many features. The most obvious one is that it's hard to visualize data the more features you have. Obviously, we can't truly visualize data in more than three dimensions, but we can often include more than three features with tricks like coloring and sizing points in a scatterplot, for instance. Additionally, sometimes there are ways of parsing data to look at only a handful of features at once, which is fine if you have 10 or 20 features, but the more you have, the less practical this is.

A second problem is *data sparsity*, where our data is spread out so much with a lot of empty space between data points or sometimes just between some. We can see how increasing dimensions can lead to sparsity in Figures 15-21 and 15-22. If we start with a one-dimensional view of data on the years of education in a fictional group of grandparents and grandkids, with just one feature on a single axis, shown in Figure 15-21, we can see it's within a small range, although the point at 20 is an outlier. (Note that I added jitter, which shifted the values slightly so ones with the same values aren't perfectly on top of each other.)

Figure 15-21. A one-dimensional view of a group of people with their total years of education

But our dataset has more features than total years of education. We can add age to the chart and create a standard two-dimensional scatterplot, as you can see in Figure 15-22. As soon as we do this, we can see two points very far away from all the others and from each other.

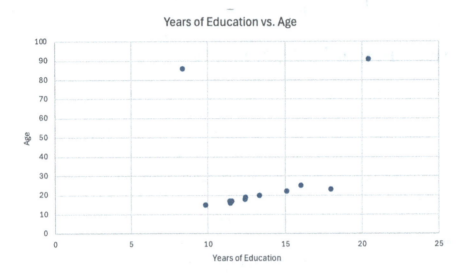

Figure 15-22. *The impact of increasing dimensions in data from one dimension to two*

These points on Figure 15-22 are so far away that it would be difficult to include these in the model without other problems, like overfitting. Also, these distant points happen to be the lowest and highest values in the whole dataset. It looks like if we used age in a model, it wouldn't know what to do—is being older associated with a high or low level of education? And what's going on in all that space between where the bulk of the points are and the two distant ones? We have no idea because we don't have enough data for this number of dimensions. If data can start to spread out with just a single dimension increase, imagine what can happen with 10 dimensions or 50 or even more. The more features we have, the more data we need to represent the real variety among each feature and how it relates to other features.

As seen in the chart, overfitting is another problem that the curse of dimensionality brings, which comes about largely from sparsity. A fourth problem is that "distance" tends to get fuzzy the more features you have. So some of the typical metrics we use to measure distance stop working well when we get into higher-dimensional data.

Two other problems relate to the amount of computation that has to be done as the number of features increase. It starts to use more and more compute resources and takes longer and longer to run. Depending on the number of features, this can reach a point of being completely impractical (or way too expensive) to run.

Data Leakage

Data leakage comes from one of the easiest mistakes to make in modeling—accidentally including data in the train set that shouldn't be there. This could be rows from your test set (easy to do) or any other information about your target that won't be available when running real forecasts.

Another thing that's easy to do, especially when you're new, is accidentally including features that are there in your train (and test) set but won't be there when you're running your model on unseen/future data. For instance, imagine you are predicting daily pizza sales at a restaurant for the next week, Monday through Sunday. You create a bunch of great features, including things like previous day's sales and previous week's sales. That works in your test set because you are able to calculate all of those since you have complete data. When it comes to running true forecasts for the upcoming week, you may have sales all the way through Sunday night, and then you're going to run the feature update for the coming week and then the forecasts early on Monday morning. No problem there, because you have the previous day's sales. What about Tuesday? You don't have Monday's sales yet because it's early Monday morning and the restaurant hasn't even opened. Most likely your feature generation or forecasts would simply fail if you tried to run it on this data because it would be missing that info, but even worse would be if your model just takes the null value and uses it (to unknown impact, but not good).

Data science requires attention to detail and always thinking about what you're doing. If you're getting very high accuracies or performance scores, you probably have data leakage. Any time you have surprisingly good performance on your test set, check for this type of mistake. First, check that there is no overlap in your train and test sets. Then check for unreasonable features. Additionally, if you're testing time-based data, you almost always want to make your test set chronologically at the very end of your data and the train before that. This is a requirement if you've included features that rely on previous values, such as the previous month's sales.

Key Takeaways and Next Up

In this chapter, we covered a lot of machine learning topics. We first focused on defining the three key types of machine learning, supervised, unsupervised, and reinforcement, and also addressed semi-supervised learning and ensemble learning, both of which add a lot to the field. We then looked at several supervised learning algorithms, some in depth, including time series analysis, generalized linear models, trees, naïve Bayes, neural networks, k-nearest neighbors, and support vector machines. We then talked about the two key unsupervised learning techniques, clustering and association rule mining, touching on both types of clustering, k-means and hierarchical. I talked about model explainers that can help explain forecasts from black box models and some of the practical aspects of doing machine learning in code. Finally, I addressed some of the major challenges we face when doing machine learning modeling. These included overfitting and underfitting, when the model either won't generalize well or misses important patterns. Then I discussed imbalanced data, when the counts of different labels are very different, and the curse of dimensionality, which describes some of the problems that come from having a large number of features in a model. Finally, I explained data leakage, when information gets into your trained model that shouldn't be there.

In the next chapter, we'll be looking at how to see how well your models have done after you've trained them. We'll talk about several metrics that can be used for measuring the performance of models in classification tasks and models in regression tasks, because they can be quite different. I'll also look at some metrics that we use to measure the performance of clustering algorithms.

PRACTITIONER PROFILE: JEFF ELMORE

Name: Jeff Elmore

Current Title: Senior principal data scientist

Current Industry: Energy, also worked in education tech, literacy research, and writing

Years of Experience: 20 in tech, 13 of that in data science

Education:

- BS Applied Mathematics

The opinions expressed here are Jeff's and not any of his employers, past or present.

Background

Jeff fell in love with computers when he was 12, and that has never wavered. He started working as a programmer after high school and enjoyed it, but eventually decided to get a degree. He started as a computer engineering major, but after taking calculus, he realized he loved the math more and switched his major to applied math. He loves how you can have a problem that's really hard and just apply a mathematical transformation to it, and suddenly it's easy. He learned about machine learning from both college and his experienced boss. There were many opportunities at the company he was working at, and he started looking at where exactly it could be used.

Work

His first major project at his company was working with a researcher on improving one of their products, a tool that analyzed books to assign a difficulty-like score to it, which could be used to match kids to the right books. He needed to understand how kids learn to read, and he realized he didn't have the right domain knowledge, so he started learning but it was a slow process. Ultimately, he and the researcher came up with a product that solved the problem. He learned a lot during his work, especially about how valuable ensembling can be in machined learning because of the way it takes a holistic view of the problem. He continued making cool stuff at that company, but a lot of his work wasn't understood and it was frustrating to see it not used, so he decided to leave. He founded a company with a couple of people he knew, and they created a product designed to help writers write their novels. That venture didn't work out and it was really disappointing. But it also shifted his perspective to valuing the life part of the work/life balance more. He found a new job that he likes but doesn't take over his life, so he's happy. He's still interested in finding ways to use AI to help people, but he feels more like an AI hobbyist at this point.

Sound Bites

Favorite Parts of the Job: Data science is a bridge role and very satisfying. Jeff loves be able to solve people's problems when they can't do it themselves. He also likes the way you can bring new tools (like ML) to a field that's never used it and improve the work in that field.

Least Favorite Parts of the Job: Data created by humans that lives in a spreadsheet. It's rarely good. He also doesn't like how 80% of data science work is still cleaning it—he wonders how this can possibly still be true, but it is. It's also frustrating when you're working within a paradigm or framework that significantly limits what you can do.

Favorite Project: He developed a new assessment for oral reading (reading out loud), which was interesting because it involved understanding a variety of areas: psychometric educational assessment (validity and reliability), literacy development (learning to read), computational linguistics, and machine learning. He was able to find a legitimate assessment of the connection between reading ability and oral reading—they were able to predict with high accuracy the reader's words correct per minute based on their silent reading ability. He was really proud of this work, but unfortunately nothing came out of it because the company didn't understand it or how to market it.

Skills Used Most: Ruminating on data—being able to think about things to solve problems. Communication is critical. Visualization skills, in terms of both conceptualizing and creating the tangible charts and so on. You can convey so much in a picture, which is great for interaction. One other thing he's learned to do is create reusable pipelines for a lot of the more repetitive work he does to avoid having to do manual work over and over.

Primary Tools Used Currently: Python, Jupyter Notebooks, LLMs, Retool (a drag-and-drop UI builder), generally uses large open source ecosystems

Future of Data Science: New AI advances are changing things dramatically for people wanting to automate parts of their work. This will continue to grow as more people are empowered to do the automations. But AI is rarely perfect because there's always more info that could have been included in a model because there's never enough time and usually there's missing knowledge.

What Makes a Good Data Scientist: Humility and an ability to listen and reserve judgment, especially when you're going into a domain you don't know. If you don't stay open, you won't understand things and will make wrong assumptions. You also need to be flexible. Visualization skills, bot technical and nontechnical.

His Tip for Prospective Data Scientists: Don't tie yourself to a tool that is only commercially available because there are so many good open source tools out there. In general, always learn from other people's work in addition to your own. Learn to read academic papers even if you don't think you'll be writing them. Or at least be able to skim them to glean what's possible. Sometimes solving a problem is simply a matter of doing something someone else has already done.

Jeff is an experienced data scientist with deep experience in literacy and psychometrics.

CHAPTER 16

How'd We Do? Measuring the Performance of ML Techniques

Introduction

It's great to do machine learning, building models, getting predictions, and finding interesting associations. But how do we know if what we've done is actually right? This is where performance metrics come in. There are a variety of ways of measuring characteristics of ML models that can help us understand whether they're doing what we want and if they're right. If our results aren't actually meaningful, we don't want to use them. Sometimes it's tempting to just take the results at face value and assume they're good. See the sidebar for more on this bad habit.

Measuring the performance of models is done mostly by calculating specific metric scores. There are different metrics for the three kinds of machine learning tasks: classification, regression, and clustering. I'll go

K. P. Vincent, *A Friendly Guide to Data Science*, Friendly Guides to Technology,
https://doi.org/10.1007/979-8-8688-1169-2_16

over the variety of metrics currently in use for each. Note that the other unsupervised method we've talked about is association rules, and there aren't metrics used to evaluate the resulting rules overall, as each rule is judged by specific metrics during creation, so there is nothing to measure afterward.

When we talk about evaluating model performance, we're generally talking about calculating a metric on the test set, and the validation set if one was used, after the model has been trained on the train set with supervised learning. With unsupervised learning, there are metrics that can be calculated that give us a sense of the quality of the model, even though we can't measure "rightness" with all of them. In cases where we've used an unsupervised algorithm but also have labels, there are additional metrics that can be used.

I'll first talk about a couple of examples of how metrics can be important to real-world scenarios. Then I'll give an overview of internal model-building metrics before talking about classification, regression, and clustering metrics to be looked at after a model has been built.

FAILING TO EVALUATE MACHINE LEARNING RESULTS

Most professional data scientists are responsible enough to ensure their models are doing a good job. But often data science is being done by people who are just learning and may not know the right way to evaluate their models (which is why a chapter like this is important). It's always critical to understand how well your model did. It's also important to ensuring that no mistakes were made, basically by checking your work. I worked with an inexperienced data scientist who provided an argument in days rather than the required months in a tool he was using, and he didn't catch the mistake. A quick check would have told him that his forecasts were 30 times bigger than expected.

Another mistake inexperienced data scientists make is to evaluate their models with performance metrics like they should, but not realizing that their great results are a result of data leakage, not a fantastic model. Any time you get really high accuracy or other metrics, check for this mistake.

But an even bigger problem is that many companies don't really evaluate how good their results are down the road. They may do some clustering to identify customer types, but two years later they don't really check to see if this has helped them in any way (especially to see if it brought improvement over the old way of doing things). They might do some machine learning to help decide where to open new stores, but several years later, they may not compare the performance of stores that were recommended by the model and other stores that weren't. Part of the reason for this is that it's difficult to follow up on things that happened years ago, and you can't really evaluate major decisions right away.

Additionally, the attitude that comes from technochauvinism (the idea that machine learning is inherently fair, right, and generally superior to human work, discussed in Chapter 9) means people don't even question the results. Data science can be incredibly powerful, but are the results actually valuable? It takes some work to know for sure.

Examples of Performance Evaluation in the Real World

A lot of the time, data scientists have their favorite metrics, but it's always good to consider different ones and how well they fit your current problem. These two examples show the importance of metrics, especially picking the right ones.

Example 1: Metrics-Driven Development

The customer relationship management (CRM) platform Salesforce has thousands of customers. These customers have different needs and use cases, and many of them want to do machine learning on their Salesforce data. It's easiest for them if they can do this within the platform, but the myriad of different use cases present a problem for Salesforce. Sometimes companies will provide customized help for customers, but this is expensive and they're limited in how many customers they can help. So Salesforce created a tool called Einstein Builder that allows their customers to generate customized self-service predictions.[1] The tool needed to support all the different use cases but also be easy to use, so Einstein Builder is a web app that doesn't require coding but still supports a machine learning pipeline that allows automation. They have tools for regression and binary and multi-class classification, which cover a variety of situations.

You may wonder, how can they dive into modeling without all the data work we've spent chapters learning about here? This is a quandary—you can't offer self-service tools that do this kind of data prep unless the users themselves know how to do it and you give them data prep tools. The whole point of self-service tools is that the users don't have these skills. You know by now that this means that the results won't be very good. This is a real problem with a variety of self-service machine learning tools that are hitting the market because of the popularity of data science.

So they face a difficult task of giving customers the ability to do useful machine learning without the data prep. They do have a decent solution that relies on an effective use of different performance metrics. Also, remember that the data the customers are using is data in Salesforce's

---

[1] "Metrics-Driven Machine Learning Development at Salesforce Einstein" by Eric Wayman, available at https://www.infoq.com/presentations/ml-salesforce-einstein/

own system, so they have knowledge of what it looks like and how it's structured. There may be additional custom tables customers have, but at least some of it is a known entity.

Still, the heart of their solution is to measure the performance of the models customers build so they can be tweaked as necessary. The system observes characteristics of the models to see if improvements are possible. As an example, in a linear regression, the system can check for overfitting by looking for near-zero coefficients or cases with way too many variables. They can then add regularization to reduce the feature space. They also have the ability to run experiments with different features and parameters (feature selection and hyperparameter tuning), measure the performance of each experimental model, and carefully track everything. One thing they're able to do is compute various metrics during the different steps in the experiments, something that we don't usually do in day-to-day data science. The observed metrics and overall results of these experiments help them fine-tune models.

Einstein Builder identifies the right metrics to monitor by looking at what models are trying to accomplish and what the characteristics of the data involved are. All of this drives the improvement of this self-service machine learning tool.

Example 2: Model Testing and QA

In most companies, quality assurance (QA) is something that's done on software and reports, but rarely on data science code or models themselves. There's some awareness that this probably isn't ideal, and some places are trying to correct this. Scott Logic, a consulting company, has integrated machine model testing into their process.[2] One of their test

---

[2] "Evolving with AI from Traditional Testing to Model Evaluation I" by Shikha Nandal, September 13, 2024, available at `https://blog.scottlogic.com/2024/09/13/Evolving-with-AI-From-Traditional-Testing-to-Model-Evaluation-I.html`

engineers described their process and approach, in an article targeting test engineers to help them understand how to make the switch from software to model testing.

She mentions the many things that make model testing different from software testing, including the fact that machine learning models do not have outputs that are known in advance and fully rely on what exactly is in the data. Additionally, the output of models changes on every run with new data, but usually their performance degrades over time (from model drift). Finally, the traditional approach to software testing of simple tests that pass or fail isn't sufficient, and instead they need to rely on the various machine learning performance metrics, which is why this is interesting for this chapter.

For this type of testing, they must understand the data being used very well. It needs to be analyzed and validated, especially checked for missing data, duplicates, and any bias. They are also doing model testing, rather than model evaluation. Model evaluation is what data scientists do when building their models and selecting the best ones by testing on the test set. But in QA, they are trying to look at how the model does in the real world. In their testing, they use cross-validation. They use A/B testing with other known solutions where possible. But the most important part is their selection of performance metrics, which they select according to the problem, classification or regression. One thing they do is not rely on a single metric, but instead use several so they can get a more complete picture. Not all metrics work on all data.

Classification Metrics

When people talk about classification in machine learning, they are usually referring to binary classification, usually understood as true or false—yes, the patient has cancer or, no, they don't; yes, this is a spam email or, no, it's not—but there are multilabel classification problems. We'll cover measures for both here.

In binary classification, the heart of performance measurement is the confusion matrix, which is an intuitive way to visualize the performance of a model. Most of the main metrics are in some way related to the confusion matrix, so we'll talk about that first and then the metrics that are based on the matrix. We'll talk about a few more metrics used in binary classification and then talk about some metrics used in multi-class classification.

Binary Classification Metrics

Most classification is binary, meaning there are two possible labels. These can be anything, but as a rule, we consider one positive and one negative. We'll talk about the various metrics that can be used in binary classification. The next section will address how to handle measuring the performance of multi-class problems.

The Confusion Matrix

When it comes to binary classification, each prediction has one of four situations, each with a name:

- **True Positive (TP)**: The predicted label is true, and it's correct because the actual instance is also true.

- **True Negative (TN)**: The predicted label is false, and it's correct because the actual instance is also false.

- **False Positive (FP)**: The predicted label is true, but it's incorrect because the actual instance is false (this is also called a Type I error in statistics).

- **False Negative (FN)**: The predicted label is false, but it's incorrect because the actual instance is true (this is also called a Type II error in statistics).

In general, we obviously want high values for TP and TN and low values for FP and FN. But it's natural to turn these values into a matrix view that makes it easy to see what's going on. The *confusion matrix* shows the counts of the four values along with the labels for predictions and reality. Figure 16-1 shows the basic structure for a cancer diagnosis, where having cancer is positive and not having it is negative.

		Predicted Condition	
	Total population	**Predicted Positive (Has Cancer)**	**Predicted Negative (Does Not Have Cancer)**
Actual Condition	**Actually Positive (Has Cancer)**	TP total	FN total
	Actually Negative (Does Not Have Cancer)	FP total	TN total

Figure 16-1. *The basic confusion matrix structure*

Usually when we create the matrix, the only labeling we have is the True and False across the top and down the side, and we know that the top is the prediction and the side is the actual. In this view, we usually want to see high values along the diagonal from top left to bottom right and low on the other corners.

Note that in cases where the dataset isn't balanced, where positives are rare (such as in fraud detection or a disease diagnosis), we would see low values in the TP box and very high in the TN box. We'd still want low values in FP and FN.

Remember that while we always want FP and FN to be low, in some cases, we want to minimize the number of false positives. For example, in cases like fraud detection, a high FN means we have missed some cases of fraud, so that transaction won't be investigated. We'd much rather have to deal with more FP than a high FN. There are many areas where this is the situation, including disease detection, where a FN means a patient won't be sent for further testing and will remain untreated for a condition they have, and security screening, where a FN might mean someone is bringing a bomb into a building. On the other hand, there are cases where FPs are very costly and a FN isn't that bad, so we'd rather err on the side of missing a positive than unnecessarily labeling something positive that isn't. For instance, in product recommendation, no one is really hurt if something they would like is not recommended to them (FN), but if too many products they don't like are recommended (FP), they will lose faith in the recommendations. Other cases where high FPs are bad are in sentencing for the death penalty (executing someone who's not guilty is much worse than not executing someone who is guilty) and spam detection (getting a few spammy messages (FNs) is better than missing an important email (FP)).

The confusion matrix is a great starting point in binary classification, but it doesn't give us much tangible on its own. Instead, we use a variety of measures calculated based on the values it holds.

Confusion Matrix–Based Metrics

The metric everyone first thinks of is accuracy, a term that is often used in the generic sense of "how well the model did," but it is also the name of a simple and specific metric. *Accuracy* is simply all the correctly labeled counts divided by the total number of records. This can be a very skewed metric, however, especially in the case of an imbalanced dataset where the total count of actual positives differs from the total count of actual negatives. In extremely imbalanced datasets, we can have very high accuracy while misclassifying every actually positive value. For instance, if we have 10,000 credit card transaction instances where only 50 are fraudulent, if the classifier simply classified every single one as negative (not fraudulent), the classifier would have a 99.5% accuracy. This obviously is absurdly misleading and useless.

Precision is another important measure, and it looks at how well the classifier did with its positive labels, or what proportion of its positively labeled predictions are actually positive. So it focuses on FPs. This is an important metric when FPs are expensive and FNs aren't that bad, like in a recommendation system. But it's not the best metric when FNs are dangerous.

Recall is considered the opposite of precision, because it emphasizes FNs. *Recall* measures the proportion of total positives were detected out of the whole dataset. This is the measure we care about in cases like fraud and disease detection. Recall is also sometimes called sensitivity or the true positive rate. The specific calculation is given in Table 16-1.

Precision and recall usually have an inverse relationship, and frequently we want to find a balance between them. Also, we can get misleading results if our model has behaved very badly. For instance, if a fraud detector labeled every single transaction as fraud, we would have a very high recall but a very low precision (and accuracy). Similarly, in a recommendation system, if we only recommended a tiny number of products but they were good recommendations, we'd have a high precision—but we'd also have missed out on all those sales of products that went un-recommended. You can see the calculation in Table 16-1.

So whichever of the two is deemed most important in a scenario, we still want to know about the balance between them. Fortunately, we have just the metric for this—the *F-score,* which is basically a modified average (the harmonic mean) of precision and recall. It's a little more complicated to calculate than the others, but you can see the calculation in Table 16-1. So, when precision or recall is more important, it's common to look at it and then also the F-score to make sure the model isn't doing something useless.

Table 16-1. *Calculations for accuracy, precision, recall, and F-score in binary classification*

Metric	Calculation from Confusion Matrix
Accuracy	(TP + TN) / (TP + TN + FP + FN)
Precision	TP / (TP + FP)
Recall	TP / (TP + FN)
F-Score	2*TP / (2*TP + FP + FN)

Although the F-score is good, it has a limitation in how it deals with TNs, so sometimes a couple of other measures are used, the Matthews correlation coefficient or Cohen's Kappa. Both are a bit more complicated to calculate, but it can be done from confusion matrix values. You can find more information by Googling.

I've talked about the most common classification metrics that are based on the confusion matrix, but there are several others that are useful and used in practice, including the false positive rate and specificity (the true negative rate). Wikipedia's page[3] on precision and recall has an excellent expanded confusion matrix that summarizes all of them.

Other Metrics

Although the confusion matrix and its associated metrics are the heart of binary classification measurement, there are some other measures that data scientists use. The most common are the related receiver-operating characteristic curve (ROC) and the area under the curve (AUC).

The *ROC* plots the true positive rate (recall) and false positive rate against each other, where in a perfect classifier, the plotted line would be a flat line across the top of the chart, like what you see on the left in Figure 16-2. Classifiers are never perfect, so a more realistic one can be seen on the right, which shows both what a randomly guessing model would produce and what a real one did on the kids height data.

---

[3] https://en.wikipedia.org/wiki/Precision_and_recall

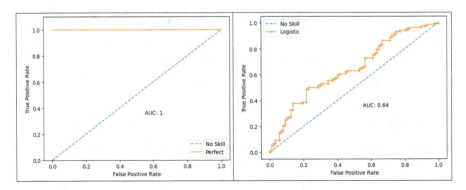

Figure 16-2. *Two ROC plots, a "perfect" classifier on the left and more realistic one on the right*

The *AUC* is literally the area of what appears under the ROC line. It's included in the charts in Figure 16-2. When comparing different runs of a model on the same data, the one with a higher AUC is considered better.

One more metric that's useful in models is where there is a probability given for how likely a prediction is likely to be right. For instance, in logistic regression, we get a probability and usually take a cutoff—0.5, normally—and anything above that is labeled True and anything below False. There's a metric called *log-loss* that works with that probability directly and calculates a score based on the difference between the probability and the actual value for each instance (the actual calculation involves some log values and can be looked up online). With this metric, we're looking for low values.

Multi-class Classification Metrics

All the metrics discussed above can be adapted to multi-class classification problems. A confusion matrix can still be created, but it will have more than four squares because it will have more than True and False labels across the top and down the side.

Accuracy is calculated the same, with all correct predictions divided by all predictions. Most of the other metrics require some modification, specifically by calculating the macro, weighted, or micro versions. With the *macro* approach, you calculate the metric by looking only at a single class at a time and calculating the average of all of those. With the *weighted* approach, you do something similar, but you multiply each individual metric by the proportion of total examples that have that label.

With the *micro* approach, you generate overall TPs, FPs, etc. by summing all the ones for individual classes to represent the overall TP and then plug those sums into the regular metric calculation. But the catch with this approach is that accuracy, precision, recall, and F-score will all be the same value. This is because of the way that for any given class, TP is the total for that class only, but any example that is correctly labeled as not being part of that class will be a "TN" for that class. So everything sums up the same in the end.

You can also generate ROC and AUC for multi-class problems, and it's common to plot each class as a separate curve on the ROC plot. Then an AUC can be calculated for each class. Figure 16-3 shows such an ROC plot for the iris data.

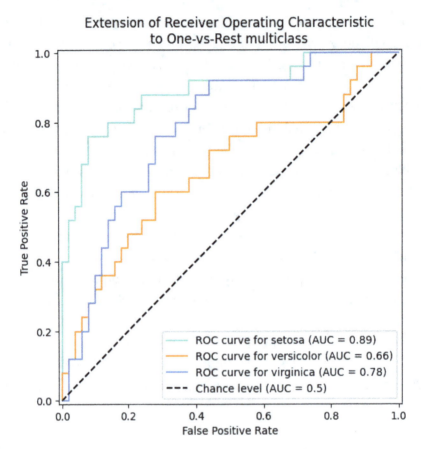

Figure 16-3.   *An ROC plot for a multi-class problem*

Regression Metrics

Measuring the performance of regression results is different because the outputs are numeric and, in most models, can be any value. Consequently, a given value isn't simply right or wrong—we want to measure how close it is to the right number.

The first thing to know about regression metrics is that there is no such thing as an accuracy metric for regression, as the accuracy metric is specific to classification. We do sometimes talk about accuracy in regression, but we are using the word in its general sense, not as a specific technical term.

A common metric that pops up with generalized linear models is R-squared score. It's technically a goodness-of-fit measure—*R-squared score* says how well the generated model fits the data and how much of the variance in the target variable is explained by the features. With linear models, it effectively compares the actual regression line to the mean line, a horizontal line across the chart through the Y-axis at the value of the mean of all the *Y* values in the data, but it can also be used with some nonlinear models. In the linear case, a terrible model would have a line close to the mean line and a very low R-squared score. R-squared can be good for different models created on the same data because it's considered relative. That means that it's not as suited to comparison across different data.

One downside of R-squared is that it assumes that all the features are important to the model. Additionally, adding a feature tends to increase the R-squared, which can lead people to add more features than necessary and end up with an overfitted model. There's a modification of R-squared score that better handles features of differing value. *Adjusted R-squared score* looks at what each feature brings in terms of added value, and the score only increases if the features improve the model more than what chance would. Because of this improvement, adjusted R-squared is almost always less than R-squared.

There are several other metrics often used in lieu of R-squared and adjusted R-squared because they aren't always the best for determining model performance on the test set.

One popular regression measure is *mean squared error* (MSE), which involves squaring the difference between the actual value and the predicted value, summing all of those, and dividing the sum by the total number of predicted, thus giving us an "averaged" value. The squaring of the difference is important in terms of keeping negative values from countering positive ones, but it also has the problem of easily becoming huge because of the squaring. A perfect model will have a MSE of 0, but it can grow very large. Additionally, it can be hard to interpret because the units of the value are the square of the units of your target variable.

A variation of MSE is *root mean squared error* (RMSE), which is just the square root of the MSE, which brings the metric's units back to the same as the target, so it's much easier to interpret. The square root has to be taken after the differences have been squared to avoid the problem of positive and negative errors canceling each other out. Just like MSE, a perfect RMSE is 0, but it too can get large.

Another popular metric is *mean absolute error* (MAE), which is similar to MSE but instead of squaring the differences between the actual and predicted values to deal with the positive and negative cancel effect, it takes the absolute value of the difference instead. The values are summed and divided by the total number of predictions, just like MSE. But this measure avoids the inflation of errors (especially larger ones) that comes from MSE's squaring them, so it's often preferred as more realistic. Additionally, it's in the target variable's units, like RMSE. Also, like RMSE and MSE, a perfect model has an MAE of 0 and can get large.

Another metric that's used a lot is *mean absolute percentage error* (MAPE), which is a variation of MAE that expresses the error as a percentage, making it instantly understandable to a wide range of people. MAPE divides the difference between the actual and predicted by the actual before taking the absolute value and summing everything (and dividing by the total number of predictions again). Percentages are very intuitive, and we don't have to worry about thinking about what the target variable units mean. One thing to be aware of is that the "percentage" can also blow up and be much, much higher than 100% when the model is pretty bad. Another thing is that when calculating the error percentage, if the actual value is 0, it can't be calculated because that would mean dividing by 0 (although there are workarounds). Additionally, the metric is relative to the actuals. If an actual is 125 and the prediction is 100, we have an error of 25 and that would turn into 25%, but with an actual of 75 and a prediction of 100 (the same error difference as the previous example), the percentage would be 33%. So predictions that are higher than the actual are penalized more than those that are lower.

Clustering Metrics

Because clustering is generally unsupervised, most of the metrics are internal and look for general characteristics of the clusters rather than compare the results to known values. However, sometimes clustering is done when labels are known, and there are some specific metrics for that case (these are considered external measures). We'll go over both types here.

Internal Measures

Internal clustering metrics are those that require no knowledge of actual labels. Because these metrics are simply looking at characteristics of the cluster, like how well they're separated from each other and how tightly packed they are, this says nothing about how correct they might be. They just tell us that the clusters seem like a good model.

One common such metric that is used to evaluate clustering results is called the *silhouette score*, which measures the separation between and cohesiveness within clusters. These things are important in a cluster model because we want clusters to be clearly distinct from each other (separation) and not have a lot of differences between the points in each cluster (cohesiveness). The calculation involves two averages for each data point: the average distance between the point and the other points within the same cluster and the smallest average distance between the point and the data points in different clusters. The value can be anything from –1 to 1, with a larger number being better clustering because the points are well-matched to their own clusters and lower numbers indicating overlap between clusters and ambiguity.

A second metric for clustering is the *Davies–Bouldin index*, which measures how similar each cluster is to its most similar cluster. It relies on taking the average for all clusters of a dissimilarity measure given between a given cluster and the cluster considered most similar to it. Lower numbers are better in this metric.

Another metric called the *Calinski–Harabasz index* considers the variance both within clusters and between clusters. The calculation involves the sums of squares between and within clusters, the number of clusters, and the count of data points. A higher value is better.

The last internal metric I'll mention is *inertia*, which looks at sum of squares within clusters. It sums the squared distances between each point's location and the centroid of its cluster. Low scores are better, but if they are too low, it can indicate overfitting. Additionally, inertia tends to decrease as more clusters are added, which is why picking an optimal number of clusters is important.

External Measures

All the metrics we've looked at so far are simply evaluating the outcome of clustering (as in unsupervised learning), but there are other metrics that are external, because they can only be used in cases where we do have labels for the data (as we would in supervised learning). The clustering itself is still run unsupervised, but a couple metrics called the *Rand Index* (RI) and the variant *Adjusted Rand Index* (ARI) allow us to compare the output of the clustering algorithm to the labels we have. The calculation of RI involves the total number of agreeing pairs of labels and the total number of pairs. It can be from 0 to 1, with higher numbers better. The Adjusted Rand Index uses the computed RI and adjusts for chance by using the expected RI value. It has the same range as RI.

One other metric that can be used when labels are known is mutual information (MI). *Mutual information* measures the similarity between a point's actual and predicted labels (clusters). Calculating it involves the actual and predicted labels and a couple of probabilities. The lowest possible score is 0, which means there's no relation, and higher values indicate the predictions align with the known labels better.

Key Takeaways and Next Up

In this chapter, we've talked about measuring performance of classification, regression, and clustering models. There are very different metrics for each of these types of machine learning. In classification, most of the measures are based on the confusion matrix, which is a way of representing counts of true positives, false positives, true negatives, and false negatives. In binary classification, "positive" vs. "negative" indicates the label we're testing for (like positive is likely a fraudulent transaction vs. negative likely not a fraudulent transaction). "True" and "false" indicate whether the predicted label matches the actual label. Some of these metrics include accuracy (what proportion it got right), precision (whether it was overzealous in making positive predictions), recall (how well it did finding the actual positives), and F-score (a balance between precision and recall). I also talked about ROC curves and the AUC calculation, which look at the true positive and false positive rates.

I then talked about a variety of regression metrics. R-squared is used with linear models and gives us an indication of how well the model fits the data. MSE looks at the error between the actual and predicted, squaring it to remove the sign. RMSE takes the square root of that to bring the units back down to be the same as the target variable. MAE takes the absolute value of the difference between actual and predicted and makes a calculation of that, also in the units of the target variable. MAPE is a version of MAE turned into a percentage, often more intuitive for less technical people.

Finally, we looked at several clustering metrics, some of which quantify the general quality of a cluster model without knowing actual labels, and others can measure models against known labels. The first group of metrics includes the silhouette score (looks at how cohesive and separated the clusters are), the Davies–Boulding index (how similar a cluster is to its most similar neighbor), the Calinski–Harabasz index (looks at the variance between and within clusters), and inertia (another look at sum of squares within clusters). Two metrics of the latter type, RI (the proportion of correctly labeled pairs and total pairs) and MI (measures the similarity between actual and predicted labels), are both also useful.

In the next chapter, we'll be looking at working with language. Data scientists don't always get involved with natural language processing or speech processing, but it's becoming more common, especially with the popularity of large language models. We'll first cover the basics of NLP including parsing language, segmenting sentences, turning text into word-like pieces called tokens, stemming (taking words down to their common base forms) and tagging parts of speech. Then we'll cover language understanding and generation. Finally, we'll spend a little time looking at speech processing (both recognition and synthesis).

PRACTITIONER PROFILE: DIEGO TANGASSI

Name: Diego Tangassi

Current Title: Advanced analytics manager

Current Industry: Retail

Years of Experience: 8 as a data scientist, previously in actuarial science and finance

Education:

- Licentiatura en ciencias actuariales (actuarial science undergrad degree)

The opinions expressed here are Diego's and not any of his employers, past or present.

Background

Diego always loved engineering, robots, and systems in general. Once he got his first PC, he downloaded things to tinker with, started messing around with software, and even jailbroke his iPhone. He originally planned to study mechatronics at university, but when he visited the school, the professor hosting his visit was uninspiring, which made Diego change his mind and look into other things. He found something new that also intrigued him—actuarial science—which he studied at university. He absolutely loved it after his first semester, and after a class in finance with a fascinating professor who had a lot of real-world experience, Diego was sure he wanted to work in finance.

Work

Diego landed a job in finance like he'd wanted, in a financial risk department (working on pensions), but it turned out that the work was boring. It was just following the same steps every day, day after day—getting output, feeding

587

it into another model, repeat. He was so bored after six months. When a consultancy firm reached out to him, looking for someone with stats and math skills, he made the jump. He loved the whole experience, including learning more about data and data engineering with ETL, databases, information cubes, and systems assurance. He also learned more about data science as a field and about revenue mix models (a way of determining which products should get boosted). The tools he was using were really basic, like Microsoft Excel Solver to optimize equations, but stakeholders were still deeply impressed. The work had two basic phases, where the first was doing the data science and the second was communicating the results. One of the things he liked at the consultancy was how he was exposed to working with high-level people at retail companies and he got really good at communicating. He would often get hand-picked for roles because of these communication skills. He left consulting but stayed in retail for his current role, where he still loves the work.

Sound Bites

Favorite Parts of the Job: Diego loves all the learning you have to do in data science. Learning all kinds of things energizes him. He also loves working with all kinds of different people. Finally, one of the most satisfying things about any kind of coding is when you write some challenging code, and it works.

Least Favorite Parts of the Job: Bureaucracy. Sometimes there can be so many obstacles to finishing things that stakeholders lose interest. Sometimes company cultures can be challenging.

Favorite Project: During his consultancy days, there was a request from a government institution to create a database of drug seizures in Mexico. They wanted a lot of details, including where it happened, the cartel involved, the number of people and cars involved, where the drugs were hidden, and more. They had to get creative with the solution, using a variety of data sources, but primarily using newspapers. They scraped the newspaper web sites, carried out NLP to extract all the details, and stored it in a database. It was a very satisfying and successful project.

Skills Used Most: An analytical mind. Being able to convert techspeak to business speak and vice versa. Being good at self-learning.

Primary Tools Used Currently: R, Python, SQL, Snowflake, GitHub, Google Suite (APIs and more), Visual Studio

Future of Data Science: It used to be that most data scientists did a huge range of work, including data engineering, data management, and more. But things have gotten more focused now so that data scientists can focus on more typical data science work, where the other tasks are done by other people with specific job titles. This trend on focusing and getting more specific will continue.

What Makes a Good Data Scientist: Understanding that data science does involve the technical stuff, but the most effort is needed in talking to nontechnical people, understanding their needs, and communicating what they can expect from you.

His Tip for Prospective Data Scientists: Try to make a plan on what area you want to focus on in your career, so you understand different domains and where you might want to work.

Diego is an experienced data scientist with expertise in finance and retail.

CHAPTER 17

Making the Computer Literate: Text and Speech Processing

Introduction

Everyone has become really aware of AI in the language context lately, mostly because of work with LLMs like ChatGPT. This is real NLP, and it's blown up in the last couple decades. The things computers can do with language nowadays are massively impressive, and it's a very exciting time to be working in the field.

Many of us also know that there have been improvements in speech technology in the last couple of decades, especially anyone who's had to call a company for help. I personally want to go back to pressing numbers again, because what I'm calling about never seems to match the options, but that's not going to happen.

For many years, it was primarily electrical engineers who did speech processing work, as it was treated as a signal processing problem. So not much knowledge of linguistics was incorporated into the efforts. In contrast, NLP was often based in linguistics university departments along

© Kelly P. Vincent 2025
K. P. Vincent, *A Friendly Guide to Data Science*, Friendly Guides to Technology,
https://doi.org/10.1007/979-8-8688-1169-2_17

with computational linguistics (a similar discipline but based more in linguistics theory), sometimes in conjunction with computer science, so processing text has involved linguistics knowledge much longer. But the field of speech processing has changed as they finally realized that speech isn't just sound—the way we say and hear words depends on many factors, including context, tone of voice, and other linguistic characteristics.

There's a lot of terminology around natural language processing and speech processing, especially around linguistics. See the sidebar for an explanation of the many linguistics areas relevant. We'll see more terms below.

LINGUISTICS SUBFIELDS

Phonetics: The study of speech sounds, including the physiological mechanism and acoustic properties

Phonology: The study of speech sounds, generally at a higher level than phonetics

Semantics: The study of meaning in language

Syntax: The study of how words and phrases are arranged in order to be grammatically correct and meaningful

Prosody: The study of higher-level sound characteristics of language, including intonation, stress, and rhythm

---

Examples of NLP and Speech Processing in the Real World

NLP is increasingly used in industry, and here we'll look at a couple of examples, first from a Scottish shoe company and then from a French hospital using speech recognition to dictate patient records.

Example 1: NLP Fills Some Big Shoes

The Scottish company schuh, a shoe company well-known in the United Kingdom and Ireland and currently expanding, was having trouble managing customer interactions and sought to speed up and improve that process.[1] They prided themselves on having good customer support, and they were specifically looking for ways to handle customer complaints in a way that would leave the customer satisfied—and still a future customer.

They started using Amazon Comprehend's NLP and ML tools to identify the sentiment in customer emails. The system looks at whether a given new email has positive or negative sentiment, which can help get brand-new emails routed to the support team properly. Different customer support agents are specialists in different kinds of support, so this routing of emails to the right person faster saves valuable time. This prioritization and routing was previously done manually, which was time-consuming. They also use the tools to analyze outgoing emails to ensure quality in responses coming from the agents.

They found that the system brought great benefits, mostly in saving time and improving the interactions with the customer, both in quality and timeliness, meaning they're more likely to retain them.

Example 2: Dictating Patient Medical Records

Doctors are notoriously busy, and paperwork is one of the things they dislike the most, but it's crucial to update patient records in a timely way. In 2016, a Paris hospital, Saint Joseph Hospital, prided itself on being paperless as much as possible. As part of that effort, they started using a speech recognition system from Nuance, a company that's been in the

---

[1] "schuh Steps Up its Customer Experience with AWS," 2020, available at https://aws.amazon.com/solutions/case-studies/schuh-case-study/

speech recognition space for almost three decades.[2] Nuance has a variety of specialized products targeting healthcare, and the Paris hospital used their product Dragon Medical Direct and integrated speech recognition into their electronic medical records system.

The main problem they were trying to solve with this solution was the speed of hospital patient records completion and delivery. A healthcare regulator required them to provide patients with their records upon release and get the record to their GPs within a week of release, which was not consistently happening. Records were handled by the doctor taking notes and sending them to a secretary, who had to retype everything. This required a lot of back-and-forth, and having a second party key in the data is rife with risk of errors being made. It's always difficult to read someone else's quick notes, and any typos the doctor made could easily be misinterpreted by another person (even if the doctors themselves would recognize and be able to correct their own mistakes). But most importantly, this process was slow. If they could cut out the need for the secretary to retype everything without putting extra work on doctors, that would be a win for everyone.

Once they started testing it, there was resistance (par for the course with any system change). Doctors didn't think it could work and didn't want to change their processes. However, early testing doctors were impressed with the dictation software because it was both accurate and quick. Soon, it was deployed in many units, with others requesting it. They found it took a few weeks for individual doctors to get used to the systems. For instance, it was a little awkward to be dictating patient records in front of the patient, as most doctors opted to do. But once used to it, they liked

---

[2] "Voice recognition, innovation to improve the healthcare process," 2016, available at https://www.nuance.com/content/dam/nuance/en_uk/collateral/healthcare/case-study/ss-saint-joseph-hospital-in-paris-epr-en-uk.pdf

being able to get that done during the visit, instead of having to work with the secretary hours or days later on it, and there was also the opportunity for the patient to correct anything wrong.

The hospital considers the project a big success. Prior to the software adoption, 32% of records going to GPs missed the one-week deadline. After the dictation system was adopted, it dropped to 5%, which is a huge improvement. Almost a third of the records are sent within a day. They've heard from the GPs on the record receiving end that the records are clear and precise, partially because using the dictation systems makes doctors speak more clearly and simply. Patients also appreciate the new transparency, as they know what's going into their records. All doctors agree that it's time-saving as well.

Language Data Sources

I first talked about language data in Chapter 1 when talking about unstructured data. Text data from things like news articles, tweets, and customer email complaints all fits that bill. But we can't just feed that data into any ML system and get something useful out. We must prepare the data in some way. We'll talk about the various ways that can be done in the next section.

Writing Systems

The way language is written is hugely important. Historically, a large chunk of work in NLP has been done with English and other European languages, but there's a growing body of Chinese and other work. One reason for this is the challenges of working with different kinds of writing systems.

English isn't the only game in town, even if it's the most popular in NLP, at least for now. I'm going to use a few linguistic terms in this section, so check the sidebar for some definitions. English and most Western

European languages are what's called an alphabetic system, which means that it contains letters that represent (generally) a specific sound, and these letters are used to make up words. It's not technically quite as simple as this, as sounds and spelling can be dependent on context and language origin, but conceptually, alphabets use letters representing sounds to make up words. They have consonants and vowels. A similar type of system is a syllabary, which generally has a character that represents a particular syllable, so a consonant and a vowel together. Most Indian languages are written this way. A final type of writing system in modern languages is logographic, which represents an entire word (or phrase) in a single character. The only one still in use is Chinese characters, but it has been incorporated into several languages in East Asia. Both alphabets and syllabaries include information about pronunciation when written, but logographic systems do not. Interestingly, most East Asian languages also have a secondary writing system that does represent sound and is used in different contexts from the main system.

Another important element of text of different languages is the script. English and most European languages are written horizontally, left to right, in the Latin alphabet script. Other alphabetic scripts include Cyrillic (Russian and others), Ancient Greek, and Arabic (Arabic, Urdu, and others). Languages that use the Arabic script are written horizontally right to left. Some alphabet-like languages include only consonants with a different script, including Hebrew, also written right to left and horizontally. Most Indian languages use a syllabary and are written horizontally and left to right. East Asian languages are often written vertically and to be read top to bottom.

One last important aspect of the way language is written is all the ancillary aspects, like punctuation (commas, periods, etc.), diacritics (modifiers for letters like the umlaut or accent marks), and white space. We often take these for granted, but Ancient Latin was written without spaces between the words. All of these things will factor in when we're trying to do NLP.

LINGUISTIC TERMS FOR WRITING SYSTEMS

Writing System: Any system used to convey words for a particular language in written form

Script: The style of the written part of language (multiple languages can use the same script)

Letter: A single written letter in an alphabetic writing system ("alpha" includes five letters)

Digit: A single written number ("973" includes three digits)

Vowel: A pronounceable language sound that can be made and held, usually with an open mouth (vowels in English include the variety of sounds that are made with "a," "e," "i," "o," and "u")

Consonant: A pronounceable language sound that is generally made with a partially or fully closed mouth (in English, "t", "m," "f,", "s," and "k" are all consonants)

Click: A specific type of consonant made with various tongue configurations that usually include popping or clicking sounds (these are common in some African languages)

Syllable: A single unit of pronounceable language that includes a vowel sound and optionally a consonant

Word: A single distinct unit of meaning in language, whether represented as a single character or a sequence of letters or syllables

Logogram: A single character, symbol, or sign that represents a word or phrase (such as Chinese characters, Egyptian hieroglyphs, and shorthand in English)

Diacritic: A sign added to a letter above or below it to indicate a pronunciation difference (common ones are accents and cedillas)

Character: Any part of written language that is a single and distinct unit, usually also including diacritics if present (in Western languages, each letter, digit, and punctuation mark is a single character; in logographic languages like Chinese, it can be a single word comprised of many marks or a part of a word representing a sound; in other languages it can represent a syllable)

Text Data

Actual text data can come from almost infinite sources. This is one of the types of unstructured data we talked about in Chapter 1. It's just any text in any form that's been digitized. I worked at one company where we analyzed book text, so we had full books scanned in, each stored in one field in one row in a SQL Server database. It might be the full text of student essays stored as files in a folder on a server. It might be tweets streaming in, in real time. Anything that is human language and represented digitally as text qualifies.

One thing to note is that whatever it is, it's almost guaranteed that some preprocessing will be required before any NLP can be started. This includes handling any character encoding issues (discussed next) and potentially tweaking line breaks and any leftover cruft (this can be weird characters from formatting in a Word file if it was converted from Word to raw text or any other detritus).

One of the things that can be a headache with text data is encoding (though this isn't as difficult as it used to be). There are several different text encodings (this is on all text files, not specific to human language text), with UTF-8 largely the favorite. Other common encodings you'll still see for Western languages are Windows-1252 and Latin-1. Most programming languages handle this rather elegantly now, but if you run into your code spitting out gobbledygook when you look at what it read in from something you've provided, it's almost definitely an encoding problem. You'll need to hit up Google.

Text Data from Audio

Speech is even more complicated because it effectively adds an extra layer on top of what NLP is handling. Generally, speech recognition involves transcribing recorded speech into text that can then be processed using the same NLP techniques that are used on text. This is true conceptually, even if a system never generates an intermediate text file going from speech input to a speech response. An automatic transcription process is error-prone.

Speech synthesis also generally uses text to create audio, because that's the starting point for the generation. It also uses some NLP techniques to determine pronunciation. Much of the speech data preparation we do is going to be similar to text, which we'll discuss below.

Natural Language Processing

Natural language processing is basically the science (and art) of taking textual language data with a recognized writing system and breaking it into small pieces so it can be handled as data. Let's just take that previous sentence and see how we'd make that data. See the sidebar for some definitions that will come up in this section.

NLP TERMS

Text: The written version of human language using a recognized writing system (in the computer world, this also means the simplest form of textual data without the formatting you get in Word or other word processing programs, like what you find in Microsoft's Notepad or Mac's TextEdit)

Document: A single object containing a cohesive unit of text that will be processed as one (in tweet analysis, usually each tweet is a single document, while working with student essays, each complete essay would be a single document)

Corpus: A collection of documents

NLP encompasses a number of different techniques, many of which can be used together. I'm going to cover many of them, but there are a lot of other things you can do, which you can find out about online. There's also one technique that is used a lot when working with language called regular expressions. It's not technically part of NLP but it's very useful, and you can read about them in the sidebar.

REGULAR EXPRESSIONS

There's more to text processing than NLP, most of which is just working with different programming languages' text processing libraries. But there's one fundamental text processing tool that isn't NLP but is still incredibly useful in the right circumstances and is often used in conjunction with NLP projects, so it's good to know about it. *Regular expressions* (often abbreviated *regex* or even just *re*) allow you to specify strings of patterns to match in text. Common uses are to create patterns to match URLs or emails when you want to find them in text you're working with. They can actually get pretty complicated fast, but you just need to know enough to look up how to write the one you need rather than having all the syntax memorized.

As an example, here's what a pattern matching an email identifier might look like:

```
\b[\w.\-]+@\w[\w]*\.[a-zA-Z]+\b
```

This is something you'll learn after working with them, so don't worry about it looking intimidating. We're going to break it down one piece at a time. It helps to understand that the backslash is an escaping character that basically changes what the next character means. Here, \b ensures it has a word boundary at the beginning. Values inside square brackets indicate a single character of any listed within the brackets. [\w.\-] matches one instance of three possible things:

- Any letter or digit (that's the \w)

- A period (the \.)

- A hyphen (the \-)

The + indicates one or more of the previous character (in this case, any of the ones specified in the brackets group). The @ just matches that character one time. The \w matches a single letter or number. Then the [\w]* matches zero or more letters or numbers (this combined with the previous and the previous \w means that there must be at least one character after the at cue). The \. matches a single period. [a-zA-Z]+ matches one or more letters. The final \b indicates a word boundary again.

I know it looks pretty ugly, but you'll surprise yourself by starting to remember it all if you ever start working with regular expressions. But even if you do, you'll probably find yourself looking up specifics all the time, which is fine.

Parsing Language

Most people intuitively know that languages have rules, even if they can't identify what they are. Yet, they follow most of them when speaking, and most when writing, even though few people get it right all the time. The only universal rule about language rules is that there is always an exception. Exceptions make learning languages difficult, and a language like English is especially difficult because it pulls from many different languages so "exceptions to the rule" are incredibly common. Additionally, there are different rules that apply in different contexts, like when dialects use different words and pronunciations from other dialects of the same language or when different groups use language differently (like teenagers vs. retirees). Even within one language, the way it's written can be radically different in different contexts. Think about looking at articles in academic journals vs. tweets.

Despite these important differences, there are still common aspects of the language across all forms. There are several standard ways of breaking language down. See the sidebar for some more linguistic definitions that I'll be using here. Breaking down language into linguistically meaningful parts is generally called parsing language. See the sidebar for some of the linguistic and NLP terms I'll talk about.

MORE NLP TERMS

Token: Any distinct unit of text as being processed, like a word, logograph, number (including multi-digit), or punctuation (can be relative, with something like $25.78 being a single token)

Affix: A prefix (a sub-word part that is added to the beginning of a word like "un-" to make "unreliable"), suffix (a sub-word part that is added to the end of a word like "-ness" to make "completeness"), or infix (a sub-word part that is inserted in the middle of a word, not too common in English)

Stem: The base form of a word with some affixes removed (does not always correspond with an actual word)

Lemma: The canonical form of a root word with morphological features like some affixes removed (always corresponds with an actual word)

Punctuation: Marks that indicate grammatical aspects (like commas, periods, semicolons, and quote marks)

Part of Speech: The type of syntactic function a word is performing in a sentence (common ones are nouns (naming words), verbs (action words), and adjectives (describing words))

Sentence: A sequence of words acting as a single meaningful unit that contains a noun and a verb and other types of words

---

Parsing has a flow and it's generally done in roughly the same order, but not always. Additionally, not every step is required for all analyses. I'm outlining the basic tasks here but will explain them below. The first parsing task is usually either tokenization (basically, breaking the text into smaller pieces, usually "words" and punctuation) or sentence segmentation (identifying where one sentence ends and another begins). Part-of-speech (POS) tagging comes next, where we identify what grammatical role the word has in a sentence (for instance, noun vs. verb). Then we do stemming or lemmatizing, which basically involves turning a word (token) into a base form when the word can appear in different forms. The order these are done in is not set in stone, but I'll present them in a common order and address when you may want to go in a different order.

Both R and Python have specialized libraries to do most of the tasks discussed here. Depending on your domain and context, you probably will be able to use them as is, or you might need to make some adjustments.

Tokenization

Most of the time when data scientists talk about *tokenization*, they are talking about breaking the text into what we usually call "words" and punctuation. This makes sense when we're working with language and want to know its meaning. But sometimes there might be a reason to break it into other pieces. For instance, it's common to identify noun phrases (something like the green tree) as a distinct unit. Alternatively, someone doing phonetic research might want to break words into syllables or to make prefixes and suffixes separate. Or they might want to divide a sentence into clauses. The type of tokenization depends entirely on the way it will be analyzed. Most of these are more complex than what a data scientist would need to be doing, but it's good to understand what's out there.

With word-style tokenization, what seems obvious when we're thinking about it isn't always so obvious, just as with sentence segmentation. The roughest tokenization involves splitting on white space, but that would leave us with things like there; where the semicolon should not be attached to the word. So we also have to pull punctuation out, too. But take a word like don't, which we know is a shortened form of do not. Should it be one token (don't) or two (do and n't, or do and not) or even three (do, ', and not)? The answer depends a lot on what you're going to do next with your tokens. Sometimes having "not" called out as a separate is really valuable, since it makes whatever word it goes with mean the opposite thing. As an example, if we are trying to figure out what the text is about, the fact that a word is negated might not be that important. However, if we are trying to figure out the exact meaning of the point the author is trying to make (like in sentiment analysis where we want to know if it's positive or negative), negation is hugely important.

Sentence Segmentation

Sentence segmentation is simply identifying the start and end of a sentence. This seems trivially easy to most literate people. But it turns out to not be so obvious when we're dealing with real text. A period (.) is the most common indicator of a sentence break. But it isn't always. Take the following text: "That will be $6.32. Cash or credit?" There are two periods there, and only one indicates a sentence break. This particular case isn't too difficult to handle because we can identify any period with numbers on either side (or maybe just a number on the right) and no space between them. But once you start looking into doing this, you'll find many of these specific scenarios pop up. Additionally, how do you handle text like this: "I went to the bookstore...they didn't have it... so I left." An experienced English speaker would intuitively recognize that this is really three different sentences. But this is a special use of the ellipsis we see in text messages and online posts by a certain demographic, where in most other text, the ellipsis (...) usually doesn't represent a sentence break (it's usually a continuation of the same sentence). Additionally, a lot of social media is informal writing taken to an extreme, where there may not be sentence breaks indicated in a detectable way.

Stop Word Removal

Stop words are words (tokens) that don't really matter in a particular context. It's very common to remove them once we've tokenized. Tokens that are almost always stop words are articles like the and an; conjunctions like and and or; prepositions like of, by, and in; and various modifiers like just and so. Like implied above, negating words (no, not, and the contraction form) can be considered stop words or might be critical. There are general-purpose stop word lists online. Like most things, stop words are relative, but you almost always want to remove at least some words, even if you have to fine-tune your list yourself for your specific domain.

Processing Other Special Tokens

A lot of real text has content that isn't words. Numbers, email addresses, URLs, emojis and emoticons, and punctuation are some of the types of text we have to deal with. Each of them involves considering our specific purpose, as there isn't just one way to handle each thing. For instance, if we look at numbers, they may or may not matter to us. If we're analyzing restaurant menus to determine if this is a "fancy" restaurant or not, prices will be important to identify. But if we're looking through book reviews to try to figure out the genre of the book being reviewed, we don't care what page numbers they mention.

The particular types of non-word tokens that may be present will also depend on the kind of text we're working with. For instance, we won't see many emojis or emoticons in academic writing (except in articles talking about them), but we would see lots of them in short social media posts. We'd expect the opposite to be mostly true for punctuation.

So how to deal with these kinds of things depends on both your context and purpose. There are often established ways to handle things that you can find with a bit of Googling. Sometimes, this requires operating at a higher level than a tokenizer, which might split some text into multiple tokens. As an example, the text $65.70 might be split into four tokens, $, 65, ., and 70, whereas we might want to treat it as a single item. Sometimes you even need to do this sort of processing before tokenizing.

Part-of-Speech Tagging

Part-of-speech tagging (POS tagging) is the process of identifying the grammatical part of speech for each word in text. Every word in a sentence serves a particular grammatical function. So POS tagging generally needs to know where sentence breaks are. POS tagging will return the part of speech for each token sent to it. See Table 17-1 for some examples of several key POS tags. Note that a given word can be different parts of speech, depending on where it is in the sentence.

Table 17-1. *Common POS tags*

Part of Speech	Role	Examples
Noun	Naming word	apple, book, training, beers, David
Pronoun	General replacement of a noun	I, you, me, it, they, her
Verb	Action word	crash, running, races, read
Adjective	Modifies a noun	green, funny, tiny, expensive
Adverb	Modifies a verb	often, happily, very, slowly, never
Preposition	Defines a relationship with a noun	by, of, in, across, until
Conjunction	Joins two ideas or clauses	and, but, or, although, where

POS can be very valuable information, depending on what you're working on. For instance, if you're looking at customer complaints for a video game company and you want to know what things people complain about the most, you might look only at tokens tagged as nouns. But if you're wanting to know what people think of specific games, you might want to see what adjectives they're using when they talk about those games.

Stemming and Lemmatization

A lot of words are basically different forms of the same base word. Adding prefixes, suffixes, and some spelling changes are morphological changes that can indicate a verb tense change (like run vs. ran to go from present to past tense or run vs. running to indicate present progressive tense) or making a word plural (book vs. books or goose vs. geese). Many of these rules are fairly predictable, but it's not foolproof. Consider the word running. It could be a verb ("I was running from the flood") or a noun ("The running of the experiment went well"). This is where part of speech determined by looking at the sentence can be valuable.

There are two slightly different ways of breaking words into the base form. *Stemming* primarily involves cutting off suffixes, like taking -ing and -ed off words to get to a base form. The base form, called the *stem*, is a common conceptual base, but not necessarily a valid word on its own. For instance, we could stem bluffing to get bluff. But many stemmers would stem running to runn, obviously not a real word. Additionally, the most common stemmers also make significant mistakes, like turning nothing into noth. They also don't handle irregular verbs like to be, because stemming doesn't change forms like is, was, are, and be to the same thing.

Lemmatization is the process of taking words in different forms and returning their *lemma*, a base form that is also a linguistically valid word. It's generally considered better because the base forms are real words. Running, runs, and ran would all return run.

While lemmatization is generally considered better, it can be more difficult to do and it's usually computationally expensive, where stemmers are simpler and faster. Like with everything else in this chapter, the right one depends on your particular context. Whichever one you pick, there are several stemmers and lemmatizers out there that you can choose from. Consider word clouds, one of the most popular NLP tasks. They represent word frequency by sizing the word in proportion to the number of times it appears in a dataset. See Figure 17-1 for a comparison of word clouds with and without some normalization done (in this case, removal of the s on plural words).

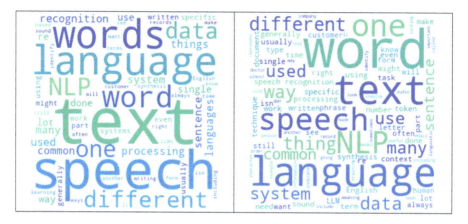

Figure 17-1. *Two word clouds with and without handling of plurals*

The word cloud on the left includes some "duplicates" because it treats "word" and "words" as separate words. Other pairs that show up in that one are "language" and "languages" and "system" and "systems." In the one on the left, "words" and "word" are both smaller than "text" and "speech," but still big. But in the right one, they're counted together as "word," and that becomes the most common word in the cloud.

Common NLP Representations and Tasks

All of the above preprocessing steps help turn text into data that can be fed into machine learning algorithms to accomplish things. In this section, I'll cover many of the different tasks that people can do with NLP. But first, a word about general tools. Large language models (LLMs) have completely revolutionized NLP. Prior to these, there were a variety of techniques used to accomplish the various tasks with text data. LLMs are not always used nowadays for a variety of reasons, and people doing NLP still use a lot of the "traditional" techniques because they're still incredibly useful and sometimes the only option for a particular domain—and they require a lot less compute resources than LLMs do. I'll talk more about LLMs specifically in a section below, but just be aware that *large language models* are models of language that have been trained on massive amounts of data. They can be used for a variety of purposes, including generating new "original" language.

Quantifying and Characterizing Words and Phrases

There are many ways to represent language in NLP, but some of the most useful techniques involve ways to represent words or short phrases. The first I'll talk about is *word embeddings*, approaches that represent words or documents as numbers in a meaningful way. The most basic approach is the *bag-of-words model*, which takes all the words in a document and tracks each individual word and the number of times it appears (the frequency), but ignores word order. This can help us understand what the document is about.

Term frequency-inverse document frequency (TF-IDF) is another word embedding approach that also counts term frequencies within documents. It relates the individual document's term frequencies to the frequency of words across all documents. It helps identify terms that separate documents from each other. If you have a database of video game reviews, words like "gameplay" or "world" might not be that informative because they will appear in many documents, but words that appear in only some documents, like names of specific games or words like "bug" or "glitch," might appear in fewer and be useful in identifying specific documents.

A fairly new word embedding approach is called Word2Vec, which looks over a set of documents and quantifies words into vectors (lines) in multidimensional space. From there, it can measure the "angle" between the two vectors to understand how related they are. This can be a great technique to identify related words that mean different things but are still connected by aspects of meaning, like "cat" and "kitten" or "bag" and "purse." Word embeddings would help us understand that the relationship between "king" and "queen" is similar to the relationship between "man" and "woman."

N-grams are groupings of *n* words (or phrases) that appear next to each other. Most commonly, we talk about bigrams (two sequential words) or trigrams (three sequential words). Bigrams and trigrams can be useful when looking for commonly mentioned things. Sometimes individual words aren't that informative, where looking at longer sequences can be. Terms like "ceiling fan" and "cracker jack" both mean something that is different from both words individually. In other cases, the pair can indicate a very specific type of one of the words, like "popcorn ceiling" or "soy milk." Looking at n-grams can therefore give information that individual words can't. Most bigrams and trigrams aren't that meaningful, either because they occur very frequently across the corpus or they occur so infrequently that they're not helpful. If you don't remove stop words, a very common bigram would be something like "in the" or "she said," which gives no useful information for most purposes.

Common NLP Tasks

The different ways of representing words and language as discussed above can be useful in a variety of tasks. Usually just counting the frequencies isn't that valuable. One exception is with *word clouds*, which are a visual representation of words and frequencies in a document or corpus, as you saw above. More frequent words are bigger in the display, so you can get a sense for the most important topics in a dataset in one quick glance. Although word clouds are usually done on individual words, you could definitely include other things, like bigrams or named entities (described next).

One of the oldest fairly basic NLP tasks is *named-entity recognition*, where "named" words or phrases are identified. These are things like people's names, company names, place names, and other terms that are generally capitalized in English. Entities could also be other things like monetary values or other domain-specific items, like order numbers or account numbers. Entities can be predefined (a requirement in a rule-based system that simply picks up when certain terms appear), but usually named-entity recognition is a supervised learning task today, where a model is trained on provided entities. The model isn't only training to recognize the provided entities, but rather to identify the patterns that make something a named entity and not just other words. Named-entity recognition isn't usually the end goal, but instead it's used widely in other NLP tasks.

Sentiment analysis is another common NLP task that involves identifying the sentiment (feeling) in some text. This is usually broken down to positive or negative, but you can also include neutral. A typical use would be to look at customer feedback and identify which are negative (complaints), positive (praise), or neutral (perhaps requesting or giving info). This is often used in monitoring things like brand reputation and topic popularity. In both cases, named-entity recognition might also be used. Some sentiment analysis is done with rule-based systems that simply identify specific words that have a known sentiment like "expensive," "broken", or "high-quality." But most utilize supervised machine learning.

Coreference is the phenomenon of words referring to other words, such as when a pronoun refers to a previously mentioned named person. It can also refer to situations in metaphors when a noun refers to something specific that isn't clear without context. *Coreference resolution* identifies when a word refers to another word, like when she refers to the word Mary in a previous sentence. *Entity linking* is the part that involves identifying which previous concept a later word or phrase refers back to. Like almost all NLP, there are rule-based and machine learning approaches to carrying it out. Supervised learning needs data with all coreferences labeled, but there are clustering techniques that don't rely on labeled data. There are also LLM and other approaches.

Word sense disambiguation (WSD) is the task of identifying the specific meaning of a word in a specific context. This is important with words that have multiple meanings in dominant dialects, but it's especially important with very informal language, such as what you find on social media. This is because such language tends to lean heavily on slang and sarcasm, which often involves using words and phrases in nonstandard, or even subversive, ways. WSD can be done with rule-based and machine learning approaches. Ideally, there's some labeled training data so supervised learning can be done, but it's also possible to use semi- or unsupervised techniques.

Text classification is the task of applying a label to a piece of text, such as identifying the subject of the piece. We might try to label what kind of essay each is in a database of high school English essays (narrative, expository, persuasive, descriptive, or argument). This relates to topic modeling, which is sometimes used as a step during classification. *Topic modeling* identifies keywords or phrases that characterize different groups of topics in a document or documents. The key difference between them is that text classification is generally done as a supervised task, where topic modeling is unsupervised.

Another common task that has been around for a while, *text summarization*, involves taking a longer piece of text and shortening it without losing the core meaning. This area has been revolutionized by large language models. There are a couple of different paradigms followed. Extractive summarization has been around a while, and it extracts the most meaningful sentences in the text and uses them as is. There are a variety of ways to score sentence importance, but it decides which to keep based on the scores. Abstractive summarization has come about since large language models. With it, new text is generated based on the original text so that it summarizes it effectively. There are three main things that can be done in abstractive summarization to generate the new text: sentence compression (shortening long sentences with rule-based techniques or supervised learning), information fusion (combining ideas from multiple sentences), and information order (putting the generated text in the right order).

Machine translation, automatically translating text from one human language to another, has also been around a long time and been revolutionized by large language models. The earliest efforts used rule-based systems and weren't that effective, but eventually statistical efforts were using huge amounts of documents and made some progress. Supervised learning techniques are used on extensive training data. Large language models are the name of the game now, but this is still an incredibly difficult task to get right.

Question answering has been around a while and was also totally changed by LLMs. Originally, *question answering* systems were fairly simple, where the goal was for short questions asked in natural language to be answered, which involved identifying what the question was asking, finding the answer (often an information retrieval task), and formulating a text response with the answer. These were originally done with rule-based systems before supervised learning approaches were used. It also involved training the system on data that could have the answers, so sites like Wikipedia and the Web as a whole have been used. Nowadays, efforts in this area are useful in areas like customer service and tech support.

Chatbots are interactive systems where a human can have a "conversation" with a computer (or, more specifically, a bot). These have been around a while using rule-based approaches before transitioning to statistical ones and now LLMs. There are many individual pieces to this. It's similar to QA, where a snippet of text that the human says needs to be understood and then an appropriate response needs to be formulated. One goal of chatbots is that it "feels" like a real conversation, which often means recognizing when it's the chatbot's turn to respond and including niceties like "thank you."

Large Language Models

A lot of people don't consider LLMs to be a part of NLP—rather it's a sister field. I think of LLMs as a specialized type of NLP, just way more advanced than the standard NLP techniques are. Most NLP uses supervised learning (but not all), where LLMs are mostly semi-supervised because the amount of data required is so immense that it's impossible to have enough labeled data.

There's so much buzz about LLMs and for good reason—the results they produce are leaps and bounds more impressive than anything that's come out of NLP after decades of work. And while they are useful for a lot of things, they're not always as substantive as we really would like. The fact that *hallucinations*—basically made-up information that a model invents and presents as fact—are so common that they make it almost unusable for a lot of purposes. I don't think this is going to (or should) keep us from using them, but they aren't the solution to everything.

There are also some serious ethical concerns about using existing LLMs trained by other companies. Much of the data they have used for training was used without permission. There are many lawsuits in progress to address this. I think that in the United States, these lawsuits may not lead to any legal changes, but that is probably not going to be the case in many other countries. The EU is taking data privacy and security very seriously, and I anticipate laws to eventually emerge related to LLMs. Depending on your use case, it may be legally risky to use a commercially available LLM. Or you may just not want to use one with dubious origins.

One option for large companies that have a lot of their own text is to train their own LLM. There are many other data sources that can be purchased or obtained for free that can supplement a company's data. But this isn't feasible for most smaller companies. Training LLMs has become easier because what were the gold standard in language processing—recurrent neural nets—have been superseded by transformers, the new neural net architecture mentioned in the previous chapter. It significantly reduces the training time. (The need for a lot of training data remains.)

The main takeaway is that LLMs can be very useful, but you should exercise caution and think about consequences any time you are looking at using one.

A Final Comment on Human Language

We all know that human languages have rules, but the #1 rule in all languages is that there's always an exception to the rule. Some forms of language are easier to work with than others. Academic writing is very formal, and most authors follow the language rules very closely—hardly any are using language in novel ways. But tweets or text messages? Those are another story altogether. Social media and all casual communication are rife with errors and the flouting of language rules, slang is all over the place, and we have all sorts of different emojis that mean different things to Gen X, Millennials, and Gen Z. Ambiguity is always there—are they using a phrase that's unique to their dialect, which you don't know? Is someone being sarcastic or passive aggressive when they say they like your hat, or do they mean it? A lot of humans can't tell, and it's almost a lost cause for computers. Add in both native speakers and non-native speakers of languages talking to each other, and it's even trickier. Subtle sarcasm is even harder for non-native speakers to detect. Sometimes non-native speakers will translate an idiom in their native language into English in a way that isn't typical in English, but everyone who speaks the other language would recognize and understand. Of all the systems humans have created, language has to be the most complex—and also the most powerful.

Speech Processing

Where NLP is the study of text, speech processing is the study of spoken language—audio. Speech recognition looks at spoken language and tries to identify what's been said, where speech synthesis generates speech. I'll talk a bit more about each.

Both speech recognition and speech synthesis generally deal with text, either as input or output. So a significant amount of the NLP work that's done on text is also done during speech processing, depending on the approach. Speech processing is best thought of as a layer on top of text processing.

Speech Recognition

Speech recognition takes audio and attempts to understand it, which historically has meant that it's transcription—turning speech into text. First, it's important to understand that speech recognition and voice recognition are not the same thing, even though they both work with audio speech. Speech recognition identifies what's been said—the words—where *voice recognition* identifies the voice as a biometric marker unique to an individual person, so it identifies the speaker. Common voice assistants like Alexa utilize both—they use voice recognition to identify different speakers in a household in case they want different things—but the main thing the assistants do is recognize speech.

The first step in speech recognition is digitizing the audio in some way. These techniques pull from signal processing, an area that falls under electrical engineering. I'm not going to talk about this part of the process, as most of the work people do when they work with speech data happens after the audio has been turned into a digital representation.

Some of the earliest work in speech recognition relied on knowledge of phonetics, the science of speech sounds (basically, the way we use our tongues, lips, and vocal cords to modify air coming out of our lungs to make speech and the specific sounds the different combinations can make). But not a lot of progress was made until hidden Markov models, a probabilistic model that allowed researchers to include a variety of linguistic and other information in their models. These were used for decades, basically until neural nets came in, in the early 2000s, and that's where we still are.

Today, most speech recognition is done with an advanced neural net called *long short-term memory* (LSTM), a type of recurrent neural net that allows "memory" of things that happened many steps back. Speech and language in general require this kind of longer memory because it's completely natural to refer to something previously mentioned after quite a bit of time. LSTMs have allowed major advances in speech recognition performance, and now there's interest in transformers, the neural net that has been used successfully in text processing.

Most speech recognition systems have several components, starting with the digitized speech input itself. Feature extraction and creation of feature vectors are the next two steps. The specific features that are extracted can vary. Once features are all prepared, they're fed into the decoder, which generates the word output (text).

Speech recognition has a lot of uses, including voice assistants, transcription, dictation, automated customer support, and real-time language translation. Often voice recognition is built into a system, as mentioned above.

Performance of speech recognition has vastly improved in the last couple decades, but it still hasn't generally reached the level of two humans speaking. General speech recognition that works for everyone talking about anything is still far in the future. I'm probably not the only frustrated person repeating the word "representative" over and over in some voice-based phone system until I get transferred to a real person. There are a lot of people who have trouble with the standard speech recognition systems. This includes non-native speakers of the language in use, those who use a dialect the system wasn't trained on, and those who have a speech impediment or speak slower than the typical person. Most English systems are trained on speech from white, middle-class American men, so sometimes even women speaking the same dialect can struggle to be understood. Additionally, most human speech contains variable elements depending on context, mood, tone, and many other things that can affect pronunciation, like stretching a vowel out or saying a syllable louder for emphasis or simply using rising intonation to indicate a question. All the "bonus" vocal things we do when speaking naturally are together called *prosody*, and it's very difficult to work with.

Speech Synthesis

Speech synthesis is effectively the opposite of speech recognition, where we start with text and create audio from it. Text-to-speech systems are the most common, and these take regular text. There are other synthesizers that can take text coded in certain ways, like phonetic transcriptions or other specialized instructions. The final step of generating the actual audio file falls back under signal processing, and I won't talk about details of that.

Because the synthesizer is starting with text, it does some of the preprocessing similar to what's done in NLP, including tokenization and other analyses. This is crucial to getting the pronunciation right, as many times the same spelling of a word can be pronounced different ways in different contexts (like "read"). This is especially true if the system's going to attempt to get intonation and other prosody right.

The next step in the process is determining how to combine the sound units together. There are different ways to do this. The most basic is *concatenation synthesis*, which is basically stringing pre-recorded sound units together one after the other. These may be phonemes (individual sound units like an "o" or "ee" vowel sound or a consonant like "n" or "s"), syllables (multiple phonemes including a single beat, like "tree," "cup", or "ah"), or even whole words, phrases, or sentences, depending on the purpose.

There are several other synthesis approaches. Another way of generating the speech audio is called *formant synthesis*, where the speech is created with additive synthesis following an acoustic model, yielding precise but mechanical speech. Another type of synthesis is called *articulatory synthesis* and is based on creating audio based on an understanding of the way speech is produced in the mouth and throat of people, so it's a physics-based model. One other type I'll mention is *deep learning–based synthesis*, which uses neural nets trained on a lot of recorded speech and associated text and labels.

Concatenation synthesis tends to be the best performing in terms of naturalness, but it can be glitchy with uncommon combinations. Formant synthesis generally yields robotic-sounding speech, but it's usually highly intelligible, requires less compute resources, and is quick to produce. For those last reasons, this is the kind of synthesis that's in screen readers for the visually impaired. Articulatory synthesis definitely has the potential to create natural-sounding speech, but modeling the human vocal tract is not an easy thing, and there is still progress to be made. Deep learning synthesis can have fairly good results in terms of sounding natural, but there are still problems when it's asked to do something it hasn't seen sufficient training examples of.

Key Takeaways and Next Up

In this chapter, we talked about the major aspects of natural language processing, which is much more than pre-built LLMs. We talked about human language and how it's written and spoken. Then we dove into the ways of parsing language with NLP, including tokenization, segmenting sentences, removing stop words, handling special tokens, tagging parts of speech, and finally stemming and lemmatizing. Then we talked about the many things you can do with language data once it's parsed and ready to go, including counting word frequencies, looking at multi-word terms, named-entity recognition, sentiment analysis, coreference resolution, word sense disambiguation, text classification, text summarization, machine translation, question answering, and chatbots. I then talked about LLMs, which have become a big part of NLP recently, even though they aren't the whole story. Speech recognition and synthesis are both important fields as conversational systems are becoming increasingly common.

In the next chapter, we'll be looking at visualization and presentation. We already saw some visualizations in Chapter 2 when we looked at descriptive statistics, but Chapter 18 will dig deeper. We'll talk about many types of visualizations as well as what makes various features good or not in different circumstances. Then we'll talk about presentation in general, beyond visualizations specifically.

PRACTITIONER PROFILE: ANDRA CIMPELLO

Name: Andra Cimpello

Job Title: Software developer/linguistic specialist, currently proofreader/editor

Current Industry: Q&A software and writing

Years of Experience: 12 years in software/linguistics, 4 years in editing

Education:

- Certificate in Machine Learning

- MSc Speech and Language Processing

- BA Modern Languages and Linguistics

The opinions expressed here are Andra's and not any of her employers', past or present.

Background

Andra always loved languages and was good at them. She learned French and German in high school, but she majored in political science in university. She found she disliked it, so it made sense for her to switch to studying linguistics. She graduated with a degree in Modern Languages and Linguistics, picking up Italian along the way. She didn't find a job in the field after graduating, and when she heard about computers doing groundbreaking work with language, she was intrigued. She started looking into that, talking with friends about this fascinating new field, and eventually she found a degree being offered that focused on the topic. She started the MSc in Speech and Language Processing later that year. Although most of it was new to her, she loved everything she studied during the degree, especially phonetics. She still loves to look at waveforms of speech. Her dissertation was on improving the computational efficiency of speaker identification, a part of biometrics focusing on voice recognition rather than speech recognition. The degree involved a lot of programming as well as linguistics.

Work

After graduating with the speech and NLP degree, she moved back home and began looking for a job involving language in some way. At first, it was tough because she was limited to one metro area because of family responsibilities, but she managed to find a software developer job with a specialty in linguistics for a Q&A company in the area, which was perfect. In the beginning, it was a lot of software development, but also a lot of linguistics, where she worked on algorithms and other NLP tasks. But after the company was bought, the culture became toxic and all she was doing was fixing bugs, so she quit and focused on her family. As her kids got older, she eased back into the workforce, now working as a freelance proofreader and editor—still working with language, which she loves.

Sound Bites

Favorite Parts of the Job: Working with language and using her expertise in it both to develop products and improve writing. She also loved trying to improve the relevancy of the answers at her Q&A job because it was challenging but so rewarding when they found ways to improve the answers being delivered.

Least Favorite Parts of the Job: At a company, politics and a toxic workplace are frustrating because you can't change them. Freelancing is better in that regard, but it's also stressful because you constantly have to be on the lookout for work.

Favorite Project: One of her favorite projects was one at the Q&A company when they were specifically working on improving relevancy of answers. The basic flow was that a user would ask a question and the system would query the knowledge base for an answer based on keywords. Andra was able to dig deep into the process and improve it by analyzing real queries and answers. She identified other important aspects that could be used to improve the answer, including coming up with a score for any given system-generated answer. In another cool project, she managed to figure something out after a light bulb went off in her head during a long night of coding, and she figured out a great way to visualize co-occurring words that clients ended up loving, making everyone happy.

Skills Used Most: At the Q&A job, she mostly used her coding skills, the NLP/linguistics knowledge somewhat less (but it was helpful when necessary). In her current work, her general language expertise is valuable. Soft skills are important everywhere, especially in freelancing. Another important skill is time management.

Primary Tools Used Currently: In the past, Java, JSP, JavaScript, NLTK (natural language toolkit), and WordNet. Currently, WordPress, Google Docs, and Microsoft Office

Future of NLP: Things are moving at the speed of light now, getting more impressive every day, but also scarier. Andra is concerned at how fast AI is advancing and worries about what it means for society in general. Now we have to be cautious when looking at anything and wonder if it's real or fake—but most people don't bother to do this. We see celebrities come out all the time saying that certain pictures are fake, but everyone believed the images were real. It's concerning.

What Makes a Good NLP Practitioner: Being willing to learn, which will be an ongoing need throughout your career. Things are always changing and improving.

Their Tip for Prospective NLP Practitioners: Don't focus too narrowly in one area when looking for a job. There aren't hundreds of these jobs open at any one time, and you may have to be flexible and go a bit outside your favorite areas to find work. You're going to be learning for the rest of your career, so look with a wider net.

Andra is a freelance editor and writer with a background in NLP and software development.

CHAPTER 18

A New Kind of Storytelling: Data Visualization and Presentation

Introduction

Visualizations are an important part of data science, even though many data scientists don't have to build a lot of them. It depends on the particular role someone has, but every data scientist will need to create a few now and again, at least. Data analysts will find themselves building many. Some teams have specialists to build dashboards based on data scientists' and data analysts' work. Whether you end up creating visualizations or not, understanding what makes good ones is crucial.

Visualizations can be powerful, frequently making things clear in a second that would take a great deal of explaining based on tables and numbers only. In Chapter 2, I talked about how the plot that the Challenger Space Shuttle engineers used did not convince leadership to postpone the

© Kelly P. Vincent 2025
K. P. Vincent, *A Friendly Guide to Data Science*, Friendly Guides to Technology,
https://doi.org/10.1007/979-8-8688-1169-2_18

launch the next day, where the right graphic might have convinced them. You'll see in the first example below how dramatic a visualization can be in showing what a disaster Napoleon's Russia campaign was. Knowing when to create a visualization and which one to create to get your point across is an invaluable skill.

In this chapter, I'll address what makes good visualizations and presentations and also cover a wide range of charts, tables, and maps. One hint is that everything is about telling a story with your data and findings. Finally, I'll talk about Tableau and Power BI, the two most popular professional visualization tools today.

Examples of Data Visualization in the Real World

I gave a couple of examples of powerful visualizations earlier in the book. The first was in Chapter 2, when I showed a simple chart that would have more clearly revealed the risk in a low-temperature Space Shuttle Challenger launch to decision-makers. Instead, engineers concerned about the performance of O-rings in cold weather shared a chart that only told part of the story, and it wasn't convincing to managers. The second, in Chapter 5, was the map of cholera deaths in 1854 London by John Snow, which helped them identify the source of the disease as the Broad Street water pump. Once they locked the pump, the epidemic ended.

Now, I'm going to talk about two more examples, one that looks back in history and one that reminds us of the importance of visualization. Both are great because they show how impactful the right visualization can be. The first is a representation of Napoleon's march to Moscow, a disastrous campaign that cost hundreds of thousands of lives. The second is a famous illustration of the importance of looking beyond basic descriptive statistics called Anscombe's Quartet.

Example 1: A Visual Record of Wartime Folly

Although Napoleon Bonaparte is considered one of the best military leaders, his attempt to conquer Moscow was acknowledged as a pointless and unequivocal disaster by Napoleon himself after he made it back to Paris. The march gained Napoleon nothing and cost the lives of hundreds of thousands of people. It's difficult to get consistent numbers on this, but most agree that Napoleon started with 450,000 troops and less than 100,000 made it home. Figure 18-1 is a famous visualization summarizing the march made in 1869 by Frenchman Charles Joseph Minard that makes clear what a disaster it was in the way that mere numbers wouldn't. The top line in tan represents the march from France (the left) to Moscow (the right), and then the return is shown in black.

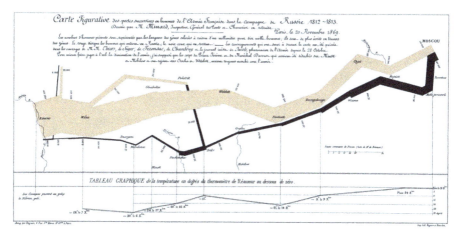

Figure 18-1. *Charles Minard's representation of Napoleon's disastrous march to Moscow. Source: From "File:Minard.png," available at* `https://en.wikipedia.org/wiki/File:Minard.png` *and in the public domain*

The army crossed into Russian Lithuania at the beginning of summer 1812. Russia was following a scorched earth retreat policy, so finding food was incredibly difficult for the army and people started dying from disease even early in the march. The presence of disease never went away, but they continued on toward Moscow. There was a battle in August that Napoleon won, but about 9,000 troops were killed. In September, they fought again on the approach to Moscow, where 35,000 French died, with even more on the Russian side (this is considered one of the bloodiest single battles in the history of modern war). Napoleon and around 100,000 soldiers took a Moscow that had been almost entirely deserted. Locals burned the city the day after, with about three-fourths of the city destroyed. Napoleon remained for a few weeks, but it was difficult to get food and starvation continued to be a huge problem, with disease still haunting them. Eventually the army started the march home in late October. This is where modern numbers and those on the graphic differ, but it's known that there was a huge loss of life on the way home as the deep winter cold descended on a worn-out army that lacked good winter clothing. Starvation, disease, and extreme weather meant that only a fraction of the soldiers returned home.

This visualization probably didn't change policy in any significant way, but it makes it extremely clear how bad that campaign was in the way that can be understood in an instant, as opposed to looking at tables and simpler charts. There usually is a perfect visualization if you're trying to show something, even though finding that isn't always easy.

Example 2: Different Looks for Statistically Identical Data

Descriptive statistics is powerful, and people have been using it for a long time to understand and describe their data. But only relying on one perspective is risky, as the statistician Francis Anscombe showed with four simple but illuminating graphs he made in the 1970s. He created four different datasets with two variables that all had the same basic statistical characteristics (mean and sample variance of the two variables, correlation between the two, linear regression line, and coefficient determination of the regression). Then he constructed a scatterplot for each of the datasets, as can be seen in Figure 18-2.

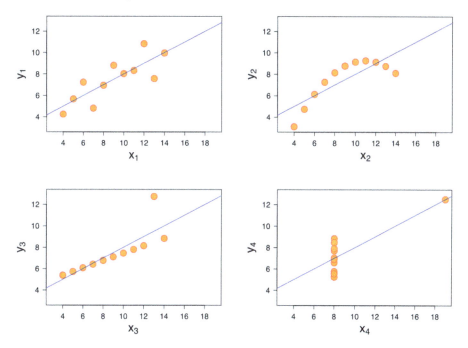

Figure 18-2. *Anscombe's Quartet, four charts showing different datasets that have the same statistical properties. Source: From "Anscombe's quartet 3.svg," available at* `https://en.wikipedia.org/wiki/File:Anscombe%27s_quartet_3.svg`, *used under the Creative Commons Attribution-Share Alike 3.0 Unported License (*`https://en.wikipedia.org/wiki/en:Creative_Commons`*)*

These let you immediately see that the datasets really aren't similar and each has its own distinctive characteristics. The upper-left chart is what we're used to seeing—data spread out with a general linear trend that we feel like we can sense. Perhaps linear regression does tell us something useful, although we really want more data before we know for sure. The upper right shows clearly nonlinear data, and the linear regression line on the plot looks ridiculous. Obviously linear regression isn't the right solution for this data. The bottom-left chart is also interesting, because the data looks incredibly linear, but one outlier throws the whole thing off. Most likely, linear regression is a good technique for this data, but that outlier would need to be thrown out before running the regression. But it shouldn't be forgotten—we need to remember that this dataset has extreme outliers. The bottom-right chart is also interesting, and an extreme outlier also throws off the chart and makes running linear regression possible. Really, the *X* variable is most likely useless in any model because it is almost always the exact same value, 8. It might be more useful to turn this into a binary feature of "is 8" vs. "is not 8."

These charts remind us both that visualization can be revealing and important and also that you should never take a myopic view of your data and just arbitrarily run correlations and linear regressions and other things without understanding what you've got first.

Storytelling with Data

It's normal to make some visualizations along the way before it's time to present your findings. But what you really need to do before you make a slide deck to present your findings is decide what story you're telling your stakeholders. You should never just drop a bunch of charts and tables in a slide deck and call it good. What's the narrative?

People sometimes get a little carried away trying to make the idea of storytelling with data analogous to fictional storytelling, and I don't think it's necessary. But the idea of having a narrative is going from one idea to another and hitting significant points along the way. In general, these are the questions you usually want to answer along the way, generally in order:

1. What is the question being asked or the problem needing to be solved?

2. What information do we have that can help us answer the question/solve the problem?

3. What kind of model or analysis can we do to answer the question/solve the problem?

4. What is the answer/solution?

5. How confident are we in the answer/solution?

6. What are the next steps?

7. What is the business impact of the solution?

These are the questions you'll be answering when you're reporting on the project, too. Start by setting up the beginning state in #1. What is the problem we're trying to solve? Always, always, always bring things back to the business case. Sometimes people have to be reminded about why they're listening to this presentation, especially if these aren't people you've been working regularly with.

In many cases, it's good to explain the materials you used to solve the problem, as mentioned in #2. The level of detail here depends on your audience, but most stakeholders will want to know what kind of data you used. They may want to know if this is a data source they're already familiar with or one they might be able to use themselves. Sometimes they want to judge how reliable or trustworthy your results will be based on their knowledge of the data sources.

It's also usually good to explain how you solved the problem, in #3. This is when you would talk about the technical stuff, but targeting the right level of technical depth depending on your audience. You might just mention that you did some cleanup and modification of the data and fed it into a machine learning model, or you might get more specific and tell them which ones or even show them the model results (like the linear

631

regression formula or trained decision tree). Remember that a big portion of the work we do remains largely invisible, so one point here is to convey that a lot of work happened, even if you don't give many details. It all depends on catering to your audience.

Next, you'll actually present the solution itself, as in #4. Explain what you found, why it matters, what assumptions were made, level of confidence, and if there are any weak spots or limitations to be aware of (this is where #5 factors in). If you are making recommendations, these would be given here. This will usually be the bulk of the presentation since this is the whole reason for the project. The specific content here will depend on the type of analytics you're doing, as well. Descriptive will be mostly charts and summaries, diagnostic will include your discovered explanations, predictive your predictions, and prescriptive your recommendations.

Finally, it's typical to identify the next steps, as mentioned in #6. If the project is ongoing, these will be what you plan to work on next (or options for discussion). If it's over, this might just be about maintenance, or there might literally be no more work that can be done. It's still typical to list some possible things that could be done in a future project.

If enough time has passed since the project was completed, #7 is a place where you can talk about the actual impact. In business, it's usually best if you can quantify the impact in some way.

One of the points of telling this as a story is not just to be organized and logical, but stories can hit people emotionally, where just a statement of facts usually does not. When you're talking about coming up with your solution, with the right stakeholders, you can even talk about the struggle of figuring things out and how you were triumphant. This wouldn't be appropriate in a short presentation to high-level leaders, but stakeholders you've been meeting with regularly might appreciate it.

Note that visualizations could appear in any step of this process, even though #4 should have the most. But remember not to overwhelm your audience with visual after visual. Stakeholders are looking for help in making better decisions. It's our job as data scientists to find the information that will help them. That can obviously take many forms. In a project that is more about understanding what has been happening and what's happening now, which will involve data analysis–style work and probably some good visualizations, data scientists deep in their EDA can easily lose sight of the end goal. Exploratory work is important, but what we are really wanting to do is create *explanatory* visualizations, which don't just give us the "what," but reveal connections with reasons—explanations—that can be understood. We need the EDA so we can understand the data and figure out the best way to find the interesting things that will explain something of value to stakeholders.

Next, I'll talk about many of the various visualizations that you can make as part of your storytelling and how you can make them most effective.

Types of Visualizations and What Makes Them Good

I already talked about several basic but solid visualizations in Chapter 2 when I covered descriptive statistics. I'll revisit some of those charts and discuss why they work here. We'll also look at many variations of charts as well as some less common ones. But first, I'll discuss some things that make charts good that are more general and some things to avoid.

What Makes a Good Visualization?

You need to know who your visualization is for and why your stakeholders need it. This means knowing your audience and making it as specific as possible. If you want to present the same basic information to your data scientist colleagues and your executive leadership, there's a good chance that, that should be two separate and different visualizations. Why they need the visualization is clearly related to what you're trying to show them. Usually when you make a visualization, it's to help people understand something better or to help them make a decision.

Another thing that you want to aim for when making a visualization is simplicity—any visualization should be as simple as it can possibly be while still conveying the information you need it to get across. As an example of a big no-no, people are often tempted to "fancy things up" by making it 3D. At best, this is unnecessarily distracting, but at worst, it distorts the actual values you are charting. To see this, take a look at Figure 18-3, which shows 2D and 3D versions of the same pie chart, equally split into thirds (every segment contains the same number of items). The proportions are easy to see in the left chart, the 2D version. It's fairly obvious that they're all one-third of the pie. But in the chart on the right, the 3D version, the slice closest to us looks bigger than the other two. I find that even though I know they're all the same size (and even though I have a pretty good math-y brain with good spatial awareness), it still *feels* bigger. Imagine if the slices aren't the same size—there's basically no way most people's brains can make the right adjustments to interpret it correctly without significant effort. And if you have to work hard to understand a chart, it's not a good chart.

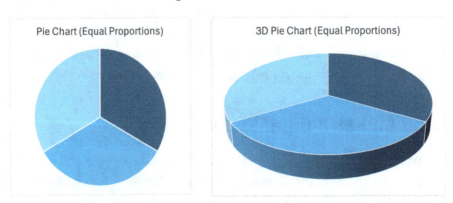

Figure 18-3. A pie chart of three equally sized segments in 2D and 3D

Scatterplots

Figure 18-4 shows an example of one of the most basic charts, a scatterplot, which we saw in Chapter 2. It simply plots two variables against each other, one on the X-axis and one on the Y-axis. It's a quick way to see how two features relate—does one get bigger when the other gets bigger, or does the opposite happen? Or is there no relationship at all? Scatterplots aren't actually used as much in business, but they are more common in science, and data scientists use them in their EDA a lot.

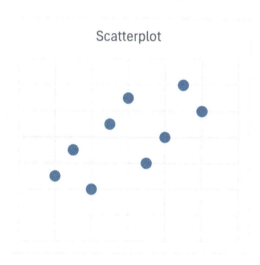

Figure 18-4. *A scatterplot*

In Figure 18-4, there seems to be a generally positive relationship, where they both increase together. Scatterplots are also a good way to see when there is no clear relationship. It's important to pay attention to scale when you're creating a scatterplot—although there can be an art to setting any chart's axes, it's usually good for the two axes to be similar in proportion to the full range of data when possible. Additionally, showing only some of the data can be misleading. Check Figure 18-5 for an example of a scatterplot that looks completely different when some of the data is excluded.

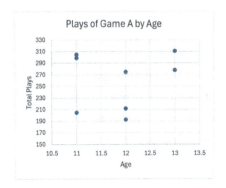 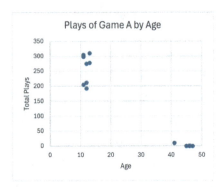

Figure 18-5. *Two scatterplots with different axes*

In the left chart, it looks like age is fairly spread out and that total plays also range from low to high. But when we add just five more data points for older players, first, it becomes clear the player ages in the first chart are low compared with the human range. Similarly, we also see that the numbers of plays the younger players had are much higher than the older players'.

There are two variants of scatterplots that I addressed in Chapter 2, which can be combined so that up to four features can be plotted on the same chart. This comes from adding color and/or size to the points. With different-sized points, we usually refer to those as bubble charts.

Color can be used with either categorical or numeric variables. In the case of categorical, each value of the variable would be a discrete color listed in a legend. In the case of a continuous variable, one color is gradated (multiple colors can also be gradated, but a single-color light-to-dark is more common). Figure 18-6 shows a bubble chart representing four variables: age (X-axis), total number of plays (Y-axis), gender (color), and rating (bubble size). The bubbles are partially transparent for clarity.

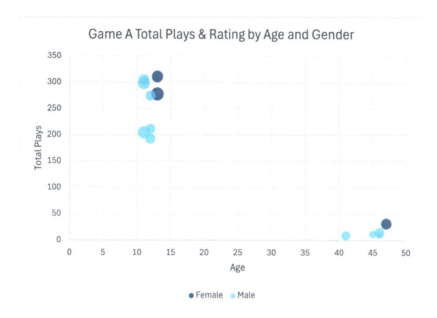

Figure 18-6. *Bubble chart with four features with color and size*

Scatterplots can be somewhat flexible, but remember that many businesspeople aren't as comfortable with them, so make sure you pay attention to how your stakeholders receive them to decide if they're the right chart for them.

As an example, we can look back at an earlier chart, Figure 2-8, included here as Figure 18-7. It shows a scatterplot of heights and genders of children. This is a very simple chart, only showing three features, age, height, and gender. When we look at this, we can see that there's an upward trend in height as age increases. From what we can see, there isn't much difference in height between girls and boys until around age 14, where the girls' heights level off but the boys continue to rise above the girls.'

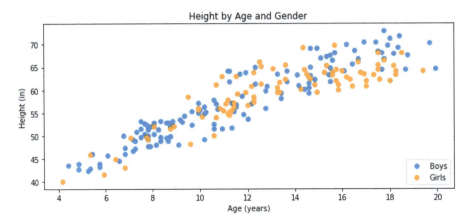

Figure 18-7. *Heights and genders of children*

Although this is potentially useful information, it doesn't really explain much—we already know that boys are taller than girls once they're teenagers. Additionally, if we look, we can also see that this dataset seems to have fewer girls under the age of 11 than boys. Is it possible that if we had more girls, we'd notice something about a difference in height among younger kids? It's hard to say. Although there's nothing wrong with this chart, there's nothing super compelling about it either. This probably shouldn't go in a report to stakeholders unless they'd specifically asked for it.

Column and Bar Charts

Another important class of charts that primarily shows a single variable in the most basic variant is the bar chart. Bar charts are one of the most intuitive charts and can be either horizontal or vertical (which is usually called a column chart). You can see the two different versions in Figure 18-8, showing the same data.

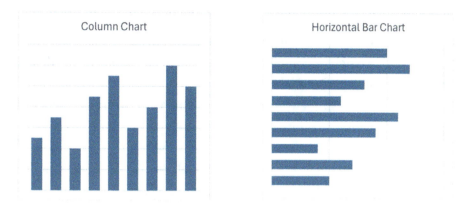

Figure 18-8. *Vertical and horizontal bar charts*

The data you're charting can often help you determine whether you want to go horizontal or vertical. It usually doesn't matter, but one case where you might choose horizontal is when your labels are particularly long. Compare the two charts in Figure 18-9. Rotating it to accommodate the long X-axis labels on the left makes for the easier-to-read chart on the right.

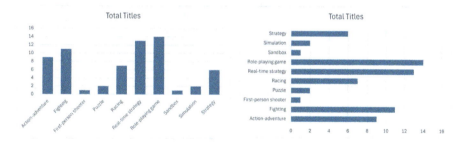

Figure 18-9. *Two views of the total number of video games by genre*

There are multiple ways of tweaking these basic bar charts, but there are some best practices to be aware of. It's generally considered best to always start bar charts' Y-axis at 0, because otherwise they can be misleading and make differences between bars hard to understand quickly. Additionally, you might have seen charts with a break (usually a jagged line) through a bar indicating that the scale has jumped—this generally is not advised because it breaks the visual comprehension. This is usually done when one bar is much larger than the others (an outlier), but it's generally considered better to leave the longer bars and add labels to the other ones so the values can be seen. You may also do a second chart that shows only the non-outlier bars, where they can be distinguished more easily.

You can get away with not having Y-axis labels in a bar chart by adding the value at the top of the bar. This can be especially effective if there aren't very many bars. For instance, see Figure 18-10 for an example of two views of the same data. These aren't hugely different, but with only three columns, you can read the one on the right faster.

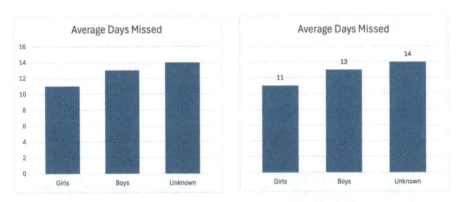

Figure 18-10. *Two views of a small column chart showing student absences from a program*

We talked in Chapter 2 about other types of bar charts. It's possible to simply show multiple groups separately or to stack them. Whether one view is more appropriate than the other depends on what is being shown. We'll start by looking at multiple bars for different groups. Imagine we use the same basic data we saw in Figure 18-10, showing student absences broken down by gender. If we add a yearly breakdown, there are a few different ways we can look at the data. See Figure 18-11 for the first two most basic ones.

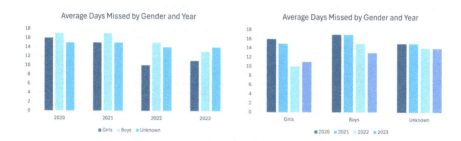

Figure 18-11. *Two ways of viewing multiple years of absences by gender and year*

Note that we've clustered groups together. We have data broken down by both gender and year, and in the left chart we put the three genders next to each other and look at each year separately. The right chart shows the opposite, with the years next to each other and each gender separate. Both charts make it fairly easy to see that there's a general decrease in absences over time, but it's much easier to understand how each gender looks on its own with the chart on the right.

Another common view is the stacked bar chart, where multiple groups are included in the same column so we can see how much we have of each, but also see how much we have in total immediately. See Figure 18-12 for an example of a basic stacked column chart. Two different groups are added together, and we can see how much each group represents as well as the overall total.

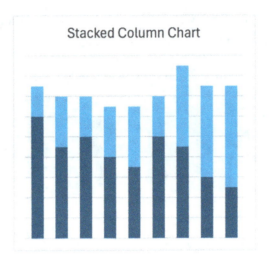

Figure 18-12. *Stacked column chart*

In our example, we can look at stacked versions of both charts in Figure 18-11. See Figure 18-13 for the stacked version of the right chart in Figure 18-11. This makes it clear that boys have the most absences in total, but it's hard to conclude much about each individual year. Line charts are often more natural for showing changes over time.

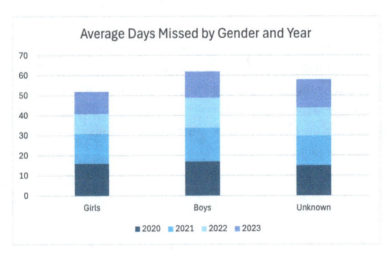

Figure 18-13. *A stacked bar chart of average absences by year and gender*

There is a better way to take this idea and make it easier to see the interior group changes. Figure 18-14 shows a view where each interior group is in its own column so it's easier to see changes over each group. This is called a small multiples chart.

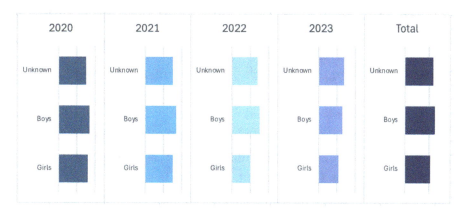

Figure 18-14. *A small multiples chart showing an alternative to the stacked bar chart making it easier to see trends across groups*

When you do plan to use a stacked bar chart, you need to be cautious that your data is appropriate for that view—make sure it makes sense to add the numbers together. Figure 18-15 shows an example of data that shouldn't be stacked. The left chart of Figure 18-15 shows the data we saw in the right chart of Figure 18-11 modified to show the percentage of missed days rather than the total number. The chart on the right in Figure 18-15 is an invalid stacked column chart—adding percentages this way is nonsensical.

Figure 18-15. *A multiple bar chart and an invalid stacked bar chart*

One more type of column chart worth mentioning is the 100% stacked column chart. This is the same as a stacked column chart, but instead of using the raw numbers, each of the numbers is represented as a percentage of the total for that point. Figure 18-16 shows a couple of charts similar to Figure 18-13, where we can see the raw counts of absences instead of the averages.

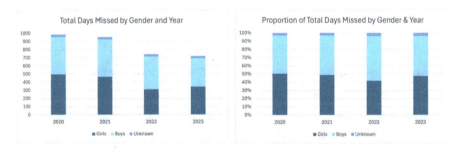

Figure 18-16. *Two charts showing both raw totals of days missed by gender and year and proportional days missed*

The left chart is simply the raw numbers broken down by year and gender. There are 31 girls in the group, 27 boys, and 2 listed as unknown. It's interesting that there are more absences from boys than girls, even though there are more girls. Looking at averages makes this clearer. But sometimes we just want to see the totals. The chart on the right shows the same data but the raw values are converted to percentages, so each column shows the proportion of girls', boys', and the unknown students' absences as a proportion of all the absences.

There are more variants on column and bar charts, as they are very versatile. You can dig deeper online if none of these works for you. You can also just play with them in a quick program like Excel or Google Sheets with some data to see all the many variations. Additionally, there are options for combining different charts, like putting a line on a bar chart that represents a significant value like an average or a target or goal.

Line and Related Charts

The standard line chart is very well-known, but there are some ways to modify it and also some related charts that can be really useful. I'll talk about line charts, area charts, and slope charts in this section.

Line Charts

Line charts are a very effective way to show changes over time. They're intuitive and easy to follow, as long as you don't try to put too many lines on one chart or do anything too fancy. The X-axis is generally time and needs to be consistently spaced (you can't show day yearly and suddenly switch to monthly in the same chart). The lines don't all have to start or end at the same point in time. As a rule of thumb, having more than four or five lines that you want to be clearly identifiable on one chart can get unwieldy. Figure 18-17 shows a couple typical line charts, with one line on the left and two on the right.

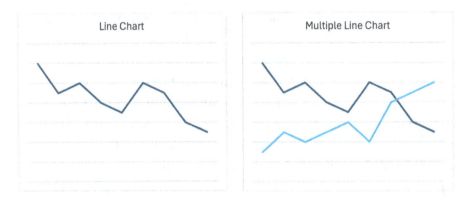

Figure 18-17. *A single line chart and a multiple line chart*

Sometimes line charts can be more intuitive than bar charts even when both are possible. Look back at the left chart in Figure 18-11. I've redone it as a line chart in Figure 18-18.

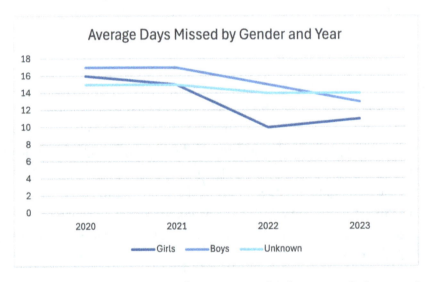

Figure 18-18. *Another way of viewing multiple years of absences by gender and year*

You'll often hear that you shouldn't have more than a handful of lines on a line chart. This is because it can get hard to follow each line when you have many, especially with lines overlapping each other. But what works depends on what you're charting, and there are ways of pulling out multiple charts to the same effect. For instance, if you're charting one or two lines that are more important than the other lines in some way, you could plot everything but make the more important lines more prominent. For example, see Figure 18-19, which shows the total number of game wins each day for each kid at a summer camp, plus the average of the two teams they've been split into.

Figure 18-19. *A chart with twelve lines, two of which are more prominent*

The camper lines are in a light color and fairly thin, so they don't stand out, but we can still see that there are different lines for the kids. The two averages are bolder, thicker, and darker dashed lines, so they are easy to see as the most important lines in the chart.

It's also possible to pull each line out separately if we want to see each camper's line. See Figure 18-20 for an example of this.

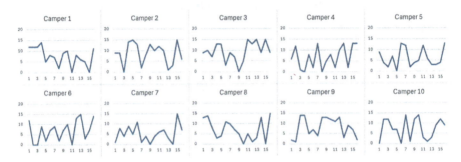

Figure 18-20. *Line charts with one line per camper, rather than all on one chart*

It's important when breaking charts down like this that all the Y-axes and X-axes are identical. This usually requires manual adjustment with whatever tool you're using. We could have also included the averages here if we wanted, but the assumption was that we were interested in the campers individually.

One modification on the line chart is adding a confidence interval. If we have used the current group of campers to estimate the average number of wins for all campers for all years, we would generate a confidence interval. How that might look on a chart can be seen in Figure 18-21.

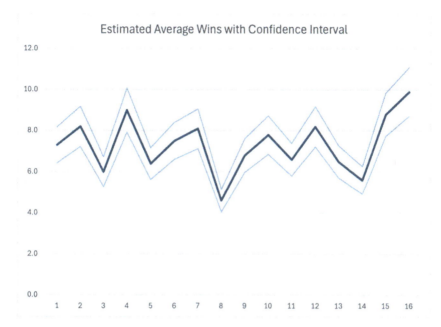

Figure 18-21. *A line chart of average number of wins showing a confidence interval*

This is a fairly intuitive chart that's easy to explain. Sometimes the area between the line and its upper and lower confidence intervals is shaded, but it's not a requirement. Generally, you want to ensure that the primary line stands out from the confidence interval lines. Usually it's made bolder and thicker like in Figure 18-21.

In Chapter 2, I talked about line charts with two Y-axes. This means that at least one line on the chart follows the Y-axis on the left side, but at least one other line goes with a separate and different Y-axis on the right. This is generally not recommended in charts you're showing to stakeholders, although it can be useful during EDA. The reason for this is that they can be confusing and hard to read. They require the viewer to stop and think, and what we really want from viewers is quick understanding. Tyler Vigen's funny website on absurd correlations,[1] which I've mentioned before and which shows how so many obviously unrelated things can appear "correlated," almost always relies on different Y-axes. It's just not as intuitive as a single Y-axis is.

Area Charts

Another type of chart that is similar to the line chart is the area chart. An area chart is simply a line chart where the area under the line is colored all the way to the X-axis. See Figure 18-22 for examples of the primary types, a simple area chart and a stacked area chart.

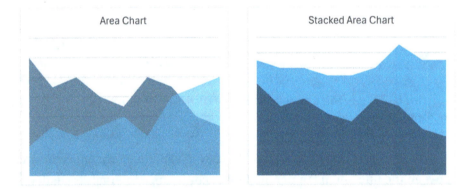

Figure 18-22. *Area charts, one with allowable overlap on the left and the other stacked on the right*

The left chart shows overlapping areas, which is conveyed by using transparency for each series. The overlapping area is a different color so you can still identify the two distinct series. This type of chart would be difficult to read if we had more than two series on it, and for this reason the most common variant of area charts is the stacked area chart, which

---

[1] "spurious correlations," available at https://www.tylervigen.com/spurious-correlations

we can see on the right of Figure 18-22. This is stacked just like the stacked column chart, with the value at each point being the sum of all the series at that point.

Note that although the right is a bit easier to read, the more lines we add, the harder it is to really understand the values. It's easiest to comprehend the lowest series since it sits on the X-axis. Because of this, it's common to make the line chart values sum to 100%, which makes each value slightly easier to understand. Compare two charts in Figure 18-23, which show a simple stacked chart on the left and a chart on the right where each day sums to 100%.

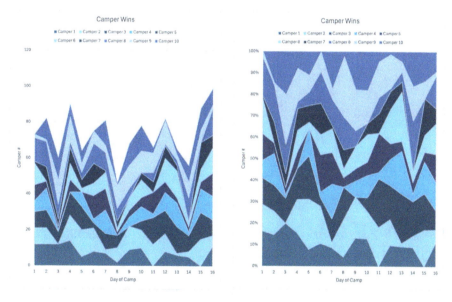

Figure 18-23. *Two stacked area charts showing camper wins each day, stacked on the left and stacked summed to 100% on the right*

Although these charts aren't particularly easy to read, it is easy to see in the left chart in Figure 18-23 that there are some days with unusually high total numbers of wins, but it's easier to see which campers were responsible in the version on the right.

Area charts are very similar to line charts, but they can sometimes be more dramatic or impactful. When to choose one or the other is a skill you'll develop over time.

Slope Charts

One last type of line chart I'll mention is the slope chart. These are used to show a two-point change, basically a before and after. Normally, slope charts have several lines, but see Figure 18-24 for an example showing the two teams' first and last day average wins at the summer camp.

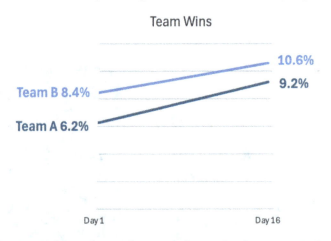

Figure 18-24. *A slope chart showing the teams' average wins at the beginning and end of camp*

It's instantly clear that both teams' average wins increased from the first day to the last day, but also that Team A increased slightly more, proportionally.

The many varieties of line charts and related ones give you so many options for charting things over time. It's usually not appropriate to use a line chart when it's not time-based, because the connections between data points indicate continuity, and discrete things don't have a natural order or anything in between them. For instance, it wouldn't make sense to make a line chart with an X-axis of Girls, Boys, and Unknown.

Pie Charts

A lot of visualization experts despise pie charts, which is understandable because they're very limited and don't often reveal much about data that can't be viewed in better ways. Also, people are notoriously bad at comparing the different slices, so they don't necessarily accomplish the goal. But sometimes stakeholders want them, and if you can't convince them otherwise, you'll have to give in and make one. You should first try to convince your stakeholders to be happy with a bar chart, but you may lose this battle. With a little knowledge you can make the best possible (least horrible?) pie chart for your particular scenario. See Figure 18-25 for an example pie chart.

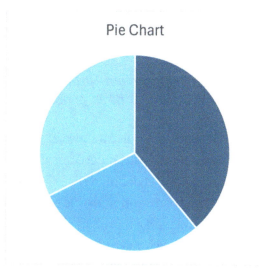

Figure 18-25. *Pie chart*

As mentioned above, the first rule is to never make a 3D version of one because it completely distorts the proportions, where the proportions are the entire purpose of the chart.

One of the hardest things to do with pie charts is comparing one to another, even though that's a common use for them. Imagine an Arkansas summer camp trying to look at where its campers come from before and after two different marketing campaigns, which targeted neighboring states in 2019 and 2022. They decide to look at the year after each campaign and compare that to 2015. Figure 18-26 shows this series of charts.

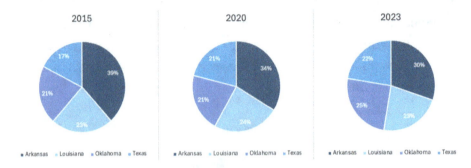

Figure 18-26. *Three pie charts showing proportional changes in the states that campers come from over three years*

This sequence of pie charts does give a sense of the change that's pretty easy to see. We can definitely see the Arkansas proportion shrinking over time, and the Louisiana and Oklahoma slices growing. Texas doesn't change much over the three years. Although this is time-based, pie charts are simpler and easier to read than a line chart showing the same data would be. Adding the percentage (or even the raw value) as text inside each slice makes it even clearer.

In summary, use pie charts sparingly, and really consider if they are the best option. Also see below to learn about an alternative to pie charts called treemaps that can work better.

Histograms and Variants

We saw histograms in Chapter 2, as it's common to generate them during EDA. They are less common in presenting to stakeholders because they require a little more explanation and a lot of people aren't familiar with them. But if you need to represent the distribution of something, they are the right go-to chart. There are some other ways to represent distributions, and these will all be discussed here.

Histograms

Figure 18-27 shows an example of a histogram. Each column is considered a bin, and it represents the count of values that are within the range defined for that bin. There are no hard-and-fast rules for defining the number of bins or bin sizes, although there are some rules of thumb based on statistical properties. Usually, the tool you use to generate the histogram can determine them automatically, but you can Google if it doesn't seem to be working well.

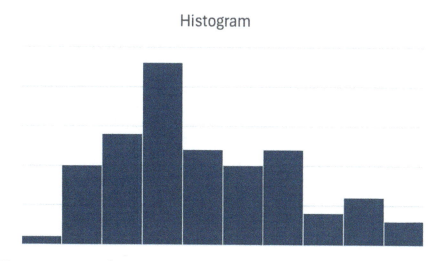

Figure 18-27. *A histogram*

Histograms have several properties worth understanding, with skew being the most important. The histogram in Figure 18-27 skews slightly to the right, because more of the data is concentrated on the left. A lot of things skew one way or another because of limitations of possible values. Grades tend to skew left because more are on the higher end of the 0–100 scale. House prices skew right because no house can be priced less than 0, but house prices can get extremely high, even though the higher it goes, the fewer "neighbors" it has on the chart. In most places, there are lots of houses that cost $100,000 but not very many that cost $4 million. This can be helpful when talking to stakeholders.

Pareto Chart

There's a variant of the standard histogram called the Pareto chart. Figure 18-28 shows one with the same histogram as above, but with labels this time. This data shows the total completed credits of 100 first- and second-year students at a college. The Pareto chart adds a cumulative percentage to a second Y-axis (one of the rare times the second axis is a good idea).

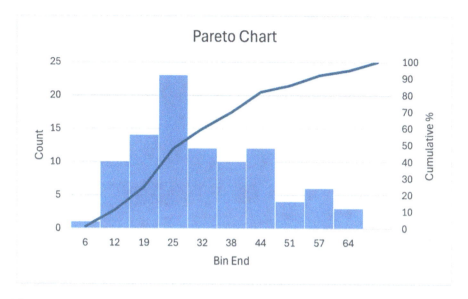

Figure 18-28. *A Pareto chart including cumulative percentage*

The cumulative line doesn't radically change the look of the chart, but it's easy to see how it emphasizes the fact that the growth slows down near the end because there are more values to the left side of the chart.

Boxplot

Another chart we saw in Chapter 2 was the boxplot, which gives us the distribution of each group among several. Boxplots rely on certain points in the distribution to show the spread of the data. They're more intuitive than some other statistically based charts and can be explained rather quickly to people who haven't seen them before. They also show outliers well. See Figure 18-29 for an example.

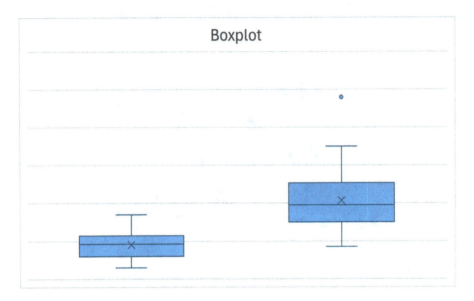

Figure 18-29. *Boxplot*

Figure 18-29 shows a two-group boxplot with the second group with values that are both higher and more spread out than the first. In this plot, the colored area represents the interquartile range (IQR; between Q1 and Q3), with the central line representing the median. In this version, the "x" represents the mean, which is noticeably higher than the median in the second group, which means the distribution is skewed. The horizontal lines ("whiskers") at the top and bottom of each group represent the min and max, with the exception of outliers. The min and max both use 1.5 times the IQR, with min being Q1 minus that calculated value and the max being Q3 plus the value. Outliers here are defined as anything greater than the calculated max or lower than the calculated min. Extreme outliers can also be included and are anything lower than Q1 minus 3 times the IQR or greater than Q3 plus 3 times the IQR.

We can look at some student credit data, this time showing first-, second-, and third-year students. See Figure 18-30.

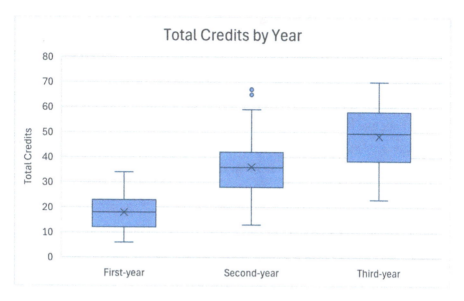

Figure 18-30. *A boxplot showing total credits completed by year of student*

The figure looks overall like we would expect, with increasing medians, mins, and maxes for each year, but also a decent spread for each year. Some students are part-time and others overload themselves or receive college credit through other methods, so not all students are going to look the same. There are a couple of overachieving second-year students, but those are the only outliers. First- and second-year students have the mean and median close together, indicating little skew, but the mean is lower than the median in the third-year group.

This chart comes up a lot in EDA and less often in charts for stakeholders, but it can still be valuable with the right data and stakeholders.

Heatmaps

A *heatmap* is an enhancement on a table where the range of values are represented by different colors and gradients, usually light to dark for small to large numbers. It can be very helpful in conveying information quickly because of the visual aspect, but still with the detail of a table. See Figure 18-31 for an example of a heatmap showing average grades of middle and high school students in a range of general subjects.

	Math	English	Science	Social Studies	Art	Music	Band	Foreign Language
6th grade	85	91	88	87	92	94	93	89
7th grade	83	90	89	88	91	96	89	90
8th grade	82	89	86	87	89	91	89	85
9th grade	79	86	82	85	90	88	88	86
10th grade	75	87	80	79	91	91	96	81
11th grade	73	85	79	82	88	87	91	77
12th grade	73	85	81	80	90	93	95	81

Figure 18-31. *A heatmap of grades of students in 6th–12th grade*

This heatmap makes it clear that as the grades progress, some subject average grades go down, but others don't have the same trajectory. The core subjects tend to follow this trend, whereas the electives don't. Without the grayscale coloring, we wouldn't see this quickly at all.

One of the many benefits of heatmaps is that you can show a lot of data while keeping it more readable than some other visualizations. This is one reason it's common when doing EDA to make correlation matrix heatmaps, especially because you can use two different colors and represent positive and negative values in both different colors and different intensities.

It's always worth checking any time you are working with tabular data to see if a heatmap would be appropriate. Stakeholders tend to like them, too.

Treemaps

I mentioned treemaps above because they can serve as an alternative to a pie chart that's appropriate when there are too many categories to make a readable pie chart, and they also allow grouping. See Figure 18-32 for an example of a treemap showing the population of US states grouped by region. The size of the rectangle represents the population of the state.

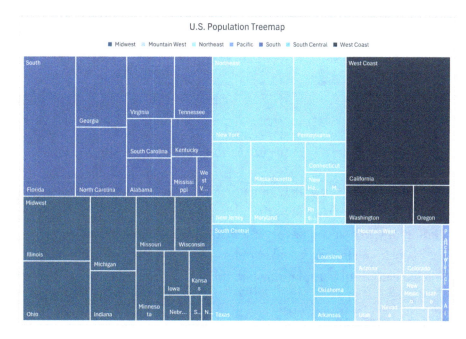

Figure 18-32. *A treemap showing US state populations grouped by region*

The grouping is a nice bonus, but despite the fact that there are 51 data points, we can still read most of them. The bottom right is Hawaii and Alaska, grouped as "Pacific." Compare that with the pie chart in Figure 18-33, which is impossible to understand without a magnifying glass and a lot of patience.

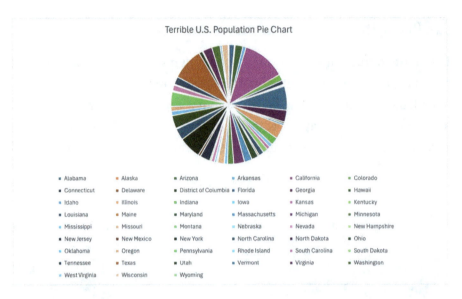

Figure 18-33. *A terrible pie chart showing US state populations*

This pie chart is obviously ridiculous, but the treemap is actually quite readable and intuitive and has the added bonus of grouping the states meaningfully. When you have numeric data with one or two categorical levels, consider a treemap. Stakeholders don't see them too often, but you'll have no trouble explaining it to them.

Choropleth Map

Everyone has seen a choropleth map, but almost no one knows the name. It's simply a geographical map of something with different areas colored according to some values. The maps all the news stations create of the United States around elections, with states colored blue or red, are all choropleth maps. See the 2008 US presidential election map in Figure 18-34, showing county- and state-level voting.

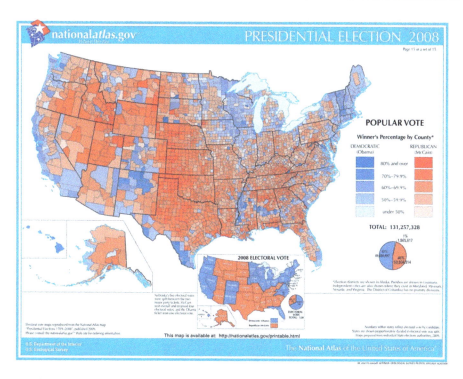

Figure 18-34. *A choropleth map showing the results of the US presidential election of 2008. Source:* `https://www.usgs.gov/media/images/printable-map-2008-presidential-election-results`

In this map, blue means the state's electoral college votes went to the Democratic candidate and red went to the Republican candidate. It's also really common to color places on a gradient based on values, as was done here at the county level. This can be done on a true gradient, where the lowest value has the lightest color and the highest value has the darkest color, with everything proportional. Alternatively, it's possible to bin the values to have a finite number of colors or shades. There are advantages and disadvantages to each. Figure 18-35 shows a map of US state populations with the shades unbinned.

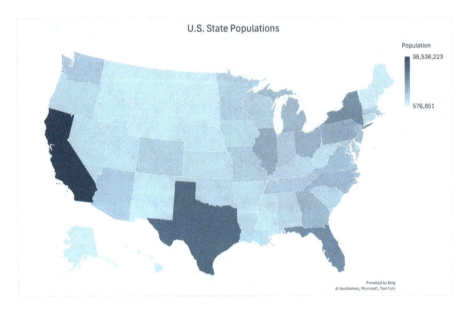

Figure 18-35. *A choropleth map showing the populations of US states*

This map is readable, but there are a lot of different shades to discern. That's why binning can be useful. We can use equally spaced bins (just divide the full range into equal-sized bins), use distribution characteristics of the data like quartiles, or simply create arbitrary bins. With such huge differences between the lowest and the highest states in this map, it might make sense to create our own bins. One disadvantage of bins is that two values could fall in different bins even if they're very close in value, which can be misleading. For instance, if we made the lowest bin cut off at 3 million, Nevada is just 105,000 over that (total 3,105,000), where Kansas is 62,000 under (total 2,938,000), but they'd be binned separately, despite being very close in actual population. Kansas would appear to have more in common with several states that have less than half a million people than it does with Nevada, which has less than 175,00 more people. The right choice will depend on exactly what you're trying to accomplish.

One other consideration when making map plots is which projection to use, especially when using a global map. The one most of us are most familiar with is the Mercator projection, which turns the globe into a rectangle with land in the north and south stretched considerably. The Robinson projection leaves the impression of a globe and doesn't stretch things out as much. A lot of tools have the Mercator projection built in and it's familiar, so for many purposes it's the easiest choice.

Presentation

At most companies, you will present your findings by creating and sharing a report (like with Microsoft Word or Google Docs) or creating and presenting a slide deck (like with Microsoft PowerPoint or Google Slides). You may also be asked to share the slide deck. I'm going to discuss what kind of info should go into your presentations based on audiences as well as medium (document or slide deck).

Audiences

There will often be reasons to create more than one version of presentations, each with different info based on who the audience is. For the sake of convenience, I'm going to name each type of general audience here to refer to below. Usually, the first people you'll present to will be the stakeholders you've done the project for ("your business stakeholders"). You may be asked to present it to their more senior leadership ("business leadership") or to your own senior leadership ("technical leadership"). If the project has gone especially well, there may be other teams like your original stakeholders who would be interested in you doing similar work ("other business teams"). Finally, it's incredibly common to be asked to present your work to your own team or other data scientists ("your peers"). In all cases, you're going to be going through your storytelling steps #1–#7 mentioned earlier in the chapter.

Every one of these audiences needs to know what the problem being solved is and how your solution helps the stakeholders do their jobs (steps #1 and #4, with a bit of #5 as needed). For both business and technical leaders, keep both of these short and sweet, but on how the solution helps, do your best to quantify that in terms of hours saved (reducing the time the stakeholders spend accomplishing the original task) or a likely increase in sales or reduction in cost. Numbers are the language of leadership. Both these groups will want a very minimal amount of info from #2 and #3. Unless you've been told otherwise, keep #2 and #3 to a bare minimum. Don't leave it out altogether, however, because some of them may have questions if you're presenting live. It can be good to put some more detail in an appendix for both a doc and a slide deck. Next steps can be good to share.

Your business stakeholders already know the problem, so you don't need to go into much detail on the problem for them, and depending on who the other business teams are, you may either need to be detailed or quickly summarize it (if it's a problem they already have, they don't need much info, but if they work in a different area, they may want much more info). For both of these groups, "it depends" is the rule on steps #2 and #3. As mentioned above, some of these groups are also working with the data themselves, so they may want to understand more about how you used it, in case that helps them. It can be good for you to spend some time on these, however, since this is the area that you are most likely to either have worked with your stakeholders or will need to work with future stakeholders (e.g., figuring out what the data means). So it's helpful to them to see that data prep is an important and time-consuming step. They're all going to be most interested in the solution, steps #4 and #5. This should be explained in detail to these audiences (even more so for any other business teams that aren't familiar with your current stakeholders' work). Just like with leaders, make sure it's clear how much this benefits them in terms of time and quality. Both of these groups will also be interested in hearing about the next steps, and with some you may even want to dive deep into these if they are on the table for immediate work.

The final common audience is your peers. For them, you need to make the problem clear, but they're usually not very interested in the details. The same is true for the solution and next steps. Don't neglect to mention them, but they're probably going to be more interested in steps #2 and #3. They will want to know about the data sources you used in order to see if they might be able to use those sources for any of their own work. Similarly, they'll be more interested in the technical info of your solution, especially if you added any code to a team repository they may use themselves or if you've solved a technical problem they also have been facing, for instance.

Mediums

The length of your documents and slide decks will vary according to your audience, as discussed above. But it can still be helpful to talk about the structure of these reports.

In a document-style report, you basically summarize everything your audience wants to know. You also should include the date, author, and people involved in the project. You'll need a table of contents and should divide the doc into sections generally corresponding with the steps outlined above for telling the story of your data. You will need to lay out all the important points you want to make at whatever level of detail is appropriate to the audience. This does mean that you might create multiple reports for different audiences. A high-level version that a lot of places want is called an executive summary, which may have an internally defined structure. Others have "one-pagers" that are a common way information is shared, obviously summarized to a single page. One advantage of a document-style report is that you can link to other documents that have more information or details, so readers can decide how in-depth to go on their own.

A slide deck will also convey everything you want your audience to know. You usually start a deck with a title slide, a date, and a list of people involved in the project. Include an agenda slide that serves as a table of contents just after this info. If you are creating a slide deck to present in a meeting, a truly good one will leave out a lot of detail that you will instead be saying out loud. This is the hallmark of a good presentation, because when you show a slide, people will either read the slide or listen to you, but generally not both. So it's usually good to have a few bullets—talking points—that you will go into, so they don't get stuck reading the slide.

However, one of the problems is that people in meetings (very) often ask you to share the deck so they can look at it later. If you've created the deck like I mentioned above, it will have very little actual content, and they won't really remember what you said. There are a couple of ways to deal with this. One is to include your major points (the key ones you're saying out loud in the presentation) in the notes field on each slide, which can't be seen when going through the deck in presentation mode. Another option is to have a separate deck that has all the explanation and details in it and share that instead.

If you've created a tool, like a dashboard, that people will use, it's also typical to do a live demo of that tool in a meeting. You can also include links to it in a document or slide deck. Make sure that you either give access to the dashboard to the people in the meeting or make it clear how they can ask for access.

Visualization Tools

There are many visualization tools out there. Data scientists frequently work with Excel and code, like matplotlib or seaborn in Python or ggplot in R. I used Excel for almost all the graphics in this chapter because it's quick, easy, and uniform in appearance. When doing EDA on real data, it's more common to use code for these (because you've usually done a lot more data manipulation in code). But for presenting to stakeholders, you may prepare data in a CSV to use in Excel or in a database or spreadsheet to use in one of the other professional visualization tools, including Tableau and Power BI.

Tableau and Power BI are the two most popular professional visualization tools in use right now. Power BI is owned by Microsoft and is often used in organizations that have gone all in on Microsoft products and the Azure cloud platform. Tableau is independent, but is integrated into a lot of other tools. Both are good to get familiar with if you're going into data science or data analysis.

Tableau and Power BI both allow you to pull data from multiple sources and combine them internally (they call this data modeling). You will essentially create a single or multiple spreadsheet-style structures within the tool that you can use to build charts and dashboards off of. The nice thing about them is that they allow you to keep the data in a fairly raw state so you can do aggregations within the tool. This way you don't have to create a bunch of spreadsheets or database tables with different aggregations and structures in order to produce multiple graphics.

Both have built-in support for all the visualizations I talked about above, and they have really flexible options for adding filters—like selecting the year to view only one year's worth of data, selecting a single product, and so on—that can impact several different charts and tables on a single dashboard. This makes it incredibly user-friendly for end users. Both tools do have a bit of a learning curve, but once you spend a little time with them and grasp the general way they work, learning new aspects of each tool becomes easier and easier.

You can learn more about Tableau and its free public option at its website.[2] This site is actually a great opportunity to both pick up some Tableau skills and create work to add to your portfolio. It's also fun to see what some really skilled people have built. Power BI also has a free version

---

[2] https://public.tableau.com/app/discover

that you can download (Windows only).[3] I highly recommend getting your toes wet with one or the other so you can list basic skills in one on your resume when you're searching for jobs. You don't need to learn both, as the skills required for one transfer fairly easily to the other.

Key Takeaways and Next Up

In this chapter, we looked at the wide world of visualization, starting with a couple of interesting examples where the visual was surprising and impactful, including a famous map from Napoleon's invasion of Russia and another showing how datasets can have several identical statistical properties but look completely different when plotted. Then I talked about how visualization and presentation of work is really all about storytelling—what story does the data and your work with it tell? I went over many different chart types and other visualizations, including scatterplots, column and bar charts, line charts, area charts, pie charts, histograms, box-and-whiskers plots, heatmaps, treemaps, and choropleth maps. I talked about ways of presenting your findings to stakeholders and others and finally about the two main professional visualization tools in use today, Tableau and Power BI.

In the next chapter, I'll be going over several modern applications of machine learning and data science in different fields, from insurance to healthcare to education, and several more. Data science is widely used, and there are no signs of things slowing down.

---

[3] https://www.microsoft.com/en-us/power-platform/products/power-bi/desktop

PRACTITIONER PROFILE: MEGHAN BERGER

Name: Meghan Berger

Current Title: Retail analytics consultant and data translator

Current Industry: Consulting at a data and analytics firm

Years of Experience: 10

Education:

- Certificates: Certified ScrumMaster (CSM) and Tableau Desktop Specialist

- MS Business Analytics

- BS Business Management

The opinions expressed here are Meghan's and not any of her employers, past or present.

Background

Meghan spent her first ten years out of college working in high-end retail management. She learned a little about analytics there and was always making charts and working up numbers because she enjoyed it and it helped with the job, even though she didn't really know anything formal about analytics. But she would look into their sales and inventory data to see if she could figure out why things weren't selling well or if there would be opportunities to trade stock with another store. She also did employee reports. It was all work that needed to be done, but she enjoyed it so she dug a bit deeper and spent more time on it than strictly required. She did get to know some of the company's analytics people and tried to learn from them, but it was all pretty basic and she didn't feel like she was getting good answers. In the end, she knew it was something she really wanted to do, so she went back to get her master's in analytics. During the degree, she learned a lot, including Python, R, and more, but especially fell in love with visualization.

Work

Meghan landed a job out of her master's with a relatively small data and analytics consulting company. She's worked with a few different clients doing different kinds of analytics and visualization, and she found that she loves being able to help clients in ways that she wanted when she was in retail, but didn't have the skills or tools back then. She's found her niche in visualization and analytics in the retail/consumer packaged goods space.

Early on, she focused on visualization development in tools like Power BI and Qlik Sense, but as she's grown in her career, she's taken on additional responsibilities, including leading data visualization teams on several projects, product management (overseeing planning and execution of large-scale analytics projects), and building relationships with business users across the organization both to help coach them in self-service visualization and to provide business insights and strategy in developing effective analytics solutions.

Sound Bites

Favorite Parts of the Job: The work is different every day so it's never boring. There are different clients and different problems to solve. She's also in a great spot because her company is pretty small, so there are a lot of opportunities for advancement.

Least Favorite Parts of the Job: Sometimes as a consultant (depending on the client), there have been times when stakeholders treated her team primarily as technical analysts and less like business partners. This limits the opportunities to share knowledge and advise on long-term solutions. It can be challenging to maintain the balance of establishing realistic development timelines and meeting stakeholder expectations, especially when she runs into unforeseen technical complexities.

Favorite Project: She got involved in a somewhat rushed project related to stock-outs in retail. She made a visualization for it that was really interesting, because it involved tying two separate sources together into one chart, and it was immediately clear what it meant. She did something similar in another project involving inventory where three sources were brought together for the first time and added some calculations that helped determine which products were likely to be delayed, which gave more time to trace the order and work on mitigating options.

Skills Used Most: Some of her earlier work was more technical, where she was doing development, but because her focus is on visualization and coaching, soft skills are more important. These include communication, management, leadership, and planning and guiding toward goals.

Primary Tools Used Currently: SQL, Snowflake, DBT, BigQuery, Power BI, Qlik Sense, DOMO, Sigma

Future of Data Science: One thing that's definitely happening is that AI is making it easier for people to do their own analytics with no-code or low-code options. This isn't inherently bad, but an understanding of the data you're working with (and how to join it with other data) can be really important. Meghan thinks that analytics professionals' roles will probably shift away from development and solution-building to more strategic oversight into projects. Self-service tools aren't going away. This all means that documentation on data is going to be increasingly important.

What Makes a Good Data Visualization Professional: Having a really good understanding of the end business use and goal. It's also important to be able to tolerate a lot of trial and error to find the right way to visualize things in a way that stakeholders will understand.

Her Tip for Prospective Data and Visualization Professionals: Learning is never over. You have to always be learning new things in the field—new tools, new ways of doing things. You may find the need to transition from tool to tool based on what is trending in the market. It's also important to try to seek partners on the business side to learn more about the domain you're working in. Trading knowledge is a great way to build relationships.

Meghan is an experienced analytics and visualization professional with a strong background in retail.

CHAPTER 19

This Ain't Our First Rodeo: ML Applications

Introduction

Both data science and machine learning in general have been around for decades now, and it's steadily increasing in popularity. With a lot of publicly available ML-based tools like ChatGPT, many companies are easily generating marketing copy, and coders are speeding up their coding. With the ever-growing popularity of analytics and ML in general, we're seeing many uses crop up in so many fields. Some companies are using third-party tools to generate their own models on their own data, like we saw with Salesforce's ML tool in Chapter 16, but plenty are also diving into creating their own data science solutions and ML tools from scratch.

Data science has helped companies improve their marketing efforts greatly, with approaches like A/B testing commonplace now. It's made it easier for companies to cater to specific groups of customers with clustering techniques, allowing hyper-personalization. It can improve sales

© Kelly P. Vincent 2025
K. P. Vincent, *A Friendly Guide to Data Science*, Friendly Guides to Technology,
https://doi.org/10.1007/979-8-8688-1169-2_19

with things like well-targeted product and service recommendation. A lot of companies are able to automate a variety of tasks that can save time and labor.

NLP and speech processing are making automating parts of customer support easier, cutting down on the number of customer support agents needed with easy-to-build chatbots.

One trend that is making it easier for companies to adopt ML is businesses shifting to the cloud for data storage and tools as they move away from hosting everything on site. Cloud providers offer a variety of services on top of the core ones that attract companies, and it's easy for them to just throw in some ML options so companies can try things out with minimal effort. If they like the results, they're already on their way. We'll be talking about the cloud in the next chapter.

One thing that's important to mention here is that you already know that a lot of uses of ML—and technology in general—come with ethical quandaries. Many of the applications I'll mention have hard-to-ignore issues even though we are ignoring them and barreling ahead. Frequently, people don't realize how things are done, so they don't know that there are ethical questions. This can include both the companies using the ML and customers using their products.

Although there are so many exciting things we can do with ML, we need to not forget the bigger picture. It's always our responsibility to think about the ethics involved, even if other people aren't.

Below, we'll look at how several industries are using data science and machine learning right now. We'll start with insurance, banking, and finance and then look into retail. Then we'll go into both healthcare and bioinformatics before jumping to the gaming industry. Finally, we'll take a look at criminal justice and education.

Insurance

Statistics has been a part of the insurance industry almost since its modern inception with rich Englishmen underwriting shipping and colonist transport to the American colonies in the 1600s (*underwriting* means taking on financial risk for a fee, with the possibility of having to make a large payout). Once Pascal and Fermat solved the problem of points, their techniques for calculating risk were picked up by underwriters, and soon insurance was an entire industry.

So statistics and statistical modeling like linear regression and generalized linear models have long been the traditional techniques used in insurance. These have helped them balance risk and profit during underwriting (like determining premium prices), but the industry hasn't yet truly embraced modern machine learning in their core business. However, there are many possible applications, especially in enhancing current methods. One possibility is to use ML to do feature engineering to create features to use in the traditional models of the field. For instance, they could prepare nonlinear features, bin existing features more effectively, or use clustering to identify entirely new features. Additionally, some ML methods like trees and regularization handle sparse data better than traditional models, so some are starting to look at those. Using ML in this way can open up ethical risks, especially because so many Western countries have significant regulations based on protected classes, which can't be included in modeling. For instance, race can't be included in determining what is offered, so data scientists have to be careful that none of the new features act like proxies for race, even if race itself is not included.

Other areas of the industry have already embraced ML more. For instance, ML can be used to speed up claims processing by looking at characteristics of the claims and routing them to analysts more effectively based on fraud risk or other aspects that require certain specialization. This speeds up processing time. ML can also be used to predict claim frequency and severity. Fraud detection and prevention are more sophisticated with ML, relying on anomaly detection approaches, catching more truly fraudulent claims and minimizing the number of non-fraudulent claims that get sent for further analysis. There are also possibilities in compliance monitoring around regulatory requirements by watching transactions, claims, or communications. Even property damage assessment in claims can be partially automated with image processing techniques. For the most part, these don't have great risk in terms of ethics, but as always, they need to be cautious to treat customers fairly, especially in light of protected classes.

Customer management and sales can also be improved with better monitoring and product recommendations. ML can inform interactions with current customers and identify methods for gaining new ones by grouping customers so different actions can be taken. Another area they're starting to apply ML is in customer churn prediction (predicting customers that leave), and it's very effective. Chatbots and other tools can assist in customer service. These are also fairly low risk in terms of ethics, but there can be privacy risks with chatbots, so they need to be secure.

Banking and Finance

ML in banking and finance has a lot in common with insurance, as they're both legacy fields that have benefited greatly from statistics and mathematics. Like in insurance, one of the most powerful and embraced areas that machine learning has revolutionized is fraud detection. Similarly, they're starting to use ML more in the underwriting phase and in credit scoring, especially in assessing credit risk and determining interest rates on loans. ML makes it easier to include more information in all these decisions, leading to more loans that are also lower risk. Like insurance, they also have to be cautious around regulations on protected classes to make sure that info isn't being included in models.

Efforts are also being poured into predicting market trends like interest rates and the stock market behavior because there is so much more data than has ever been included before. NLP makes it easier for companies to utilize more data from unstructured sources like news and other documents for use in stock market forecasting, as well as any other area that could use that data. A final area that's growing in popularity is fully automated stock trading, called algorithmic trading. The ethical risks aren't significant here, especially if the data used is publicly available. However, what is done with these forecasts can lead to ethical questions in terms of how it's presented to customers.

Money laundering has always been something banking and finance has had to watch for, and ML has made that easier, making it easier to stay in compliance with regulations with anomaly detection. Regulations in general can be especially difficult because they change periodically, but there are methods for automating compliance in the face of change.

There are also many areas related to customers where ML is being used in banking and finance. They use NLP and chatbots to automate or semi-automate customer service, especially on simpler tasks. Customer churn prediction is also done here because it's very useful to try to convince customers to stay if they're planning to leave. Personalization is invaluable for improving the customer experience on website usage, and product recommendations can both make customers happy and increase sales. Some of these things come with privacy risks that need to be properly handled.

Retail

Like other sectors, retail uses data science for improving the customer experience and sales through customer segmentation, targeted marketing, and personalized product recommendations. This is hugely important in retail as repeated selling is their main goal. Loyal customers are invaluable in retail.

There are many other areas where retail companies use ML. They maximize profits with pricing strategies that monitor competitor prices and tweak prices frequently, especially for online storefronts. Automatic pricing optimizes much more effectively than manual management can. In order to do it, a pricing system needs to be able to match products from the company's catalog to competitor sites' products, analyze the variety of prices, and set the new price. There may be ethical questions about accessing other companies' websites and prices.

Another important area of ML adoption in retail is in supply chain management. The supply chain is hugely complicated nowadays at most companies, where they have multiple suppliers, different productions or warehouses, and different shipping mechanisms. It's common to use computer systems to manage the entire supply chain, such as through enterprise resource planning (ERP) systems. There are often ML tools built into those, which are easy to use because the system already has access to much of the important data. Whether used within an ERP or not, ML can help with demand planning to maintain stock by determining when to reorder product, components, and supplies (replenishment) through forecasting usage. Other forecasting is invaluable in general for aiding business decisions. Forecasting in general helps with many aspects of business, including staffing, product placement, and product launches. These tools are also fairly safe from ethical issues.

Data science is also used heavily in marketing, as in other industries. This includes managing promotional prices or general discounts. Customer segmentation is massively important in marketing, and retail has been doing it for a long time. It helps them cater different promos to different groups. There's a concept of customer lifetime value, which allows companies to forecast a given customer's value over the long term, which can help them determine marketing or offers for different customers. Like with everyone else, customer churn prediction can also be helpful in marketing. The same ethical questions apply to any situation dealing with customer data.

Managing the website, especially descriptions and images, can be made easier with ML tools, such as using GenAI to generate the product descriptions. AI can also be used in image touch-up. Website monitoring can help the company learn about customer behavior, such as where they click and which pages are visited most, and A/B testing lets them try different things out to see how they do with customers. These are fairly low risk in terms of ethical issues.

For stores that have physical retail locations, there are tools to monitor customer behavior even in the store, like tracking how many people come in, the paths they take through the store, the displays they spend time looking at, and the products they pick up. This can be incredibly helpful in determining store layout and shelf optimization. It can even be used to help with theft prevention both by using facial recognition to identify known shoplifters and by monitoring behavior. Obviously, the minute you start monitoring people, the possibility of privacy and ethical risks is right there. Facial recognition, especially of known criminals, has a host of ethical concerns. Some risks may not be worth it.

Companies that make the products they sell either for resale or directly to customers also use data science in determining new products to make and even in tweaking ingredients in food products. Ingredients and parts can be optimized on a variety of factors, including cost, taste, health factors, sustainability benefits, and more. Ethical questions can arise depending on what is considered.

Healthcare and Pharma

Another area that's already made good headway with data science and has the potential for even greater benefits is healthcare. The promise of personalized medicine hasn't been fully realized yet, but we have made some progress and already have some major improvements in diagnosis, risk assessment, treatment recommendations, drug development/

discovery, assessing drug and other treatment trials, and automating a lot of administrative tasks. Healthcare is one area that is rife with discrimination, so it's crucial that we address these issues as they arise.

One of the biggest areas of improvement is in the analysis of medical images like MRIs, CT scans, and X-rays, something that used to be done exclusively by specially trained people. We talked about one of these tools helping to diagnose hard-to-spot breast cancer in Chapter 15—ML-based image evaluation systems are especially good at detecting very small anomalies that are usually missed by people. Additionally, relying on ML for this saves time from doctors—they do still need to look at the images, especially ones where the ML detected nothing, but time is saved when it finds something. The main ethical risk here is relying too much on ML. These tools do make mistakes, and humans still need to do their due diligence. One advantage of using these tools is that they minimize the risk of discrimination. When looking only at the inside of our bodies, things like ethnicity and socioeconomic status aren't visible.

There is also potential for ML to improve other methods of diagnosis as well as treatment plans. Tools that can recommend whether a patient should receive a particular medicine, or a particular treatment, are already in use in some hospitals. These can spot things doctors miss because of the sheer number of factors, so they can be very powerful. Some of these tools can take things a step further and actually predict outcomes for patients based on different combinations of treatments, helping doctors choose a strategy. One thing we are finding is that these can be biased because human biases are baked in. For instance, one system is known to be less likely to recommend a pain medication for Black patients because there are inaccurate beliefs that they have higher pain tolerance and are at higher risk for addiction, and these biases have been captured in the model the system uses. But the truth is, there is also great potential here to overcome doctors' own biases, if fair ML tools can be created.

There are also benefits that can be brought to patients through their medical records and monitoring technology. We've talked in this book about using automatic transcription on patient record dictation, and that and other NLP approaches can be invaluable for understanding and using patients' records to help them. They can be mined for missed diagnoses and useful in chronic disease management. A lot of patients can benefit from wearing medical devices that monitor aspects of their health, even making it easier for patients to manage their own conditions better. Some of these areas can touch on the same problems we see with discrimination in healthcare, but there are additional concerns related to the need for security in medical devices that can be hacked, and the question of who owns the data from a tracking device has so far always been the company that makes the product and not the patient.

There's potential for higher-level benefits like better hospital management and public health benefits. Hospitals can use supply chain management tools like they use in retail to plan for supply usage, stock appropriately, and order in time. Demand forecasting can be used both for supplies and scheduling procedures, and staffing can be planned more easily with ML. There are also a variety of tasks that can be automated in healthcare, including communication, billing, and appointment scheduling. This does put pressure on patients to be able to interact with the healthcare system digitally, which may be a challenge for people who don't have access to computers either at all or securely.

There are a lot of ways ML can be used in public health, including forecasting disease trends and planning for them. We can even go all the way back to patient records to identify and understand the origin of disease outbreaks, although that would require caution in terms of dealing with sensitive data.

Drug discovery and development are two other areas that are starting to use ML with great promise. Drug discovery has long been an expensive and time-consuming process. The programs can take over a decade and

cost billions of dollars and still fail the vast majority of the time.[1] There are
so many factors that can contribute to a particular chemical compound
being potentially useful that it's impossible for humans to consider them
all, so ML is ideal for this problem. It's not easy, however, as it requires
dealing with high-dimensional data, which is resource-intensive, but with
the preponderance of large chemical and biological datasets (public and
private), getting useful data is increasingly possible. ML can help with a
variety of stages in the drug discovery and development process, including
identifying potential chemical compound components that might be
useful (prognostic biomarkers), analysis of aspects of clinical trials, and
optimizing chemical properties.

One current challenge with ML in drug discovery and development
is in interpreting the results of ML, along with the lack of reproducibility
in ML outcomes (when nondeterministic methods are used). One area
that drug researchers are working on is getting better at understanding
the technical aspects of ML, making them better at picking the right
approaches and evaluating the results. In my view, one of the biggest
risks with ML in drug discovery and development is higher-level—which
diseases do companies try to find treatments for, and who is most affected
by those particular diseases? But it also will speed up the process and
decrease the failure rate. This both saves pharma money and helps
patients, even though pricing can often exclude a lot of patients—but if
companies behave ethically, they can keep the costs of new drugs down
since it costs less for them to develop them.

I mentioned that ML can be used during drug discovery to evaluate
clinical trials, but it can also be used to evaluate them for other
applications, including treatment strategies. Furthermore, it can be used to
help design new studies and choose factors to evaluate.

---

[1] "Machine learning in preclinical drug discovery" by Denise B. Catacutan, Jeremie
Alexander, Autumn Arnold, and Jonathan M. Stokes, https://www.nature.com/
articles/s41589-024-01679-1

Bioinformatics

Bioinformatics is a field focused on understanding biological systems through several traditional disciplines, including biology, computer science (and ML), chemistry, and physics. It's especially valuable when the datasets get large, which they do with biological data. Bioinformatics has a lot in common with healthcare in terms of ML uses, especially with drug discovery, as we saw above.

Prior to the use of ML, working with the massive biological datasets was difficult, and it was largely impossible to identify all of the important features. Since its introduction, feature engineering has been one area that ML has excelled in bioinformatics. The dimension reduction technique principal component analysis (PCA) has been invaluable in working with the extremely high number of features these datasets often have.

Proteomics is the study of proteins, and there have been several areas that ML is helping. Protein structure prediction is very similar to drug discovery, and similar methods have been used to identify structures that are likely viable and worth further study. It's also been used to predict the function of different proteins and potential interactions between them. It's also possible to use NLP techniques to annotate research materials related to proteins.

Genomics is another important area where ML is increasingly in use. The human genome was sequenced in 2003, which took over a decade of work—ML has enabled the process to be sped up, and it can now be done in a day. There are also genomes of a lot of different animals available, so genomics work touches all areas of biology. Genomics also shares some types of ML-aided tasks with proteomics, like predicting the structure of RNA and the function of genes. ML can also help with finding genes and identifying their coding regions. In genomics, they use ML to help identify *motifs*, subsequences of DNA with a specific structure that appear repeatedly. Motifs are assumed to serve a function, and ML can help figure that out as well as a variety of other facets of the motifs. There's also a lot of genomics work that relates to sequences, including assembling them, that ML helps with.

The study of evolution is interdisciplinary with other bioinformatics areas, but one particular area that ML helped is the building of *phylogenetic trees*, structures that represent evolutionary relationships between species. See Figure 19-1 for an example of a phylogenetic tree of the order Lepidoptera, containing butterflies and moths.

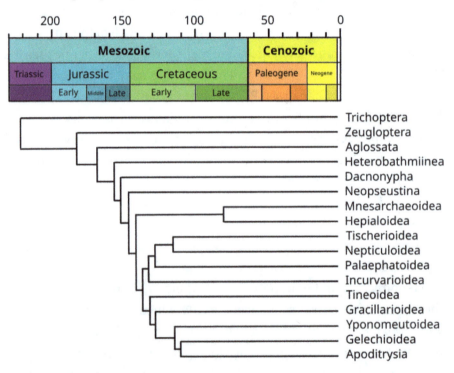

Figure 19-1. *A phylogenetic tree of the order Lepidoptera. Source: From "File:Phylogenetic chart of Lepidoptera chronogram.svg," available at* https://en.wikipedia.org/wiki/File:Phylogenetic_ chart_of_Lepidoptera_chronogram.svg, *used under the Creative Commons Attribution-Share Alike 3.0 Unported License (*https:// en.wikipedia.org/wiki/en:Creative_Commons)

Most bioinformatics work is done away from individual humans' data, so the ethics risk isn't huge, but any genomics that does deal with individual people's genomes needs to involve awareness of data privacy. There can be significant risks associated with individuals' genomes. As an example, the genome sequencing company 23andMe promised its customers it would never share their genomes, but after a bankruptcy in 2025 they were sold to another company. One of the conditions of the sale was that the new company honor that promise, but there is no real way to ensure the data isn't used differently down the road.

Gaming

ML is already highly associated with games in the public consciousness because of the famous 1996 and 1997 matches between IBM's AI agent Deep Blue and the reigning chess champion Garry Kasparov. They played multiple games, and Deep Blue won a couple in the 1996 match, but Kasparov won the match overall. But the next year, they played again and Deep Blue won. There have been further advances in computers playing chess, but more recently the big news was that an AI agent AlphaGo beat the reigning champion of Go (in 2016). Go's considered a much more challenging game than chess, so this was a big deal. However, since then, results have been mixed, and these tools haven't been able to consistently beat Go champions. It seems that it depends on style of play, and AI can't always defeat certain strategies.

But ancient board games aren't the only place ML is being used. The modern video gaming industry uses it in a variety of ways. One is to improve non-player character (NPC) interactions and behavior with players. NPCs are the characters that are in the game world while the player moves through it, which are not being controlled by other human players. These characters are often important in quests or challenges, so players need to interact with them to get information. Historically, these interactions were written with a confined script, so there were only a certain number of specific things the NPCs could say or do. As a player, you'd usually end up getting to the point where the NPC repeated itself, and you knew you'd exhausted the info. But with ML, there can be a lot more natural behavior that can respond to specific actions the player does. This makes the world far more lifelike.

NLP plays into how NPC characters have been improved by making the conversations more natural, but it also allows other improvements to gameplay, especially when combined with speech processing. Players can give commands to the game verbally and receive instructions from chatbots.

Another area is personalization, similar to what happens with customers in retail. Games can monitor player behavior and offer different game behavior based on the specific actions they make. This is especially valuable to companies who sell things within their games, because it can encourage purchasing by offering "just the right thing" the player needs in that exact moment. But it's also useful for giving a better gaming experience by offering more of what the player likes about the game and minimizing what they don't like. For instance, a lot of games have sidequests that don't have a significant impact on the overall story of the game but can be fun, and some players might like ones that involve puzzles but not fighting lesser monsters, where another player might feel the exact opposite. The game can serve up what they like. Monitoring player behavior is also a way to understand what players in general like and don't like, which can influence improvements or enhancements to the game, or even identify bugs. This level of monitoring also allows the game developers to minimize cheating.

Previously, all content in games was fully scripted by people, just like NPC behavior. But ML can allow some content to be automatically generated, a process called *procedural content generation*. During development, this can be everything from entire game levels, settings, specific in-game quests, and many other things. It's also possible to dynamically generate content during play, although this doesn't happen as much.

Another benefit to the development of games is testing and quality assurance (QA). Like other things, this used to rely entirely on humans, who would play through the game and hopefully find any particular paths that were buggy. Automated testing and QA usually look for certain patterns or anomalies in game data to identify bugs and performance problems more quickly. Automation makes it easier to test on a variety of different hardware configurations.

We've already seen how some of the innovations in AI have involved image processing and computer vision, so it's not surprising that this factors into game improvement as well. Modern graphics and animation are really high quality, especially for those of us who grew up playing 8-bit games like the original Atari systems. ML allows them to avoid drawing every single part of the game and instead create a character template that ML tools can manipulate in different actions.

Finally, gaming companies are also selling things, so they have customers like traditional retail companies. They can use ML in the same ways to manage customer interactions and speed up elements of customer service. They can also use it to help with business functions like planning new games, managing promotions, and making other business decisions. Forecasting is just as important as in retail.

Criminal Justice

We saw in Chapter 9 that some court systems are using ML-based tools to aid in sentencing convicted criminals. Proponents generally believe that this can help us get rid of bias, even though most studies that have looked at these tools have instead found that it exacerbates it. Like with any machine learning, there is the potential for being less biased, but that requires training models in a way that doesn't learn the biases that are already there in the real world. I'll go over how it's being used now, but keep in mind that most of the current uses are incredibly ethically problematic because no one has really figured out how to remove bias from training data.

Police organizations use ML in a variety of ways. One is analyzing video and images, including the relatively simple tasks of identifying people and objects, but also to detect red light violations, car accidents, and crime scenes (even in real time from cameras posted in public spaces). Crime scene images can also be analyzed to look for evidence. They're definitely known to use facial recognition technology, a thing a lot of people don't like. Similarly, they can detect gunshots with sensors placed in public spaces, which allows the police to be notified about a shooting before anyone calls in and also to determine the location of the shooter. There have been many famous cases of ML being used in crime prevention tools by determining where to place additional police and implementing certain policies. Finally, forensics departments use ML in DNA analysis now. Medical examiners are also using image processing to help determine cause and manner of death in some cases, looking at radiological images.

ML is also in the court systems. As we saw in Chapter 9, the likelihood of a convicted defendant reoffending is predicted by ML tools, and that number is used to inform sentencing decisions. This falls under risk assessment, and other related decisions are those made pre-trial about bail, post-conviction in sentencing, during incarceration about release, and post-release about probation and parole.

There are also tools available to help automate some administration tasks, including those in court systems like scheduling, managing documents, and coordinating workflows. This obviously can save time and reduce human error, but at the same time, it can introduce other types of error, which can lead to bad experiences for people. This means that it's important that humans monitor this.

Education

It's probably not surprising that ML is used in education, too. There are a lot of tools that schools use to manage all their resources and students, and those frequently have tools in them that use ML in some way. For instance, there are tools that scan students' computers looking at their writing, emails, and messages to identify those at risk of suicide.

Tracking, a traditional way of grouping students based on perceived student ability and putting them through different curriculums ("tracks"), can be partially automated. This can determine their future options. This has been done in high schools for decades, but now it is often assisted by ML tools that can look through extensive amounts of data. Tracking has always been risky for students, who may be virtually locked out of college opportunities because of decisions made by others, and now this is handled at least partially by ML.

Tracking isn't the only thing that can be done with the vast amount of student data out there. It also allows schools to identify students that could use some extra help, and they can offer that help if they have the resources. Colleges can also use these approaches, and some have found that finding struggling students and offering them support has had very positive outcomes.

The biggest problem with both tracking and cherry-picking students for extra help is that these systems again perpetuate bias because they are trained on data that has come out of an already-biased system. Additionally, both come with privacy risks.

One exciting educational area that uses ML is personalized learning. More and more educational software is coming out, especially for younger students, and it often contains ML that determines what should come next after a student completes one lesson. There are also ML tools used to identify fraudulent behavior in learning systems, which is important when it's used to assess students directly. AI tutors and chatbots are also becoming popular. All these things can definitely be positive, but we need for students to not get lost in technology. This means we need humans to be a part of this system to make sure that students are treated fairly.

Key Takeaways and Next Up

In this chapter, we looked at eight different areas where ML is being used. Almost all of these areas have customers in some form or another, and most companies are using analytics to help manage their customers and marketing in some way. Similarly, many are using it in their supply chain efforts. Insurance and banking and finance have a lot of similar problems and uses, including determining how much risk to take on and dealing with fraud and regulations. Retail uses ML in a variety of ways, including pricing strategies, website management, and product development or purchasing decisions. ML is big in healthcare, where there are efforts to improve diagnosis, treatment plans, and drug discovery. Bioinformatics has a lot in common with healthcare, including processes similar to what's done in drug discovery, but ML is used in genomics and proteomics. The video gaming industry uses ML in a variety of ways, mostly to improve gameplay and graphics in modern video games. The criminal justice system is using ML quite a bit in both law enforcement (monitoring people and looking for crimes) and the court systems (sentencing, bail, and release decisions). ML is also being used in education at all levels, from primary through higher education.

In the next chapter, we'll be looking at the cloud and scalability. This means looking at the old way of doing things with on-premise data centers and the new way with decentralized cloud computing (paying for resources hosted by another company). Cloud options bring a lot of benefits, including scalability, elasticity, resource pooling, and a pay-as-you-go model, and I'll talk about those next. Parallel and distributed computing are addressed, which help with speeding up processing of large amounts of data. We'll look at the most common cloud platforms and some of their tools that data scientists are most likely to use.

PRACTITIONER PROFILE: CAITLIN TETERS

Name: Caitlin Teters

Current Title: Senior analyst, insights and analytics (data translator)

Current Industry: Consumer packaged goods

Years of Experience: 3 years as a data analyst

Education:

- MS Business Analytics

- MBA

- MME Vocal Pedagogy

- BFA Acting

- BM Vocal Performance

The opinions expressed here are Caitlin's and not any of her employers, past or present.

Background

Like a lot of the practitioners in the book, Caitlin has loved numbers and math since she was a kid. She's always felt like it brings her energy. In high school, she initially expected to follow a path into STEM, but after falling in love with musical theater, she followed a different path and went to college for vocal performance and acting. One of the things she loved about it was the storytelling, and she deeply enjoyed studying human behavior during acting classes. When she realized she didn't want to be a performer, she pursued a master's degree in music teaching. As part of the degree, she did primary research involving a lot of quantitative analysis, which she discovered she enjoyed. After the degree, she did become an elementary music teacher while also running and managing her own music studio business. She still didn't feel like she was quite where she wanted to be, so she gravitated back toward quantitative work with a plan to enter the business world. She did an MBA to give herself wider context in the business world, enjoying the business analytics part, but that's where she started to see how it all fit together. She could see how piecing together data into stories could be powerful and drive impact in a business context. As part of a dual master's program, she pursued an MS in Business Analytics, fully committing to that path.

Work

Caitlin cut her teeth in business by doing some consulting for startups while still in grad school, which involved a lot of different activities that taught her a variety of functional skills. She learned how to set up Google Analytics effectively and analyzed website traffic, and she also did some analytics. One of the things she loved was identifying business problems or needs and doing a deep dive to understand the problem and find the best potential solutions to recommend.

After graduating, Caitlin started the job she's at now, working on an insights team, which does analytics and runs a lot of direct qualitative and quantitative studies where they analyze the results. They also use third-party data in different analyses, which is a primary focus of her current role. Currently, she works with marketing strategy and marketing innovation teams to quantify emerging categories and sub-categories, strategically identify growth areas and opportunities with holistic category analysis, and leverage a variety of data sources into cohesive stories so she can deliver actionable recommendations. At the same time, Caitlin's still figuring out exactly what her role is as her company is itself still figuring out what role analytics will play in the business. So Caitlin's in an exciting and still somewhat intimidating position to figure out how she can be the most helpful, in some ways defining her own role.

Sound Bites

Favorite Parts of the Job: Caitlin loves telling stories with numbers and data, especially when it involves problem-solving and flexing her logical reasoning skills to find creative ways to solve problems with the resources she has available. She also loves being able to collaborate with people with many different backgrounds and priorities.

Least Favorite Parts of the Job: Data quality, especially in terms of cleanliness and usability, is always a challenge. Preparation for analysis takes time and can be so frustrating. It can be especially difficult when using different data sources, which is very often what needs to be done to get at all the right info.

Favorite Project: One of Caitlin's favorite projects was one she worked on while at a startup incubator. She quantified the economic impact of the operations of the incubator in order to show the effectiveness of the funding and also generated forecasts. This analysis was directly leveraged to secure funding from local and state governments by demonstrating the return on investment (ROI) through direct and indirect impacts to the local and regional economy. The work specifically involved quantifying jobs and wages on the activities happening in the startups they funded as well as using predetermined models to determine total indirect economic impact. It put a value to the question "what is the ROI that can be generated by directing tax dollars toward this org?"

Skills Used Most: The two top skills she uses are logical reasoning and flexible storytelling, both of which rely on communication skills. Reasoning also requires deep understanding, which comes through active listening to stakeholders and other knowledgeable people. Active listening is so important for getting to the bottom of the surface question, determining how the analysis will be used, and understanding who it will impact. One thing that surprised her is how a lot of the skills she uses every day came from her experience in totally other pursuits, like performing and teaching. A final important skill that comes from business experience is simply a combination of domain knowledge and business acumen, understanding the field and the business at a high level.

Primary Tools Used Currently: Caitlin works a lot in Excel and Google Sheets because most of what she does is on the more basic side, but she's working on some automation for some of her repetitive analytics tasks in Python. She also uses Domo and Looker Studio, two very simple visualization tools.

Future of Data Science: GenAI is obviously completely changing the game and will continue to play a big role in the future. Humans aren't going to be out of the loop for a while, if ever, but GenAI will continue to make an impact. GenAI analysis can be especially impactful with unstructured data like business reports and other text documents because they can be summarized automatically, which can change the process for analysts. Also, computing and hardware development will ensure that GenAI keeps growing in ability to handle more and more.

What Makes a Good Data Analyst and Scientist: The ability to listen and ask good questions, including what the goals are, what decisions it will drive, and who the stakeholders are. Developing business acumen and domain knowledge and understanding how that ties into your specific work is important. Finally, creativity in problem-solving is so valuable. This can come from thinking broadly about different angles to take on analysis to find the most impactful one.

Her Tip for Prospective Data Scientists: Build industry knowledge in whatever area you're interested in pursuing. It makes any analysis you do better informed and will make it easier to break into the industry.

Caitlin is a data professional with experience helping companies in a range of industries understand their data and impact better.

CHAPTER 20

When Size Matters: Scalability and the Cloud

Introduction

"Cloud" is a definite buzzword right now, but what came first? Historically, companies operated with *on-prem* computers—short for on-premise computers—where they had racks full of thin and wide computers called servers slid into multiple slots in their server rooms (giant rooms with many racks full of servers) or in even bigger data centers. A *server* is a regular computer in a lot of ways, but it's bigger and has more memory and larger disks, but it doesn't have monitors or keyboards regularly connected and is instead set up to be accessed via the company's network. See Figure 20-1 for a view of several servers.

© Kelly P. Vincent 2025
K. P. Vincent, *A Friendly Guide to Data Science*, Friendly Guides to Technology,
https://doi.org/10.1007/979-8-8688-1169-2_20

Figure 20-1. *A view of several server racks. Source: "Network Servers on an Enclosure" by Sergei Starostin, available at* https://www. pexels.com/photo/network-servers-on-an-enclosure-6466141/

On-prem data centers were important for most of a business's operations. Their website and intranet would live on servers. All of their databases also would be in their data center servers, as would email and shared tools that they hosted for employees to use (for things that couldn't be installed directly on people's computers). Data centers also hosted the company's networking devices.

Servers were the only way people could process large amounts of data, so they were also critical for data workers. If programmers' and other people's personal computers couldn't handle the data, they would log onto a server, which would often be running a flavor of the Linux operating system, and do some command-line work to get their code to run. Some companies preferred Windows servers, and these could be easier to use because they had a graphical user interface virtually identical to people's PCs.

This was how things were run for decades, but data centers were costly to run for many reasons—the hardware (servers and more) were expensive, the devices produced a ton of heat so cooling the rooms was important and expensive, it took a lot of staff to manage them, and physical security was a lot of work and could be expensive. Cloud computing has therefore provided a revolutionary alternative. Companies can't completely get rid of hardware, but they can significantly scale things back.

Nowadays, *cloud computing* simply refers to the use of computers that are available via the Internet so can be located almost anywhere. We'll focus on cloud computing, parallel and distributed computing, and the importance of scalability today.

Examples of Scalability and the Cloud in the Real World

The cloud opens up many possibilities that would have been impossible or very difficult with only data centers. This is due both to location, where the things you're doing need to be on the same hardwired network, and to scalability—sometimes more compute and storage are necessary for short periods of time than are worth owning all the time, and sometimes it's hard to keep up. We'll look at two examples here, both where scalability was important and one where location also factored in.

Example 1: A Digital Tennis Tournament

The 2021 Australian Open was in the midst of the pandemic, but the organizers still wanted to run a big event with fan and vendor involvement. They use a cloud platform called Meridian (created by Infosys) to create a virtual space called Virtual Hub that gave fans and vendors an immersive experience.[1] They had actually started a digital transformation with the 2019 Open, but they really expanded everything in 2021. This was something that wouldn't have really been possible without the cloud, enabling people to watch the games and interact with others, as well as get a lot of additional info and experience both during the games and outside them. They had a shop selling clothing and other gear, with virtual models showing things from different angles. For the games, they included 3D-driven animations with data overlays on every shot. This included analytics like speed, spin, and serve placement in real time. They used machine learning to extract match highlights, with a variety of features that were both objective (physical aspects like speed of shots) and slightly more subjective (crowd reaction and player emotion). They also offered virtual experiences like tours behind the scenes, concerts, tennis clinics, and other live events. They were so happy with the results that they continued offering much of this in future Australian Opens even post-COVID, with extensive statistics and visualizations on strokes, player rally analysis, and matchbeats.

Example 2: Music for the Masses

The music platform Spotify made the jump to the cloud several years ago, but the benefits were immediate and lasting.[2] The company only launched in 2008 and cloud services were not available yet, so they were doing everything in-house, running their own on-prem data centers.

---

[1] "Reimagining The Digital Experience for Australian Open," available at https://www.infosys.com/products-and-platforms/meridian/use-cases/reimagining-digital-experience.html

[2] "Views From The Cloud: A History of Spotify's Journey to the Cloud, Part 1," available at https://engineering.atspotify.com/2019/12/views-from-the-cloud-a-history-of-spotifys-journey-to-the-cloud-part-1-2/, and "Spotify: The future of audio. Putting data to work, one listener at a time." at https://cloud.google.com/customers/spotify

They surpassed 1 million customers in 2011 and continued rapidly growing. They'd reached capacity with four data centers in 2015 and were looking at building more to support both their growing customer base and their engineers. So they started considering cloud solutions.

Even at that time, cloud platforms were available from Google, Microsoft, and Amazon but were relatively rudimentary. After some considerations, they chose to go with Google Cloud Platform (GCP). They actually had a smooth migration process that was managed by only six people but involved all the engineering teams working together to do what's called a *lift and shift* for the majority of their infrastructure, which meant they moved the data and processes basically as is rather than rewriting anything (although there were some rewrites necessary). This involved moving 1,200 online processes from 100 different teams, with over 20,000 executions of these many processes every day, so it was not a simple job.

The effects were profound, and they were able to close their data centers—the first one in 2017 and the other three in 2018. Two of the biggest benefits are how easy it is to scale as their data storage needs to grow and how GCP fully ensures the privacy of their customers' data. Their customers play music billions of times a day. Spotify engineers can focus on their core business—supporting those customers with automations and other time-savers available on GCP. They also can easily use many of the other services GCP offers, including the hugely valuable data analytics and machine learning services.

On-Prem, Virtual Machines, and the Cloud

I talked above about how people logged onto servers to do their big(ger) data work, but technically, they weren't usually logging into the servers directly. Instead, multiple *virtual machines* (VMs) were created on servers, which were just like they sound—virtual "computers," something you could log into that would look and act like a computer to you but actually lived entirely on a particular server with some of the server's memory and disk space assigned to it. It could be set up with any operating system. A user is given access to that VM, and they can log in, and it looks to them like they were logged into a regular computer. The user can install software, open programs, run it as normal, and restart it as necessary (just not with a physical power button). Practically, there are usually limitations on what can be done in a VM (usually for security reasons), and they often run slowly and have a lot of lag, but conceptually they operate like normal computers.

Because virtual machines are created by software, they can be configured to be whatever size you needed (depending on IT's willingness to give you one big enough). Although not perfect, these options did allow for some *scalability*, the ability of code to handle larger and often increasing amounts of data. But it also meant that *code optimization* (work to make programs run more efficiently and therefore take less time and resources) was quite important for most people working in data. Two other types of computing, parallel (where you break code into pieces that can run at the same time) and distributed (where you send pieces of code out to different servers to take advantage of more memory and disk space), also came into play, often working together.

This was pretty much the only way, so nobody called it on-prem 15 years ago. With the advent of the Internet, options expanded. Today, while on-prem is still around, cloud computing is getting more common every day. One important thing to note is that it is also possible to go hybrid—some companies have a data center and also use cloud services. Sometimes this is done because of regulatory issues like the ones that banks deal with. For instance, the most sensitive data or any data that relates to regulatory compliance might be kept in the company's own data centers, while everything else is in the cloud. However, this is not necessarily required as the cloud providers do offer a regulatory-related service that can help manage things.

"Big Data" and the Problem of Scale

I've mentioned "big data" before as a term used in a variety of ways because it's not a term of precision yet, but basically means a lot of data that takes a lot of processing power to work with, usually beyond what an individual computer could handle. But it is starting to get formalized more, and now there are now several Vs of big data, although these change depending on which "expert" site you're looking at: volume, velocity, variety (variability), and veracity.

Big data volume refers to the massive amount of data generated by daily life and that it's often gathered from many different sources into a more centralized place because storage is so much cheaper now. This also ties into *big data variety* because of the many sources. *Big data velocity* relates to how fast the data is generated through real-time collection of things like shopping transactions, finance trading, and social media activity. *Big data veracity* relates to accuracy and reliability because of how important those traits are. There is one more V that isn't usually considered part of the pillars, but is presumed—value. None of the others matter if the data itself doesn't have value.

The existence of big data brings up the issue of scale. This has actually been a problem for a long time—often code or dashboards are written using a subset of data (this is actually standard because it's faster) and, when connected to the real data, perform poorly. This is an example of a program that didn't scale well. Making the code more efficient through code optimization has been a part of software engineering for decades now. In the old days, it involved having intimate knowledge of the particular hardware (processor and disk) the program would be running on and managing and improving memory management directly in the programming languages. Most modern programming languages handle memory management automatically, so this isn't something programmers usually have to deal with as much today.

When you are trying to work with a lot of data, the problem will rear its ugly head, whether it's considered "big data" or not. Some of the true big data tools make this easy, but it's still important to know a little something about improving code efficiency. This generally involves understanding how ML algorithms work, along with other techniques like cross-fold validation, grid searching, and other common ML techniques. Also, knowing some of the more classical computer science data structures and algorithms can be helpful.

Distributed and Parallel Computing

Traditionally, when you run a program, it runs one step after the other in your computer's memory and disk drive. One step at a time is *sequential computing*, and running on your machine is usually referred to as running locally. Nowadays, it's common to do this same basic approach in the cloud, where you're still running a sequential program on a single machine in the cloud, rather than on your personal computer.

There is an alternative to the sequential approach called *parallel computing*, where a program is broken down into different steps and run at the same time—in parallel. In processors that have multiple cores, whether on your computer or on a cloud compute system, each core can handle a single process. A four-core processor could therefore run four parallel processes at a time. The primary advantage of this is obviously that it will take less time because we don't have to wait for each step to finish before starting the next.

Conceptually, parallel computing is great, but it requires code that can be split into steps that can run separately. So you can't just take some existing code you've written sequentially and drop it in a system that offers parallel computation. The code has to tell the computer how to split things up. There are tools that can do this somewhat automatically, but it's usually not that simple.

If you had a machine learning pipeline that cleaned and prepped video game data, trained a model, ran forecasts on thousands of games, and uploaded those to a table, you couldn't parallelize the entire pipeline. Depending on the specifics, several of the steps probably could be done that way. Cleaning and prepping the data could partially be done in parallel, with cleaning and feature generation that relies on external info or only on row values. For instance, we could break a large dataset into some number of smaller ones and run those steps. However, if we have any aggregations that need to know values for an entire column, we can't do that on a split-up dataset. Training partially in parallel might or might not be possible (depending on the algorithm), but it will all have to come together in the end for final steps. Forecasting likely can be done in parallel, where each game could be run at the same time as the others (if there were enough processors).

Parallel computing often goes hand in hand with *distributed computing*, which is when you distribute different code steps to different computers. In the context of parallel computing, this means running a program by splitting it into parts that run on different computers and then sending the results back to the original computer, where the original program pieces the results back together and completes. But anything that runs on different computers that communicates with each other counts, like the World Wide Web, distributed databases, massively multiplayer online games, and even cellular networks.

People sometimes do distinguish between parallel and distributed computing by saying that in parallel computing, all the processors that code has been sent to have the same shared memory (like a program that has taken advantage of your personal computer's four-core processor but is still using your system's single memory) and in distributed, each processor has its own distinct memory associated with it (like a virus that has been sent to run on every individual computer in the network and send back information).

These distinctions aren't hugely important in most data scientists' work lives, but understanding the basics of parallel and distributed can come in very handy when you need to use it because you're working with large data.

Virtual Machines and Containers

We talked above about what virtual machines are, but one of the key ways they are used is based on the fact that each is a known entity and considered isolated—the version of the operating system, programming languages, libraries, and so on aren't updated without serious consideration because they impact everything on that machine. One important point is that because we want to keep applications isolated in their own VMs, it's expensive to run an entire VM for each application because of all the processors and memory it had allocated to it.

One popular answer to the problem of running multiple applications on one VM is containers. *Containers* are conceptually similar to VMs in that they are set up with a specific language, libraries, etc., except they can share an underlying operating system with other containers, reducing the cost significantly because they're using the host system's resources more efficiently. This also makes them easily scalable. Normally, you wouldn't put more than one application in a container. Note that you can run a container on any computer, including a VM. The point is that each container is isolated.

Nowadays, VMs and containers are often in the cloud, which we'll talk about below.

DevOps and MLOps

"DevOps" is another buzzword you'll hear a lot in tech. *DevOps* isn't rigidly defined, but it is basically both a philosophy and a methodology that came from an Agile mentality and encourages collaboration with everyone involved with deploying software, from developers to testers, to those managing the code repository, and to those doing the actual deployment. It emphasizes automation, lean principles, measurement, collaboration, and a shared culture. This isn't always something data scientists will have to worry about, but it helps to understand what's going on, and sometimes it is something you'll have to deal with.

Automation plays a huge role in DevOps. Software testing is usually automated, along with builds and releases (when the software will be finalized in the state it will be deployed and then actually deployed). Other aspects related to development are also automated, such as creating environments for development and anything else that is prone to human error and can be automated. It usually ties in with another acronym, CI/CD (continuous integration/continuous deployment), both of which require the automation just mentioned. CI refers to regular code integration and testing and CD to code being deployed reliably and through the correct environments.

All of this amounts to faster development and production releases—and quicker user feedback. A lot of major companies are serious about this velocity, and Amazon supposedly has every new software engineer make a real change to the codebase and have it deployed it on their first day. But it isn't only faster, it's usually better—more reliable and more secure. Communication and collaboration between the dev side and the ops (operations) side is critical to this methodology.

MLOps is analogous to DevOps in the ML space. Just like DevOps, *MLOps* is about both the development of ML solutions and deployment of them and everything in between. This refers to deploying code, feature stores (more below), and models into production (not necessarily all every time). In organizations that aren't very mature, data scientists may never need full MLOps because they're not deploying into any production systems, but almost follow parts of it like version control and some automation. A lot of the pre-modeling work can also benefit under MLOps practices even if it's not in production.

Like DevOps, MLOps is powered by automation. Let's first talk about the general flow of an ML tool, usually called the *pipeline*. We start with data prep and may also create a *feature store* (a collection of defined and fine-tuned features that can be derived from common data being used), as well as the training and testing datasets we'll need further along. The next step is the training and tuning, basically preparing the model. We perform our experiments to select the algorithm and parameters and then train the final model. The final step is deployment (with monitoring an implicit fourth step in order to detect model drift and other situations requiring retraining). In deployment, the model and feature store are put in place, along with any code that must process the incoming data. That's the end to end from development to deployment, and now the system is ready for end users to use it.

Cloud Services and Computing

At its most basic, cloud computing is just computing over the Internet. Any service you use that involves using a website rather than downloading software that works whether you're online or not is cloud computing. So is software that requires you to be online to use it, even if the program lives on your computer. Gmail and most email services are cloud, even when you're using Outlook installed on your computer. Your email is actually an application running on the cloud because it lives on a server somewhere, and the website (or Outlook) just lets you see it. Slack, Steam, and Zoom are additional examples of cloud-based applications where you install software but must be online to fully use it. Web applications like online shops, online banking, and social media all run in a browser only. Products that manage licenses online but don't require you to be online all the time you are using them, like Adobe products and Microsoft Office 360, use cloud services despite functioning offline.

However, convenient Internet-based applications aren't what we care about when we talk about cloud computing in data science. Instead, we are referring to the delivery and use of computing resources like disk storage, processing power (often just called compute), and software over the Internet. This allows users to access all of these things without using a company's on-prem data center or even using the actual resources on a user's computer (beyond the web browser and what's required to run that). This is valuable because it means we gain the benefits of scalability (using large amounts of resources), elasticity (automatic adjustment of the resources being used based on current need), resource pooling (a variety of computing resources being available to multiple users), and pay-as-you-go pricing (paying only for what you use).

Cloud computing products make things easy for users. Often you don't have as much control over your data and the software and pipelines you write, which may be stored in a proprietary way, and this makes the risk of lock-in a problem. *Lock-in* is when you can't easily switch from one program or platform to another. For instance, if you have a data pipeline created in Microsoft's cloud platform, Azure, there may be no clear way to export it to run it somewhere else. Lock-in can also happen when you can't move your data out at all. It's a concern when you're using any platform, like one of the major cloud platforms we're going to talk about next, and this should be considered when choosing tools to develop software and pipelines in.

Major Cloud Platforms

We saw above that cloud platforms basically replace data centers, and data centers basically provide compute and memory. But cloud products and services are simply any software that runs over the Internet, and there are many, some of which offer only one or a handful of specific services, but there are also three large, full platforms (mentioned in the Spotify example): Microsoft Azure, Amazon Web Services (AWS), and Google Cloud Platform (GCP). Nowadays, all three provide a huge number and variety of products and services ranging from basic services like data storage and scheduled data loads to advanced AI services like chatbots and recommendation engines. Because there are so many services available, it's easy to find one you need and start using it.

It is important to note that the services I'm talking about in this section are intended more for businesses. Cloud platforms can be very costly, which is often a big barrier for companies looking to start using them (or allowing employees enough access to be able to do real work).

Cloud platforms can also be costly for individuals, and it can be difficult to figure out how to avoid racking up accidental costs. All the platforms I talk about below do have free options, so it's worth exploring if you really want to get familiar with them.

Microsoft Azure

Azure got started in 2008 and developed over the years to include the over 200 products and services it now has. They have a good list online of everything currently on offer,[3] broken down by type. These include storage, databases, media, analytics, AI + machine learning, and many more. There is overlap in what some of these tools can do, usually because they are optimized for slightly different uses.

---

[3] https://azure.microsoft.com/en-us/products

I'll talk about some of the most commonly used ones here. There are many data storage options along with migration tools to move companies' data into Azure easily. One of the most popular products for object storage is Azure Blob Storage, which is highly scalable and secure and often used in ML applications. Azure Data Lake Storage supports high-performance analytics. Azure Files is good for storage of files and is secure and scalable.

Some of the most popular services are SQL Server on Azure (which can also live in a virtual machine), Cosmos DB, and Azure's own SQL Database. SQL Server is a Microsoft database product that's been around since long before the cloud. Azure also supports MySQL and PostgreSQL, two other popular SQL databases. Cosmos DB is intended for data used in high-performance applications and offers good scalability. Azure also has a NoSQL key–value database simply called Table Storage. Azure Backup is a popular service that can be used to back up all of your data in Azure.

There are also many tools for moving and processing data (extract, transform, and load, or ETL), but the most popular is Azure Data Factory (ADF). It's a workflow-style tool, which means you drag different nodes onto the workspace and connect them in ways that data can flow through them left to right. Each node does a specific task, like pulling data from a database, performing an operation on it (generally transforming), and writing to a database. It's largely code-free, which makes it accessible to more people. It offers a number of transformations and tasks, including many analytics ones. The tool can pull in data from over 90 different sources, including data with other cloud providers.

ADF is a good low-code ETL tool. There are other popular tools, especially a much more advanced one called Databricks that makes it easy to do large-scale data processing and analytics with Spark (a distributed processing system) and also allows easy management of machine learning models. Databricks can be used to do many other things, including ETL. Power BI was mentioned earlier as a popular visualization tool, and it's available within Azure, making it easy to share dashboards with customers. Microsoft Fabric is a relatively new all-inclusive platform with AI built in for data scientists and other data professionals.

Another type of product Azure offers is under the Compute label. These are basically products that allow you to use computers—processors and memory—to run code. Azure Virtual Machines (VM) allows you to create and configure a "computer" that might be used for a particular product—for instance, a particular operating system is installed, a specific version of the programming language distribution is installed, all the required libraries are installed, the codebase lives and runs there, and any configuration files are set up. You can also set up your databases inside a VM. Virtual machines can be Windows or Linux.

As mentioned above, VMs can be expensive, so containers and other serverless tools are present in the application infrastructure world. Azure Kubernetes Service allows you to easily set up containers that hold code and more to run applications. Another popular service is Azure Functions, which lets you run code outside any sort of dedicated server (computer) that you've set up, which can be done on other existing Azure objects. One huge benefit with Azure Functions is you can bind them to these objects and assign triggers, where it runs if a certain defined event on another Azure object happens.

One more product that's popular in organizations is Azure DevOps, which is primarily for improving collaboration, planning, and managing the development process. It supports software development with code repositories, testing tools, and support for continuous integration and continuous delivery (CI/CD; a way of managing deployment of code to production). It also has Kanban boards and other Agile tools (software development concepts we'll talk about in the next chapter).

That's an introduction to some of the Azure products and services—basically the tip of the iceberg only. Also, as mentioned above, there's a ton of overlap and interconnection between these products—you might be reading data from Azure SQL Server in your Databricks code, for instance, or setting up any number of services inside a VM.

Amazon Web Services

Amazon launched their cloud storage solution Amazon Simple Storage Service (S3) in 2006 and its compute offering Amazon Elastic Compute Cloud (EC2) soon afterward. These were the first two products in the AWS platform, but there are many more now. They also have an online list[4] that includes a searchable area for all products.

Like Azure they have many services, over 200. Each product category has its own Decision Guide to help walk you through picking the right products for your situation. For example, the one for Storage[5] talks about different types of storage (block, file, and object) and the different products for each of those. Amazon S3 is an object store that can be integrated with some on-prem systems with AWS Storage Gateway, which makes storing files and migrating data easy. AWS Backup is another useful product that can be used to back up different types of files in the system.

AWS also has several databases to choose from for regular database loads, including two relational services, RDS (which supports engines like PostgreSQL, MySQL, and SQL Server) and Aurora (which offers high performance and scalability for PostgreSQL and MySQL workloads), plus several NoSQL ones. It also offers Redshift, which is for very large amounts of data being processed rapidly.

There are several analytics tools that let people work with their data to extract info from it relatively quickly, even if they're nontechnical. Amazon Athena allows users to do real-time data exploration, analytics, and visualization quickly. AWS Glue DataBrew does quick data prep, and Amazon QuickSight lets you do quick visualizations. There are also more advanced tools for data scientists, some of which require coding like

---

[4] https://docs.aws.amazon.com/

[5] https://docs.aws.amazon.com/decision-guides/latest/storage-on-aws-how-to-choose/choosing-aws-storage-service.html?icmpid=docs_homepage_storage

the comprehensive machine learning service Amazon SageMaker that supports everything from data prep to model training and deployment. It also integrates the popular frameworks PyTorch and TensorFlow for neural nets work. Other tools don't require much coding, like Amazon Comprehend for NLP, Amazon Lex for building chatbots, and Amazon Rekognition for video-based work.

In the world of compute, Amazon EC2 is the classic choice, allowing you to run virtual servers. Amazon Linux allows you to spin up Linux-based environments. Amazon Lambda offers serverless computing triggered by events. AWS Batch allows you to run large batch jobs without having a lot of infrastructure to manage.

AWS also has a lot of developer tools, including AWS App Studio for internal applications, AWS CodeBuild for writing and testing code, and several tools related to specific languages.

Google Cloud

Google launched a preview of Google App Engine in 2008, allowing people to create web apps that could easily scale. It came out of preview in 2011 and became one of the major cloud platform providers, now offering over 150 products and services, which you can find on their list of products.[6]

BigQuery is a data warehouse that can be used for many things, including analytics. Cloud Storage supports object storage for unstructured data like images and videos, and Filestore provides fully managed file storage for network file systems used in enterprise applications. They have Cloud SQL that supports three flavors of relational databases as well as some other options for SQL Server and PostgreSQL. There are also NoSQL options, Datastore and Cloud Spanner, and Cloud Bigtable. Dataform allows you to run SQL workflows in BigQuery.

---

[6] https://cloud.google.com/products/

There are a handful of products that can be used for ETL in Google Cloud, including Cloud Data Fusion and Datastream. Cloud Data Fusion is a fully managed tool that lets users build, manage, and schedule ETL through a visual interface. Datastream is a change data capture and replication service that enables moving data between databases and storage services.

Another important tool is Dataplex, which helps to manage data even if it's spread out in silos. Spanner allows you to bring several types of databases together, including relational and NoSQL.

In the analytics space, BigLake unifies data from data lakes and warehouses for analytics, and Dataflow is a serverless stream and batch processing tool with ETL and real-time analytics in mind. Document AI allows for some NLP work, and Recommendations AI is a recommendation engine. Vertex AI is a general-purpose analytics and AI platform with several specific offerings, including an agent builder.

In the compute space, there are several useful tools. Dataproc controls Spark and Hadoop clusters allowing for work with big data. Compute Engine allows the running of customizable VMs. There are also some other tools for compute for specific needs, including Cloud GPUs (for heavy-duty processing and ML training) and Cloud TPUs (for accelerated neural net computations). Cloud Workstations is another option for the development environment. Cloud Run allows for serverless deployment of containerized applications, and App Engine provides a serverless platform for hosting web apps. They also offer Kubernetes-based serverless capabilities through Knative.

Key Takeaways and Next Up

In this chapter, we talked about the cloud and scalability and everything that ties into that world. This included understanding the differences between the on-prem data centers of yesterday and modern cloud computing and the scalability, elasticity, resource pooling, and paying only for what you need and use. We talked about big data and the importance of scalability. We also addressed distributed and parallel computing, both important in improving computation time of large-scale jobs. Lastly, we looked at the three major cloud platforms and some of what they have to offer.

In the next chapter, we'll be taking a deep dive into data science project management first by understanding traditional project management and modern software development management. We'll give an overview of a Kanban approach to managing data science projects. We'll also talk about post-project steps.

PRACTITIONER PROFILE: BHUMIKA SHAH

Name: Bhumika Brijen Shah

Current Title: Container technology engineering manager

Current Industry: Retail

Years of Experience: 17 years in tech, including 3 as a data analyst, 7 years in cloud DevOps, and 4 in platform solutions

Education:

- Certifications in Kubernetes, Azure, and Google Cloud Platform

- Postgraduate in Computer Science

- Bachelor's in Computer Science

The opinions expressed here are Bhumika's and not any of her employers', past or present.

Background

As a teenager, Bhumika loved computers and formally studied computer science in school. The next logical step was to get an undergrad degree in computer science, which she did. She did an internship while still in university, where she learned about software engineering, doing testing and other important steps. She started to develop a real software developer mindset there. She got her master's in computer science, too, and kept taking courses to continue to grow.

Work

Her first job was as an ETL developer (extract, transform, load—basically moving data from one place to another), which is where she first really encountered "data" as a thing in business. She loved how valuable the transformation step could be, taking something hard to use and making it useful. She moved into a data analyst role and continued to grow her data skills. She enjoyed the visualization and how it enabled her to show things that were difficult to see in the data. She continued down this path toward data science, learning about data obfuscation, and then she became interested in data security, which led to another small shift in her career. She moved into a cloud DevOps role, where she was developing infrastructure and solutions doing automations. She loved scripting all of that. She grew in that area before moving into container platform engineering, where she now manages a team.

Sound Bites

Favorite Parts of the Job: Bhumika loves the problem-solving aspect of all the jobs she's done, as well as constantly learning. She also loves collaborating with different teams and making solutions work. She especially enjoys creating solutions that are scalable.

Least Favorite Parts of the Job: Trying to balance competing priorities can be frustrating. As a manager, she has to make sure all of her team's work aligns with company goals and vision but still make stakeholders happy. This can be hard at times. One thing that can be frustrating is when other people make decisions about things you need them to do for your work.

Favorite Project: One project she liked was to create a secure and scalable Kubernetes environment that would improve on deployment time. She was able to reduce deployment time by 40%, which helped developers avoid downtime. This was especially useful to data scientists because they are starting to use Kubernetes. This gives them data tools, including ones to obfuscate their data. Some of the aspects of the project that were interesting were the autoscaling, optimization, reducing operational costs, and being able to migrate monolithic tools to microservices.

Skills Used Most: Problem-solving, critical thinking, communication, leadership, and her technical expertise, in that order

Primary Tools Used Currently: Kubernetes, Terraform, Jenkins CI/CD pipelines, cloud services (GCP and Azure), Golang, and SQL

Future of Cloud: All industries are moving toward cloud computing and automation. This makes it easier for everyone to deploy applications and means AI tools are being used both as part of the cloud and automation tools and as part of the applications being deployed.

What Makes a Good Technology Professional: Curiosity and creativity with strong technical skills. A solid understanding of statistics. Communicating at the right time (not too soon, not too late). They also need to know how to benefit from AI tools—know what you're trying to get out of it rather than just using it willy-nilly.

Her Tip for Prospective Technology Professionals: Go to open houses to meet people and learn about companies and the kinds of roles out there. Then focus on building strong technical foundations in whatever area you intend to focus on.

Bhumika is a container platform engineering manager with a varied background in data and other technologies.

CHAPTER 21

Putting It All Together: Data Science Solution Management

Introduction

One of the key questions I asked in every interview I got during my last job search was "What development methodology do you use?" I asked this because at my previous company, they were forcing us to use Scrum on the same board as the software developers, data developers, and BI developers, and it did not work.

But the reality is that there isn't a known, established way to manage data science projects yet. People just make do with different approaches, including Scrum. I'm going to go over the various aspects that are important in managing data science projects, and then I'll share a management approach that can work for some data science teams.

© Kelly P. Vincent 2025
K. P. Vincent, *A Friendly Guide to Data Science*, Friendly Guides to Technology,
https://doi.org/10.1007/979-8-8688-1169-2_21

Project Management

Project management is an entire discipline that's been around for a very long time, with many fairly standard approaches in software development and engineering. All of them aim to define what needs to get done, give a timeline to achieve it, and then monitor progress as the project proceeds, all to make sure everything gets done. Traditionally, there was only one way to manage a project—through a methodology we now call waterfall. With the waterfall paradigm, everything is carefully planned in advance. This makes sense when change is expensive or time-consuming, so large-scale construction and engineering projects definitely need it.

There's a story from the annals of "mistakes were made" from my college days that still makes me laugh. A rumor held that the relatively new building that housed the engineering college I attended had an eternally empty fountain on the top. The tale was that before they filled it, they checked the plans and apparently engineers had forgotten to calculate the weight of the water when designing the floor, and the filled fountain would have exceeded the weight limit. This isn't like most software bugs— you can't just throw down a few more I-beams to patch it up. This is why planning and management are important.

General Project Management Methodologies

Project Management Body of Knowledge (PMBOK) is a collection of knowledge, terminology, and guidelines that many project managers (PMs) follow. The canonical source for this is a book put out by the Project Management Institute, with the seventh edition the most recent at the time of writing, published in 2021. The institute offers two certifications that provide a common route for people getting into project management. PMBOK is fairly traditional and rigid, but more flexible practices were added in the second most recent version, in 2017. Another project management methodology is Projects in Controlled Environments

(PRINCE2). This one focuses on organization and control, but it does also emphasize continuous improvement, making it more flexible than traditional PMBOK. Yet it is also still rigid in structure and ends up requiring a lot of documentation.

Both of these approaches break the project into phases. They both define five of them, and these are PMBOK's:

1. **Project Initiation**: Define the business case and project goals and identify stakeholders.

2. **Project Planning**: Define the scope, identify tasks and subtasks, create schedule, identify resources (people and tools to do the work), create a budget, identify third-party vendor or tool needs, identify risks and mitigation plans, define quality and success measures, and determine the method of communication with stakeholders.

3. **Project Execution**: Do the work.

4. **Project Monitoring** (runs concurrently with execution): Ensure the objectives are being met and work is proceeding as necessary, based on established factors and measures of success.

5. **Project Closure**: Write up lessons learned and report on the overall project.

These approaches both came out of the waterfall project management methodology, which is front-heavy, with extensive planning before work can even get started. In PMBOK, it's said that about 50% of any project's time is spent in the second phase, project planning. Much of the rest is in project execution. We'll talk more about those below.

In traditional project management, the *project manager* (PM) is the person who oversees the entire process and is critical to the project's success, including getting external resources for the project to succeed, whether people, tools, or data. The PM is the person who represents the project to the rest of the company, and the project planning step is where they present their case in order to get funding and resources.

Project Planning

In traditional project planning, the "what" and "how" are paramount with "what" giving the whole purpose of the project and "how" the way it will be done. Here are the basic ten steps, to be discussed below:

1. Define the project objective and scope.

2. Describe the project.

3. Create a detailed work plan.

4. Define milestones and deliverables.

5. Define personnel needed.

6. Define schedule.

7. Define budget.

8. Define change control processes.

9. Identify risks.

10. Define policies and guidelines.

The first real step in the process is to define a clear project objective, along with defining the boundaries of that object (the scope). This includes getting the requirements of the stakeholders. Next, the project should be described. This involves asking several questions, including (among others) What will be done and when? What will be delivered?

Who is responsible for accepting the project as complete? What is the responsibility of the developer and the user? How will progress be measured? At this point, we have the requirements of the customer defined as well as the scope of the project. Next, an actual work plan must be created with details, including milestones and deliverables.

The next steps in the process involve defining different aspects of the project, including personnel needs, a schedule, a budget, change control processes, risks, and policies and guidelines. This is generally done by the project manager reaching out to the company for information, and a lot of companies have tools that will handle some of this automatically. Once personnel are assigned, it plus the detailed work plan can help define the schedule. From there, the budget can be worked out. Policies and guidelines are necessary to lay out the roles and responsibilities of people on the project. Finally, change control processes are defined. This one involves any kinds of changes, including personnel, role, and modifications to the project's product.

In the traditional approach, the design of any system or object being created would be done during the planning phase because it's needed to create the schedule and budget and identify resources, among other things.

Project Execution

Project execution starts when everything is planned and resources are gathered. This is the actual work, whether it's design, coding, or something else. People will be working on reaching milestones and preparing deliverables. At this point it's the PM's job to keep everything on track, and if additional support is needed (for instance, more people or tools), they will advocate for that. The PM will also keep stakeholders informed. Changes may be requested here, which the PM will take back to the team.

Project execution stops when everything has reached the defined point of completion (generally this is at the decision of one person or team defined during planning).

Project Management in Software

Although both approaches to general project management are relatively rigid, they have been adapted to Agile approaches (new and far more adaptable than waterfall, discussed below), so they're fairly common in software development. Although I feel quite strongly that project management methodologies that work for software development do not work unmodified for data science, it's really common for data science teams to be forced to follow them. The field hasn't settled on good methodologies for managing data science projects, which is why they fall back on others. Still, the Agile software development methodologies do provide a good reference point for data science project management.

For decades, waterfall was the way software projects were also managed. But a completely different methodological paradigm called Agile has mostly supplanted it. Agile's key aspects are iterative development driven by customer needs with continuous feedback from customers. A focus on having working software as early as possible and improving it by adding features is a fundamental part of the approach. The minimum viable product, or MVP, is a functional piece of software that has some of the requested functionality, with more functionality added as the product is developed. These aspects make the process incredibly flexible. It's the polar opposite of waterfall, which is incredibly front-heavy. You can start development in an Agile project even before you have all the requirements of a project, as long as you have the foundation and beginning concepts nailed down. The two most common specific Agile methodologies are Scrum and Kanban.

Project Planning

Even though you can get going on an Agile project faster, there is still project pre-work to be done. It's still critical to determine the project objective and scope, which involves getting key requirements. The project needs to be described, including the final gathering of requirements, and a work plan needs to be created. In the case of Agile, the work plan will be lists of the specific piece of work to do, kept in what's called the "backlog" in the most common type of Agile management, Scrum (which we'll talk about below). This backlog is not necessarily a complete record of everything that has to be done because of the intentional agile nature of the project. More needs may crop up during execution, which can be added to the backlog. Important milestones and deliverables will be identified, but those can also change during execution. The remainder are documents that wouldn't be as specific as in a traditionally managed project, but they should exist in some form. Risks and personnel are identified and a schedule and budget created. Finally, a change control process is defined, as are policies and guidelines.

Project Execution

In a software project, further system design might happen during the execution stage, as would all coding, basic testing, performance testing, quality assurance, and implementation.

For project execution, Scrum is characterized by short development cycles called sprints (two to four weeks), the backlog, several specific meetings at various stages in the sprint, and a visual board, either physical or virtual, where all active stories are tracked. Developers pick stories from the backlog at the beginning of each sprint, and those are what they work on during that sprint. The various meetings are called "ceremonies" and include daily standups (very short check-in meetings on progress), a sprint retrospective (looking at how the last sprint went), backlog refinement (adding stories to the backlog and improving the information on them as more things are known), sprint review (where development progress is demoed), and sprint planning. The ethos of Scrum is empowering developers and self-organization.

Kanban is a bit looser, but also has a visual board for tracking work on stickies (virtual or physical) that are moved from left to right through the development stages. It's also iterative and is intended to encourage focused work on a smaller number of tasks (usually with limits to how many can be worked at once). It is also flexible, empowers developers, encourages collaboration, and makes it easier to see obstacles quicker.

Both Kanban and Scrum boards are updated regularly, ideally daily. Many teams have daily meetings where they update the board together by moving stickies on a physical or virtual board or simply go over the change they made since the previous meeting.

Figure 21-1 shows a typical Kanban board. The status sections can be almost anything, depending on the type of work it's used for. This one is for software development. In this one, a sticky/task moves from the Stories section (backlog) to To Do when it is going to be worked on soon, then to In Progress once it's started, then into Testing when that work has started, and finally to Done when complete. A Scrum board looks very similar. The only required statuses in Scrum are To do, In Progress, and Done, but other common ones are On Hold, Testing, and Blocked.

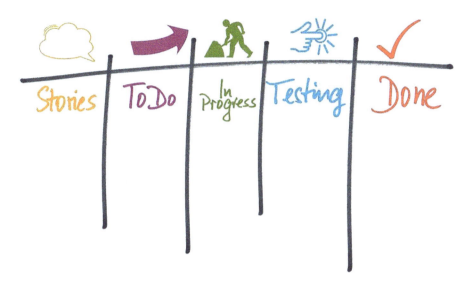

Figure 21-1. *An example Kanban board. Source: "Work, Work Process, To Organize royalty-free stock illustration" from* `https://pixabay. com/illustrations/work-work-process-to-organize-4057334/`

To understand how things would work on the software development Kanban shown above, we'll look at a restaurant point-of-sale system being developed from scratch. A point-of-sale system is the system that restaurants use to take orders and payments. The first thing that would happen is that the requirements would be gathered, which involves developers and/or project managers talking to the stakeholders the system is being built for. This would include identifying needs like the ability to select food items from a menu, which would require prices to be entered in advance, and the ability to create an order containing multiple items, subtotal it, add appropriate tax, allow custom tips, get the overall total, accept cash and/or run credit cards for the total amount, process returns, and print and/or email receipts.

Once the developers have the stakeholder requirements, they dig deeper and come up with a solution detailing the technical needs. From there, they can create a backlog of the specific things that need to be done for the system. These are usually called stories, and they should be as small and finite as possible, with the goal of having as few dependencies on other people or stories as possible. Here are some examples that would likely be in the backlog:

- **Database**: Create a database to store orders and more.

- **Database**: Create a table to store items.

- **Database**: Create a table to store item prices.

- ...

- **Front End**: Create a display of items.

- **Front End**: Make displayed items clickable to pop up more info.

- **Front End**: Add a clickable button and quantity to each displayed item to add it to the order in the database.

- **Front End**: Add a clickable button to check out an order that opens an order summary screen.

- ...

Note that the backlog usually isn't included on the board because there are usually a lot more backlog stories than space on the board. The above list is obviously tiny compared with what needs to be done, and it's also pretty clear that some stories depend on others being completed before they can be started. This is unavoidable on most projects. There are always many stories that don't depend on each other and can be worked at the same time by different developers, which is why it works so well in software development.

Once there's a backlog of stories, developers choose stories to start work on soon, and those go in the To Do in the chart above. This section gives a good sense of what's coming. Once a developer starts working on it, it gets moved to In Progress. When ready, it gets moved to Testing and, when that's complete along with any necessary updates, into Done. It's common for a story to get moved from Testing back to In Progress when bugs are found, and this can repeat until all testing is good. It can then be moved to Done.

It's worth mentioning that the goal is always for the backlog to be "complete"—have everything that needs to be done listed—but this is difficult even in software development, and the process is flexible enough to allow stories to be added to the backlog at any time.

Because both Scrum and Kanban are very good for software development, people have naturally tried to apply these approaches to data science projects, and it hasn't always gone very well. In the next section, we'll talk more about ways to work on data science projects.

Data Science Solution Development Methodologies

Data science seems like software development in a lot of ways, but most of the work is quite different. Sometimes data scientists are creating software to deploy, so in that situation, it is similar to software development. But the extensive data work, exploratory work, feature engineering, development and testing of different models, selecting and preparing visualizations, and preparing of results just don't fit into software development approaches.

You may be forced to follow Scrum or a software development Kanban in a job, even though it's not ideal, so it's helpful to understand how they work. But what if your team doesn't have a methodology picked? In case you're lucky enough to be able to help your team determine the best way to run data science projects, read on.

The hard truth is that there isn't any one methodology to rule them all for data science. That doesn't mean that there aren't good ways to do things.

Project Planning

Even though you can get going on an Agile-style project faster, there is still project pre-work to be done. Although project planning is well-established in software projects, it's not in data science projects. Still, many of the steps are the same or similar. If we look again at the CRISP-DM for a reminder, the first two steps are related to the planning phases of a project.

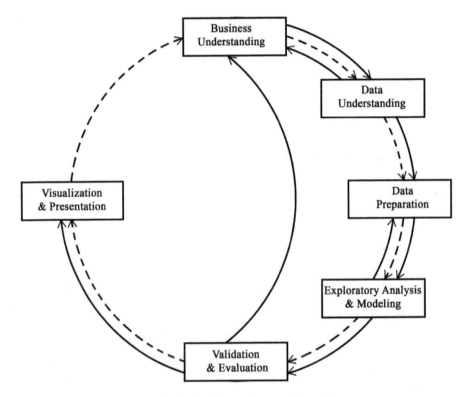

Figure 21-2. *The CRISP-DM data science lifecycle*

Business understanding is part of planning. Data understanding is also partially included in planning, but it can also be a part of execution. The planning steps are outlined here:

1. Define the project objective, scope, and high-level requirements (business understanding).

2. Determine candidate data sources, methods for obtaining access, and data owners where possible (data understanding).

3. Carry out pre-EDA to determine candidate fields in data and if they will provide the necessary info (data understanding) (go back to Step 2 as needed).

4. Create a backlog of tasks known at this point, primarily EDA.

5. Define milestones and deliverables.

6. Define personnel needed.

7. Define schedule.

8. Define budget.

9. Define change control processes.

10. Identify risks.

11. Define policies and guidelines.

The first step is to define the project objective, scope, and high-level requirements. If the work involves a dashboard, a mockup of this is also created. In a data science project, the data is critical, so this needs to be considered in planning. Likely data sources need to be identified and explored enough to see that they will be sufficient to fulfill the stakeholders' requirements. The rest of the tasks should also be done, but like with software, they aren't as strict or as complete as in traditional project planning. The backlog is created next, but because only a fraction of all tasks is known at this point, that is the majority of what will be added. Key milestones and deliverables can be defined. Personnel can be identified, a schedule created, a budget determined, and risks identified at this point, but all may need to be revised based on what happens during execution of the project. A process for change management and any necessary policies and guidelines can also be created here.

Project Execution

For project execution, neither Kanban nor Scrum as used for software development is good for data science, but Kanban is a lot more feasible, with adjustments. Fortunately, it's flexible. When using a Kanban board, the idea is there is one "thing" that can be moved from left to right through the different statuses until it can go into the Done section. In software development, that thing is an individual software component or task that can be worked on in parallel with other software components or tasks, but data science tasks are not like that. So much of what gets done in data science is dependent on findings in earlier work, and it can't really be planned in advance the way software can. A lot of the time, we have an idea about what should be done, but we find after working on it a bit that it's not a good path, so we don't take it any further. Or we find out that there's a lot more to be done than we realized, and it generates a bunch more work.

As an example, consider a task like preparing a data source to use in the model. It's not straightforward or like all other tasks. Imagine it's sales of all products at store B. The basic tasks related to that would be as follows (some relate to other things, and not all may need to be done):

1.   Find all relevant data sources.

2.   Get access to store B sales data (database access).

3.   EDA on store B sales data.

4.   Explore any anomalies identified.

5. Determine if store B sales data is sufficient for modeling or if anything is needed to supplement it.

6. Store B sales data cleanup.

7. More EDA on store B sales data.

8. Data preprocessing on store B sales data.

9. Combine store B sales data with other data.

10. Feature engineering on data.

At this point, the data source is usable in a model. Now consider the task of creating a model to make forecasts. This can't even start until there is feature-engineered data available, so it's way down the road from the start of the project. Data prep regularly takes 80–90% of project time. Additionally, once we're ready to start building the model, it might not really be a single task. Normally there is experimentation—we try several different algorithms, different ways of splitting the data, different parameters. It might make sense to make a single story for each algorithm. The work that would be done on that story would be preparing the data if any additional changes are necessary for the particular algorithm, setting up the code to run cross-validation and hyperparameter tuning on the specific algorithm, running it, and collecting performance metrics. At that point, that story might be done. A follow-up story might be to evaluate the results of all the stories encompassing the different algorithms tried.

On the other hand, a mature team may be able to treat "building the model" as a single task through automation. Code can be written that will try every specified algorithm, with specified cross-validation and hyperparameter values. This might actually take a relatively small amount of time to run—maybe even only a few minutes on a smaller dataset. Writing the code that runs these experiments will take a much longer time. That code also needs to be tested, just like in software development. In the previous way, with a story for each algorithm being tried, that code will also need to be written for each story.

It's clear that the modeling tasks are different from the data source preparation tasks. It's not easy to see what specific statuses they have in common. But they do have the basics: To Do, In Progress, and Done. If we consider that and look again at the CRISP-DM lifecycle again for guidance, there's a way forward. See Figure 21-2 for one more look at CRISP-DM.

The Stepwise Data Science Kanban Board

What we can do is make each of these stages in the lifecycle a *swimlane*, a horizontal grouping on the simple Kanban board. You know there's a lot of interaction in the process, going back to a previous stage to revisit something or process something new. The bulk of that occurs in the data understanding, data preparation, and exploratory analysis and modeling steps. Business understanding is usually completed at the beginning of the project, and visualization and presentation and validation and evaluation may not happen until closer to the end. Even that's not a guarantee, and this proposed board structure would accommodate anything. We'll stick with the basic statuses but add a Blocked column for more visibility when there are challenges. See Figure 21-3 for what the proposed board looks like.

	To Do	In Progress	Blocked	Done
Business Understanding				
Data Understanding				
Data Preparation				
Analysis & Modeling				
Validation & Evaluation				
Visualization & Presentation				

Figure 21-3. *The stepwise data science Kanban board*

Example Project on the Board

To illustrate how the stepwise data science Kanban board might work and how we'd run through a data science project, I'm going to go talk through it with a recent real project I worked on. In the project, we wanted to do some calculations to determine *incrementality*, the amount of truly new sales that came from a new product launch. Whenever we launch a new product, it's usually similar to another product or products in terms of potential customers. We know some customers will switch—they'll buy the new product but stop buying the old product. This situation is called *cannibalization* of the old product, when we lose sales to another of our own products. It's not beneficial if those are the only people buying the new product. What we want to see is that overall sales—from both the old and new added together—are higher than what we would have sold if the new product had never launched. So we forecast the sales of the old product after the new product launches based on its historical sales. Cannibalization and incrementality are quick calculations based on the forecast and real sales values. The data science part of this project is the forecasting of the original product.

I'm going to go through the work we did on the project over several months and show snapshots of our Kanban board once a month. The board would be changing much more often than that, but this way you can see how things progress over time. This is very detailed, and you might find that just glancing through the monthly boards gives you enough of a picture of the flow of a project. On the other hand, if you've never done data science work in a company, this can help you understand what working on a real project is like.

We started with a kick-off meeting with the stakeholders to gather the basic requirements, so that's in the Done column of Figure 21-4, at the end of week 1 of the project. Their need was pretty straightforward, but the data situation was far more complicated. I always start a project doc to outline our overall plan and a rough estimate on a timeline, which captures the stakeholders' needs, as well as serving as a working doc we can modify as we move along, saving data sources, etc. We were also exploring possible data sources based on what they knew as well as reaching out to other people (it always feels a bit like a wild goose chase with only occasional wins). Those were both In Progress. We had two things we knew about coming up and listed in the To Do column: getting signoff on the completed draft of the project doc and making a final list of the data sources we'd be using (we wouldn't be adding any more). Note that everything is still in the Business Understanding swimlane because we were just getting started.

738

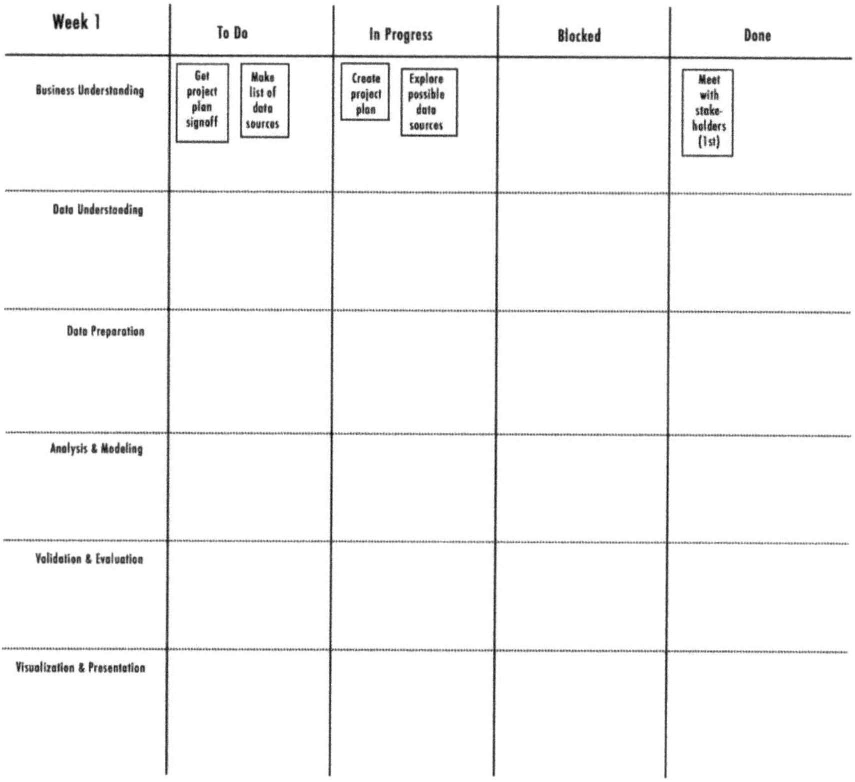

Week 1	To Do		In Progress		Blocked	Done
Business Understanding	Get project plan signoff	Make list of data sources	Create project plan	Explore possible data sources		Meet with stake-holders (1st)
Data Understanding						
Data Preparation						
Analysis & Modeling						
Validation & Evaluation						
Visualization & Presentation						

Figure 21-4. *The stepwise data science Kanban board at the end of week 1 on the incrementality project*

Four weeks later, we'd made good progress, which you can see in Figure 21-5. This time we have stories in both the Business Understanding and Data Understanding swimlanes. Most of the Business Understanding stories were already done, however—the data for this project was uncertain, and we kept hearing about other possibilities. I'd technically said we wouldn't include any new sources, but the stakeholders wanted us to use as much as possible, so we had our ears to the ground. We were mostly working in the Data Understanding swimlane, still doing EDA on the first sales data source after having a meeting with someone on that data source to understand some of the columns better. We started doing

739

the EDA on the market share data but discovered that it was missing the SKU, so we couldn't join it to our list of SKUs to work with. We figured out how to use another table to get the SKU from the UPC, only to discover that the UPC that was in that source was wrong, with the final number chopped off. A data engineer was working on fixing it, so we had to pause that work waiting on them. We did a cursory glance at the second sales data source and realized we needed more knowledge, so we were trying to set up a meeting with a SME on that data source.

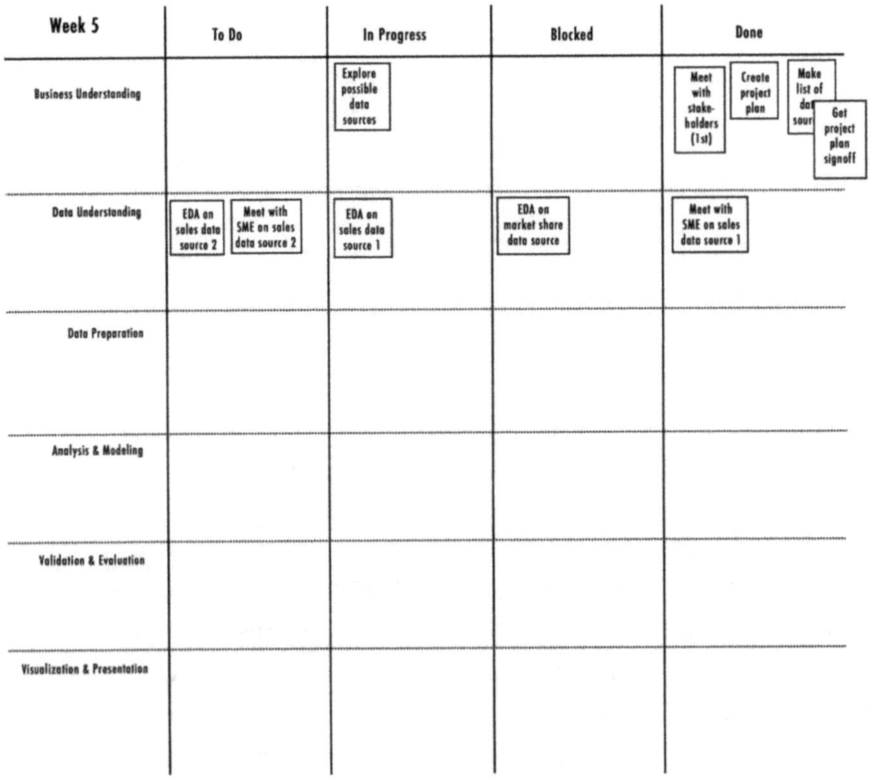

Week 5	To Do	In Progress	Blocked	Done
Business Understanding		Explore possible data sources		Meet with stake-holders (1st)　Create project plan　Make list of data sour　Get project plan signoff
Data Understanding	EDA on sales data source 2　Meet with SME on sales data source 2	EDA on sales data source 1	EDA on market share data source	Meet with SME on sales data source 1
Data Preparation				
Analysis & Modeling				
Validation & Evaluation				
Visualization & Presentation				

Figure 21-5.   *The Kanban board at the end of week 5 on the incrementality project*

At the end of week 9, we'd made a lot of progress and were now working in the Data Preparation swimlane. See Figure 21-6. We were still hypothetically waiting for info on further data sources, though that was wrapping up because we couldn't take anything new this far into the project. But one thing that came out of the last month was a new data source that we had to add, so doing the EDA for that got added to the To Do column in Data Understanding. We already completed three Data Preparation stories, finalizing the queries for the market share data and first sales data source and finishing the preprocessing on the market share data because it was pretty simple and clean. We were actively working on the preprocessing of the first sales data source and had three stories on the docket in To Do for Data Preparation.

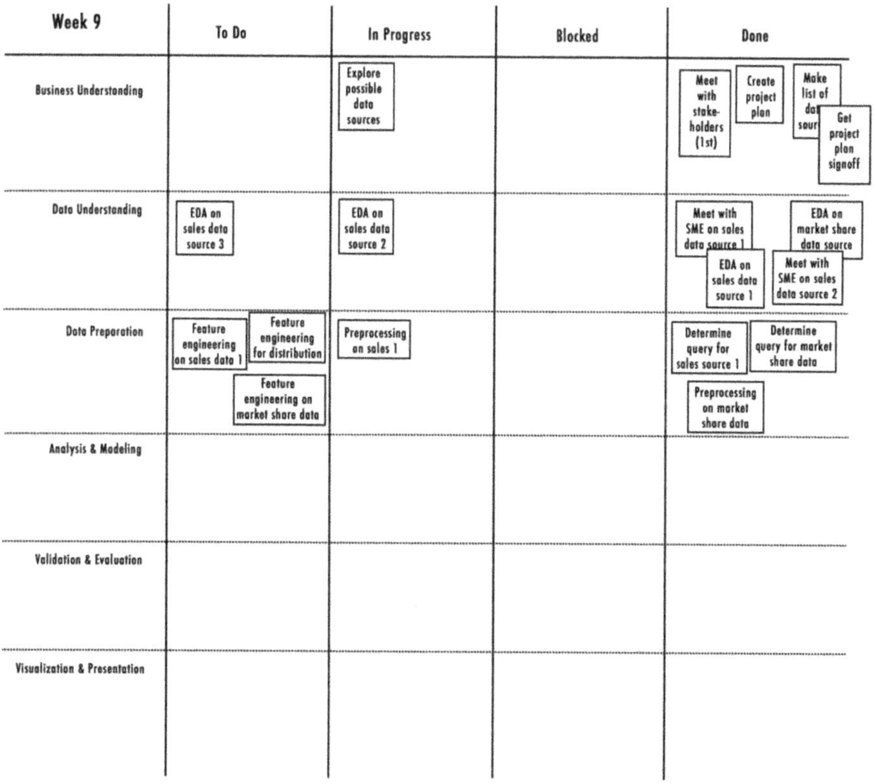

Figure 21-6. The stepwise data science Kanban board at the end of week 9 on the incrementality project

By the end of week 13, shown in Figure 21-7, quite a bit of progress had been made. I'd said we definitely wouldn't be taking new data sources at that point, so the last Business Understanding story got moved to Done, and so did all the Data Understanding stories. We started on the feature engineering in the Data Preparation swimlane and got it done for market share and distribution (an additional feature we got from the first sales source). We also got the preprocessing done on the first sales source, and we realized we wanted to combine the three data sources together and do the feature engineering on that combined dataset, so we added a story and modified the feature engineering one for the first sales data. We added another story to To Do about refactoring (redesigning) the code that did the data prep so it would be easier to use on additional products. (This would be done later.)

We started in the Analysis & Modeling swimlane with writing the code that would run all our experiments (different algorithms, cross-validation, and hyperparameter tuning). We also did some planning on what remained and additional work we would need to do. Once all the feature engineering was done, we'd have to modify the experiment code and then we could run it, so that story was added to the To Do in Analysis & Modeling. Two other new stories were to write and run the code on the best-performing models per SKU and to write and run the code for calculating incrementality and cannibalization. A final new story in the swimlane was to refactor the code that did the modeling, also to be done later.

We then added stories to the Validation & Evaluation and Visualization & Presentation swimlanes' To Do column for the things we already knew would have to happen.

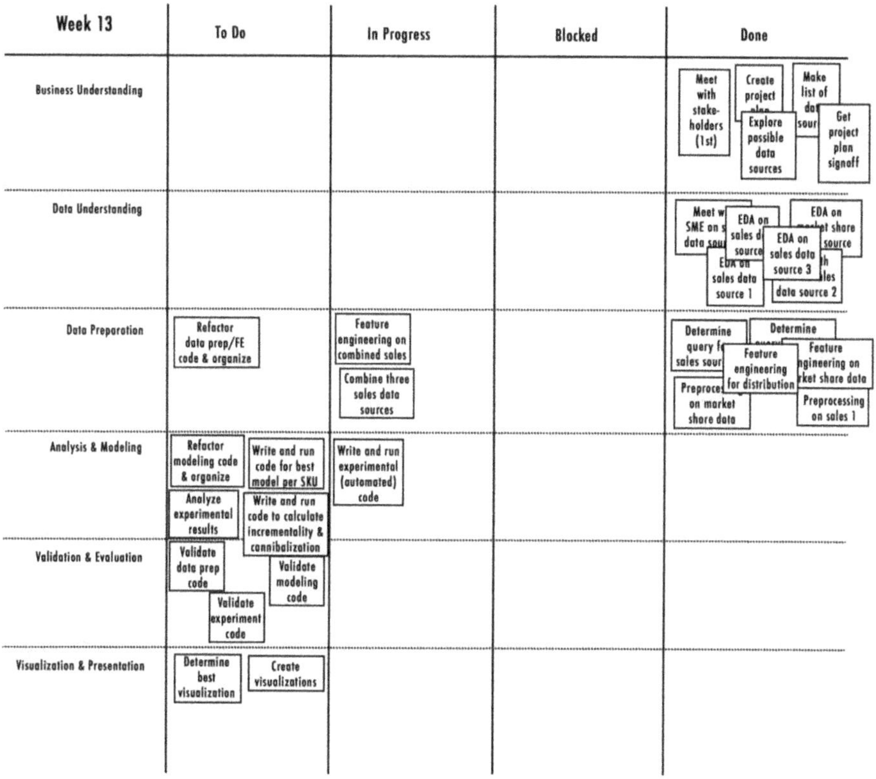

Week 13	To Do	In Progress	Blocked	Done
Business Understanding				Meet with stakeholders (1st); Create project plan; Make list of data; Explore possible data sources; Get source; Get project plan signoff
Data Understanding				Meet w/ SME on sales data sour; EDA on sales d; EDA on market share source; EDA on sales data source 1; EDA on sales data source 3; EDA on sales data source 2
Data Preparation	Refactor data prep/FE code & organize	Feature engineering on combined sales; Combine three sales data sources		Determine query for sales sour; Determine Feature engineering on market share data; Feature engineering for distribution; Preprocess on market share data; Preprocessing on sales 1
Analysis & Modeling	Refactor modeling code & organize; Write and run code for best model per SKU; Analyze experimental results; Write and run code to calculate incrementality & cannibalization	Write and run experimental (automated) code		
Validation & Evaluation	Validate data prep code; Validate experiment code; Validate modeling code			
Visualization & Presentation	Determine best visualization; Create visualizations			

Figure 21-7. *The stepwise data science Kanban board at the end of week 13 on the incrementality project*

A major wrench got thrown in after week 13, because we were asked to include two new market share data sources in week 11. Sometimes it's hard to say no, so we had to accommodate the request. See the week 17 board in Figure 21-8. We were working on the EDA for both sources, back in the Data Understanding swimlane, and at the same time starting the preprocessing and the feature engineering for both (we moved the prior feature engineering story for the market share data back from Done to In Progress) in the Data Preparation swimlane. This is not generally what you want to do, but we needed to be able to include those features ASAP, so we worked on the code knowing that we'd probably have to make some tweaks based on what came out of the EDA. The experiment code in the Analysis & Modeling swimlane was done except for adding the market share features, so it was stuck in In Progress. We were still able to get started on some of the code and fine-tune our process for evaluating the experiment results, so those were moved to In Progress. We were also able to start validating the data prep code (all parts except what related to the new data sources) in the Validation & Evaluation swimlane. We added one more story to the Visualization & Presentation swimlane, for creating the final slide deck, and got started on the first slides that just explained the project and other preliminary things.

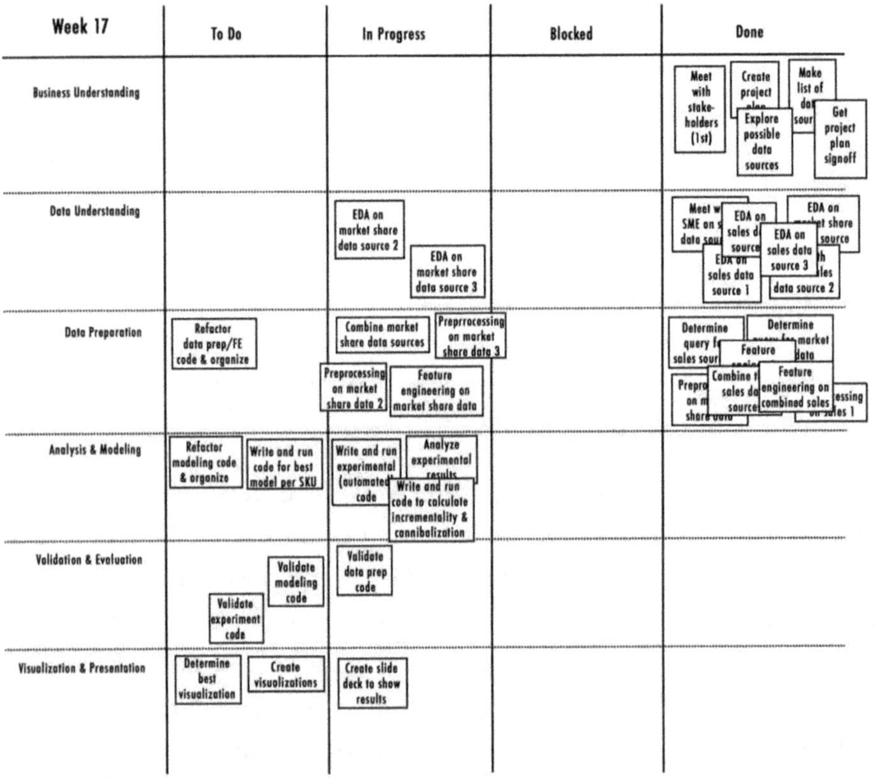

Figure 21-8. *The stepwise data science Kanban board at the end of week 17 on the incrementality project*

After the new data sources came in, we scrambled and managed to get everything done and with some long days got everything done in time for a deadline that had been looming. See Figure 21-9 for what the board looked like. We did finish the EDA on the two new sources in the Data Understanding swimlane. Then, in the Data Preparation swimlane, we got the preprocessing done on both sources and combined them and finalized the feature engineering code for it. Only the refactoring story remained in that swimlane, and it was saved for later. In the Analysis & Modeling swimlane, we finished all the In Progress stories and only the refactor for later remained. In the Validation & Evaluation swimlane, we did enough validation of the code to feel like it was accurate, although we didn't have time to do the rigorous code reviews we normally do. Finally, we finished all the Visualization & Presentation stories and presented the deck to the stakeholders.

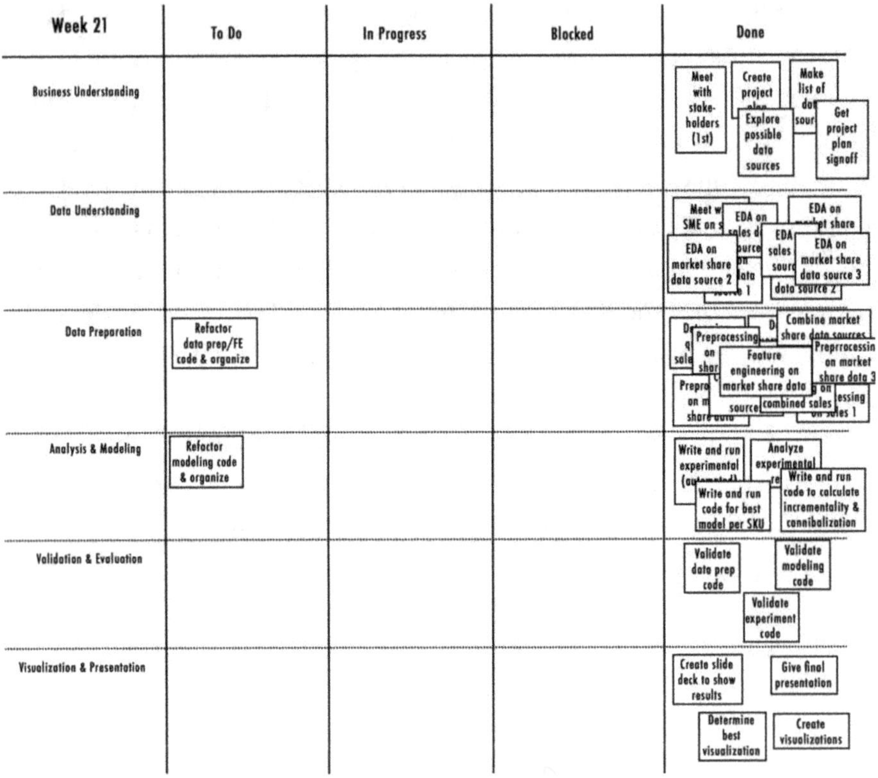

Figure 21-9. *The stepwise data science Kanban board at the end of week 21 on the incrementality project*

Hopefully, this gives you a sense of how the Kanban board can work for a data science project. Note that it's common to leave all the stickies in the Done column on a physical board, but it obviously can get crowded so sometimes they do get moved. On a virtual board, they will stay there if set to visible.

Post-project Work

Projects of all types usually have a start and end. This is true for data science projects, too. We talked about the lifecycle of a data science project, but there's nothing addressing what happens after the project is finished. With software, there's an acknowledged post-completion phase, maintenance. In software, that just means ensuring that it's still working. Things are more complicated in data science.

In terms of post-project work, some data analysis and data science projects don't have anything based directly on the project work. This is the situation when you've been asked to perform some analysis and have delivered a report. You might be asked to repeat or update the analysis, but that involves rerunning things (hopefully your code makes it as simple as that), not really tweaking the work from the actual project. In two other typical cases, there are post-project responsibilities. We'll talk about these next.

Code Maintenance

One scenario when there is work to do after the project is when a dashboard has been created as part of the project that needs to be regularly refreshed, so code is in place to prepare the data for the dashboard, either in the BI tool or outside of it. A second is when a model has been created that will continue to be in use (whether it's queried via an API or is used in code you've also written that refreshes regularly or some other scenario).

In both of these cases, any scheduled code or data transformations in a BI tool need to run as expected. The advantage of code is that it will always run the same unless something changes or a transient random computer fail occurs. A random computer fail could be something like the network connection being lost while code relying on data in the cloud ran or a computer simply being down so the job isn't triggered. The point is that the fault lies not with your code, but on the mechanism it relies on to run.

Usually, the fix is to just manually kick off the process this one time after the other system is back up and running, and it should go back to working again the next scheduled time.

The "something changes" scenario is more difficult. This can be a whole host of things, including database structure changes (removed or added columns in one of your tables, for instance), a change in permissions on the account that's used to connect to a database, a refresh to a table you use moving to a later time than you've allowed for. There are basically an infinite number of possibilities for your code to be borked. This is actually a huge pain, because now you have to figure it out, and if you're unlucky enough to have only an unhelpful error message (very common, unfortunately), it can be anything—so it's time-consuming to troubleshoot. Sometimes you may find it's something you need to ask someone else about, and if they're busy, you may be stuck until they respond. If you need to get it working, you'll have to keep going until you figure it out.

Model Drift and Maintenance

One other area that's important when you have a deployed model, whether it's trained and static or retrained regularly, is the idea of model drift. *Model drift* is the situation where a model's performance changes over time, usually for the worse. So, checking for that, we monitor the performance of the model over time. As an example, let's say you have a model that's in production and forecasts sales at all stores for the next seven days, running every night. The model itself isn't retrained, so the longer it's in place, the more "out of date" the data is. We may see good performance metric values (for instance, MAPE scores less than 10%) on most days. But after it's been running for a year, a lot more of the MAPE scores are in the teens. The solution is to retrain the model, but that will mean you have to evaluate the results and likely test different algorithms and parameters, so this isn't a quick fix.

Another option that's usually better for sales forecasting is to retrain the model regularly, even every night if it's feasible. This will reduce the likelihood of model drift. However, it doesn't remove it completely. In our example, imagine a regular promotion is introduced that wasn't there when you first wrote the code that you used to generate the model. That info is probably going to help the model a lot because promotions usually change purchase patterns, but until you modify your code to add the feature, that information is lost and the model will underperform.

One other situation requiring adjustments is if something new is added that your model didn't consider but that must be there now. For instance, if you're forecasting sales of specific products differently, a new product won't immediately be included. You will need to do the work to include it.

Post-project Time Management

The important thing to remember here is that if the need for intervention comes up after the project is officially over, you won't necessarily have "free" time to work on it. You'll be working on something else at that point. It probably won't be up to you to prioritize the broken code or dashboard over your current work, but it will likely take time away from your current work, which is frustrating for you and everyone else interested in that project.

Key Takeaways and Next Up

In this chapter I talked about the challenges of managing data science projects. There aren't established ways and people generally cobble together ways of managing projects. I talked about traditional project management, especially in relation to project planning and execution, and then about the software variations on that. I proposed a way of managing data science projects with the stepwise data science Kanban board, which adds swimlanes to define project stages to a typical Kanban board. I then talked about some of the post-project work that has to be done.

In the next chapter, I'll be talking about the various known biases and fallacies that can affect data scientists and data science projects. Knowing about them will help you avoid them or at least mitigate them.

PRACTITIONER PROFILE: HADLEY KOMBRINK

Name: Hadley Kombrink

Current Title: Technical project manager

Current Industry: Retail, previous healthcare and IT consulting

Years of Experience: 6 in managing data science

Education:

- Certified Scrum-master

- Master of Public Administration

- BA English and Sociology

The opinions expressed here are Hadley's and not any of her employers, past or present.

Background

After Hadley finished her undergraduate degree, she went on to do a master's and ended up teaching in the program for several years afterward. But she was also exposed to analytics during that time and enjoyed it. She gravitated toward IT in general and worked for a hospital with a database administrator, where she learned even more. She soon moved into a new role on another IT team, infrastructure project management, where she became a technical project manager. In that role, she managed technical rollouts that could still affect the business, like SharePoint and Internet Explorer. She got to where she felt she knew a little about everything. She worked in a similar role at another job before landing a new role at the hospital as a business solutions analyst.

Work

In Hadley's role as a business solutions analyst, she had a variety of responsibilities and acted as a product owner, Scrum-master, business analyst, and more and managed many types of projects. It also involved a lot of interaction with customers, and she also dabbled in some coding at this point. But the most interesting thing was that there were several positions in the team, including a data scientist, which is where she started learning more about that field. She soon found herself specializing in managing data science projects, which she still does today.

Sound Bites

Favorite Parts of Managing Data Science Projects: Variety in the work day to day and working with different kinds of people with different backgrounds and specialties. She also loves helping to set priorities and guide business value without being a people manager.

Least Favorite Parts of Managing Data Science Projects: Bureaucracy and when she can't see how to make an impact, like when something is beyond her control.

Favorite Project: She worked on a mobile device management project at the hospital that involved a lot of people skills. They were trying to get doctors to use their phones for work, but had to convince them that the hospital wouldn't steal their photos or get into the other areas of their phones. She had to really flex her communication skills on that project, which was ultimately successful.

Skills Used Most in Managing Data Science Projects: Cat herding and various organization techniques like managing emails, to-do lists, and systems to manage and organize project work (like Agile management tools).

Primary Tools Used Currently: Hadley still loves pen and paper, especially for to-do lists. For project management, she's used Jira, ServiceNow, and other Agile tools. Also, Google Suite and Canva.

Future of Data Science: It's still the future, along with having a strong data foundation for all work. There is still work to be done on data in general. But good communicators will also remain very important.

What Makes a Good Data Scientist and DS Project Manager: Data scientists should be very inquisitive, be willing to work with the business and stakeholders, have technical skills, and always consider different options. DS PMs need to be flexible because it's not like management of other technical projects.

Her Tip for Prospective Data Scientists: Ask lots of questions and don't have a big ego. Network and forge relationships, and especially find a mentor if possible, as the data science community is small.

Hadley is an experienced project manager with experience managing data science and other IT projects.

CHAPTER 22

Errors in Judgment: Biases, Fallacies, and Paradoxes

Introduction

We all know that people have cognitive biases and fallible memories and beliefs—or at least we know that *other people* do. It's hard to remember that we ourselves have them. There's actually a name for that—*bias blind spot*, where we assume that we're less biased than other people. There's no real way to get rid of this lizard brain irrationality altogether, but we can be aware of these inaccurate beliefs and correct or compensate for them consciously.

Obviously in a book about data science, we care more about ones that affect our ability to do quality and fair data science, but a lot of them can affect our data science work in surprising ways.

In this chapter, I'm going to cover a variety of biases, fallacies, and paradoxes, how they can manifest in data science work, and what we can do about it. First, we'll look at a couple of examples where bias or fallacies caused problems.

K. P. Vincent, *A Friendly Guide to Data Science*, Friendly Guides to Technology, https://doi.org/10.1007/979-8-8688-1169-2_22

Examples of Biases and Fallacies in the Real World

The two examples we'll look at include Amazon's sexist resume screener, which we originally saw in Chapter 9. This example exhibits several different biases that we'll talk about in the chapter. The second example is of a tool used in healthcare that was accidentally biased against Black people because it used an economic measure that mistakenly made Black people seem less ill than they actually were. It also shows several problems, but especially one paradox that we'll also talk about below.

Example 1: Amazon's Sexist Automatic Resume Filter

In 2014, Amazon started work on a tool to help them identify quality applicants by automatically reviewing their resumes.[1] It was supposed to crawl the Web to find these excellent candidates and try to bring them into the hiring process. By 2015, Amazon realized that their screener was really sexist, heavily favoring male applicants. This was not the intent, and the company scrapped the project (although they did apparently start another attempt). How did the system turn sexist?

This screener was intended to be a classifier that would identify candidates that were worth starting through the evaluation process. The team used ten years of applicant resumes as the training set and created a

---

[1] "Insight—Amazon scraps secret AI recruiting tool that showed bias against women" by Jeffrey Dastin on Reuters, https://www.reuters.com/article/us-amazon-com-jobs-automation-insight/amazon-scraps-secret-ai-recruiting-tool-that-showed-bias-against-women-idUSKCN1MK08G/, and "Why Amazon's Automated Hiring Tool Discriminated Against Women" by Rachel Goodman on ACLU.org, https://www.aclu.org/news/womens-rights/why-amazons-automated-hiring-tool-discriminated-against

model for each job function and location, around 500 in total. Then they used NLP to parse the resumes and identify tens of thousands of potential features, basically words and word combinations. The models first learned that most of the features weren't useful, and they disregarded certain characteristics that were common across all the resumes under that job function (like programming skills for software development engineers). So far, so good.

At that point, things started to go south. The models picked up and started favoring "strong verbs"—words like "executed" and "captured" that appeared far more often on men's resumes than women's and because so many more of the hired people were men. It also penalized terms that were clearly associated with women's resumes, like "women's" (as in something like "women's rugby") and the names of certain women-only colleges.

What basically went wrong is that the team fell for technochauvinism— believing that if they automated it, it would be better than doing it manually. They wanted to mine the Web to find the diamonds in the rough. But what they forgot was that all they were doing was training a system to do exactly what they'd always done—hire with gender bias.

There are a couple of other biases that play in here, both of which we'll talk about more below. One is selection bias—their training set consisted of people who had applied to Amazon. Amazon is known as tough to break into, so anyone who doubts themselves will be less likely to apply, and that imposter syndrome is in a lot more women than men. It's known that while men tend to apply for jobs even if they don't meet all the required skills, women generally don't. The other is survivorship bias, which is when only those who've made it past a certain point are considered. In this case, so few women were hired over the ten-year training period that information about them (and their quality) isn't a consideration. They can determine if the algorithm is mostly letting only candidates who will be successful through, but they have no idea if other, way better, candidates were rejected. They have no idea about the quality of those applicants.

Using AI in recruiting across industry is full-steam ahead right now, even though these kinds of examples show how difficult it is. Additionally, it all seems to hinge on the idea that a resume gives you valuable information about a candidate. Anyone who's been involved in hiring knows candidates' resumes may or may not match their actual experience and skills.

Example 2: Racist Healthcare Risk Forecasting

Another area that is rife with discrimination is healthcare, with significant biases against Black people in particular perpetuated by textbooks, practices, and the people in the profession. Their pain management is pretty bad because of false beliefs that they have a higher pain threshold than white people and are additionally drug-seeking, but in general they're not listened to. Even tennis superstar Serena Williams nearly died after giving birth because her reports of what turned out to be a serious condition were dismissed (being Black and also a woman is especially dangerous in the American healthcare system).

A tool that had been in common use by the mid-2010s was found to lead to discriminatory care of Black people because it relied on biased data and features.[2] A tool existed to help hospitals and insurance companies identify which chronically ill patients to put into "high-risk care management" programs. Patients in these programs tend to get better because they're provided extra care and specialized nurses. The idea is that treating sicker patients here prevents more serious complications from coming up, which lowers costs and keeps patients healthier and happier.

---

[2] "Racial Bias Found in a Major Health Care Risk Algorithm" by Starre Vartan in *Scientific American*, available at https://www.scientificamerican.com/article/racial-bias-found-in-a-major-health-care-risk-algorithm/

This sounds like a good plan, but they used a feature that was intended to provide insight into a patient's medical need, previous spending—in other words, what that patient had spent on their medical care in the past. Apparently, this is a common metric used in ML in healthcare, and your alarm bells should be ringing if you know anything about healthcare in America. This feature would be more equitable in countries with socialized medicine, but everyone knows that American healthcare is expensive and cost-prohibitive to many people. People avoid going to the doctor regularly because they can't afford it, so issues that are relatively minor and treatable get so severe over time without treatment that people end up in the ER, hospital, or even dead. All because of money. So comparing previous medical costs of different patients is going to tell you more about their economic class than their actual health. Additionally, because of the discrimination they experience, Black people are in general less likely to trust the healthcare system, another reason they don't go to the doctor as often. So this feature is inherently biased and should not be used.

Researchers looking at this particular tool found that the tool assigned lower risk levels to Black patients than it should have. On average, Black people have lower incomes than white people, so it follows that they will spend less on healthcare because they won't be able to afford it as easily. When researchers looked at white patients and Black patients at the same spending level, the Black patient would be in poorer health. Among patients with very high risk scores, Black patients had 26.3% more chronic illnesses than their white counterparts. This all meant that Black patients who would have benefitted from the additional care were not recommended for it by this tool as often as white patients with the same level of need were because their spend was less. So the cycle of discriminatory care continued.

There are obviously several issues at play here, including lack of data, but one is Berkson's paradox, which I'll talk more about below. It comes about when a system or people come to a false conclusion because they didn't consider certain conditional information. In this case, the information was about the actual need of Black patients, which was masked by using the misleading metric of prior healthcare spend.

A Curious Mind

One of the most important things about being a data scientist or data analyst is that you have to listen to the data—be a data whisperer, as it were. We have to want to know what the data is going to tell us. This means that while you can go in with knowledge, techniques to help bring meaning out, and understanding of things we think are similar, you must keep from having preconceived ideas and jumping to conclusions too quickly.

There are quite a few cognitive biases that we have that undermine this curiosity. *Anchoring bias* happens when we fixate on one idea or characteristic of something and disregard other aspects. It's usually the first thing we learn about the thing. The classic example is if you are out shopping and you see a shirt that's $300 and then another that's $150, you're prone to think the second is a good deal since the first one you saw was $300. If you'd seen them in the other order, you likely wouldn't think $150 was a good deal—instead, you'd think $300 was a rip-off. In data science, this might be stopping your EDA too early because you think you've found something interesting. Imagine you have a dataset of reading habits of teenagers from ten local high schools. You start your EDA and see quickly that most of the books read were written by male authors and that thrillers and sci-fi are the two most popular genres. If you've read this far into the book, you know better than to stop here and come to strong conclusions. But if you really find the fact that most of the authors are male—and you feel that means that kids today still read mostly male authors—this can taint how you see everything else.

Anchoring bias relates to *conservatism bias*, where we're unwilling to change our belief even when presented with new information that contradicts it. Imagine you break down the readers into girls and boys and see that, actually, girls tend to read more female authors. But if you stick to this idea that teens overall read more male authors, this is conservatism bias. If you eventually discover that your dataset contained three boys-only high schools and the rest were mixed, that should rid you of your belief, because the data itself was clearly biased, but conservatism bias could still keep you from changing your belief.

If you look at more data about teen readers and do some exploratory work on the new dataset (which you've confirmed is all from mixed high schools), if you can't get the idea out of your head, you might see a slight preference for male readers as more significant than it is. This is *confirmation bias*, where we tend to focus on and interpret things presented in front of us (and also remember them) so that they confirm what we already believe. Another aspect of confirmation bias is that people seek out more information that will support their beliefs, which can influence the hypothesis driving the work.

It's also possible to fall into a bias called *belief bias*, where people are more likely to think an argument is logical if the conclusion feels "believable." A similar one is the *illusory truth effect*, where we're more likely to believe something if it's easier to understand or is more familiar. If we already think that people in general read male authors more than female, this initial finding might feel believable, sensible, and familiar—and accept it too readily.

All of these biases can compromise our ability to do good data science. We're more likely to fall into one if what we see is what we were hoping to see. To deal with anchoring bias and conservatism, force yourself to ask: Are you sure? Could there be anything else? How would I prove that this is true? Dealing with confirmation bias, belief bias, and the illusory truth effect is similar. You need to identify other truly possible explanations and rule them out. But you have to do this with integrity—don't just make up faux "alternative explanations" and test them and rule them out. Always try to take a step back and keep your personal feelings tamped down as much as possible.

The basic takeaway is that you need to stay curious and don't jump to conclusions or accept the easy answers.

Data Science Practice

There are many biases and fallacies that are important to think about and deal with when doing data science and interpreting and reporting results. I'll first talk about ones that are important when planning and carrying out the data science work and then move on to ones that are important when interpreting results and finally to ones that we need to keep in mind when communicating our results to stakeholders and others.

Planning and Doing

Depending on the data science you're doing, you may be creating experiments or studies or simply working with existing data. Some biases and fallacies are specific to planning experiments, while others are relevant in other situations where we're planning and carrying out analysis and modeling.

We've talked before about *selection bias* in the context of experiment and study design, which is when an ostensibly random statistical sample is taken without true randomness, perhaps affected by the experimenter's cognitive biases or simply because of systemic or technical issues in sampling. These are all things that can be dealt with if we follow the standard selection methods, such as those outlined in Chapter 3.

There are a couple of other things to consider when designing experiments and studies. *Expectation bias* is a type of confirmation bias where someone running an experiment is only looking for what they want to see based on their preconceived notions, and the *observer-expectancy effect* is where an experimenter manipulates (usually unconsciously) an experiment or their interpretation of data in order to see the result they expect, generally by influencing participant behavior. Questionnaires with leading questions are a typical example of the observer-expectancy effect. It's important to look for these in your design. You want to do the work in such a way that your research arch-nemesis can't find fault in your experiments.

The *observer's paradox* is a phenomenon that comes up in social sciences, basically in studies of humans, where the mere act of the researcher observing something changes that thing. It first came up in sociolinguistics where when linguists tried to study the way people spoke, their presence as an observer changed the way people spoke. It's something that should be considered whenever humans are being studied.

As we saw in the previous section, keeping an open mind is important. There are some phenomena that can act against that. *Groupthink* is a phenomenon that occurs when decision-making in a group of people is illogical or dysfunctional because harmony in the group is considered paramount. This can be important in the way data science teams operate. If you know something is wrong, try to be the one brave enough to speak up.

Being open doesn't mean ignoring important information. A framing bias called *domain neglect* leads us to ignore important knowledge in a domain when we're working in an interdisciplinary context. It's your responsibility to ensure that you have gathered all the relevant info on something, especially when there might be ideas that you don't know about because they come from other fields or parts of the business. This problem can be accomplished by thorough research or talking to colleagues from the other disciplines and teams.

It's also important to ensure you're on the right path. First, always ensure that your project is worth doing and won't have unintended negative consequences. *Automation bias* is another term for technochauvinism, where we have too much faith in automated systems at the expense of human decisions. Simply because the project is cool doesn't mean we should do it. Once the project is begun, it's important to look at progress periodically. *Sunk-cost fallacy* is where people stick to the original plan even when it starts to look like it was wrong, all because of the investment they've already put into it. If you realize partially into a project that it isn't going to work, it might be better to start over rather than try to reshape what's done so far into something that will work. This is especially difficult in the work world when you have deadlines. Nobody wants to start over. But sometimes it's the best way.

It's important to always think about what you're doing when you're planning and carrying out any data science project, whether it involves designing an experiment or study, analyzing some data, or using existing data to create a forecasting model.

Interpreting

There are many biases and fallacies that are important to think about when interpreting and analyzing data.

Domain neglect, mentioned above as the tendency to not look outside your own immediate discipline for ideas, is also important at this stage. You might see an effect in your results that you don't know how to interpret, but someone in psychology or on the finance team might. It would be better to have already dealt with this (and established a relationship with your colleagues) in the planning stage, but you may need to revisit your original contacts or search some more.

One important paradox is *Berkson's paradox*, which affects data science work directly and occurs when people come to false conclusions because they've disregarded important conditional information. A typical scenario can happen when a sample is biased. For instance, imagine a parent who concludes that novels written for teens are much longer than novels written for adults, since their sample is the books in their house, where the book collection consists of their cozy mysteries and romances and their teenager's young adult fantasy books. In the young adult category overall, fantasy novels are much longer than contemporary or romance novels, and in the adult category, cozy mysteries and many romances tend to be on the shorter side. This is not a representative sample of "YA books" and "adult books."

There are several biases that fall under *extension neglect*, which happens when experimenters don't consider the sample size when making judgments about the outcome. The most important here is *base rate fallacy*, which crops up when people only focus on information relating to a specific case and disregard more important general information. The term "base rate" basically means the general case, and this is sometimes disregarded in favor of something more interesting. An example particularly relevant to data scientists is in situations where we have a classifier that's testing for a relatively rare thing and has low precision (a lot of false positives relative to true positives); we have to not fall into the trap of believing all the positives. This is called the *false positive paradox* and is a specific type of base rate fallacy.

Apophenia is how we tend to see connections and relationships between things that are unrelated. This is why the warning "correlation is not causation" is so important—it's a fundamentally human thing to identify connections in our world. That means we have to be careful. If you think you see a connection, you need to confirm it's real. Don't jump to conclusions just because it feels like something interesting. This is one reason that statistical testing is used. Just because one number is higher than another doesn't mean it's a true reflection of reality and not just a result of randomness.

There's a famous (but apocryphal) story that illustrates another bias, *survivorship bias*, which is when we look at the results of something and only think about the things that made it through ("survived") a filter of some sort, disregarding the things that didn't. The example concerns damage to American bomber planes returning from bombing campaigns during World War II. They wanted to add armor to the planes but couldn't put it everywhere, so they needed to protect the most important areas. The military aggregated the damage to the planes by showing all of it in one diagram, as can be seen by the red dots on the plane in Figure 22-1.

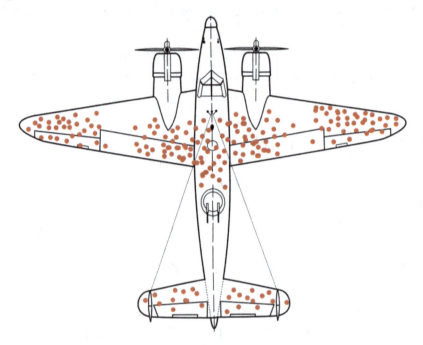

Figure 22-1. An aggregation of damage to planes in World War II illustrating survivorship bias. Source: From "File:Survivorship-bias.svg" available at https://en.wikipedia.org/wiki/File:Survivorship-bias.svg, used under the Creative Commons Attribution-Share Alike 4.0 International License (https://creativecommons.org/licenses/by-sa/4.0/deed.en)

A lot of people when first looking at this think that the areas with the concentration of red dots should be reinforced with armor. But actually, the exact opposite is true—these are the planes that made it back (survived). Planes with damage to the engines, the nose of the plane, central spots on the wings, or the area of the body from the gunner to just in front of the tail didn't. That's because when planes were hit in these areas, they would crash. The apocryphal story has it that one statistician named Abraham Wald pointed this out when everyone held had fallen for the bias. This was also seen in the Amazon recruiting tool example, where they looked only at the people who were hired and didn't consider whether any of the ones who weren't hired were actually good. Always think about the data you're not seeing in a problem.

Another really interesting problem is *Simpson's paradox*, which occurs when trends that appear in individual subsets of data are different, even opposite, when the subsets are combined with each other. This sounds impossible, but it often crops up in real data. This is one reason it's important to slice and dice the data in different ways. See Figure 22-2 for an example of what this can look like visually.

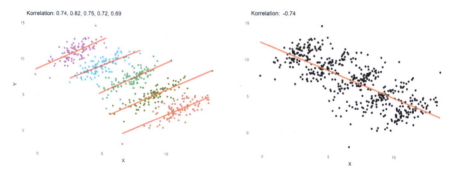

Figure 22-2. An example of Simpson's paradox, showing positive correlation in individual groups (the left chart) and negative correlation across the whole dataset (the right chart). Source: From "File:Simpsons paradox—animation.gif" available at https://en.wikipedia.org/wiki/File:Simpsons_paradox_-_animation.gif, used under the Creative Commons Attribution-Share Alike 4.0 International License (https://creativecommons.org/licenses/by-sa/4.0/deed.en)

You can see in the figure how strong these effects in each group look, but also how important it is to not miss the overall trend. This does not mean that the in-group trends don't matter, but the entire picture is important.

One final fallacy worth mentioning is the *McNamara fallacy*, which occurs when people only measure success by quantitative measures and disregard other important factors (qualitative ones). This is especially important when the data science work being done affects people through public policy, healthcare decisions, and so on. Imagine a popular video game company looking only at sales of a new game and disregarding that everyone hates it, reviews are terrible, and people are saying it's bug-ridden. If they're *only* looking at sales, they're missing hugely important info—which is saying that sales will soon drop off. The initial sales numbers are propped up by the company's reputation, not the popularity of the game itself. The only way to deal with this is to force yourself to look beyond the numbers.

It can be easy to fall into traps when analyzing results in data science, but if you pay attention and are aware of what these traps are, you can escape them.

Communicating

Many of the biases and fallacies already mentioned are important during communication, partially because your stakeholders may have the same biases and you may have to address them to explain why they aren't accurate. But there are a handful of others that are fairly specific to communication.

The *curse of knowledge* occurs when experts forget how to look at things from a regular (non-expert) person's perspective. This is something I've harped on over and over—you have to be good at explaining technical things to nontechnical audiences in a way that they can understand. Forgetting to consider their backgrounds and knowledge is never going to be good. Before communicating with stakeholders, do your best to understand their technical level and the things they understand and believe.

There's also an interesting self-assessment phenomenon called the *Dunning–Kruger effect*, where experts underestimate their skill level and novices or completely unskilled people think their ability is significantly better than it is. It doesn't seem hard to you, so you can forget how technical your knowledge is and how inaccessible that is to other people. Another challenge in communication can be the other half of that effect, the situation where your nontechnical stakeholders think they know more than they do. This can be challenging to deal with because they can say things that are flat out wrong, and it's difficult to correct people in meetings, especially when they're more senior than you. The prudent thing to do will depend on many factors. The best thing to do is observe how your other, more experienced colleagues deal with this. Honestly, sometimes you just have to ignore it and move on.

One final bias that is important to communication is *pro-innovation bias*, which is related to technochauvinism and occurs when people view things that are innovative or new as being more beneficial to society than they may be, largely by disregarding the negative aspects. Don't let yourself fall into this trap, and don't let your stakeholders fall into it either. This means thinking about the wider consequences and not disregarding the negatives or risks.

Career

Being a data scientist generally means having a long-term career as a data scientist or in adjacent fields. There are a lot of ways we sabotage ourselves or at least miss opportunities to improve and grow. There are two key areas of biases that can impact our careers. Most are related to self-perception and getting that wrong. There are a couple more that relate to planning, which influences how successful you can be in your career. Growing your career is your responsibility, and knowing yourself and your strengths and limitations is the best way to know how to improve.

Self-Perception

One of the most famous biases is *imposter syndrome*, where you feel like you're a fraud—you're not actually competent and everyone around you *is* competent and pretty soon they're going to figure out you're worthless. Obviously, this is a scenario that could theoretically happen, but it's rarely the reality. Most people act more confident than they feel and have lots of self-doubt, so you're not alone. The best way to deal with this one is to really reflect on your skills and see how you've worked to develop them so you know they're real. Identify the skills you have that your colleagues don't have, even if your instinct is to discount them as not important. If you're a good networker, that's a skill a lot of people don't have. If you speak several languages and are more comfortable with NLP than some of your colleagues, that's a win, too.

Self-awareness is a valuable thing to have in any career (in life, too). Having a realistic view of your strengths and weaknesses makes it easier for you to choose good paths and grow yourself. But there are several biases that can affect your ability to see yourself accurately, all of which are in contrast to imposter syndrome and fall under *egocentric bias*, where we overvalue our own perspective and think we're generally better than other people think we are. I mentioned blind spot bias above, where we assume we're less biased than others. There also is the *false uniqueness bias*, where we view ourselves as unusually unique and more special than we actually are, perhaps thinking we're unusually good at writing optimized code because our immediate colleagues aren't as good at it. The *false consensus effect*, where we think others agree with us more than they actually do, can lead us to think everyone agrees with our solution even when some people have doubts. There can be a lot of reasons people might not speak up in the moment, not only because they agree with you. Finally, the *illusion of validity* is where people think their opinions and judgments are more accurate than they really are. With the *illusion of explanatory depth*, we think we understand something much better than we do. Both of these can lead us to being more confident in our solutions than we should be.

All of these are natural. We live in the world in our own minds, so of course we feel that our perspective is the best one. But knowing better can help us avoid letting this self-centeredness get in the way of improving ourselves. To deal with these, you have to first realize that everyone is biased in various ways, including you. Knowing this is power, and if you listen to other people and watch for their biases, this can provide a mirror into your own. You don't have to go around telling everyone what your biases are, but if you know you have them, you can adjust for them in your head and choose different actions. If one of your biases is discovered publicly, don't get defensive. Instead, listen, acknowledge the mistake, and do better.

Uniqueness bias and false consensus bias are actually somewhat opposite, so it's kind of funny that we can carry around both. These two along with the illusion of validity all are important to be aware of when you're working with other people. Uniqueness bias makes us think that we are just generally better than other people, and the illusion of validity makes us think that our ideas are better. When you're working on a team, these aren't universally true. Even if you do have unique experience or knowledge, you still don't know everything. Additionally, it can be important to remember that other people sometimes have experience or knowledge you don't know about. It's important to not fall for the false consensus effect when you're on a team trying to make a decision. Make sure you are all in agreement by restating the thing you've just agreed on, and invite critique.

I mentioned the Dunning–Kruger effect above, where experts downplay their skills and non-experts presume they know more than they do. Don't let yourself undervalue your skills. You've worked hard to develop them. This is obviously related to imposter syndrome, but it's not always about feeling good about yourself, but actually just not recognizing what a big deal some of your skills are. If someone looks to you as a mentor, you may think you don't have much to offer; the mere fact that they have approached you shows you do.

A final self-perception bias worth mentioning is the *hot–cold empathy gap*, which exists when we don't acknowledge how much emotion influences our beliefs and behaviors. We are all human and none of us is 100% dispassionate and rational, despite how much we think we are. It's okay to be human, but this just means you should try to identify any emotions affecting your decisions and evaluate whether they are relevant to decisions you're making.

Planning

One of the skills everyone in the work world needs to develop is the ability to plan, including estimating effort and time. This is notoriously difficult, and two fallacies are primarily to blame. One is the *hard–easy effect*, which occurs when people overestimate their ability to complete hard tasks but underestimate their ability to complete easy tasks. This feeds into the *planning fallacy*, which is a belief people have that things will take them less time to complete than they really will. These two go hand in hand and really do come up all the time in the tech world. Providing estimates is fundamentally hard, and the only way to get good at it is to pay attention to your own estimates and learn from how wrong they are.

Other Biases and Fallacies

There are many other biases and fallacies that affect us, even if they don't always apply directly to data science work. Any sort of human failing can affect work, however, so here are a few more to think about.

We are really bad at thinking accurately about the future, especially when it involves uncertainty. *Gambler's fallacy* causes people to think that future likelihood is altered by past events even when there's no connection. This name comes from the fact that people gambling will consider what has happened so far when betting on future things. For instance, someone playing a slot machine might think, "I've lost 20 games in a row—I'm due for a win soon." This thought process also relates to *optimism bias*, where people overestimate the likelihood of positive outcomes and underestimate the likelihood of negative outcomes. *Neglect of probability* is also common, where we just flat out ignore probability when looking at situations that have uncertainty. The slot machine player is forgetting that each pull of the handle has the same likelihood of winning. In data science, these can relate to our ability to plan projects.

We often have trouble understanding complicated situations, especially (but not only) when uncertainty is involved. *Zero-sum bias* occurs when people believe that a situation is a zero-sum game when it isn't (a *zero-sum game* is one where one person only gains when someone else loses or vice versa). It is simply not true that something always needs to be taken away from someone else to give another person something. We also often see situations that are ambiguous as inherently threatening or negative, which is called *interpretation bias*. If you look more closely, it may be possible to understand it differently. However, we often want more information about situations even when that info will add nothing to our understanding of it, a bias called *information bias*.

Other biases related to trying to understand situations are *scope neglect*, which occurs when people disregard the size of a problem or level of impact when thinking about solutions, and *zero-risk bias*, when people want to remove a small risk entirely rather than simply lowering risk over a wider range. Scope neglect means that people might not immediately grasp that sending 10,000 desks to poor schools in an African country is going to cost a lot more than sending 1,000. With zero-risk bias, people are often uncomfortable simply lowering risk, meaning that there still is some risk across a wide range of conditions, and prefer to completely remove it in some of the specific conditions. The examples of this come from surveys done on people looking at hypothetical hazardous site cleanup.[3] Imagine two hazardous waste sites, A and B. A causes eight cases of cancer and B four every year (on average). There are three cleanup options. The first two bring the total number of cases down to six, but there are some from both sites. The third completely eliminates cases at B but only one at site A, meaning the total overall cases are higher, at seven. Almost half of the survey respondents preferred the third option over at least one of the first two, even though the overall reduction was less.

We're generally bad when it comes to looking into the past. The cognitive bias *recency bias* causes people to favor memories that are more recent. The *recency illusion* makes people think that something they've only just learned about is itself relatively new. We also really want the world to make sense, and *hindsight bias* causes people to think that events in the past were predictable in advance. When we look back at a past event, it's easy to say, "Oh, of course that happened—it's obvious after the way he was acting." But at the time, we had no idea, and we couldn't have because it wasn't actually predictable.

We are also bad at waiting for good things. *Present bias* occurs when people overvalue payoffs that are in the near future vs. undervaluing those further out. This is why so many people struggle to invest money—they'd rather have that new hoverboard than be able to buy five of the future equivalent of hoverboards down the road.

Another area where we struggle is getting distracted by subtle differences between things. There are several biases that fall under the *framing effect*, where how an idea is presented affects how we perceive it. One is the *default effect*, which occurs when people are presented with several options but prefer the one presented as the default. The *distinction*

---

[3] *ProjectThink: Why Good Managers Make Poor Project Choices* by Lev Virine and Michael Trumper

bias is the tendency to see things as more different from each other when we look at them at the same time as opposed to separately. The *decoy effect* is a phenomenon exploited in marketing where when someone is looking at two things and favoring one, introducing a third option (the decoy) with certain characteristics can cause the person to switch to favoring the second option. The decoy is clearly the worse option than one of the original two, but mixed when compared to the other one (better in some ways and worse than others).

Regression to the mean is a statistical concept that we talked about earlier in the book, but it's good to keep in mind. It holds that when we see an extreme value, the next instance is likely to be less extreme. It's directly opposite of the *hot-hand fallacy*, where people believe that someone who's had success will continue to do so. The name comes from beliefs of fans of sports that somebody has a "hot-hand"—like when they've sunk an unusual number of basketball goals in a row in a game because they're on a streak. However, it's worth mentioning that because of psychology, there's some belief that the hot-hand fallacy may not always be a fallacy—a basketball player who's in the middle of a streak may believe they can do no wrong and is therefore more confident, and this causes them to actually do better. But this wouldn't apply to someone on a winning streak at a slot machine, because winning is entirely determined by chance regardless of the person's confidence.

These biases and fallacies all affect our ability to function and make decisions on a day-to-day basis, which can also influence how we function and make decisions in our work and careers.

Key Takeaways and Next Up

We've talked about a montage of biases, fallacies, and paradoxes in this chapter. I first gave a couple of examples where biases had a negative impact on an ML system. Then I talked about several biases and fallacies that can keep us from having the open mind that we need to be able to do good data science and generally be good data scientists. I then talked about several that directly relate to planning, doing, interpreting, and communicating data science work. I then talked about biases and fallacies that can affect your career and finally covered some others that are more general, but can cause trouble in work and life.

In the next chapter, I'll be talking about various ways you can "get your hands dirty" by doing some real data science. Getting practical experience is important both for skill growth and for building a portfolio. I also talk about the various tools and platforms you can use to get this experience, as well as how to find data sources you can use.

PRACTITIONER PROFILE: DANIEL TARAZÓN BARBERÁ

Name: Daniel Tarazón Barberá

Current Title: Senior data scientist

Current Industry: Numerous through consulting, including retail, banking, and gas

Years of Experience: 6, plus 2 as a data engineer

Education:

- Título de grado en física (physics undergrad degree)

The opinions expressed here are Daniel's and not any of his employers', past or present.

Background

Daniel enjoyed physics so that's what he majored in, in college. But as he got close to the end of the degree, he wasn't sure what he wanted to do when he finished. He was interested in either nuclear physics or working in medicine, but those jobs were hard to get. There were only a few of relevant jobs in medicine every year, and thousands of people took the exam to qualify for the 12 or so positions available. He was uncertain and getting nervous as graduation neared as he knew that path was unlikely. A family member stumbled across a course on big data and told him about it, so he started researching that. He knew nothing about it and hadn't heard about data science, but when he stumbled across it in his research, he knew that was the right path. He had always loved programming and working with data even during his physics degree. He started a master's in data science and completed the coursework with only his thesis project left, which he didn't finish because he managed to land a job in data.

Work

Daniel worked for a couple of years as a data engineer before moving into a data science role. The data engineering wasn't too interesting, but he did learn a lot about working with data on the job. He really likes being a data scientist because of his love of programming. It suits him because he has good attention to detail and believes in rigor and doing things right, including checking for statistical significance. His data science work has all been with a consulting company, so he's been able to do a lot of different types of work at different companies and in different industries, so it's been interesting.

Sound Bites

Favorite Parts of the Job: His favorite thing in data science is working with models that use optimization techniques, like SIMPLEX and genetic algorithms. One other thing he likes about data science is that it can have a positive impact on a lot of aspects of society.

Least Favorite Parts of the Job: Especially working as a consultant, you don't always have control over what companies you work for or what kind of work, and Daniel sometimes feels moral conflict over some of the work being done in data science. He also doesn't like how there can be so much uncertainty in EDA and other aspects of data science. He often finds himself questioning his work afterward, thinking maybe it would have been better if he'd done it a different way or spent some more time exploring.

Favorite Project: Daniel has a couple of projects he's particularly proud of. One was creating a recommender system for books in his area's public library. They only did a pilot using collaborative filtering on the top 5,000 most borrowed books (as opposed to millions in the library system), but it actually performed really well. There wasn't enough money to take the project beyond the pilot, but it was clear it would have been a good system if they could have. The second project was one done for a gas retailer on a major time crunch. They wanted to optimize delivery of gas, which had two aspects—one was deciding which customer would be supplied by which plant, and the second was optimizing the routes from plants to customers. They managed to get everything done within a month, and the company adopted the findings so it was a big success.

Skills Used Most: Practicality and prioritizing transparency and explainability. Think of Occam's Razor—the point as it relates to data science is that if you have two solutions, the simpler one is probably the best. Simpler solutions are usually better in terms of transparency, explainability to stakeholders, and understanding for data scientists (so they know how it worked, like understanding which features it used). Daniel always remembers a Spanish saying that translates to "Don't kill flies with cannonballs." You probably don't need GenAI to classify text into subjects, for example, because we've known how to do that for a decade.

Primary Tools Used Currently: Python or R (depending on the customer), SQL, and Excel (sometimes it's easier to make a quick chart in Excel instead of cracking open matplotlib or ggplot)

Future of Data Science: Unfortunately, the future is GenAI, even though he feels it largely solves a "problem" nobody has. He doesn't trust it because it makes things up while still "sounding good." He believes there are going to be some bad years where everyone is trying to use GenAI, but they'll figure out it's actually not that valuable and we'll go back to traditional ways of working in data science.

What Makes a Good Data Scientist: Understanding statistics is important, as a lot of people don't (even data scientists). Sticking to your principles is important because otherwise the business can make bad decisions if they're based on wrong or misleading results. This can happen because frequently businesspeople don't understand things like statistical significance, so they don't think they matter. It's important to have standards in terms of quality and rigor and to stand up for what's right when people want you to compromise (the funny thing is that it's for their own good). One more point is that you shouldn't look for specific results because you can always manipulate data to make it "say" what you want—instead you need to keep an open mind and let yourself discover what's really in the data.

His Tip for Prospective Data Scientists: Learn statistics well. Most people (even many data scientists) have only a surface understanding of statistics. Also, don't be overly concerned with having perfect technical skills like programming. You can develop your technical skills on the job.

Dani is a data scientist with experience in a variety of industries. He's also done some data engineering work and is training to be a teacher.

PART III

The Future

CHAPTER 23

Getting Your Hands Dirty: How to Get Involved in Data Science

Introduction

Data science is not one of those things you can just read about and be ready to do successfully, despite what a lot of online scammers would tell you. This is true of most tech fields, but it's absolutely required that you practice your data science skills in order to develop them. Yes, you can develop your skills on the job—this will always happen—but it's far more difficult to walk into an entry-level data science job without any experience than it used to be. The good news is that your experience doesn't have to be working for a company (although that's usually the best)—nowadays almost everyone, experienced or not, needs to have a portfolio that they can share with potential employers. This does mean that even if you have

© Kelly P. Vincent 2025
K. P. Vincent, *A Friendly Guide to Data Science*, Friendly Guides to Technology,
https://doi.org/10.1007/979-8-8688-1169-2_23

work experience, if it's proprietary, you may not be able to share your work without modification. You can always talk *about* projects you've done, but usually it's good to have something tangible to show people.

How do you make a portfolio? You have to do some projects, whether those are personal projects, further developed projects from school, or part of competitions or public challenges like through Kaggle. I'll talk about how you might go about that in this chapter. I'll start with the basics of how to develop your skills (and which need developing), move on to the tools and platforms available to help you develop your portfolio and skills, and finally discuss the many available data sources out there.

Skill Development

As mentioned above the most tangible goal of actually doing some real data science is to create a portfolio you can use in a job search. But the real goal is to develop your data science skills so you can go into your shiny new data scientist position and hit the ground running.

Data Mindset

I've mentioned the data mindset before. It's basically a perspective of looking at data and the world with curiosity and without assumptions, allowing the data to show you what it holds, rather than forcing your beliefs on it to make it say what you want. But it also relates to having good instincts with data, which means knowing how to look at data and when and where to dig deeper. The curiosity aspect is hugely important. Data science really is part art, because finding creative ways to reveal secrets in the data takes more than knowing how to code up a few simple charts. Wanting to know what it really says is critical to being a good data scientist.

A good data science mantra is: peruse with purpose and don't presume.

Real-World Data

One of the points that came up in almost all the practitioner interviews I did was the difference between the data people used in their analytics coursework and what's in the real world. The key difference is messiness. I've talked about it throughout, and real-world data truly is very, very messy. The 80–20 split on time spent doing data prep vs. modeling is not an exaggeration.

So why don't you get experience with this in your college courses? Well, to be honest, every analytics degree program should have at least one semester dedicated to learning how to work with messy data, even though most don't. The main reason you don't learn it in other classes is that they're trying to teach you something specific, and to spend only 20% of your time learning that specific thing because you've spent 80% of your time making the data ready doesn't make a lot of sense. But it downplays the importance of data prep and gives you a false impression of what doing data science is really like.

So just bear in mind that real data will have missing data, weird outliers, and values that are simply wrong for any number of reasons— none of which you will be able to identify easily. Or you'll have one table that you want to join to another table that has really valuable data, only to find that only a fraction of the records exist in both tables. There will always be challenges. You've been warned.

Problem-Solving and Creativity

Some people are naturally better problem-solvers than others, but it is also entirely possible to learn how to solve problems effectively. One of the best ways to learn to solve problems (or just to get better at it) is to practice doing it. Find some data analysis and data science problems (ideally real), and walk through the steps below to figure them out. Every time you figure one out, that's knowledge and experience that will help you with another, future problem.

There are many ways to find projects you can practice on. It's generally best to pick something you're interested in or curious about and even know something about (i.e., have domain knowledge in). It will make the learning more enjoyable. It will also make it easier for you to have real insights into how to proceed. You can always do your own, personal project on data from your life in some way. Or you can find a common problem online like the Titanic problem on Kaggle or another less common one on Kaggle or other sites. I would warn against picking something complicated when you're first starting out because you don't want to get bogged down in details specific to one problem when you're supposed to be developing your overall skills. I also would really recommend starting with some data analysis projects before getting into modeling in order to develop your EDA and analysis chops.

Once you've picked a problem and done some high-level feasibility checking (basically, making sure there's some data for you to work with), start by following CRISP-DM. That means you should start with business understanding, which refers to what is wanted by whoever will benefit from the analysis you're planning to work on. If it's a personal project, that's you, and you should ask yourself what your goals are. Whether you're doing this for practice or have found a problem online, imagine what could come out of this work and identify some questions to answer.

From that point, continue to follow CRISP-DM and you'll work your way through a solution, spending most of your time in data understanding, data preparation, and exploratory analysis and modeling. Validation and evaluation is another important step where you basically check your work to ensure you're confident in your results. Finally, the visualization and presentation you do will depend on what your goals are in terms of practicing. I'll talk about that below.

If you want to get better, do a project, rinse, and repeat. Try to find different kinds of problems or increasingly difficult ones in one domain, especially if you want to go into that for your career.

Presentation and Visualization Skills

We've already talked extensively about the importance of presentation and visualization skills. Here, I'm talking more about presenting your findings rather than charts and other visualizations you make during your EDA as part of a portfolio or presentation in an interview. You should make your conclusions and findings clear and as succinct as possible.

You'll want to make sure all of your visualizations look clean, axes are well-labeled, and you have titles and other appropriate labeling (legends and others). Make sure text is readable (large enough and a non-fancy font) and try to be consistent in both style and color schemes throughout. If you created a dashboard with several components, it would be a good idea to take screenshots zoomed in, in case you aren't able to demo it live. Make sure everything is easy to read.

In general, but especially when you are exercising these skills to create a portfolio, you'll need to create different ones targeting different types of audiences, from nontechnical leaders to data science peers. This is one of the places you can demonstrate some of your communication skills.

Portfolio

For a portfolio, you'll want to clean up your work so you can show it and create some good visualizations and a nice, clean presentation (in whatever format you want). You can absolutely use projects you did in classes, but you'll probably need to tweak and enhance it to make it look more professional.

787

You'll want to have a little bit of everything, from EDA to some code, some visualizations, and some explanatory text along the way, plus your conclusions. A common way to present this is in a notebook like a JupyterLab notebook, which is very appropriate for technical peers. But you may want to also have a slide deck prepared that would be targeting less technical people. You can save any style out to PDF. But the best way to share your portfolio is to have it on a public platform. This could be your website or any of the many data science and visualization public shares, like Kaggle or Tableau (more below on both).

Tools and Platforms

There are a variety of tools and platforms you can use to do and share data science. I'll talk first about coding options and then about visualization, presentation, and sharing.

Coding

You have three basic options for where you are going to write your code and compile your results. One is personal, whether your own computer, a computer from work, or a VM on a shared computing space like your college network. A second is on a public data science platform like Kaggle. And the third is on a general cloud platform, such as Azure, AWS, or GCP. For an individual, the first two are most likely the only truly feasible options because they're free, whereas cloud platforms can be really expensive (it's also easy to accidentally rack up charges, even if it sounds affordable).

Personal

If you are going to write your code on your personal computer or a VM, unless it is already set up, you will need to prepare it to do data science work. That means picking a language (R or Python) and installing it and then installing all the basic libraries needed for doing data science. See the Appendix for details on setting up your personal data science environment, but I'll cover the basics here.

You can install R from the language's project page.[1] There are different distributions of Python you can choose from. For Python, the gold standard has been Anaconda,[2] a full distribution of Python that includes many of the important libraries pre-installed. It used to be common in businesses, but there have been some issues that means companies have moved away from it, but it is still safe for individuals. As an alternative, you can install the basic Python distribution from Python.org.[3]

Chapter 11 lists some of the most common libraries for both R and Python. You can wait to install some of them until you need them, but most people go ahead and install all the common ones at a minimum. If you plan to work notebook style in Python, you will also need to install JupyterLab.

If you are working in R or you want to have more options in Python, you will also need to install an *integrated development environment* (IDE). IDEs are text editors where code is written, and most of them let you write in almost any language after you install a plugin in the IDE. IDEs are great because they have lots of convenient and helpful features like color formatting of code that makes it clear what kind of "thing" a particular bit of text is—like a function, a variable, or some text. It also will include line numbers (very helpful for debugging) and other useful functionality and tools. For R, the preferred one is RStudio Desktop.[4] Python can be used in many IDEs, and currently Visual Studio Code[5] (VS Code) is popular. One nice thing about VS Code is that it will support working with JupyterLab

---

[1] https://www.r-project.org/
[2] https://www.anaconda.com/download
[3] https://www.python.org/downloads/
[4] https://posit.co/download/rstudio-desktop/
[5] https://code.visualstudio.com/

inside it, and you can also work with other languages and formats. Additionally, you can use it for R. Once you've installed your IDE, you can install the necessary plugins (again, see the Appendix for more details).

One of the most important skills for data scientists is SQL, and getting SQL on your personal computer is a little more complicated than R or Python. You could install a free database engine and GUI tool to interact with it, which would allow you to create and query databases. A couple of these are MySQL[6] and MySQL Workbench[7] or PostgreSQL[8] and pgAdmin.[9] There's a bit of a learning curve with these, but it might be worth it to get some more practice. Alternatively, you might just look for online SQL courses and practice tools. These will be discussed in the next chapter.

Free Platforms

Kaggle[10] is another excellent free option, especially if you want to share your work easily. Kaggle also is a place where you can participate in contests to test your data science chops, like the Titanic project. It has a huge number of datasets available that you can play with, which we'll talk about below.

You'll need to create an account if you want to access the notebook tools where you can write R or Python. I recommend heading over to the Code section and exploring the many public notebooks that are there. You can search for things that interest you and see what other people have done. One nice thing you can do in Kaggle is *fork* a notebook, which means to make a copy of an existing notebook into your own space, where you can make changes to it as you want (without affecting the original).

---

[6] https://www.mysql.com/
[7] https://www.mysql.com/products/workbench/
[8] https://www.postgresql.org/
[9] https://www.pgadmin.org/
[10] https://www.kaggle.com/

If you're feeling like you don't know where to begin, Kaggle has some free courses available in the Code section, where you can earn certificates that you can share even outside the platform. These courses are very hands-on and will give you a great starting point.

Once you've gotten comfortable, check out the Titanic challenge, look for data in the Datasets section, or even look at some of the competitions. Create a notebook and get started.

Cloud Platforms

I'm not going to go too much into the cloud options because they're complicated and unlikely to be the right choice for you, mostly because of cost. But Azure, AWS, and GCP are out there, and you can sign up fairly easily. They will all give you a free trial and there are some plans that do offer free service up to a certain amount of usage, but you just need to be very careful to stick with the free options. Azure has a page[11] listing the free options, as do AWS[12] and GCP.[13]

Obviously, they offer these free options as a way to convince you to start spending money, so proceed with caution. At one of my jobs, we did a hackathon to learn Azure, and my colleague created a SQL Server instance, which he never turned on. He read all the docs to ensure we didn't spend any money on it, and when we came in the next day, it had still racked up $75 just sitting there doing nothing. In another instance, my team had some servers created in GCP that we could run VMs on, but nothing was running, and it still hit $300 in less than a week. The docs related to pricing tend to be byzantine. Businesses have people in charge of managing these services who understand the rules, so they avoid accidental massive charges, but it's a risk as an individual. You have to know what you're doing.

---

[11] https://azure.microsoft.com/en-us/pricing/purchase-options/
azure-account?icid=azurefreeaccount
[12] https://aws.amazon.com/free/
[13] https://cloud.google.com/free

Visualization, Presentation, and Sharing

All the coding options discussed above include options for visualization, with some presentation options, but they may not be the best. Additionally, not all allow for easy sharing. In this section I'll talk about the overall options for all of these needs.

Personal

If you're doing everything on a personal computer or space, the most obvious methods of sharing your work including visualizations and presentation are notebooks. You can potentially share these directly in an interview (sharing your screen), or you could download them as PDFs or even raw notebook files (.ipynb) to share with potential employers.

If you're using a visualization tool that you've downloaded yourself, you will have similar options. But there may be public spaces you can upload dashboards you've created on your computer to share widely.

Another option for sharing your code and notebooks publicly, even if all your development is on your computer or a VM, is GitHub, which I'll talk about in the next section.

A final option for sharing your code is your own dedicated website or blog. There is a lot that's possible with this option, including free or low-cost sites like Squarespace and Wix. You don't need to do anything really fancy, but the option would be there to do anything you want. You probably would want to pay for a domain name for this, especially in case you ever think this might be something you'd like to pursue in a larger way.

Free Platforms

As mentioned, Kaggle is an obvious place to develop your code in public notebooks, where you can include visualizations and include nicely formatted notebooks intended to present your project. A lot of people use Tableau Public to share their live and interactive dashboards publicly.

The famous platform GitHub is fundamentally a place to store code with *version control* (changes to code are tracked so you can quickly roll back to an older version if you introduce a bug or something you don't want to keep). But it's also great for sharing and collaborating. The way it works is that the code has to live somewhere on a computer, where it's tied to a copy of the code on the GitHub website called a *repository*. When you update it on your computer, you can push the code to the repository website. Other people can push to the same repository from their own computers if they have access. You can include your notebooks with your results. Unlike Kaggle, you can't run anything in GitHub, but people can view your code on the GitHub site directly. Additionally, you can share your results via text files using a simple language called markdown to create even more nicely formatted results and READMEs. GitHub also has the ability to create what are called branches, which involves taking a snapshot of a repository at a point in time and then editing it only within that branch. This is how most coding is done nowadays, with everyone creating their own branches and working in them before merging the branches back into the original repository when a unit of work is completed. Getting some experience with GitHub wouldn't be bad for your career, either.

Another less common option similar to running your own website or blog is to use a writing platform like Medium or Substack to do blog-like writing with access to a built-in potential audience. This involves writing articles that people can subscribe to. If you do find yourself producing enough content this way, you can even monetize it and get articles published in some of the big data science sites like Towards Data Science.

Cloud Platforms

Cloud is always an option, but I still don't recommend it for individuals. But if you've gone this route, look through the free options for good tools for visualization and sharing your presentations.

Data Sources

Finding data to work with in your projects can be a challenge. But at the same time, there are an almost unlimited number of datasets out there if you know where to look. You have to do your own due diligence in determining quality and potential for bias. Some datasets have good documentation where they're hosted or somewhere else, while others have so little that it's hard to justify using the data (since you won't know what it really means).

Kaggle is an obvious one we've already talked about. There are thousands of uploaded datasets, some of which have several notebooks attached to them, some no one has ever looked at since they were uploaded. They're of varying quality so you should consider that when you look at one. There are also a lot of datasets on GitHub. Sometimes the cloud platforms and learning platforms make data available in their courses, and you can always use that in ways beyond what's taught in the courses.

Various levels of governments and nonprofits often have data stores where they share data that can be downloaded. A large amount of government data is public, but there are also potential sources you can request access to. In the United States, the Freedom of Information Act requires entities to provide certain data on request, such as court records that may not be online but are considered public.

See Table 23-1 for a list of quite a few dataset sources. The website KDnuggets maintains a list of data sources,[14] and you can find more if you Google, especially if you have special interest in certain data. There are also data providers that aren't free but may have exactly what you're looking for if you're willing to pony up the cash. For instance, if you're really interested in sports, there are many companies that have sports data for fantasy sports or real sports. Somebody will sell a mound of that data to you.

---

[14] https://www.kdnuggets.com/datasets/index.html

Table 23-1. *A handful of dataset sources*

Name	Type	URL
Kaggle	Many types	`https://www.kaggle.com/datasets`
GitHub	Many types	`https://github.com/search?q=datasets&type=repositories`
UCI Machine Learning Depository	Many types	`https://archive.ics.uci.edu/ml/datasets.php`
Google Dataset Search	Many types	`https://datasetsearch.research.google.com/`
Data.gov	US demographic and social data	`https://data.gov/`
Data.world	Many types	`https://data.world/`
World Bank Open Data	Worldwide demographic and social data	`https://data.worldbank.org/`
DataHub	Many types	`https://datahub.io/collections`
Humanitarian Data Exchange	Worldwide demographic and social data	`https://data.humdata.org/dataset`
FiveThirtyEight	Many types	`https://data.fivethirtyeight.com/`
Academic Torrents Data	Data from scientific papers	`https://academictorrents.com/browse.php?cat=6`
AWS Datasets	Many types	`https://registry.opendata.aws/`

(continued)

Table 23-1. (*continued*)

Name	Type	URL
Pew Research Center	Many types	https://www.pewresearch.org/datasets/
Nasdaq Data Link	Financial and economic data	https://data.nasdaq.com/
The Global Health Observatory	World health data	https://www.who.int/data/gho/
UNICEF Datasets	World social data	https://data.unicef.org/resources/resource-type/datasets/

Key Takeaways and Next Up

In this chapter we looked at the need to start doing data science if you want to make it a career. You will develop your skills most effectively if you actually practice on real projects. It's also important in any modern career search that you have a portfolio so you can demonstrate your skills beyond talking about them. There are many ways you can start coding and doing data science, from using your own computer to finding an online platform that will let you code there, like Kaggle. Finally, there are so many sources for datasets that you can surely find some about something that interests you.

In the next chapter, I'll be talking more about learning and developing your skills outside of doing data science directly. The best learning resources depend on what you want to learn and how you learn. There are a huge variety of options, including blogs, joining organizations, social media, in-person events, and more.

PRACTITIONER PROFILE: CAROLINE WEI

Name: Caroline Wei

Current Title: Advanced analytics manager

Current Industry: Retail, previously finance (risk management)

Years of Experience: 7

Education:

- MSc Analytics

- Honours Business Administration

The opinions expressed here are Caroline's and not any of her employers, past or present.

Background

Caroline first learned about analytics through an internship during her undergrad business degree, where she was analyzing Amazon data in Excel. She instinctively knew that there had to be a better way. So she started looking into the field and learned some VBA and ways of automating things. She loved analyzing data and especially when it helped companies make more informed decisions. When she graduated, she wasn't interested in the typical path business students take (into consulting or investment banking). Analytics was still fairly new (there were only three degree programs at the time), but it intrigued her and she found a fast-track analytics degree and enrolled in that. She enjoyed learning although it wasn't easy.

Work

Caroline's first job out of her degree was with a bank working on in-depth analytics. She was intimidated at first because her colleagues were mostly PhD's and most from computer science backgrounds and she was having to learn everything (especially math and machine learning), which she ended up loving. She had a great mentor at that job who helped her understand you can learn almost anything on the job if you have soft skills and put the work in. Her imposter syndrome faded after she finally realized she was as valuable as her more qualified colleagues were. There were many interesting problems to solve in that and her current job, where she gets to build solutions end to end and also see the impact of her work.

Sound Bites

Favorite Parts of the Job: Caroline is an extrovert so she loves all the stakeholder engagement in data science. Listening to their questions and the issues they're facing is interesting, as is coming up with ways to help them solve their problems. She also likes the ever-changing nature of the field, requiring constant learning, so nothing ever gets boring.

Least Favorite Parts of the Job:

Caroline's frustrations primarily revolve around two key challenges. The first is the prevalence of data silos and legacy systems, which often hinder seamless data integration and limit the potential of analytics projects. These outdated infrastructures make accessing and leveraging data a time-consuming and complex process. The second is the lack of robust data governance. Without clear ownership, standardized processes, and quality control measures, data often becomes fragmented and unreliable, making it difficult to generate accurate and actionable insights. These issues highlight the foundational gaps that need to be addressed to unlock the true value of data science.

Favorite Project: Caroline's favorite project was a recommendation system she built end to end after learning how a group she worked with was handling recommendations. She pitched the project to several teams because the issue was actually a pain point for multiple teams. The project was her idea after she learned about recommender systems from a friend at another company, and she took ownership and built the system with over 2,000 features and an ML model in 90 categories of products across more than 400 stores in Canada. It was so successful that several other teams were interested in it.

Skills Used Most: Problem-solving and practicality—you don't need the fanciest model, just something that works for the current problem. Expectation alignment—it's important to make sure your stakeholders understand how long everything will take and what deliverables to expect and then keep them up to date with the status of things. Communication in general is important, including learning the jargon and lingo of the domain/business you're working in. A final important one is resilience because not everything is successful—sometimes a project fails, but that doesn't mean you failed.

Primary Tools Used Currently: Python, SQL currently, previously PySpark, SparkSQL, Airflow, Google Cloud Platform, Azure

Future of Data Science: The current data scientist role will probably be rooted out. Simply building models and doing analysis won't be valuable enough soon. A successful data scientist has two paths: (1) get deeply embedded in the business and become the domain expert in that business area, or (2) become a full-stack data scientist with end-to-end skills (which is more like an ML engineer right now).

What Makes a Good Data Scientist: Be a curious and practical problem-solver. Be curious about what the real business problem is. Be practical to find the best solution (not just using your favorite algorithm) and ensure you're delivering business value.

Her Tip for Prospective Data Scientists: You don't necessarily need the degree if you're going to be a typical data scientist, because it's the skills you need, so learn those and make a portfolio that demonstrates you have the skills. Don't forget about soft skills—a lot of people can understand the technical stuff, and soft skills can make you stand out from the crowd both in a job search and in a work setting. Don't forget about the big picture and demonstrating that you can think at that level (strategy and domain knowledge and needs).

Caroline is an experienced data scientist with wide experience in different types of ML for finance and retail.

CHAPTER 24

Learning and Growing: Expanding Your Skillset and Knowledge

Introduction

One aspect of data science is that the field is constantly changing and there's an almost infinite amount of information one can learn. It's not only about having technical skills, or even soft skills, as I've talked about before. It's also important to understand the bigger picture ideas around the place of data science in the world, its limitations, and its ethical consequences.

The practitioners I interviewed for the profiles in the book mentioned a huge variety of resources they've used in their own learning and growth. In this chapter, I'll talk about the many resources you can utilize for learning about every aspect of data science. No one will use all of these, but you can decide which ones will benefit you the most.

K. P. Vincent, *A Friendly Guide to Data Science*, Friendly Guides to Technology, https://doi.org/10.1007/979-8-8688-1169-2_24

A general tip is to look for resources about specific things you're interested in, not general data science skill-building. You can find out about what skills you might want to target by looking at online job descriptions of the kinds of jobs you're interested in.

Social Media

Although it can be a time suck, social media actually has a lot to offer if you look in the right places. Specifics ebb and flow, but many people I interviewed recommended YouTube as a good resource, especially for seeing different perspectives on how to do things and for picking up jargon. It has a lot of career advice coming from individuals in the field as well as more formal channels. #datasciencetiktok TikTok has a lot of resources for beginners and people looking for practical career advice. One practitioner I interviewed said it was following Data Science Twitter that got him excited about data science and exposed him to ideas of what can be done. That community is no more, but there are always going to be new ones. One of the nice things about social media is that you can hear from individuals in short doses day to day rather than larger messages that come in articles. You can learn what kind of personalities go into data science.

Blogs and Articles

Blogs and online articles can be a great source of information for learning specific skills, the state of data science and AI, practical career info, and really anything else.

Medium is an online platform where anyone can publish articles. There's a really strong data science community on there, especially as can be seen by the Medium publication Towards Data Science,[1] one that several of the practitioners I interviewed specifically mentioned. There are many writers on there who publish regular data science content and are worth following, but you'll find that a lot of them produce for a while and then stop. But all of their old content is still out there. A good way to find people is reading Towards Data Science and following the writers of the best articles you find there. One nice thing about Medium is the whole range of topics are out there—technical guides, tutorials, career tips, big picture ideas, and more. Some of Medium's content is behind a paywall (it's up to the creator to set this), but the many good articles on the site are worth the small monthly fee to me. One thing also worth noting is that because anyone can publish there, Medium also has a lot of so-so content or articles that are rehashing the same basic stuff without adding much to the body of knowledge. So proceed with caution.

LinkedIn came up several times for the articles some people publish. There are a lot of data scientists out there writing content for it, and a good thing to do is find companies you're interested in (those that work in domains you're interested in) and start following some of their data scientists to see what they have to say. It's also a good idea to follow big industry leaders.

There are also some independent blogs and sites that people recommended. Several practitioners mentioned Machine Learning Mastery[2] with Jason Brownlee. Daily Dose of Data Science[3] posts weekly articles that dive deep into various topics. KDnuggets is a site that's been

---

[1] https://towardsdatascience.com/
[2] https://machinelearningmastery.com/
[3] https://www.dailydoseofds.com/

around a long time and runs an informative blog[4] and a repository of articles.[5] The Sequence[6] has several newsletters offering different takes of the AI world, some paid and some free.

Courses and Training

I mentioned that Kaggle has some free courses you can take, but there are lots of other learning opportunities available online, many of them free or very low cost.

If you're already working somewhere, your company may provide access to on-demand learning resources like LinkedIn Learning, DataCamp, or Udemy. All of these have some options for learning data science, and there are also options for individuals to pay a relatively low cost. DataCamp in particular comes recommended from a lot of the practitioners I interviewed (especially for career changers). Another good option is Coursera, which has several data science courses. Most of these offer certifications, some of which you pay extra for. These generally don't hold much value in terms of qualifying you for jobs, especially if you have a degree (or are working toward one) in data science, but the learning opportunities are good if you like structure. Certifications and courses are more likely to be valuable if they are in very specific skills—like some of the cloud platforms or deep learning in Python. MIT OpenCourseWare and Stanford also have some good courses.

The major cloud platforms and associated tools often have free trainings available to encourage interest in their platforms. Tableau has training videos,[7] and Microsoft offers a Power BI training track.[8] Google

---

[4] https://www.kdnuggets.com/news/index.html

[5] https://www.kdnuggets.com/topic

[6] https://thesequence.substack.com/

[7] https://www.tableau.com/learn/training

[8] https://learn.microsoft.com/en-us/training/powerplatform/power-bi

for Developers has a site[9] with many courses on a variety of topics. Google Cloud Skills Boost[10] has cloud-specific courses. Microsoft offers Azure training,[11] Fabric training,[12] and data scientist track training,[13] and they also sometimes run learning challenges you can participate in. AWS also has free training and certifications.[14] Certifications in these tools tend to be a little more valuable than the general ones you can get through Coursera and the like because they represent more specific skills. But you should pick these based on where you want to go in your career rather than just getting all of them.

One final note on courses: I highly recommend against signing up for any of the "courses" individual content creators on social media offer. These are largely people who have minimal experience at best and are true scammers at worst. Look for recognized courses and resources.

Books and Papers

Some people recommended books and papers, especially for bigger picture ideas rather than skill-building. Some specific book recommendations include *The Alignment Problem* by Brian Christian, *The Master Algorithm* by Pedro Domingos, *The Signal and the Noise* by Nate Silver, and *The Black Swan* by Nassim Nicholas Taleb. The books *Weapons of Math Destruction* by Cathy O'Neil and *Algorithms of Oppression* by Safiya Umoja Noble are also both very good. Some of these are rather old

---

[9] https://developers.google.com/learn
[10] https://www.cloudskillsboost.google/
[11] https://learn.microsoft.com/en-us/training/azure/
[12] https://learn.microsoft.com/en-us/training/fabric/
[13] https://learn.microsoft.com/en-us/training/career-paths/data-scientist
[14] https://aws.amazon.com/training/

in the fresh-faced world of data science, but these and others like them are important for understanding how data science impacts the world. You can search on Amazon and peruse bookstores for more current books.

A couple of people recommended reading academic papers, including "The Unreasonable Effectiveness of Data" by E. Wigner and "Statistical Modeling: The Two Cultures" by L. Breiman. Watching for whitepapers, articles, and presentations on data science from major companies like Google (Google Research[15] and Google Cloud Whitepapers[16] are both available), Microsoft,[17] Netflix,[18] and Uber[19] can be good. Search for a company and either "technical papers" or "whitepapers" to find more. Whitepapers or white papers are just papers written by an organization and released without being officially published. But many companies have research arms that also do publish in technical journals. You can find a lot. Additionally, if you're still a student, take advantage of the journal subscriptions your school libraries make available to you for rigorous reading.

Research, Google, and ChatGPT

Something that's obvious but that you can forget is an option is just research topics you're interested in learning more about. Google and ChatGPT can both be great for this. Especially with ChatGPT, you can be very specific and drill down, building off of one answer to ask your next question. As always, make sure what you find is accurate.

---

[15] https://research.google/pubs/
[16] https://cloud.google.com/whitepapers
[17] https://www.microsoft.com/en-us/research/publications/?
[18] https://research.netflix.com/
[19] https://www.uber.com/blog/seattle/engineering/

Language, Library, and Tool Documentation

It's also a great idea to look at documentation of the languages, libraries, and other tools you are interested in using. You can find both R[20] and Python[21] documentation online, along with virtually any library or tool you'll use. It's just generally good to get familiar with documentation because you'll be referring to it a lot when you write code. But also, a lot of tool documentation includes tutorials and examples, and it's worth perusing those to see what can be done and how. You'll find that these are very specific to a particular tool or even function or class, whereas you can also find the aforementioned tutorials on Medium that may include several tools to accomplish a wider goal. Both are useful.

Events

In-person conferences and user group meetings can be great for both learning and networking. Industry conferences can be very expensive because usually companies are paying for attendees. If you are already working, you might be able to convince your employer to pay for you to attend if you can convince them it's related to your current job in some way. However, there are also conferences that aren't so expensive or are even free. Additionally, it's worth checking to see if any of the expensive ones have cheaper rates for individuals or even specifically for students. If you're a very outgoing person who gets a lot out of networking and personal relationships, it might even be worth paying a big chunk of your own money to attend one of the expensive ones. I wouldn't recommend this if you're not a great networker.

---

[20] https://www.r-project.org/other-docs.html
[21] https://docs.python.org/3/

A lot of bigger cities have user groups of many of the tools that meet in person or virtually monthly or occasionally, like Tableau, Databricks, and AWS, all of which have groups in my area. This is great for meeting other people in the same area (networking really does lead to getting jobs). But also, you'll learn a lot.

As far as finding things to attend, hit up Google. Look for the things that interest you and that you think you want to eventually work in or local events since more locals will be there to network with. There are also organizations that maintain lists of events[22] that can be helpful. Some conferences have more of a business focus, while others are much more technical. So make sure to check out the agenda on any event you're considering to make sure you will be able to get a lot out of it. Some of the ones to look for include Data Summit, Analytics and Data Summit, ICML (International Conference on Machine Learning), NeurIPS (Conference on Neural Information Processing Systems), and ICLR (International Conference on Learning Representations). There are conferences all over the world.

Note that there are many events that are available virtually. Some of the expensive conferences offer virtual attendance for free. Of course these can be very educational, but you miss out on the networking. But it's a great way to learn what's currently going on in the data science world.

One last comment about all of these kinds of events: Many of them are sponsored by vendors or allow vendors to participate, so they can be pushy and full of hard sells. There's nothing inherently wrong with this, but be aware and adjust your understanding accordingly.

---

[22] https://www.kdnuggets.com/meetings/index.html

Joining Organizations

Another way to learn is to get involved with professional organizations related to data science and other areas. Many organizations offer mentoring programs, seminars, conferences, and other events to their members. Some have certifications you can earn. These are not free, but they are intended for individuals and often have student rates. Some general ones include the Association of Data Scientists[23] (ADaSci), Data Science Association,[24] American Statistical Association,[25] and Institute of Analytics.[26]

There are also many groups intended for under-represented groups. Examples include Women in Data Science Worldwide,[27] Black in AI,[28] R-Ladies,[29] LatinX in AI,[30] and Out In Tech.[31] There are more for different communities—just spend some time in Google to find these niche things. These groups are also likely to offer scholarships and opportunities to give back to the community through volunteering or mentoring.

Working and Volunteering

Obviously one of the best ways to get experience is to do an internship in data science, the absolute best thing to have on your resume when you're applying for jobs. They are out there and they usually pay reasonably well

---

[23] https://adasci.org/
[24] https://www.datascienceassn.org/
[25] https://www.amstat.org/
[26] https://ioaglobal.org/
[27] https://www.widsworldwide.org/
[28] https://www.blackinai.org/
[29] https://rladies.org/
[30] https://www.latinxinai.org/
[31] https://outintech.com/

in the tech world, but they're also very competitive, so don't despair if you can't land one. The process of getting one is pretty much the same as getting a full-time job—an application with a resume, an interview or interviews, and possibly a take-home assignment or sharing your portfolio. So even if you don't end up with an offer, going through the process is good practice for your eventual job search. If you do get an internship, make sure to try to get as much out of it as possible—try to network and learn about other work being done at the company outside of your particular assignments. Informational interviews are commonly encouraged during internships (and regular jobs), where you talk to managers or other people on different teams to learn about other work being done at the company. This is more networking. If you do well, this can also turn into a full-time job offer.

Another option to get some experience is by volunteering for a nonprofit or other organizations looking for someone to do some data work for them. Be cautious here not to overpromise, and be aware that a lot of these kinds of organizations really have no idea what they want or what is possible, so there may be more work for you to figure out what they need and what is possible. But this can be good in a portfolio.

Others

There are other ways of learning. One that comes up a lot in visualization specifically is challenges like Tableau's Makeover Monday, where they share a chart and invite people to redesign and submit it.

Other fun sites are those that show bad examples of charts or other things and then explain why it's bad. Junk Charts[32] does this for visualization and Tyler Vigen's aforementioned Spurious Correlations[33] for correlation.

---

[32] https://junkcharts.typepad.com/
[33] https://www.tylervigen.com/spurious-correlations

Storytelling with Data is a whole brand with a website,[34] book (it's concise and good—*Storytelling with Data* by Cole Nussbaumer Knaflic), a podcast, a YouTube channel, seminars, and more.

Key Takeaways and Next Up

This chapter looked at the huge variety of resources available for learning more about data science, from technical skills to career info to how it affects society. The more you know, the better data scientist you will be. Social media provides opportunities to learn about the community, blogs and articles cover a wide range of valuable topics, books and papers are especially good for big picture ideas, researching on the Internet is tried and true, reading documentation from tools you'll be using is useful, events can be great for both learning and networking, joining organizations can provide many opportunities, and gaining real-world experience through internships and volunteer work can be great.

In the next chapter, I'll talk specifically about how you can break into data science as a career. I'll talk about the many different types of data science and related jobs out there and then the important considerations when searching for a job, including picking the right industry and domain, right kind of role, and right kind of organization. I'll give some practical tips and address salary.

---

[34] https://www.storytellingwithdata.com/

PRACTITIONER PROFILE: RACHEL BAKER

Name: Rachel Baker

Job Title: Academic consultant, language learning products

Industry: Language education

Years of Experience: 20 years, including her PhD (14 years in industry)

Education:

- PhD Linguistics

- MSc Speech and Language Processing

- BS Cognitive Science

Background

Rachel always liked language, but her undergrad college didn't have a linguistics major, so she chose to study cognitive science. Cognitive science is a combination of linguistics, psychology, and computer science. It turned out to be perfect for her because she loves interdisciplinary work. She found that working across disciplines is a great way of avoiding going down research rabbit holes tied to individual disciplines' assumptions, a belief that stayed with her into her career. After finishing her bachelor's, she taught English in France for a year, which was her first hands-on experience with language teaching. She continued her education with a one-year master's in Speech and Language Processing, still in linguistics but leaning a bit more toward computers. During that degree, she learned that she loved phonetics, phonology, and prosody (areas of linguistics that focus on sound and the way things are said). As she was finishing the degree, she felt like she still didn't know enough to start working in the field, so she went on to a PhD in Linguistics, where she studied prosody and language acquisition.

Work

During her degree, she loved doing research and analyzing the results of her studies. But after graduating, Rachel wanted to do more applied work, so she went into industry. She was hired at a major company in the language learning space, where she stayed for ten years. She quickly realized that her belief that she "didn't know enough to start working" after her MSc really wasn't true— she totally could have. But the PhD qualified her for certain jobs that the MSc wouldn't have and helped her develop skills she uses daily, so it was still quite valuable to her. Her first job was working on language learning products, which involved analyzing data to support product development. In general, being an academic working in industry meant wearing a lot of hats. Her primary role involved three main areas: planning and developing education materials, assisting with the development of digital products, and analyzing data from teachers and students. The data analysis portion of the job was fairly involved, as she carried out lesson observations, interviews, and surveys with teachers. Most of the education products she worked on were used by teachers in their classes rather than by students directly. Additionally, she analyzed teacher and student usage and performance data that came directly from the products. Wearing many hats meant working with very different kinds of people, from educators to product designers to software developers. Different people had different perspectives, and she had to learn to work with people who weren't like academic researchers—deeply methodical and careful. Industry is much faster-paced, so teams often use less precise methodologies in order to get the information they need faster. One of her additional roles at the company was to act as a liaison with university researchers that the company collaborated with, something she found very rewarding.

Since that first job, she has worked for several other education companies and is now working as an expert consultant on language learning products. She loves the freedom and flexibility it gives her because she works remotely and part-time, so she sets her hours and schedule, which suits her family responsibilities right now. Additionally, she likes the chance to work on a wider variety of projects for different people and types of organizations. It also gives her time to do some of her own projects, such as collaborating with a former colleague to bring a language test preparation course to a larger market. She is also looking into creating products based on language learning techniques that have been shown to be effective in research but have not been incorporated into commercial products.

Sound Bites

Favorite Parts of the Job: She liked working on a team to make products that tried to solve real problems and engage their users. It was satisfying to improve the products over time. She also liked encouraging teachers to use more communicative lessons and more evidence-based teaching methodologies.

Least Favorite Parts of the Job: Sometimes, conflict at work can be very stressful. She disliked it when someone in a position of authority would make decisions for the team without good reason or data, overriding team members with more knowledge or experience. It was especially difficult to try to make a successful product in that kind of environment.

Favorite Project: Rachel had two projects she's really proud of. One was fairly early in her career when she worked with university researchers to develop a corpus of language learner writing. This was exciting because it filled a gap in the data available on language learners. The corpus they created had different levels of students (including lower levels) and came from an education setting rather than from test-takers. The other project was developing a digital product for teaching English in high schools. It offered teachers real-time data on their students' performance, which they could use to make decisions in and out of the classroom.

Skills Used Most: The first skill is her deep knowledge of language learning. She also uses her data analysis skills, including structuring data for visualization, running statistics on it when appropriate, and being able to explain her findings as well as the limitations of the data. She also uses more advanced analysis skills, especially the ability to take lots of different types of data and information from various sources and bring it all together to come to meaningful, accurate, actionable recommendations that can be prioritized.

Primary Tools Used Currently: She mostly uses Excel for simpler analysis and R for more complex work.

Future of Data Science: Rachel thinks GenAI is a big curveball, but it's here to stay, and we'll have to figure out how to work with it being around. Her biggest complaint about it is how it's deceptive—it makes it easy for non-experts to produce something that "seems" good but actually isn't. Other non-experts think it's great, but the second an expert lays eyes on it, they know it's wrong. For factual texts and texts with specific language requirements, it may not save time because of the time you have to spend on fact-checking and editing. So she has only found limited uses for it in her work, but she has begun advising companies on how to improve their own GenAI products.

What Makes a Good Data Analyst: The most fundamental skill is logical thinking, which leads to solid data analysis skills. You need to be able to see the wood from the trees because it's easy to get lost in the numbers, trust them blindly, and forget to think about how the data came to be. Always wonder if the data is capturing what you think it is or if it could actually be measuring something else, based on how it was collected. The same is true for the results you find—are they really what you think they are? Additionally, it's important to be humble and able to talk to all kinds of people, some of whom may have a different understanding of the data or how it was collected than you have. Reaching out across disciplines can be a great way to find a different perspective, which might improve your understanding of the data.

Her Tip for Prospective Data Analysts and Scientists: For anyone close to starting a career, think about the problems you want to solve in the world and let that guide your job search (or academic career). Rachel feels lucky that she somewhat stumbled into the right career for her, but there were lots of jobs that she wasn't aware of when she was first looking. Also, pay attention to whether you want to be domain-specific or domain-general. Academics have usually studied something very deeply to develop their expertise, so it can be hard for them to work outside of that area. But data scientists, software engineers, and many other data jobs can be generalists—you could work for a banking company and then move on to a retail company with minimal fuss. Neither way is right or wrong, but it's good to think about it when looking at starting a career.

Publications That She Would Recommend for Learning More Data Science: Rachel would recommend the *More or Less* podcast, hosted by the economist Tim Harford. The show digs into statistics in the news, determining whether they're reliable and providing much-needed background and context on the source data and its analysis. The show offers a useful and accessible perspective to anyone interested in data science by illustrating common pitfalls in data collection, analysis, and communication that happen in real life. She'd also recommend his books, such as *How to Make the World Add Up,* which explores similar topics in more depth.

Rachel is a language pedagogy and technology professional with over two decades of experience working to make language education better.

CHAPTER 25

Is It Your Future? Pursuing a Career in Data Science

Introduction

If you're reading this book, you probably are interested in becoming a data scientist. But there are actually a lot of roles around data science, and you might find you're as interested—or even more interested—in those, once you learn a bit about them. I'm talking about three different types of jobs in this chapter. The first group are data-focused, and these include data scientist, data analyst, and BI engineer. The next group is engineering, and there are quite a few jobs under that umbrella, including machine learning engineer, software engineer, and data engineer. Finally, there are sales, business, and management positions, which include sales engineer, business analyst, and project manager among others. I will also address how to pick the right positions to apply for beyond the particular role and how to actually get the job.

© Kelly P. Vincent 2025
K. P. Vincent, *A Friendly Guide to Data Science*, Friendly Guides to Technology,
https://doi.org/10.1007/979-8-8688-1169-2_25

Data Science and Related Jobs

There are a lot of job titles out there and a lot of companies using them in different ways. If you start looking through job boards and compare job titles with their job descriptions, you'll see things are all over the place. In this section, I'm going to try to talk about each job title and its "Goldilocks" description—what most people mean when they use that term. But bear in mind that these things are also changing all the time. The jobs world is one you're just going to have to constantly research to stay on top of.

An Aside: Further Education

One thing I will mention is that if you're finishing up a bachelor's in data science, you don't necessarily have to go into the work world. You could always continue to grad school, a master's or even a PhD, or other doctoral degrees. You might even find the job search difficult after only a bachelor's and think a master's would make you more competitive. Data science is a fairly applied field, so most people who study it do go into the workforce, but the skills you've learned qualify you to study other subjects, too. I would recommend going deeper into a particular subject if you're going for a master's, rather than doing another data science degree. For instance, if you loved the stats portion, you might pursue a statistics master's or PhD. Applied statistics is another option for a master's, which tends to be more industry-focused and most people do it as a terminal degree. You may want to go deeper into machine learning or deep learning specifically, and you may find degrees in those disciplines. You may want to get more technical and do a computer science degree or skill up as a programmer and do a software engineering degree. You may want to pivot a bit and go into more general AI or dive deeper into NLP. Or you could focus on a particular industry and find a degree in healthcare analytics or finance. There are many options, especially if you're planning to pay for any further degree.

If you want to go onto a funded PhD, you'll need to prep a bit more in advance for that. You'll want to get some research experience and ideally establish a relationship with a professor who will serve as a mentor to help you get accepted and funded somewhere, and you'll need other professors who can enthusiastically recommend you. If you think this might be a path you want to follow, try to get started getting research experience as early as possible in your degree.

If you feel yourself drawn to teaching either at secondary or higher education level, you could pursue a teaching master's (for secondary) or another master's for higher education. There's a growing demand for teachers of STEM subjects in secondary education. Tenure-track positions aren't usually available to people with only a bachelor's, but adjunct positions and community college positions are available to people with a master's. A lot of people in technical fields supplement their incomes with teaching in continuing education programs, adjunct positions, or at community colleges.

Data-Focused Jobs

There are three jobs I'm including under the data-focused category: data scientist, data analyst, and business intelligence (BI) engineer. I think of these as being on a continuum, because there are a lot of overlapping skills and responsibilities, which can vary at different companies. At some places, a data scientist would never make a dashboard to be shared with a stakeholder, but at others they might. Figure 25-1 shows the continuum as it exists now.

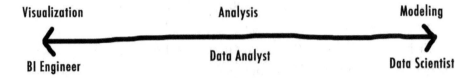

Figure 25-1. *The BI engineering–data analysis–data science continuum*

Because of the fluidity of these titles, see Table 25-1 at the end of this section for some of the other job titles you might find these three roles listed under.

I'll start in the middle, with *data analyst*. I've talked previously about how there's a lot of overlap between data analysis and data science, but they are distinct things. A Goldilocks data analyst will take various data sources and perform EDA and other types of analysis to identify trends in the data and draw conclusions. They're usually focused on descriptive and diagnostic analytics. They might also perform statistical analysis on data and even help design and run statistically sound experiments. Although they often will use SQL to query data, they typically do not write a lot of code or use very advanced techniques beyond statistical testing, although some may do linear regressions. When they are doing either of those, this is when they are leaning more toward data science work. On the other end of the spectrum, most data analysts will make visualizations as part of their analysis using basic tools, and some may create dashboards in enterprise tools that are refreshed on a schedule and will be available to stakeholders for regular use. When they're making dashboards using company visualization tools, they are leaning toward the BI side of the spectrum. Data analysts sometimes do peer reviews like BI engineers and data scientists, but it tends to be less structured than many roles. They will also spend a decent amount of time validating their work, especially looking for data and analysis quality and statistical assumptions.

Following from that, a *business intelligence engineer* is someone who primarily makes visualizations for customers to use. Generally, they do not have to do a lot of data preparation because usually they're using data sources that have been prepared by other teams for use in many areas. So BI engineers usually wouldn't do extensive EDA or other analyses, but they might do some exploration and validations to make sure the data they have meets the expectations based on what they've been told. They will be using the company's enterprise (professional, official, and usually company-wide) visualization tools, often Tableau or Power BI, but there are some other cheaper ones that companies try to get away with or specific ones embedded in other tools the company has committed to. Most BI work involves making fairly complicated dashboards that will contain several individual charts, tables, and other components. If they do anything beyond building dashboards, they're leaning toward the center of the continuum. BI engineers will test their systems and do peer (code) reviews with others on their team and have their own reviewed. They're generally focused on dashboard accuracy and up-to-date information.

At the other end of the spectrum is *data scientist*. This book has focused on the responsibilities of this role, so you already have a pretty good idea of what a data scientist does. They work with data to find trends and draw conclusions the way data analysts do, but they tend to use more sophisticated tools like machine learning, and they also take things further by doing predictive and prescriptive analytics. Data scientists use code to do most of their work, usually SQL and Python or R, although sometimes they'll work with other languages. When they're doing the more basic work or even creating visualizations to share with stakeholders, they're definitely leaning toward the middle of the continuum. In a lot of less-mature companies, a data scientist really does cover the full continuum. One last thing worth mentioning is that a data scientist could be responsible for productionizing their models (putting them somewhere someone else (or other software) can use them). This is ideally done by an ML engineer (see the next section), but data scientists often end up responsible for this.

It can be easy or difficult, depending on the platforms and infrastructure available at the company. Data scientists also must validate their work and test and review their own and other people's code. They will be looking to ensure their models are rigorously tested, performant, and reproducible.

The lesson here is that you're going to have to work to figure out what kind of job a particular job listing represents and look widely when you're looking for a specific type of job.

Table 25-1. *Some of the other job titles you'll see data-focused jobs listed under (not exhaustive, and some of these more commonly mean something else, like business analyst)*

Business Intelligence (BI) Engineer	Data Analyst	Data Scientist
BI analyst	Analytics manager	Analytics manager
BI architect	BI developer	Business analyst
BI developer	BI engineer	Data analyst
Business analyst	Business analyst	Data engineer
Data analyst	Data manager	Data modeler
	Data scientist	Data storyteller
	Research analyst	Machine learning engineer
	Research scientist	Statistician

Engineering Jobs

There are quite a few engineering roles that support data science work. Like with the data-focused jobs, these can involve work falling under various other labels, and some jobs involve wearing many hats. But I'll talk about four different engineering jobs that someone interested in data science might also be interested in. Note that all of these involve doing lots of peer (code) reviews and testing of code and systems they develop, along with validation when data's involved. See Table 25-2 for some of the other job titles you might find each of these listed under.

A *data engineer* is basically someone who prepares data for use by other people. It's a pretty technical role and does require expertise in several tools and usually coding. Their primary tasks are data modeling and ETL (extract-transform–load) or ELT (extract–load–transform), depending on platform options. Data modeling is designing tables or other storage locations for data and then also preparing them, such as creating a table with all the right data types. ETL is basically pulling data from one place, transforming it as appropriate, and loading it into the new location they have designed and prepared in the data modeling step. The transformations in ETL can be anything from small and trivial to massive, complex, and time-consuming. There's a shift occurring to ELT over ETL, which is available with many cloud platforms and allows the data to first be loaded into a central location and then the transformations can be done. The tools used to accomplish this will vary from company to company. Many use tools designed for this that handle a lot of it without having to write extensive code, where others still require coding. Most will require SQL even if they are otherwise low-code.

An *analytics engineer* is really just a specialized data engineer, with deeper knowledge of analytics and the specific needs of data analysts, data scientists, and even BI engineers. The tasks really aren't very different, except they would know more specialized transformation and potentially different data storage methods. They sometimes use specific tools to prepare data for self-service analytics.

A *machine learning engineer* is someone who works fairly closely with data scientists to deploy their models. This can be as simple as putting a model in a production location that other systems can access or taking the code data scientists have written and putting it into a production system or be much more involved, like reworking or even completely rewriting the code. In rewriting the code, an ML engineer might write it in the same language but optimize it to run faster and more efficiently at scale, they might rewrite it in another more efficient language (with the same goals), or they might implement it in another tool. ML engineers are very skilled at writing efficient code, and they also know a lot about the needs of analytics. Many of them have skills in deep learning (neural nets) beyond what many data scientists have. Usually, ML engineers are skilled with the entire data science pipeline, from data ingestion (usually data engineers' domain) through data prep and the other data science steps and then deployment. They usually follow MLOps and frequently use specialized tools for containerization, orchestration, and more. Data scientists can also own the whole pipeline at some places, but the difference is that ML engineers tend to be more technical with coding and infrastructure, while data scientists tend to be more comfortable in the data prep and modeling steps. What determines an ML engineer's specific tasks is the team they're on.

A *software engineer* is a classic role that isn't inherently related to data science or machine learning, but sometimes these roles can exist on a data science team, and other times the role involves doing more than software engineering, which can include data science. On a data science team, their responsibilities would possibly include both data engineering and ML engineering. They'd likely support data scientists in many ways, including building APIs that expose machine learning models on the Internet, designing infrastructure for data pipelines, and ensuring that all team code follows CI/CD workflows. Software engineers fundamentally write software, usually by writing code on their computers or within a specific platform. But some may use other tools to develop software that will run within another system, such as with a click-and-drag interface (though usually with these, there's still some code to write). The realities of a given software engineer's daily work vary based on company. Software engineers use version control systems like Git and often use integrated testing and deployment infrastructure. This involves writing tests and doing various amounts of testing on the software they write. A key feature of a software engineering role is that they really do spend the majority of their time writing code, where all the other engineer (and data scientist) roles have a bigger mix of tasks and won't spend as much time coding.

Just like with data-focused jobs, you'll have to do some digging to figure out roles that might interest you.

Table 25-2. *Some of the other job titles you'll see engineering jobs listed under (not exhaustive)*

Data Engineer and Analytics Engineer	Machine Learning Engineer	Software Engineer
Big data Engineer	AI engineer	Application engineer
Data architect	Analytics engineer	Back-end developer
Data scientist	Data scientist	Developer
Data warehouse engineer	Deep learning engineer	Front-end developer
ETL developer	ML analyst	Full-stack developer
	MLOps engineer	Programmer
	ML scientist	Software developer
		Software development engineer

Business Analyst Jobs

There are roles that are good for people who understand and enjoy the world of data science but may not be as interested in doing the day-to-day technical work. They require technical expertise, but not at the level of data scientists' or engineers'. There's one job title in particular that means a hundred different things at a hundred different companies, but I'm using *business analyst* here in the sense of someone who knows the business (certain areas within a company) and analyzes data and processes to figure out better ways of doing things. They can also understand and work with data and analytics people. This role can be similar to a data analyst role, but most people with this job title aren't quite as technical. They definitely use Excel widely and may even use SQL for getting at the data they need. Business analysts propose solutions to improve the things they're responsible for. They are usually involved in project planning in the areas they're assigned to. This job involves a lot of communicating with different people in the business, sometimes involving conveying technical information. This role appears under many different job titles. Ones that relate to data include data analyst and analytics translator.

Sales Jobs

If you love convincing people to do things, tech sales might be a good place for you. There are several different roles in sales that involve different levels of technical know-how. Note that these positions are usually partially commission-based, which does mean that you can make a lot of money if you're good, but much less if you're not as good at sales as you thought you'd be. All of these jobs require excellent people skills, and you will frequently be presenting to or talking with very senior-level people, especially when a customer is new. These roles require a lot of confidence and assertiveness and perhaps a willingness to stretch the truth a bit. There may be travel involved, depending on the company, so sometimes you're responsible for schmoozing and entertaining potential or existing customers. And most importantly, you wouldn't be doing much actual data science in these roles, but your knowledge of the field is useful.

There are several job roles that require very little technical knowledge, which you'd probably want to avoid if you're trying to capitalize on your new expertise in data science. Sales (or business) development representatives reach out to potential customers but then pass their leads on to an account executive. An account executive is basically responsible for making the sale, and then the account gets passed to an account manager, who manages the relationship with the help of other, more technical people.

The *sales engineer* position works with other sales staff to develop solutions that help potential customers, all in service of closing the sale. This role requires a good technical understanding of the company's products specifically, as well as a broad understanding of the industry and landscape of products like these. It also requires great people skills and a persuasive personality.

The *customer success manager* role is one that works closely with account managers and their customers to help them adopt and use the company's products after the sale has closed. Selling is a part of this especially (additional products or services), but that is mostly the account manager's responsibility. This role requires a medium level of technical expertise because they need to understand the company's products well enough that they can guide customers' usage. So the knowledge they need is not just general, but they'll need to learn about the company's products on the job. Having a good technical foundation is a good way to jump-start that. This still requires a lot of people skills, as the relationships they have with customers should be close and long-term. They also are expected to advocate on behalf of customers, which means they can bring customer needs back to the company and subsequently guide development of the products.

At some companies, the *solutions architect* role is virtually identical to the customer success manager role, but at others it's far more technical. As the name implies, this role involves actually creating solutions for (or with) customers. The goal is of course sales—keeping customers and selling them more things—but it happens through technical expertise more than wining and dining. It still requires good skills for communicating with customers and understanding what they really need, even if they themselves don't know, and then knowing what to do with that info.

A final sales role I'll mention is *partner sales manager*, which is somewhat technical. But rather than working with customers directly, this role involves working with intermediaries, other companies that are using the company's products to serve their own customers. It's similar to the customer success manager role except the relationship is with these intermediaries rather than the companies directly using the products. But they still can be involved in bringing customer needs back to the company. Table 25-3 at the end of the section summarizes some of the available positions.

Table 25-3. *Sales job titles and level of technical knowledge (not exhaustive)*

Lower Technical Knowledge	Medium Technical Knowledge	Higher Technical Knowledge
Sales development representative	Customer success manager	Sales engineer
Business development representative	Partner sales manager	Solutions architect
Inbound sales representative		
Account executive		
Account manager		

Management Jobs

A final category of jobs related to data science are related to product, project, or people management. Like the last ones, if you like data science but don't want to do it all day, these might be good to look at. Historically, technical expertise was generally a nice-to-have with these jobs, but technical expertise is becoming increasingly valued. These roles don't have a lot of different job titles they'll appear under. There aren't as many entry-level opportunities with these roles, but someone with experience could easily move into one of them.

Project managers are the people who are in charge of getting a project going, keeping it on track, and seeing it finished. The specifics can vary, but usually they're assigned to a new or existing project with people assigned for the various roles on the project. They're generally responsible for planning; creating and managing budgets and schedules; establishing appropriate regular and special meetings; monitoring work progress to ensure work is progressing appropriately; helping to clear roadblocks as they come up; communicating with senior leadership, stakeholders, and project members; and ensuring that once all the work is done, all closing tasks are completed. They use appropriate enterprise tools to help them manage projects. As implied above, in a lot of companies, project managers are generalists and don't really have any technical expertise (or even specific expertise in anything beyond project management) and can be assigned to any project. But at other places, having that technical expertise could qualify you for specific projects or just make the job much easier. This is definitely a role that takes leadership, communication, and general people skills, especially the ability to motivate people. Sometimes the role also requires being a bit pushy, both on people on the project and on others outside the project when roadblocks need to be cleared. Sometimes teams have their own project managers, so if you're specifically interested in managing data science projects, you could look for a role like that. Additionally, if this is something that interests you, you can get a certification in project management to help you stand out.

A related position is *program manager*, which is basically an advanced project manager role—program managers usually manage several projects that fall under a larger effort (a program) and manage their strategic alignment. A lot of places don't have these roles because they only exist at larger companies. The role is basically the same as project managers but comes with more responsibility, and they speak to senior leaders more than a typical project manager. This is generally a senior position for experienced project managers.

Another position similar to project manager is *product manager*, with a lot of overlap between the two roles. But a product manager is more focused on a specific product and manages the lifecycle of that, rather than individual, isolated projects. Like with program management, there may be multiple projects going, all related to a specific product. Many of the actual tasks are very similar to project management, but with a wider scope. However, product managers have expertise and technical understanding of the particular product they're responsible for and will help even with ideation of the product itself. They're often responsible for putting the product team together in the first place, or finding additional people for it, and planning immediate and long-term work. They handle much of the communication with stakeholders. They're also responsible for paying attention to competitor products, managing strategy, and planning launches. They also look at performance and reception of the product, collecting and analyzing feedback. Like with project managers, leadership and soft skills are crucial.

A role that shares some responsibilities with project manager is *Scrum master*, but they're actually quite different in important ways. First, a Scrum master is a specific role within the Scrum methodology, which as I've previously said is not ideal for data science. So this is probably not a role to pursue if you want to stick with the data science world, but if you are interested in going more generally into tech, it might be worth considering. One of the key traits of Scrum is that teams should mostly self-organize and operate, which makes this role different from a traditional project manager. The Scrum master mostly ensures that the Scrum methodology is being followed correctly by everyone, and they facilitate and organize the daily standup meeting and all of the ceremony meetings (retrospectives, sprint planning, and sprint demos). They help protect the team from distracting work, ensure that the work tracking tools are being updated regularly by team members, and report on progress to leadership. They also help clear blockers and manage conflict within the team as needed. The skills needed here are similar to the previous roles'—leadership, communication, and empathy. Although not always required, Scrum masters generally need technical skills related to the work they're managing because they need a fine level of understanding to manage the work effectively. This is another role you might consider getting a certificate if you want to break into it.

The final role I'll talk about is *people manager*, usually just called manager (but so many roles nowadays have the term "manager" in them even though they aren't true management roles). A manager generally leads a team of people, usually referred to as individual contributors to distinguish their roles from leader positions. The manager is responsible for the general running of the team. This involves working with each team member on things like performance, goals, career growth, and disciplinary issues. They provide guidance to the team as a whole and usually help define a vision for it. They delegate assignments, specify deadlines, and assign training as appropriate. They also handle interpersonal and other conflicts as necessary. People skills are the most important for good managers. Managers spend a lot of time in meetings on higher-level topics like strategy, goals and timelines, project selection and planning, and many other things. Core skills are leadership, communication, decisiveness, and (ideally) technical expertise. At a lot of companies, people managers aren't expected to have any technical knowledge, but many places understand that a manager with an understanding of the work being done is a better manager. This is probably not a role that someone with only a bachelor's in data science could get straight out of school, but it could be something you strive for if it appeals to you.

Job Levels

Almost all jobs have different levels. You'll see these in job listings, and they can be largely incomprehensible because there's no rhyme or reason to most of them. You have to look at the listings and the years of experience they're looking for to figure it out. Some places use three levels, with "junior" at the bottom, "senior" at the top, and no label for the middle roles. These don't have any meaning on their own either—senior might be five years of experience at one place and fifteen at another. "Principal" is another common label along with those, indicating someone with extensive expertise in their field. It's usually the top of the chain in a job hierarchy. Some places use "lead" as similar to senior or between senior and principal. You'll also see the term "associate" added before a job title to indicate a junior-level position. Other places use numbers to indicate the level (often Roman numerals to add to the craziness), but there can be any number of levels, defined per company. Other companies use the term "staff" to indicate more experienced people. Other places do the same with the term "scientist." One thing that's frustrating when you're looking for entry-level positions is that almost nobody labels their listings that way. You'll have to click in to find if they are requiring experience or not. Even more frustrating is when they do label it entry-level and then you click in and they're requiring a year of experience. As you search for jobs, you'll start to learn how different companies use the levels, so if you find you're qualified as a level II data scientist at one company, you can start to filter on that label in other jobs at that same company.

Picking the Right Job(s)

Finding a job you love isn't an easy task. This is especially true if you've never worked before in the corporate world because it can be hard to imagine what it's really like day to day. I'll talk about some of the most important considerations, but I recommend you dive into job boards and start seeing what's out there (LinkedIn and Indeed are a couple places to start).

Note that there's a basic structure to a job listing that's pretty commonly followed. They start with a little info about the company and the job, trying to convince you why it's a great place to work and how great this opportunity is. Then they'll list the basic responsibilities of the job usually in paragraph form, and then there will be a bulleted list of required skills and experience. There may be some more info, and then sometimes they'll list preferred skills either in a paragraph or another bulleted list. Usually, they'll include a bit about benefits next. Finally, they'll include the legal verbiage at the bottom of the listing (equal-opportunity employer and so on).

The Type of Role

Trying to identify what you really want to do for eight-plus hours a day five days a week is admittedly difficult. If you read about the many possible jobs in the previous section, did anything appeal to you more than the others? Think about how you felt when you were deep in coding for a project or when you were doing EDA, trying to determine the best algorithms to use, creating visualizations, or writing up your results. Which parts did you like the most? Did you like collaborating with other people when you worked on group projects? (However, it is worth mentioning college group projects aren't that similar to working with other people in a job, although there usually still is that one person who doesn't do much.) Did you have opportunities to present your work, and if so, did you like planning the presentation, writing the slides, and giving the presentation? Do you like explaining complicated things? Do you like planning in general? Do you love wrangling other people in group projects?

These are all things to consider when you're thinking about the kind of role. Remember, it's not only what particular job role you want to identify, but the kind of role within that label. If you really don't think you can ever learn to talk with stakeholders, you're really limiting yourself, but that means you need to look carefully at job descriptions and see if you can suss out how much you would work with stakeholders. Some roles definitely have more of this than others.

It's also worth mentioning that it can be difficult to get a sense of what a job will be like from a job description, because a lot of them are ridiculous, pie-in-the-sky wishlists of skills. At one job, the leaders were so excited about AI (It's amazing! It can do anything!) that when we needed to hire just a regular data scientist working on retail data, they listed robotics, speech tech, and image processing as desired skills. And, no, they didn't have some secret project in the works; they just didn't know what they were doing. This is common with recruiters and even (unfortunately) sometimes with hiring managers.

The problem of recruiters not knowing what they're doing also shows up in the inconsistent use of job titles that were mentioned above, like with a data analyst position requiring deep learning skills. There can be many reasons a job is listed under an inaccurate job title. Sometimes it's so the company can pay less for a lower-status job title, sometimes it's because the people listing the job don't know what they're doing, and other times it's because there are weird legacy job titles internally.

One thing to definitely consider when you're trying to pick a role is career growth. They may talk about it in the job listing itself, which is a good sign. But you can also figure it out by looking at other similar jobs, especially at different levels, at the same company. If there are clearly named levels, that's a good sign that you will be able to grow through the ranks. If you're looking at a position that will be the only data scientist at a small nonprofit, there isn't as much clear growth potential there. That doesn't mean it's not a good job for you, but it is something you might talk about in an interview or at least consider if you want to advance quickly. But also note that taking an isolated role like that, getting some experience, and then moving onto a new company is totally normal in today's work world.

Consulting or Direct Hire

I've talked about all the jobs above as though they were direct positions with a given company. But almost all of them can also be done through consulting companies. You should definitely consider whether you want to try consulting or not. Sometimes it can be hard to tell if a job in a listing is through a consulting company or not.

Working for a consulting company can be a good way to get a wide range of experience, and if you are struggling to figure out which direction you want to go, it might be worth exploring as an option because you can get exposure to different kinds of work. There are positives and negatives in working as a consultant.

Consulting is different depending on which consulting company you work for. Your assignments could be relatively short, or you could end up assigned to the same company for a while, even years. Your experience on assignment can vary widely as well. At some companies, consultants are looked down on and treated as second rate, especially when companies hire long-term consultants because they don't want to pay enough to cover the benefits they'd have to provide if they made them permanent staff. Other consultancies are huge and relatively well-regarded by the industry as a whole and by leadership.

The Industry

As we've seen in previous chapters, there's virtually no field that hasn't at least dabbled in data science. There may be certain ones you're drawn to or find tedious. It can help you be happy at work if you have at least a passing interest in your company's core work. But at the same time, it's not required.

Of interest to those looking to go into data science, different industries have different maturity in that regard. Insurance and finance have been using statisticians for decades, so everyone trusts the work and you're not constantly trying to prove that your work is valuable, useful, and to be trusted. This is something you'll face in a lot of other industries that are only starting to get on the ML bandwagon.

If you are interested in an industry that is newer to data science, you might look specifically at marketing jobs at companies in these industries, because a lot of them have been doing more advanced analytics for longer than the rest of the company.

Thinking About Your Values

But picking an industry to work in can also come with implications and choices about your values and beliefs. Not everyone is suited to every industry because of these, and you should think about these things before applying for jobs.

I'll give a couple examples from my own career. I personally have qualms with the insurance industry as a whole, but especially the health insurance industry, so I will never apply for those kinds of jobs. I once applied to a job at a religious hospital, and during the interview I was told that they pray at the beginning of all their meetings and everyone has to say a prayer. I was surprised, but it was good info because I don't want to mix personal and work life to that degree, so I did not apply for any more jobs at that hospital.

In other examples, maybe you don't want to work for a pot company, a gun company, a sugary drinks company, or a defense contractor. Many large tech companies use their significant wealth to sway political leaders to make laws that benefit themselves and their shareholders. These same companies often play fast and loose with their customers' data. Some companies knowingly cause harm to groups of people who are already at a disadvantage. If any of these issues are things that you don't like, you should avoid those companies when applying for jobs. It's up to you to do the research before applying.

Additionally, sometimes a company overall is fine, but some of the jobs within that company might give you pause. For instance, being a data scientist for a human resources team might involve "optimizing jobs," another way of saying call for layoffs. Somebody will do that job, but if you don't want it to be you, you can avoid it. Alternatively, some teams are trying to be good in what other people might view as a bad situation. For instance, in the pre-Musk era of Twitter, despite the fact that a lot of people were troubled by some of the content on the site, there were teams that were responsible for monitoring and trying to stamp out hate, misinformation, and other harmful content. You might have been comfortable working on a team like that, but not for some other area of the company.

It is also worth pointing out that not everyone has strong opinions about these kinds of things, and that's fine. If these aren't going to be things that niggle at you, then don't worry about it. But if you do care, don't get yourself into the position of doing a job or working for a company that you're uncomfortable with or ashamed of.

The Type of Organization

There are a lot of different types of organizations out there, and not everyone will enjoy or thrive in all of them. You can get a sense of what different companies are like on the site Glassdoor,[1] where people review their employers, jobs, and share their salaries.

Many technical students dream of working for one of the big west coast US tech companies (often called FAANG jobs, for Facebook, Amazon, Apple, Netflix, and Google). There's nothing wrong with these. They're hard to get into, but they come with big salaries, prestige, and name recognition when people ask where you work. You'll usually be doing interesting and challenging work, and it can be great for career growth

---

[1] https://www.glassdoor.com/index.htm

and developing expertise. They're also competitive internally, some even cutthroat, and they demand a lot from their workers. They reorganize frequently, so even if you land on a great team, a year later you could be somewhere else with a terrible boss. You have to be flexible and put your job first. These companies can be especially hard for women, especially if they go through a pregnancy. They all say your job is safe, but most of them find ways to punish women for prioritizing their kids over their careers. These companies also lay people off periodically, and nobody's really safe. The good news is that getting laid off or fired from places like this isn't really a black mark on your resume because it happens.

Startups can exist in many industries, but the data science jobs are mostly at the tech ones. Working for a startup can be exciting—being part of something new is very cool, especially if it's successful. There's variety here, but salaries are sometimes middle-of-the-road because you often get a bunch of equity (stock) that has a chance of making you rich if you're really lucky and the company makes it big. The company culture is often scrappy, and everyone is scrambling to do as much as possible quickly and with fewer resources than the tech giants. But sometimes you're expected to work long hours because everyone else is working late, and there's no guarantee the company won't go belly up.

Then there are large corporations with names people recognize and that have been around a long time, like Hilton, Colgate-Palmolive, American Airlines, and Harley-Davidson. There's a lot of variety among these, but companies that have been around longer than personal computers often have distinct and sometimes old-fashioned company cultures, where tenure and hierarchy are highly valued. These can be slow to change and often have very unrealistic views of what is possible with data science (or how quickly). Old companies also tend to have nightmarish legacy data systems that they've possibly been trying to clean up with attempts that frequently end in failure because it's very complicated. Depending on your personality, it can be easy to get lost in a giant company. It depends a lot on what kind of team you land on, which can be difficult to know in advance. Salaries are usually decent at these kinds of companies, and there are often significant annual bonuses.

Smaller corporations vary quite a bit in terms of company culture. They're usually younger than the big companies, so not as old-fashioned, and it's easier to get to know leaders and people at different levels of the hierarchy, which can be useful for some people (especially you good networkers). Sometimes you can find a company that's in the process of transitioning from a startup to a corporation. If they're going to go public soon, you may be able to land some good stock options that might pay off long-term. Salaries can be good at these, but you're less likely to receive annual bonuses.

You can also find opportunities at very small nontech companies, those with fewer than 50 people or so. At these companies, everybody sort of knows everybody else, like living in a small town. Sometimes they intentionally encourage a "family" culture, which may or may not work. This can be good for getting noticed. If they're new to data science, there's a lot of opportunity for you to make a big splash. But you're also more visible if you have trouble accomplishing anything really helpful, which may simply be because they don't have enough of the right data for you to do much. Companies like this often don't have good data and usually have very little idea how data science works. Salaries are usually on the lower end, and you won't generally find bonuses.

Universities and other educational institutions can be large and impersonal or small and cozy. There's usually pretty good job security at these, and they are usually far less competitive than the corporate world is. They can move slowly but at the same time be hugely innovative, depending on where you land. Salaries tend to be mediocre, and there won't be bonuses.

Nonprofit organizations do hire data scientists, but they often are very immature in the space. They're usually small so you can make a visible impact if you do well, but also risk a visible failure. Salaries are usually mediocre (which you'd expect), but the work can be very rewarding, especially if they work in a space you're passionate about.

There are a huge variety of options out there, more than I've listed here, but this is good for you to think about. Definitely do your research on different types of companies to see where you might fit.

It's Not a Lifelong Commitment

One important thing to remember is that although you want your first data science job to be good, if it turns out to not be the right fit for you, that's fine—you're not stuck there for life or forever confined to the industry or type of company. If you talk to Boomers, they'll tell you how important company loyalty is, but that is completely gone in today's world. You owe your employer nothing more than the work they're paying you to do. They feel the same about you. In the tech world, it's considered healthy to change jobs every two to three years. Usually this is within the same company, but it's also fine if your resume shows you at three companies in seven years. Most people won't bat an eye. You do need to have a reason for leaving each one, and the answer should always be something along the lines of "There weren't good growth opportunities for me there." Note that it's also okay to stay at one place even in one job for several years, but a lot of potential employers will wonder if you are interested in career growth (pro tip: they want you to be).

Salary

Although salary is important and I mentioned it above, it should never be the most important thing in your job search. Yes, you want to make sure you're getting a good salary. Most places will allow you to negotiate once they've decided to give you an offer. Figuring out the right salary for the job you're looking at is notoriously difficult. This is especially true if you'd be moving to a new location. The best thing to do is go on a site like salary.com[2] and Glassdoor (mentioned above) and search for the job title, location, and experience level. You'll see ranges, and you might also try related job titles to get a better sense of the real ranges. You also should look into cost of living by looking up apartments to rent and other expenses you will have, like a car, parking, or public transport.

You can also ask about annual bonuses, as many companies have them and they can be everywhere from small to very large. But do factor that in when you're looking at the salary in an offer.

The other benefits—vacation, sick time, health insurance, retirement savings matches—and sometimes others are also important. A lot of career advice makes it sound like you can negotiate for different things here. Although that's theoretically possible, at most places you get what you get. It can vary a lot from company to company, and sometimes you'll get additional benefits (especially more vacation or a higher retirement savings match) after you've been there a number of years. Make sure to pay attention to these and consider what matters to you. Some places allow you to roll vacation or sick time over (keep what's unused one year for the next year); others don't. Some companies even have theoretically unlimited vacation.

How to Land a Job

As I've implied above, you should be researching things even before you start applying for jobs so it's not a willy-nilly process. This means researching the industries, companies, and jobs you're interested in. Start studying the job boards, too, so you know what's out there. Save jobs you're interested in.

---

[2] https://www.salary.com/

Make sure you're continuing to work on your portfolio, making it as impressive as possible. Work on your resume, making it visually appealing (but nothing fancy) with a decent amount of white space and ensuring there are no typos or formatting inconsistencies (have someone else who's good with language check it—if you're still in school, take it to your school's writing or career center).

Applying

Once you're ready to start applying, make sure you have your resume and portfolio in order. Look back at the jobs you've saved. One thing I recommend is to avoid applying through the job sites and instead go directly to the job listing's company and find the job on their careers site and apply there. This way you avoid getting stuck with a middleman recruiter. In some cases, you won't be able to figure out what the company is and you can just apply through the job listing site. One disadvantage with working with an independent recruiter is that they get a finder's fee if the company hires you, so they're only interested in getting that fee, rather than whether you're a good fit or not. One warning about companies' careers sites: They have these huge byzantine application processes where you have to manually fill out all the information that's on your resume into these clunky forms, which takes a ridiculous amount of time. But you still have to do it the first time you apply at that company.

Note that frequently sites will make it appear that a cover letter is optional. Always include one. A lot of companies won't care, but you might miss out on the perfect job because a recruiter is offended you didn't include one. A cover letter basically just highlights the most important features they'll find on your resume that relate to the job requirements. Try to have someone read over this as well, especially if you're not a clean writer. Typos look really bad to people who notice them. Apply, and then apply for more, and keep going.

Note that what you will mostly hear in response to dozens of applications are crickets. Even if you literally meet every single requirement and even all the desired skills, you may hear nothing. Don't despair—this is just how it is. Companies often get hundreds of applications for a single position. One important thing to know is that resume screeners look for specific keywords on your resume related to the job. Google to figure out which ones you need to make sure appear somewhere on yours.

Interview Process

Congratulations! You've gotten your first interview. Now you start studying up on the company and see what you can find out about the position itself.

Usually, the first step is a quick interview with the company's recruiter, generally called a screen (a screening interview). This is mostly to ensure that there's not a glaring problem and to make sure you know a little more about the position and are definitely still interested. These are usually not technical at all (recruiters famously have very little technical knowledge), but it gives you a good chance to ask questions about the job and the company. You should always be ready to demonstrate your knowledge of the company and also ask questions. Just avoid looking greedy, and bringing up salary early in the process is considered bad form (unless they bring it up themselves).

If you make it past the initial screen, you'll find different companies do different things. Some of the things you may face, in no particular order, are a screen with the hiring manager, a group interview with the team, some number of interviews with individual team members, a brainteaser interview, a take-home assignment (timed or not, large or small), a live coding session, a technical interview without live coding, a behavior interview, a lunch with the team and/or hiring manager, a full day of back-to-back interviews in front of a whiteboard, and so on. Usually, the recruiter will tell you the process if you make it past the screen—make sure to ask if they don't. You can research online to find out more about all these types of hiring activities, as they're fairly standard.

I'll give a few tips. Always have questions for every single interview you have, and make sure they show that you've done your research (for instance, don't ask when the company was founded—anybody could find that kind of info online). Your questions are supposed to impress people, but also give you info. Try to learn more about the job and the team.

Never say anything negative. Even if your last boss was a mean, vindictive jerk who took a special interest in making your life hell for no reason, don't tell anybody that. You always left a job because there weren't enough opportunities to grow your career there. Or you were looking for something more challenging and interesting.

For the behavior interviews, make sure you've done your research on these and you have prepared answers for the type of questions they ask. These are the "Tell us about a time …" you had to deal with conflict with a colleague, you saw somebody doing something unethical, you made a mistake that no one noticed, and so on. The more specific your examples are, the better. But aim to make the answers succinct. Always stay positive. If you have to admit to having some failing, make sure to talk about how you learned from the error. Also, it is okay to not know something—you're better off admitting you don't know and, if appropriate, asking what the answer is. I was once asked what the difference between the SQL commands UNION and UNION ALL is, and I didn't know. I said that and asked what it was, they told me, and I got the job and have never forgotten the difference. Pretending you know something you don't is very risky because it makes you look untrustworthy, and they will wonder if you can take critique and grow or if you'll never admit you're wrong (a bad trait).

When you get asked about ways you know you need to grow, you can always go with the old standby, "Sometimes I'm a bit of a perfectionist and am hard on myself." But that's rather trite now, so see if you can come up with something that's at least mostly true, but is still a backhanded brag. In general, lying in interviews is bad, but spinning everything positively is the norm.

The Job Offer

Congratulations! They want to hire you! This can be a very exciting time, especially if you've been looking for a while. This is probably a great thing for you if you've done your research and ensured the team and job feel like a good fit for you. You don't have to accept right away. Sometimes they'll want to talk to you first before making a salary offer, so try not to give away how excited you are at this stage (but don't sound uninterested— aim for keen and pleased). This is like any negotiation. Ask questions to understand exactly what they're offering. Sometimes they may ask you if you have other offers pending. You can be honest, or you could imply that you are far in the process with one or two other companies and are waiting while they finish up other interviews. Once you're done talking to them, you can hang up or shut your laptop and *then* jump up and down a few times. But you need to sit down and decide what you want to do. If you really do have other irons in the fire, this would be a good time to reach out to the other recruiters for an update. You can explain that you have an offer and would like to be able to make a decision soon. This is especially important if one of the other companies is a job you're convinced you want more. But you also don't want to keep the original company's offer waiting. Just be respectful and rational, and you'll make the right decision.

Best of luck.

Key Takeaways and Next Up

In this chapter, I talked about the many types of jobs that are out there and related to data science, from the obvious data scientist, data analyst, and BI engineer to different engineering roles, to project management, to business analyst, and even to sales. Then I addressed the types of roles there are beyond job titles, industry considerations, and making sure to stick to your values in your job search. I also reminded you that a job is not a lifetime commitment and that salary matters but isn't the most important thing. Finally, I talked a bit about the job search, application, and interviewing process, finishing up with a discussion of how to handle a job offer.

PRACTITIONER PROFILE: ALEX BAUR

Name: Alex Baur

Current Title: Senior specialist solutions architect (full-stack data scientist)

Current Industry: Tech, with stints in several others including cannabis, chemicals, and telecommunications

Years of Experience: 8

Education:

- BA Chemistry, minor in Entrepreneurship

The opinions expressed here are Alex's and not any of his employers, past or present.

Background

Alex always loved chemistry so that's what he majored in, in college, but he also has an entrepreneurial spirit so he minored in that as well. His first job was in chemical sales, but he really wanted to do more of the science part of chemistry, and he started learning about other fields and found data science. He worked his way through the Johns Hopkins Data Science certificate on Coursera and then networked like crazy before landing his first job as a data scientist at a cannabis company.

Work

Alex's new job wasn't quite a full data science role, as he was doing a lot of marketing, sales, and website analytics, but he did start doing some modeling in R. One of his earliest projects involved creating a network node graph showing the heritage of all strains of cannabis. He was still doing a lot of advertising for their principal revenue stream, but also started working on classifying the users coming into their website. He left that company to form a

847

consulting company doing analytics and marketing for a variety of customers (including from the cannabis and finance industries as well as an English Duke) and a second consulting company that worked on helping cannabis dispensaries manage inventory better.

After Alex had trouble with his business partners, he ended up landing a job at a retail corporation as a data scientist. He did a lot of forecasting and classification initially and then worked on a project that dealt with a supply chain problem, ships of goods being stuck in ports. He created a very successful algorithm that sorted the inventory effectively. Then he got pulled into some projects where he learned about cloud services in general and Databricks specifically and ultimately ended up in a job that involves wearing a lot of hats. He helps create long-term deployable bundles that automate MLOps so teams can get started right away.

Sound Bites

Favorite Parts of the Job: The fact that when you're doing good data science, it's a result of marrying true technical insights with actual communications with human beings. He is good at both and enjoys the range of challenges. Successfully translating technical work to nontechnical people feels like magic sometimes.

Least Favorite Parts of the Job: Data cleaning. Also, the lag in data science being truly applied as part of business. Conceptually, people appreciate the idea of data science, but they don't understand that it's a process and they have to be patient for it to come to fruition. They also have trouble understanding that we're never going to have a 100% definitive answer, but can still be useful.

Favorite Project: The supply chain problem of optimizing shipping containers for a huge global retail organization was both difficult and fascinating. His worked ensured that things like Christmas trees and highly valued toys would arrive before Christmas, not in January. It saved $3.35 billion in wasted inventory.

Skills Used Most: Architecture and design of systems, as well as visuals. His broad knowledge of existing, new, and emerging ML tech and how that fits in with cloud infrastructure, as well as his understanding of the nuances of successful ML and AI. Being able to communicate with anyone at any level on both tech and nontech topics.

Primary Tools Used Currently: Databricks, Python, PySpark, SQL, Azure, AWS, GCP, and Fabric

Future of Data Science: What will never change is the need for high-quality, timely, and relevant data. There will be less focus on data scientists understanding the nuances of specific ML algorithms. It seems like we're in a more compound era of AI, where we care more about how the output helps the business than about the internals of it.

What Makes a Good Data Scientist: Practical stuff like strong coding skills in Python (C++ and C# can also be helpful in some roles), strong SQL, and solid statistical knowledge (so you can make sure the data is really saying what you think it is). Soft skills like working with and communicating with all sorts of different kinds of people, both technical and nontechnical. As something to strive for, being given a problem and being able to immediately see ways ML or AI can do to achieve stakeholders' goals and drive business value.

His Tip for Prospective Data Scientists: Get good at cleaning data. Colleges should have a full year teaching how to clean data. They usually don't, but it's going to be a huge part of the job, so get good at it on your own.

Alex is a data scientist and MLOps leader with experience in a variety of industries.

Setting Up Your Computer to Do Data Science

Introduction

In Chapter 23, we talked about getting your hands dirty by getting some data and starting to do actual data analysis and data science on it. There are online platforms you can use, but a lot of people like working on their own computers, so I'm going to go over the basics of setting your computer up to do data science.

Installing the Language

Most computers come with Python installed at the system level, but if you're going to use it for data science, you'll want to install your own version. And if you want to use R, you'll have to install that.

© Kelly P. Vincent 2025
K. P. Vincent, *A Friendly Guide to Data Science*, Friendly Guides to Technology,
https://doi.org/10.1007/979-8-8688-1169-2

Installing Python

Traditionally, data scientists installed a particular distribution of Python called Anaconda because it contained both the base Python language and also many of the libraries that data scientists need. However, concerns have been raised about using Anaconda within companies, so if you're planning to install it on your work computer, I'd recommend not using Anaconda.

Installing Anaconda

To install Anaconda, go to their download page[1] (you can skip providing your email). There are actually two different distributions, based on how much space you want to use up. The regular installation is just called Anaconda, and those are the first options on the page. Miniconda is the alternative, which installs fewer libraries, but is a much smaller installation, so if space is a consideration, it might be the right choice. It does have some other limitations, so do your research first.

The options for the full installation will be available in a dropdown under Anaconda Installers. For Windows, there's just one installer. For Macs, you'll have to choose Apple Silicon or Intel for the processor type on your machine. You can figure out which you have by going to About This Mac and seeing what it says next to Chip. You can download either the graphical installer or command line install for each chip type (the graphical installer is the easiest). For Linux, there are three options, also dependent on processor type.

If you do want to install Miniconda, the process is the same, but it's the lower row of options, Windows, Mac, and Linux again.

Once you've downloaded the package you want to install, double-click it and follow the prompts.

---

[1] https://www.anaconda.com/download/success

Installing Python.org Python

If you want to install the canonical Python, you can get it from the python.org downloads page.[2] You'll have the same options in terms of operating systems. It will default to offering the one that's appropriate for your machine at the top of the page, but you can click the other options under that. For each operating system, there's only one option, so it's straightforward.

Follow the prompts after double-clicking the package you downloaded.

Installing R

Installing R is straightforward because there's really only one real option: The Comprehensive R Archive Network page.[3] At the top of that page, you'll see the three options for downloading R: Windows, Linux, and Mac. Click the appropriate one for your machine, and it will take you to another page where you may have to pick the right installation package based on your computer specs. For Windows, you just want to install the base binary distribution. For Macs, pick the Apple chip (M1, M2, etc.) or the Intel version. You can find out which type you have by going to About This Mac and seeing what it says next to Chip. It's a little more involved for Linux users, as they have versions for five flavors of Linux (they show Debian, Fedora, RedHat, SUSE, and Ubuntu, although SUSE seems defunct) and you'll need to pick the right one for you and then follow the instructions. Once it's downloaded, double-click the package and follow the instructions.

---

[2] https://www.python.org/downloads/
[3] https://cran.r-project.org/

Installing and Setting Up an Integrated Development Environment (IDE)

IDEs are the text editors that programmers use because they have extra features over something like Microsoft's Notepad or Apple's TextEdit. IDEs have a lot of built-in functionality that helps development. Some of the basic things are the ability to run code within the IDE, color-coding your code (like making variables one color, functions another, and so on, so that it's easy to tell what's what at a glance), including line numbers (which is very helpful for debugging), and code folding (where you can collapse a block of code like a for loop so you don't have to wade through all that code every time you're going through the code file).

R programmers usually use RStudio Desktop. You can also write Python code within RStudio, but Python programmers usually use others. Right now many are using Visual Studio Code (VS Code). VS Code is a general IDE and can be used for virtually any programming language (including R). You also can use Jupyter Notebook within VS Code, which can be useful. You'll just need to install an extension or two for each language you want to use within VS Code.

VS Code

You can download the version of VS Code for your operating system at the downloads page.[4] There's one primary option for each of Windows and Mac, and there are two options for Linux depending on the flavor. There are also more specific download types, but you don't need to worry about those unless you already know what you're doing with them. Download the package and double-click to follow the instructions.

---

[4]`https://code.visualstudio.com/download`

Once installed, you'll need to add extensions for the languages you want to work in. You don't have to add everything now, but definitely add extensions for Python and/or R and the Jupyter ones now. Start by clicking the extensions tab on the left bar. It looks like Figure A-1.

Figure A-1. *The extensions tab icon*

You'll have to search for the ones you want to install in the bar at the top. These are recommended:

- Python
- Python Debugger
- Jupyter
- R
- R Debugger

There's really no reason not to install all of them even if you're not sure you'll use them. But you can always wait. Additionally, there are many more Python-, Jupyter-, and R-related extensions you can also install if you want to try them out.

R Studio

Download RStudio Desktop from the tools page.[5] Under step 2: Install RStudio, you'll find a button that should show the appropriate option for your operating system. Note that there may be minimum system requirements for the current version of the tool, so follow the instructions

---

[5] https://posit.co/download/rstudio-desktop/

on the page to get an earlier version for an older computer. You can also scroll down to see all the installers. Once you've downloaded it, double-click it and follow the instructions. Everything is ready to go for R.

Installing Data Science Packages

If you code for any length of time, you'll eventually come across a package you want that you don't already have installed. It's common to install several before you even get started just because you know you'll probably need them eventually.

Note that I tend to use the terms "library" and "package" interchangeably in reference to Python. They are subtly different, but it's not important for this appendix. However, in R, they are more different, and the term "package" is the right one.

Python Packages

As mentioned, the Anaconda distribution already has most of the libraries you need to do data science already installed. If you're going to be doing some specialized stuff like NLP or deep learning, it won't have packages for those, so you might want to go ahead figure out which ones you need and install them. Table A-1 shows good data science libraries to start with, indicating which aren't included in the Anaconda distribution.

If you've installed the official Python distribution instead, you will likely want to install the libraries listed in Table A-1.

Table A-1. *Recommended Python packages*

Package	Use	In Anaconda
pandas	Data frames	✓
numpy	Numeric data manipulation	✓
scipy	Builds on numpy	✓
plotly	Visualization	✓
matplotlib	Visualization	✓
seaborn	Visualization	✓
statsmodels	Statistics	✓
scikit-learn	ML modeling and performance	✓
xgboost	ML modeling	No
pytorch	ML modeling (deep learning)	✓
keras	ML modeling (deep learning)	No
nltk	NLP modeling	✓
spacy	NLP modeling	No
jupyter	Jupyter Notebooks	✓

The most common way to install libraries is at the command line with the command conda (with Anaconda) or pip. Note that this is not within the Python interpreter itself, but rather at the system level. With Anaconda, you have an additional option. You can run the Anaconda Navigator GUI (available in all operating systems) and select libraries to install. Otherwise, use the Anaconda Prompt in Windows and the system terminal

on Mac and in Linux (in VS Code, you can open a terminal in the IDE and
type the command there).

The full command to install a library, let's say XGBoost, is `conda`
`install xgboost` or `pip install xgboost`. Usually, if you have Anaconda,
you'll try with `conda` first, and if that doesn't work, use `pip`. Otherwise start
with `pip`. `conda` requires some integration with the distribution to work,
and not all libraries have been set up for this, which is why sometimes you
have to use `pip`. But they both accomplish the same thing. Note that often
you'll need to use `pip3` rather than `pip`. That just ensures you're using the
Python version 3 installer.

R Packages

In R, you install packages from within the R interpreter itself, rather than
at the system level like you do with Python. A new installation of R only
comes with 15 base packages installed, so you will need to install some.
Open RStudio and run the function `install.packages()` with the name
of the package you want installed inside single quotes in the parentheses,
like `install.packages('lubridate')`. You can also install multiple in one
line, which looks like this: `install.packages(c('lubridate', 'dplyr',`
`'stats'))`. Table A-2 shows some of the packages you might want to
install.

Table A-2. *Recommended R packages*

Package	Use	Base Install
data.table	More data frame manipulation	No
dplyr	Data manipulation	No
lubridate	Date manipulation	No
ggplot2	Visualization	No
stats	Statistics	✓
e1071	ML modeling	No
xgboost	ML modeling	No
randomForest	ML modeling	No
tensorflow	ML modeling	No
naivebayes	ML modeling	No
metrics	ML performance	No
tidymodels	ML support	No
shiny	Presentation	No

System vs. Virtual Environment Installation

The methods described above are how you would install these packages at the system (or the logged-in user) level. It's long been common to work with what are called virtual environments in Python, where you basically create an encapsulated space on your computer that has a particular version of Python and specific packages installed. R has a package called renv that allows you to manage virtual environments in the same way.

For a given virtual environment, you would specify which versions of each package should be installed, and that environment won't be modified by any other changes at the system level. This can be useful if you need to use an older version of some package because you're using another one that isn't compatible with the most recent version. It's also a way of keeping things static, as sometimes installing a package on your computer will automatically update another package, and if you had code working on a specific version of that second package, it could break.

When using virtual environments, you normally manage the packages with a list that is used to configure the environment. Usually, this is done with a YAML file (extension `.yml`) that lists all the packages and their version number in a particular format. This can be done with both Python and R environments.

Start Coding

Everything's set up and ready to go. Now you just need to open up your IDE and start writing code.

Coding in Python in VS Code

Now that you've got your extensions in place, you'll need to ensure that you can access your Python installation (and R if appropriate) from within the IDE. Often it works out of the box, but if you have trouble, go to Settings in VS Code and find the option "Python: Default Interpreter Path" and point that to the python.exe file in your new Python installation folder.

If you didn't install Jupyter, you will need to create a file with a `.py` extension. This is called a script. You can put whatever code in this file, but to run any of it, you'll have to run the whole file. If you have a step that takes a long time, it will still have to run every time you want to run any of the code. This is how it used to be done before notebooks revolutionized coding. You run a script by clicking the right-pointing arrow in the upper right next to the open file names.

If you installed Jupyter, there are a couple options for writing your code. You can still do it the old way, but better is to create a notebook. Create a file with the extension `.ipynb`, and VS Code will treat it as a Jupyter Notebook. This is the best way to get started writing code. A notebook has individual cells one after the other, and you can put code or text (markdown) in each cell. This allows you to run each cell separately. If you run a big data processing step that takes several minutes and you want to do use the data frame (the table) that it produces, you won't have to rerun that step even if you change code that's after it. This really speeds up experimentation. To run a cell in a notebook, click the right-pointing arrow to the left of the cell. There are other options above and in the menus and keyboard shortcuts (Ctrl-Enter).

Coding in R in RStudio Desktop

When you first open RStudio, it shows a console, which is basically the R interpreter. You can run commands here. You can also create files with lines of code just like Python, but you can run each line independently, giving you the flexibility that Jupyter Notebooks do for Python. It's recommended to create files because otherwise you haven't saved what you've run. When you create a file, it opens in the top left. You can start typing your code there. Each line can be run with Ctrl-Enter or clicking the Run button at the top of the file. The Source button runs the entire file.

Conclusion

Now you've got your computer set up, and you're ready to start doing data science, whether you're using R or Python (or a mix of both). Make sure that if you run into any trouble installing anything, check in with Google because somebody somewhere has probably had the same trouble.

Index

A

Actuarial science, 50
Aggregations/interactions, 30–32
Amazon Simple Storage Service
   (S3), 713
Amazon Web Services (AWS), 403,
   710, 713, 714
Anaconda, 852
Analytics program
   advanced techniques, 250, 251
   aggregation/grain, 246–248
   approaches and tools, 177
   basketball franchise, 241, 242
   business intelligence, 248, 249
   business reporting, 236
   cholera epidemic level, 183–185
   classes
      descriptive analytics, 255
      diagnostic analytics, 255
      maturity, 256, 257
      predictive/prescriptive, 256
   communication skills, 188
   computer, 179–181
   concepts, 176, 235
   CRISP-DM (*see* CRoss Industry
      Standard Process for Data
      Mining (CRISP-DM))
   customer relationship
      management, 239–241
   data/finding insights, 250
   DataOps, 244
   development environment, 243
   disciplines, 209
   domain knowledge, 188–190
   foundation, 245–248
   functional skills, 185
   Goldilocks, 176
   high-level view, 238–240
   history, 177–179
   maturity concept, 237
   maturity model
      analytics initiative, 253, 254
      business intelligence
       effort, 252
      classes, 251, 256, 257
      foundation, 251
      implicit requirement, 252
      levels, 251
      program, 258
   mining, 209
   Moneyball, 182–184
   new analytics, 243, 244
   Ngram viewer, 210
   production/deployment
      support/tools, 236

© Kelly P. Vincent 2025
K. P. Vincent, *A Friendly Guide to Data Science*, Friendly Guides to Technology,
https://doi.org/10.1007/979-8-8688-1169-2

GPSR Compliance

The European Union's (EU) General Product Safety Regulation (GPSR) is a set of rules that requires consumer products to be safe and our obligations to ensure this.

If you have any concerns about our products, you can contact us on

ProductSafety@springernature.com

In case Publisher is established outside the EU, the EU authorized representative is:

Springer Nature Customer Service Center GmbH
Europaplatz 3
69115 Heidelberg, Germany